WORLD HERITAGE SITES

FIREFLY BOOKS

A FIREFLY BOOK

Published jointly by the United Nations Educational, Scientific and Cultural Organization (UNESCO), 7, place de Fontenoy, 75352 Paris 07 sp, France, and Firefly Books Ltd. 2009.

Text © UNESCO 2008
Maps © Collins Bartholomew Ltd 2009
Photographs © as per credits on page 830

First printing

Publisher Cataloging-in-Publication Data (U.S.)
World heritage sites : a complete guide to 878 UNESCO
world heritage sites / UNESCO. [832] p. : col. photos., maps ; cm.
Includes index. Summary: Each site has an entry explaining
its historical and cultural significance, with a description and location map.
ISBN-13: 978-1-55407-463-1 (pbk.)
ISBN-10: 1-55407-463-0 (pbk.)
1. World Heritage areas--Pictorial works. 2. World Heritage areas – Guidebooks.
I. Title. 910.2 dc22 G140.5U547 2009

Library and Archives Canada Cataloguing in Publication
World heritage sites : a complete guide to 878 UNESCO world heritage sites / UNESCO.
Includes index.
ISBN-13: 978-1-55407-463-1
ISBN-10: 1-55407-463-0
1. World Heritage areas--Guidebooks. I. UNESCO
G140.5.W68 2009   910.2'02   C2009-901986-8

Published in the United States by
Firefly Books (U.S.) Inc.
P.O. Box 1338, Ellicott Station
Buffalo, New York 14205

Published in Canada by
Firefly Books Ltd.
66 Leek Crescent
Richmond Hill, Ontario L4B 1H1

Printed in China

For more information on World Heritage, please contact:
UNESCO World Heritage Centre
7 place de Fontenoy
75352 Paris 07 SP, France
Tel: (33) 01 45 68 15 71
e-mail: wh-info@unesco.org
http://whc.unesco.org

WORLD HERITAGE SITES

A Complete Guide to 878 UNESCO
World Heritage Sites

United Nations
Educational, Scientific and
Cultural Organization

UNESCO
Publishing

How to use this book

The page on which the information on a World Heritage site can be found is accessed in a number of ways – by consulting the continent maps on which all the sites are located, or by reference to either the alphabetical or country by country index. All entries are presented in a similar manner and are arranged chronologically by the year in which they were first inscribed on the World Heritage List.

The diagram below indicates the individual components of each entry and explains the colour coding used to distinguish whether a site is classified as natural, cultural or mixed.

Site title
gives the official UNESCO World Heritage title for each entry.

Red band
represents entries classified as cultural sites.

Locator map
shows the location of the site in a wider region.

Blue band
represents entries classified as mixed sites.

Timeline
on every page highlights the year in which the sites were first inscribed.

Site location
indicates the country where the site can be found.

Green band
represents entries classified as natural sites.

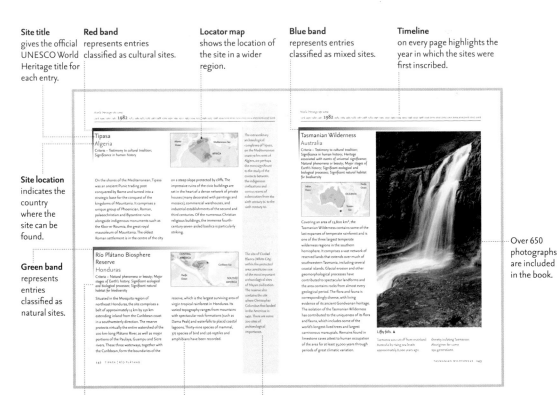

Criteria summary
To be included on the World Heritage List, sites must be of outstanding universal value and meet at least one out of ten selection criteria. Full criteria explanation can be found on pages **800–1**.

Main text
gives concise descriptions and information about each site.

Extra information
about each site supplements the details in the main text.

Over 650 photographs are included in the book.

Contents

6 Foreword

8 World Heritage sites – mapped by continent

25 World Heritage sites – descriptions, locations and photographs

800 UNESCO's World Heritage Mission Statement, inscription criteria and information about UNESCO worldwide

804 Country by country index

814 Index

830 Acknowledgements

Foreword
by Mr Koïchiro Matsuura,
Director-General of UNESCO

Fifty years ago, the construction of the Aswan High Dam resulted in the flooding of an extensive stretch of the Nile Valley, home to numerous ancient Nubian treasures, such as the Abu Simbel temples. Mindful that the threatened temples and artefacts were an urgent priority transcending national interests and pride, the United Nations Educational, Scientific and Cultural Organization (UNESCO) launched its first international safeguarding campaign. Funds and expertise were mobilized to dismantle and reassemble the monuments in new locations. UNESCO's appeal to save the truly outstanding vestiges of one of the world's richest and oldest civilizations made people all over the planet appreciate the universal dimension of cultural heritage. Thus, in addition to its triumph as a technical feat of unprecedented scale, this hugely successful campaign paved the way for the key notion of the common heritage of humankind that underpins the Convention concerning the Protection of the World Cultural and Natural Heritage, developed by UNESCO and adopted by its Member States in 1972.

Commonly known as the World Heritage Convention, this international treaty has been ratified by 186 countries to date, and the famous World Heritage List now includes 878 sites in 145 countries across the planet. For over thirty-five years, the Convention and its List have proved invaluable tools in UNESCO's constant efforts to encourage the identification, protection and preservation of cultural and natural heritage around the world considered to be of outstanding value to humanity. Moreover, they contribute significantly to advancing UNESCO's mission to safeguard the world's precious cultural and biodiversity.

Although every year new sites are inscribed by the World Heritage Committee, many sites of outstanding universal value have yet to be included on the List, which strives to ensure a true representation of the full diversity of all types of tangible heritage. The open-ended nature of the List is precisely what makes it such a vibrant and attractive instrument for preservation.

The prestigious List includes some of the most famous places in the world, such as the ancient Nabataean city of Petra in Jordan, the legendary Acropolis in Athens, the Great Barrier Reef in Australia and Machu Picchu, the 'Lost City of the Incas', in Peru. These renowned breathtaking sites are obvious inclusions on the List as they represent extraordinary examples of our cultural and natural heritage. However, many are in need of extensive conservation efforts, partly as a consequence of their immense popularity. Fortunately, inclusion on the World Heritage List and the resulting economic benefits can give a huge boost to these conservation efforts so that future generations can enjoy this irreplaceable heritage.

Along with the well-known sites on the List are a number of lesser-known places. In reading this handsome book, you will gain a wealth of information about our common heritage: that the largest coliseum in North Africa is located in the small village of El Jem in Tunisia; that Ethiopian emperors were crowned at Aksum; that one of the most influential cultural centres of antiquity was Palmyra in present-day Syria; and that the world's largest free-roaming bison herd can be found in Canada's Wood Buffalo National Park. Other more unusual features revealed in these pages include the massive *moai* (carved heads) found in Rapa Nui National Park in Chile, the monastery-crowned rock pinnacles of Meteora in Greece, and the 'dragon' whose home is at the Komodo National Park in Indonesia. These are just a few of the many highlights to be discovered in this publication.

All currently inscribed UNESCO World Heritage sites are described in this single volume of *The World's Heritage*, which is illustrated with over 650 stunning full-colour photographs. It is my sincere hope that you will enjoy this unique guide to the planet's outstanding cultural and natural treasures, and that it will contribute to the dissemination of the universal ideals and values that UNESCO endeavours to uphold in its worldwide action.

World Heritage Sites
Europe

Belarus
Architectural, Residential and Cultural Complex of the Radziwill Family at Nesvizh p748▼; Belovezhskaya Pushcha / Białowieża Forest p61; Mir Castle Complex p635; Struve Geodetic Arc p747▲.

Cyprus
Choirokoitia p554▼; Painted Churches in the Troodos Region p210▼; Paphos p101.

Denmark
Jelling Mounds, Runic Stones and Church p426▼; Ilulissat Icefjord p708 (see map on p15); Kronborg Castle p608; Roskilde Cathedral p452.

Estonia
Historic Centre (Old Town) of Tallinn p509; Struve Geodetic Arc p747▲.

Finland
Bronze Age Burial Site of Sammallahdenmäki p599▲; Fortress of Suomenlinna p378; Old Rauma p368▼; High Coast / Kvarken Archipelago p630▲; Petäjävesi Old Church p436▼; Struve Geodetic Arc p747▲; Verla Groundwood and Board Mill p486▼.

Germany
Aachen Cathedral p33; Abbey and Altenmünster of Lorsch p379▲; Bauhaus and its Sites in Weimar and Dessau p484▲; Berlin Modernism Housing Estates p784▼; Castles of Augustusburg and Falkenlust at Brühl p193▲; Classical Weimar p549; Collegiate Church, Castle and Old Town of Quedlinburg p441▲; Cologne Cathedral p474; Dresden Elbe Valley p709; Frontiers of the Roman Empire: Upper German-Raetian Limes p273; Garden Kingdom of Dessau-Wörlitz p610▲; Hanseatic City of Lübeck p281; Historic Centres of Stralsund and Wismar p690▲; Luther Memorials in Eisleben and Wittenberg p503▲; Maulbronn Monastery Complex p407▲; Messel Pit Fossil Site p456▼; Mines of Rammelsberg and Historic Town of Goslar p387; Monastic Island of Reichenau p640▼; Museumsinsel (Museum Island), Berlin p578; Muskauer Park / Park Muzakowski p719▼; Old town of Regensburg with Stadtamhof p753; Palaces and Parks of Potsdam and Berlin p346; Pilgrimage Church of Wies p176▼; Roman Monuments, Cathedral of St Peter and Church of Our Lady in Trier p251; Speyer Cathedral p128▼; St Mary's Cathedral and St Michael's Church at Hildesheim p225▼; Town Hall and Roland on the Marketplace of Bremen p727; Town of Bamberg p406; Upper Middle Rhine Valley p684; Völklingen Ironworks p441▼; Wartburg Castle p589; Würzburg Residence with the Court Gardens and Residence Square p132; Zollverein Coal Mine Industrial Complex in Essen p667▼.

continued on page 12

Azores (Portugal)
731 162▼

Key to maps

● Cultural site

● Natural site

● Mixed site

234 Page number reference

▲▼ top/bottom of page

The maps in this section are laid out by geographical continent. Please see the maps on page 802 for UNESCO's regional areas.

602
Madeira (Portugal)

587▼
250▲ 766 **Canary Islands (Spain)**

ICELAND
710▼
789▼

241 570

244 273 470 **UNITED KINGDOM**
659
401▲ 273 246
IRELAND 257▼ 707▼ 248
492 257▼ 239 665
620▼ 675▼
757 290 270 312
254 507
671 692 272 327 485▲
441
NET

PORTUGAL
211 649▲
397 217 220▲
535▲ 758
682▲ 662 647▲
482 563▼ 200 535▼ 508▼
337 177 328 209 644▼
163 175 256
447 258 253 233 193▼ 366 182
410 416▲ 242 568 505 621▼ 528▲
266 198 562▲
430▲ 697▼ 639 572
202

SPAIN

Only countries States Parties to the World Heritage Convention are labelled on these maps. United Nations countries boundaries shown as of March 2009.

8

NORWAY

SWEDEN

FINLAND

RUSSIAN FEDERATION

DENMARK

ESTONIA

LATVIA

LITHUANIA

RUS. FED.

GERMANY

POLAND

BELARUS

UKRAINE

TURKEY

CYPRUS

232▼
747▲
497
747▲
504▲
716▲
392▼
449▼
449▼
742
104
630▲
630▲
436▼
356▲
81
747▲
43
742
368▼
599▲
486▼
643▼
667▲
378
744
408▲
371
509
747▲
344
646
432▼
404▲
435▲
747▲
383
725▲
459
426▼
608
640▲
544
420
396
452
553
747▲
350
281
690▲
690▲
615
747▼
728
427
727
503▲
578
547
421
526
635
SR 387
441▲
146
610▲
719
752
748▼
225▲
484▲
503▲
675▲
61
456▼
132
549
709
675▲
61
747▲
379▲
406
753
40
30
382▼
347
407▼
273
594
645
702
564
640▼
176▼
772
747▲

large-scale map on pages 10–11

747▲

226
442
550
237▼
225▲
292
230
329
333

210▼
101▲
554▼

Scale 1 : 20 000 000

9

650 648▼ 745▼ 562▼
562▼ 562▲ 562▲ 562▼
575 655▲ 554▲
652▼ 555
738▲ BELGIUM
117

LUXEMBOURG
424

360 372
622 62 169
65 113 669 334

168▼ 381 66 121▲
F R A N C E

153▲ 176▼ 498 531
157 SWITZERLAND 585
777 660 610▼ 787 159▼
784▲ 704▲ 76▲ AUSTRIA
775 600 695▲ 695▲ 465 SLOVENIA
44▼ 561 541 95 565▲ 261▲
557▼ 695▲ 618 428 268 516▲
695▲ 797 532▲ 57 CRO
508▼ 229 120 759▲ 797 520 453
480 444 513 487
514 131 513 782▲ SAN
732▲ 265 134 MARINO 627
ANDORRA 358 567 82
MONACO 460 533 79
174 495 632
Corsica 724
(France) 729▼ 729▲ 673 506▲
180 92 595 399
HOLY 527
SEE 448 524
Sardinia 538 558▲
(Italy) 546 I T A L Y

625

542 546▲ 690▼
Sicily 690▼ 690▼ 750▼
(Italy) 690▼ 750▼
690▼ 690▼

96
93
MALTA 98▲

Only countries States Parties to the World Heritage Convention are
labelled on these maps. United Nations countries boundaries shown
as of March 2009.

World Heritage Sites
Europe

Albania
Butrint p389; Historic Centres of Berat and Gjirokastra p732▼.

Andorra
Madriu-Perafita-Claror Valley p732▲.

Austria
City of Graz - Historic Centre p585; Fertö / Neusiedlersee Cultural Landscape p664▼; Hallstatt-Dachstein / Salzkammergut Cultural Landscape p531; Historic Centre of the City of Salzburg p498; Historic Centre of Vienna p678; Palace and Gardens of Schönbrunn p502; Semmering Railway p551▲; Wachau Cultural Landscape p641.

Belgium
Belfries of Belgium and France p575; Flemish Béguinages p562▼; Historic Centre of Brugge p650; La Grand-Place, Brussels p555; Major Town Houses of the Architect Victor Horta (Brussels) p648▼; Neolithic Flint Mines at Spiennes (Mons) p652; Notre-Dame Cathedral in Tournai p655▲; Plantin-Moretus House-Workshops-Museum Complex p745▼; The Four Lifts on the Canal du Centre and their Environs, La Louvière and Le Roeulx (Hainault) p554▲.

Bosnia-Herzegovina
Mehmed Paša Sokolović Bridge in Višegrad p779; Old Bridge Area of the Old City of Mostar p734.

Bulgaria
Ancient City of Nessebar p165; Boyana Church p60▼; Madara Rider p69▲; Pirin National Park p179, Rilu Monastery p178; Rock-Hewn Churches of Ivanovo p69▼; Srebarna Nature Reserve p168▲; Thracian Tomb of Kazanlak p49▼; Thracian Tomb of Sveshtari p234▲.

Croatia
Cathedral of St James in Šibenik p627; Episcopal Complex of the Euphrasian Basilica in the Historic Centre of Poreč p516▼; Historic City of Trogir p533; Historical Complex of Split with the Palace of Diocletian p82; Old City of Dubrovnik p86; Plitvice Lakes National Park p57; Stari Grad Plain p790▼.

Czech Republic
Gardens and Castle at Kroměříž p565▼; Historic Centre of Český Krumlov p386; Historic Centre of Prague p384; Historic Centre of Telč p392▲; Holašovice Historical Village Reservation p551▼; Holy Trinity Column in Olomouc p637; Jewish Quarter and St Procopius' Basilica in Třebíč p704▼; Kutná Hora: Historical Town Centre with the Church of St Barbara and the Cathedral of Our Lady at Sedlec p469; Lednice-Valtice Cultural Landscape p493; Litomyšl Castle p588▲; Pilgrimage Church of St John of Nepomuk at Zelená Hora p443; Tugendhat Villa in Brno p681▼.

continued on page 13

Scale 1 : 10 000 000   11

Europe (continued from page 8)

Iceland
Surtsey p789▼; Þingvellir National Park p710▲.

Ireland
Archaeological Ensemble of the Bend of the Boyne p401▲; Skellig Michael p492.

Latvia
Historic Centre of Riga p544; Struve Geodetic Arc p747▲.

Lithuania
Curonian Spit p615; Kernavė Archaeological Site (Cultural Reserve of Kernavė) p715▼; Struve Geodetic Arc p747▲; Vilnius Historic Centre p421.

Netherlands
Defence Line of Amsterdam p488; Droogmakerij de Beemster (Beemster Polder) p580; Historic Area of Willemstad, Inner City and Harbour, Netherlands Antilles p528▼ (see map p16); Ir. D.F. Woudagemaal (D.F. Wouda Steam Pumping Station) p566; Mill Network at Kinderdijk-Elshout p523; Rietveld Schröderhuis (Rietveld Schröder House) p636▲; Schokland and Surroundings p449▲.

Norway
Bryggen p43; Rock Art of Alta p232▼; Røros Mining Town p104; Struve Geodetic Arc p747▲; Urnes Stave Church p81; Vegaøyan – the Vega Archipelago p716▲; West Norwegian Fjords – Geirangerfjord and Nærøyfjord p742.

Poland
Auschwitz Birkenau German Nazi Concentration and Extermination Camp (1940-1945) p40; Castle of the Teutonic Order in Malbork p547; Centennial Hall in Wrocław p752▲; Belovezhskaya Pushcha / Białowieża Forest p61; Churches of Peace in Jawor and Swidnica p675▲; Cracow's Historic Centre p30; Historic Centre of Warsaw p89; Kalwaria Zebrzydowska: the Mannerist Architectural and Park Landscape Complex and Pilgrimage Park p594; Medieval Town of Toruń p526; Muskauer Park / Park Muzakowski p719▼; Old City of Zamość p382▼; Wieliczka Salt Mine p36; Wooden Churches of Southern Little Poland p702.

Portugal
Alto Douro Wine Region p662; Central Zone of the Town of Angra do Heroísmo in the Azores p162▼; Convent of Christ in Tomar p175; Cultural Landscape of Sintra p447; Historic Centre of Évora p258; Historic Centre of Guimarães p682▲; Historic Centre of Oporto p482; Landscape of the Pico Island Vineyard Culture p731; Laurisilva of Madeira p602; Monastery of Alcobaça p337; Monastery of Batalha p177; Monastery of the Hieronymites and Tower of Belém in Lisbon p163; Prehistoric Rock-Art Sites in the Côa Valley p563▼.

Russian Federation (see also p24)
Architectural Ensemble of the Trinity Sergius Lavra in Sergiev Posad p420; Church of the Ascension, Kolomenskoye p427; Cultural and Historic Ensemble of the Solovetsky Islands p392▼;

Curonian Spit p615; Ensemble of the Ferrapontov Monastery p643▼; Ensemble of the Novodevichy Convent p728; Historic and Architectural Complex of the Kazan Kremlin p646; Historic Centre of Saint Petersburg and Related Groups of Monuments p344; Historic Monuments of Novgorod and Surroundings p383; Historical Centre of the City of Yaroslavl p744; Kizhi Pogost p356▲; Kremlin and Red Square, Moscow p350; Struve Geodetic Arc p747▲; Uvs Nuur Basin p693▼; Virgin Komi Forests p449▼; White Monuments of Vladimir and Suzdal p396.

Spain
Alhambra, Generalife and Albayzín, Granada p202; Aranjuez Cultural Landscape p658▼; Archaeological Ensemble of Mérida p410; Archaeological Ensemble of Tárraco p621▼; Archaeological Site of Atapuerca p647▲; Burgos Cathedral p200; Catalan Romanesque Churches of the Vall de Boí p644▼; Cathedral, Alcázar and Archivo de Indias in Seville p266; Cave of Altamira and Paleolithic Cave Art of Northern Spain p220▲; Doñana National Park p430▲; Garajonay National Park p250▲; Historic Centre of Cordoba p198; Historic City of Toledo p242; Historic Walled Town of Cuenca p505; Ibiza, Biodiversity and Culture p572; La Lonja de la Seda of Valencia p485▲; Las Médulas p535▲; Monastery and Site of the Escurial, Madrid p193▼; Monuments of Oviedo and the Kingdom of the Asturias p217; Mudéjar Architecture of Aragon p256; Old City of Salamanca p328; Old Town of Ávila with its Extra-Muros Churches p233; Old Town of Cáceres p253; Old Town of Segovia and its Aqueduct p209; Palau de la Música Catalana and Hospital de Sant Pau, Barcelona p528▲; Palmeral of Elche p639; Poblet Monastery p366; Pyrénées - Mont Perdu p508▼; Renaissance Monumental Ensembles of Úbeda and Baeza p697▼; Rock Art of the Mediterranean Basin on the Iberian Peninsula p562▲; Roman Walls of Lugo p649▲; Route of Santiago de Compostela p397; Royal Monastery of Santa María de Guadalupe p416▲; San Cristóbal de La Laguna p587▼; San Millán Yuso and Suso Monasteries p535▼; Santiago de Compostela (Old Town) p211; Teide National Park p766; University and Historic Precinct of Alcalá de Henares p568; Vizcaya Bridge p758; Works of Antoni Gaudí p182.

Sweden
Agricultural Landscape of Southern Öland p640▲; Birka and Hovgården p404▲; Church Village of Gammelstad, Luleå p504▼; Engelsberg Ironworks p408▲; Hanseatic Town of Visby p459; High Coast / Kvarken Archipelago p630▲; Laponian Area p497; Mining Area of the Great Copper Mountain in Falun p667▲; Naval Port of Karlskrona p553; Rock Carvings in Tanum p432▼; Royal Domain of Drottningholm p371; Skogskyrkogården p435▲; Struve Geodetic Arc p747▲; Varberg Radio Station p725▲.

Turkey
Archaeological Site of Troy p550; City of Safranbolu p442; Göreme National Park and the Rock Sites of Cappadocia p230; Great Mosque and Hospital of Divriği p225▲; Hattusha: the Hittite Capital p237▼; Hierapolis-Pamukkale p329; Historic Areas of Istanbul p226; Nemrut Dağ p292; Xanthos-Letoon p333▼.

Ukraine
Kiev: Saint-Sophia Cathedral and Related Monastic Buildings, Kiev-Pechersk Laura p347; L'viv – the Ensemble of the Historic Centre p564; Primeval Beech Forests of the Carpathians p772; Struve Geodetic Arc p747▲.

United Kingdom
Blaenavon Industrial Landscape p620▼; Blenheim Palace p270; Canterbury Cathedral, St Augustine's Abbey and St Martin's Church p327; Castles and Town Walls of King Edward in Gwynedd p257▼; City of Bath p290; Cornwall and West Devon Mining Landscape p757; Derwent Valley Mills p675▼; Dorset and East Devon Coast p671; Durham Castle and Cathedral p246; Frontiers of the Roman Empire: Antonine Wall and Hadrian's Wall p273; Giant's Causeway and Causeway Coast p244; Gough and Inaccessible Islands p468▼ (see map on p18); Heart of Neolithic Orkney p570; Henderson Island p321▼ (see map p21); Historic Town of St George and Related Fortifications, Bermuda p621▲ (see map on p15); Ironbridge Gorge p239; Liverpool – Maritime Mercantile City p707▼; Maritime Greenwich p507; New Lanark p659; Old and New Towns of Edinburgh p470; Royal Botanic Gardens, Kew p692; Saltaire p665; St Kilda p241; Stonehenge, Avebury and Associated Sites p254; Studley Royal Park including the Ruins of Fountains Abbey p248; Tower of London p312; Westminster Palace, Westminster Abbey and Saint Margaret's Church p272.

Europe (continued from page 11)

France

Abbey Church of Saint-Savin sur Gartempe p168▼; Amiens Cathedral p117; Arles, Roman and Romanesque Monuments p131; Belfries of Belgium and France p575; Bordeaux, Port of the Moon p775; Bourges Cathedral p381; Canal du Midi p480; Cathedral of Notre-Dame, Former Abbey of Saint-Rémi and Palace of Tau, Reims p372; Chartres Cathedral p65; Cistercian Abbey of Fontenay p121▲; Fortifications of Vauban p794; Gulf of Porto: Calanche of Piana, Gulf of Girolata, Scandola Reserve p174; Historic Centre of Avignon: Papal Palace, Episcopal Ensemble and Avignon Bridge p444; Historic Fortified City of Carcassonne p514; Historic Site of Lyons p561; Jurisdiction of Saint-Emilion p600; Lagoons of New Caledonia: Reef Diversity and Associated Ecosystems p788▼ (see map on page 71); Le Havre, the city rebuilt by Auguste Perret p738▲; Mont-Saint-Michel and its Bay p58; Palace and Park of Fontainebleau p113; Palace and Park of Versailles p62; Paris, Banks of the Seine p360; Place Stanislas, Place de la Carrière and Place d'Alliance in Nancy p169; Pont du Gard (Roman Aqueduct) p229; Prehistoric Sites and Decorated Caves of the Vézère Valley p44▼; Provins, Town of Medieval Fairs p669; Pyrénées - Mont Perdu p508▼; Roman Theatre and its Surroundings and the 'Triumphal Arch' of Orange p120; Routes of Santiago de Compostela in France p557▼; Royal Saltworks of Arc-et-Senans p153▲; Strasbourg – Grande île p334; The Loire Valley between Sully sur-Loire and Chalonnes p622; Vézelay, Church and Hill p66.

Greece

Acropolis, Athens p282; Archaeological Site of Aigai (modern name Vergina) p504▲; Archaeological Site of Delphi p262; Archaeological Site of Mystras p341; Archaeological Site of Olympia p336; Archaeological Sites of Mycenae and Tiryns p574; Delos p353; Historic Centre (Chorá) with the Monastery of Saint John, the Theologian, and the Cave of the Apocalypse on the Island of Pátmos p603; Medieval City of Rhodes p324; Meteora p332; Monasteries of Daphni, Hosios Loukas and Nea Moni of Chios p342▼; Mount Athos p309; Old Town of Corfu p771; Paleochristian and Byzantine Monuments of Thessalonika p313▼; Pythagoreion and Heraion of Samos p388▼; Sanctuary of Asklepios at Epidaurus p315; Temple of Apollo Epicurius at Bassae p240▼.

Holy See

Historic Centre of Rome, the Properties of the Holy See in that City Enjoying Extraterritorial Rights and San Paolo Fuori le Mura p92; Vatican City p180.

Hungary

Budapest, including the Banks of the Danube, the Buda Castle Quarter and Andrássy Avenue p274; Caves of Aggtelek Karst and Slovak Karst p456▲; Early Christian Necropolis of Pécs (Sopianae) p619▼; Fertö / Neusiedlersee Cultural Landscape p664▼; Hortobágy National Park - the Puszta p583▼; Millenary Benedictine Abbey of Pannonhalma and its Natural Environment p479; Old Village of Hollókő and its Surroundings p293; Tokaj Wine Region Historic Cultural Landscape p686.

Italy

Archaeological Area and the Patriarchal Basilica of Aquileia p565▲; Archaeological Area of Agrigento p542; Archaeological Areas of Pompeii, Herculaneum and Torre Annunziata p524; Assisi, the Basilica of San Francesco and Other Franciscan Sites p632; Botanical Garden (Orto Botanico), Padua p532▲; Castel del Monte p506▲; Cathedral, Torre Civica and Piazza Grande, Modena p520, Church and Dominican Convent of Santa Maria delle Grazie with 'The Last Supper' by Leonardo da Vinci p95; Cilento and Vallo di Diano National Park with the Archeological sites of Paestum and Velia, and the Certosa di Padula p558▲; City of Verona p618; City of Vicenza and the Palladian Villas of the Veneto p428; Costiera Amalfitana p538; Crespi d'Adda p465; Early Christian Monuments of Ravenna p487; Eighteenth-Century Royal Palace at Caserta with the Park, the Aqueduct of Vanvitelli, and the San Leucio Complex p527; Etruscan Necropolises of Cerveteri and Tarquinia p729▼; Ferrara, City of the Renaissance, and its Po Delta p453; Genoa: Le Strade Nuove and the system of the Palazzi dei Rolli p759▲; Historic Centre of Florence p134; Historic Centre of Naples p448; Historic Centre of Rome, the Properties of the Holy See in that City Enjoying Extraterritorial Rights and San Paolo Fuori le Mura p92; Historic Centre of San Gimignano p358; Historic Centre of Siena p460; Historic Centre of the City of Pienza p495; Historic Centre of Urbino p567; Isole Eolie (Aeolian Islands) p625; Late Baroque Towns of the Val di Noto (South-Eastern Sicily) p690▼; Mantua and Sabbioneta p797; Piazza del Duomo, Pisa p265; Portovenere, Cinque Terre, and the Islands (Palmaria, Tino and Tinetto) p513; Residences of the Royal House of Savoy p541; Rhaetian Railway in the Albula / Bernina Landscapes p784▲; Rock Drawings in Valcamonica p76▲; Sacri Monti of Piedmont and Lombardy p695▲; Su Nuraxi di Barumini p546▼; Syracuse and the Rocky Necropolis of Pantalica p750▼; The Sassi and the park of the Rupestrian Churches of Matera p399; The Trulli of Alberobello p501; Val d'Orcia p724; Venice and its Lagoon p268; Villa Adriana (Tivoli) p595; Villa d'Este, Tivoli p673; Villa Romana del Casale p546▲.

Luxembourg

City of Luxembourg: its Old Quarters and Fortifications p424.

Malta

City of Valletta p93; Hal Saflieni Hypogeum p98▲; Megalithic Temples of Malta p96.

Montenegro

Durmitor National Park p107; Natural and Culturo-Historical Region of Kotor p84.

Republic of Moldova

Struve Geodetic Arc p747▲.

Romania

Churches of Moldavia p412▲; Dacian Fortresses of the Orastie Mountains p597; Danube Delta p362; Historic Centre of Sighişoara p592; Monastery of Horezu p408▼; Villages with Fortified Churches in Transylvania p404▼; Wooden Churches of Maramureş p581.

San Marino

San Marino Historic Centre and Mount Titano p782.

Serbia

Gamzigrad-Romuliana, Palace of Galerius p776; Medieval Monuments in Kosovo p717▼; Stari Ras and Sopoćani p76▼; Studenica Monastery p250▼.

Slovakia

Bardejov Town Conservation Reserve p645; Caves of Aggtelek Karst and Slovak Karst p456▲; Historic Town of Banská Štiavnica and the Technical Monuments in its Vicinity p414; Primeval Beech Forests of the Carpathians p772; Spišský Hrad and its Associated Cultural Monuments p417; Vlkolínec p418; Wooden Churches of the Slovak part of the Carpathian Mountain Area p795▲.

Slovenia

Škocjan Caves p261▲.

Switzerland

Benedictine Convent of St John at Müstair p159▼; Convent of St Gall p176▲; Lavaux Vineyard Terraces p777; Monte San Giorgio p704▲; Old City of Berne p157; Rhaetian Railway in the Albula/Bernina Landscapes p784▲; Swiss Alps Jungfrau-Aletsch p660; Swiss Tectonic Arena Sardona p787; Three Castles, Defensive Wall and Ramparts of the Market-Town of Bellinzone p610▼.

The former Yugoslav Republic of Macedonia

Natural and Cultural Heritage of the Ohrid Region p77.

World Heritage Sites
North America and the Caribbean

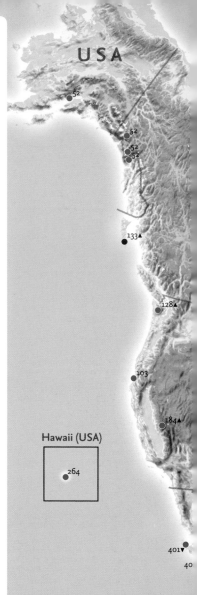

USA

Belize
Belize Barrier Reef Reserve System p483.

Canada
Canadian Rocky Mountain Parks p194; Dinosaur Provincial Park p48; Gros Morne National Park p302; Head-Smashed-In Buffalo Jump p118▲; Historic District of Old Québec p206; Joggins Fossil Cliffs p796▲; Kluane / Wrangell-St Elias / Glacier Bay / Tatshenshini-Alsek p52; L'Anse aux Meadows National Historic Site p37▼; Miguasha National Park p599▼; Nahanni National Park p38▲; Old Town Lunenburg p463; Rideau Canal p773; SGang Gwaay p133▲; Waterton Glacier International Peace Park p450; Wood Buffalo National Park p170.

Costa Rica
Area de Conservación Guanacaste p596▼; Cocos Island National Park p516▲; Talamanca Range-La Amistad Reserves / La Amistad National Park p162▲.

Cuba
Alejandro de Humboldt National Park p672▲; Archaeological Landscape of the First Coffee Plantations in the South-East of Cuba p649▼; Desembarco del Granma National Park p583▲; Historic Centre of Camagüey p785▼; Old Havana and its Fortifications p146; San Pedro de la Roca Castle, Santiago de Cuba p548▲; Trinidad and the Valley de los Ingenios p323; Urban Historic Centre of Cienfuegos p749; Viñales Valley p593▼.

Dominica
Morne Trois Pitons National Park p515.

Dominican Republic
Colonial City of Santo Domingo p343.

El Salvador
Joya de Cerén Archaeological Site p416▼.

Guatemala
Antigua Guatemala p73; Archaeological Park and Ruins of Quirigua p130▼; Tikal National Park p88▼.

Honduras
Maya Site of Copán p90; Río Plátano Biosphere Reserve p142▼.

Haiti
National History Park – Citadel, Sans Souci, Ramiers p139▲.

Mexico
Agave Landscape and Ancient Industrial Facilities of Tequila p754; Ancient Maya City of Calakmul, Campeche p687▲; Archaeological Monuments Zone of Xochicalco p605; Archeological Zone of Paquimé, Casas Grandes p563▲; Central University City Campus of the Universidad Nacional Autónoma de México (UNAM) p770▲; Earliest 16th-Century Monasteries on the Slopes of Popocatepetl p432▲; El Tajin, Pre-Hispanic

City p394; Franciscan Missions in the Sierra Gorda of Querétaro p703▲; Historic Centre of Mexico City and Xochimilco p284▲; Historic Centre of Morelia p36g; Historic Centre of Oaxaca and Archaeological Site of Monte Albán p286; Historic Centre of Puebla p284▼; Historic Centre of Zacatecas p419▲; Historic Fortified Town of Campeche p577; Historic Monuments Zone of Querétaro p496; Historic Monuments Zone of Tlacotalpan p548▼; Historic Town of Guanajuato and Adjacent Mines p318; Hospicio Cabañas, Guadalajara p534; Islands and Protected Areas of the Gulf of California p733; Luis Barragán House and Studio p722▲; Monarch Butterfly Biosphere Reserve p791; Pre-Hispanic City and National Park of Palenque p299; Pre-Hispanic City of Chichen-Itza p330; Pre-Hispanic City of Teotihuacan p288; Pre-Hispanic Town of Uxmal p489; Protective town of San Miguel and the Sanctuary of Jesús Nazareno de Atotonilco p781▼; Rock Paintings of the Sierra de San Francisco p419▼; Sian Ka'an p267; Whale Sanctuary of El Vizcaino p401▼.

Nicaragua
Ruins of León Viejo p619▲.

Panama
Archaeological Site of Panamá Viejo and Historic District of Panamá p529; Coiba National Park and its Special Zone of Marine Protection p745▲; Darien National Park p130▲; Fortifications on the Caribbean Side of Panama: Portobelo-San Lorenzo p99; Talamanca Range-La Amistad Reserves / La Amistad National Park p162▲.

Saint Kitts and Nevis
Brimstone Hill Fortress National Park p596.

Saint Lucia
Pitons Management Area p711.

United States of America
Cahokia Mounds State Historic Site p151▲; Carlsbad Caverns National Park p446; Chaco Culture p285; Everglades National Park p53; Grand Canyon National Park p54; Great Smoky Mountains National Park p164; Hawaii Volcanoes National Park p264; Independence Hall p64; La Fortaleza and San Juan National Historic Site in Puerto Rico p166; Kluane / Wrangell-St Elias / Glacier Bay / Tatshenshini-Alsek p52; Mammoth Cave National Park p126▲; Mesa Verde National Park p28; Monticello and the University of Virginia in Charlottesville p304; Olympic National Park p128▲; Pueblo de Taos p380▲; Redwood National Park p103; Statue of Liberty p187; Waterton Glacier International Peace Park p450; Yellowstone National Park p34; Yosemite National Park p184▲.

Hawaii (USA)
264

133▲
128▲
103
184▲
401▼
40
52

708
Greenland
(Denmark)

Only countries States Parties to the World Heritage Convention are labelled on these maps. United Nations countries boundaries shown as of March 2009.

170

C A N A D A

37▼

302

48
118▲
○

599▼

796

206
463

34

773

UNITED STATES
OF AMERICA

187
64

151▲

304

126▲

621▲
Bermuda (UK)

28
285
380▲

164

446

563▲

53

MEXICO

146
749
785▼
Puerto Rico (USA)

ANTIGUA AND
BARBUDA

593▼
323
649▼
672▲
139▲
166

781▼
419▲

CUBA
583▼
548▲
343
SAINT KITTS AND NEVIS

733

703▲
770▲, 722▲

330
HAITI
DOMINICAN
REPUBLIC

596
515
DOMINICA

754
318
394

JAMAICA

SAINT LUCIA
711

534
496
288

489
267

BARBADOS

369
791
432▲

577
687▲

SAINT VINCENT AND
THE GRENADINES

284▲
284▼
548▼

299
88▼
483

BELIZE
GRENADA

605
286

130▼

HONDURAS
142▲

TRINIDAD
AND
TOBAGO

73
90

416▼
619▲
NICARAGUA

GUATEMALA
EL SALVADOR

Cocos Island
(COSTA RICA)

596▼
162▲
99

516▲
162▲
529
745
130▲

COSTA RICA
PANAMA

Scale 1 : 37 500 000

15

528▼
398▼ 644▲
184▼
473▼
426▲
687▼
431 GUYANA
658▲
VENEZUELA
756▼
SURINAME
COLOMBIA
473▼
458
624
537
32▼
171▲
ECUADOR
576▼
668
374▲
356▼
238
BRAZIL
214▲
210▲ PERU 296
326
172 158
208
681▲ 579▼
434
636▼
674 280
617
634 BOLIVIA
681▲ 588▼
556▼
656
348▲ 606 223 108 579▼
367▲
289
606
736
606
705 CHILE PARAGUAY
235
188 407▲
159▲ 159▲
628
628 647▼
472
URUGUAY
696▲
762▲
ARGENTINA
620▲
590
607▲
111

Galápagos Islands
(Ecuador)
26

Rapa Nui (Chile)
454

Scale 1 : 37 000 000

World Heritage Sites
South America

Argentina
Cueva de las Manos, Río Pinturas p607▲; Iguazu National Park p188; Ischigualasto / Talampaya Natural Parks p628; Jesuit Block and Estancias of Córdoba p647▼; Jesuit Missions of the Guaranis: San Ignacio Miní, Santa Ana, Nuestra Señora de Loreto and Santa María Mayor (Argentina), Ruins of Saõ Miguel das Missões (Brazil) p159▲; Los Glaciares p111; Península Valdés p590; Quebrada de Humahuaca p705.

Bolivia
City of Potosí p289; Fuerte de Samaipata p556▼; Historic City of Sucre p367▲; Jesuit Missions of the Chiquitos p348▲; Noel Kempff Mercado National Park p636▼, Tiwanaku: Spiritual and Political Centre of the Tiwanaku Culture p634.

Brazil
Atlantic Forest South-East Reserves p606; Brasilia p280; Brazilian Atlantic Islands: Fernando de Noronha and Atol das Rocas Reserves p668▲; Central Amazon Conservation Complex p624; Cerrado Protected Areas: Chapada dos Veadeiros and Emas National Parks p681▲; Discovery Coast Atlantic Forest Reserves p579▼; Historic Centre of Salvador de Bahia p208; Historic Centre of São Luís p537; Historic Centre of the Town of Diamantina p588▼; Historic Centre of the Town of Goiás p674; Historic Centre of the Town of Olinda p150; Historic Town of Ouro Preto p108; Iguaçu National Park p235; Jesuit Missions of the Guaranis: San Ignacio Miní, Santa Ana, Nuestra Señora de Loreto and Santa María Mayor (Argentina), Ruins of Saõ Miguel das Missões (Brazil) p159▲; Pantanal Conservation Area p656; Sanctuary of Bom Jesus do Congonhas p223; Serra da Capivara National Park p374▲.

Chile
Churches of Chiloé p620▲; Historic Quarter of the Seaport City of Valparaíso p696▲; Humberstone and Santa Laura Saltpeter Works p736; Rapa Nui National Park p454; Sewell Mining Town p762▲.

Colombia
Historic Centre of Santa Cruz de Mompox p473▲; Los Katíos National Park p426▲; Malpelo Fauna and Flora Sanctuary p756▼; National Archeological Park of Tierradentro p473▼; Port, Fortresses and Group of Monuments, Cartagena p184▼; San Agustín Archeological Park p458.

Ecuador
City of Quito p32▼; Galápagos Islands p26; Historic Centre of Santa Ana de los Ríos de Cuenca p576▼; Sangay National Park p171▲.

Paraguay
Jesuit Missions of La Santísima Trinidad de Paraná and Jesús de Tavarangue p407▲.

Peru
Chan Chan Archaeological Zone p238; Chavín (Archaeological site) p210▲; City of Cuzco p158; Historic Centre of Lima p326; Historic Sanctuary of Machu Picchu p172; Historical Centre of the City of Arequipa p617; Huascarán National Park p214; Lines and Geoglyphs of Nasca and Pampas de Jumana p434; Manú National Park p296; Río Abiseo National Park p356▼.

Suriname
Central Suriname Nature Reserve p658▲; Historic Inner City of Paramaribo p687▼.

Uruguay
Historic Quarter of the City of Colonia del Sacramento p472.

Venezuela
Canaima National Park p431; Ciudad Universitaria de Caracas p644▲; Coro and its Port p398▼.

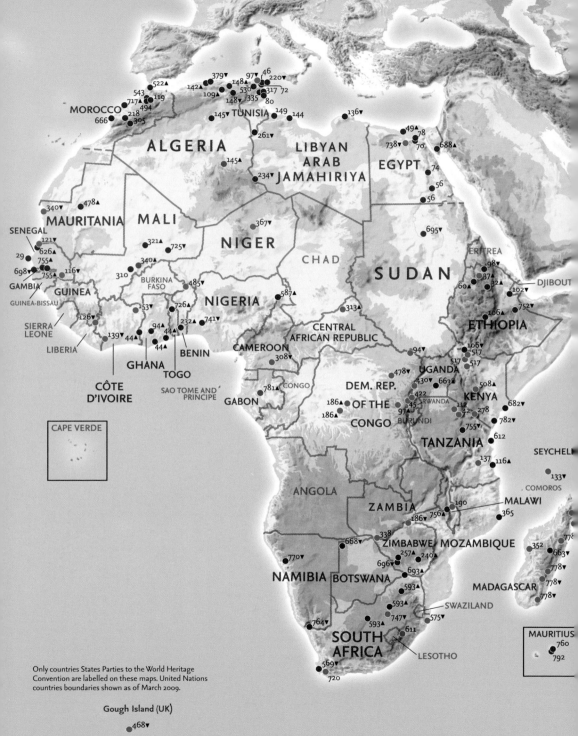

MOROCCO

543
717▲
494
666
218
305
119
522▲
142▲
379▼
109▲
148▼
530
148▼
335
97▲
46
220▼
317
80
72

TUNISIA
145▲
261▼
149▼
144

ALGERIA

LIBYAN
ARAB
JAMAHIRIYA

145▲
234▼

EGYPT
136▼
49▲
78
738▼
70
688▲
74
56
56
695▼

SENEGAL
121▼
29
626▲
755▼
698▼
755▲
116▼
340▼
478▲

MAURITANIA

MALI
321▼
725▼
340▼
310
485▲
587▲

NIGER

CHAD

SUDAN

ERITREA
98▼
37▼
32▼
60▼
102▼
DJIBOUT
106▲
752▼
106▲
517▼
517▲
508▲

ETHIOPIA

GAMBIA
GUINEA-BISSAU
GUINEA
SIERRA
LEONE
LIBERIA
126▼
139▼
44▲
153▼
726▲
232▼
741▼
94▲
44▲
44▼

Burkina
Faso

NIGERIA
313▼

CENTRAL
AFRICAN REPUBLIC

94▼
478▼
430▼
663▲
422
45
97▼
112
42▼
278
755▼
782▼
UGANDA
517▲
508▲
KENYA
682▼

GHANA
TOGO
BENIN
CAMEROON
308▼
781▼
CONGO
186▲
186▼
GABON

DEM. REP.
OF THE
CONGO
RWANDA
BURUNDI

TANZANIA
612
137▼
116▲

SEYCHEL

133▼

COMOROS

ANGOLA

ZAMBIA
186▼
756▲
190
MALAWI
365

352
778▼
663▲
778▼
778▼
778▼

MADAGASCAR

668▲
338
257▲
240▼
696▼
693▼
593▲
593▲
593▲
747▼
575▼
764▼
611
ZIMBABWE
MOZAMBIQUE

NAMIBIA
770▼

BOTSWANA

SWAZILAND

MAURITIUS
760
792

SOUTH
AFRICA

LESOTHO

569▼
720

CAPE VERDE

Only countries States Parties to the World Heritage
Convention are labelled on these maps. United Nations
countries boundaries shown as of March 2009.

Gough Island (UK)
468▼

CÔTE
D'IVOIRE

SAO TOME AND
PRINCIPE

Scale 1 : 50 000 000

World Heritage Sites
Africa

Algeria
Al Qal'a of Beni Hammad p109▲; Djémila p148▲; Kasbah of Algiers p379▼; M'Zab Valley p145▼; Tassili n'Ajjer p145▲; Timgad p148▼; Tipasa p142▲.

Benin
Royal Palaces of Abomey p232▲.

Botswana
Tsodilo p668▼.

Cameroon
Dja Faunal Reserve p308▼.

Central African Republic
Manovo-Gounda St Floris National Park p313▲.

Congo, Democratic Republic of the
Garamba National Park p94▼; Kahuzi-Biega National Park p97▲; Okapi Wildlife Reserve p478▼; Salonga National Park p186▲; Virunga National Park p45.

Côte d'Ivoire
Comoé National Park p153▼; Mount Nimba Strict Nature Reserve p126▼; Taï National Park p139▼.

Egypt
Abu Mena p49▲; Ancient Thebes with its Necropolis p74; Historic Cairo p78; Memphis and its Necropolis – the Pyramid Fields from Giza to Dahshur p70; Nubian Monuments from Abu Simbel to Philae p56; Saint Catherine Area p688▲; Wadi Al-Hitan (Whale Valley) p738▼.

Ethiopia
Aksum p98▼; Fasil Ghebbi, Gondar Region p60▲; Harar Jugol, the Fortified Historic Town p752▼; Lower Valley of the Awash p102▼; Lower Valley of the Omo p106▼; Rock-Hewn Churches, Lalibela p32▲; Simien National Park p37▲; Tiya p106▲.

Gabon
Ecosystem and Relict Cultural Landscape of Lopé-Okanda p781▲.

Gambia
James Island and Related Sites p698▼; Stone Circles of Senegambia p755▲.

Ghana
Asante Traditional Buildings p94▲; Forts and Castles, Volta, Greater Accra, Central and Western Regions p44▲.

Guinea
Mount Nimba Strict Nature Reserve p126▼.

Kenya
Lake Turkana National Parks p517; Lamu Old Town p682▼; Mount Kenya National Park/Natural Forest p508▲; Sacred Mijikenda Kaya Forests p782▼.

Libyan Arab Jamahiriya (Libya)
Archaeological Site of Cyrene p136▼; Archaeological Site of Leptis Magna p144; Archaeological Site of Sabratha p149; Old Town of Ghadamès p261▼; Rock-Art Sites of Tadrart Acacus p234▼.

Madagascar
Rainforests of the Atsinanana p778▼; The Royal Hill of Ambohimanga p663▼; Tsingy de Bemaraha Strict Nature Reserve p352.

Malawi
Chongoni Rock-Art Area p756▲; Lake Malawi National Park p190.

Mali
Cliff of Bandiagara (Land of the Dogons) p340▲; Old Towns of Djenné p310; Timbuktu p321▲; Tomb of Askia p725▼.

Mauritania
Ancient Ksour of Ouadane, Chinguetti, Tichitt and Oualata p478▲; Banc d'Arguin National Park p340▼.

Mauritius
Aapravasi Ghat p760▼; Le Morne Cultural Landscape p792▲.

Morocco
Archaeological Site of Volubilis p543; Historic City of Meknes p494; Ksar of Ait-Ben-Haddou p305; Medina of Essaouira (formerly Mogador) p666; Medina of Fez p119; Medina of Marrakesh p218; Medina of Tétouan (formerly known as Titawin) p522▼; Portuguese City of Mazagan (El Jadida) p717▲.

Mozambique
Island of Mozambique p365.

Namibia
Twyfelfontein or /Ui-//aes p770▼.

Niger
Aïr and Ténéré Natural Reserves p367▼; W National Park of Niger p485▼.

Nigeria
Osun-Osogbo Sacred Grove p741▼; Sukur Cultural Landscape p587▲.

Senegal
Djoudj National Bird Sanctuary p121▼; Island of Gorée p29; Island of Saint-Louis p626▲; Niokolo-Koba National Park p116▼; Stone Circles of Senegambia p755▲.

Seychelles
Aldabra Atoll p133▼; Vallée de Mai Nature Reserve p171▼.

South Africa
Cape Floral Region Protected Areas p720; Fossil Hominid Sites of Sterkfontein, Swartkrans, Kromdraai, and Environs p593▲; iSimangaliso Wetland Park p575▼; Mapungubwe Cultural Landscape p693▲; Richtersveld Cultural and Botanical Landscape p764▼; Robben Island p569▼; uKhahlamba / Drakensberg Park p611; Vredefort Dome p747▼.

Sudan
Gebel Barkal and the Sites of the Napatan Region p695▼.

Tanzania
Kilimanjaro National Park p278; Kondoa Rock-Art Sites p755▼; Ngorongoro Conservation Area p42; Ruins of Kilwa Kisiwani and Ruins of Songo Mnara p116▲; Selous Game Reserve p137; Serengeti National Park p112; Stone Town of Zanzibar p612.

Togo
Koutammakou, the Land of the Batammariba p726▲.

Tunisia
Amphitheatre of El Jem p80; Dougga / Thugga p530; Ichkeul National Park p97▼; Kairouan p335; Medina of Sousse p317; Medina of Tunis p72; Punic Town of Kerkuane and its Necropolis p220▼; Site of Carthage p46.

Uganda
Bwindi Impenetrable National Park p422; Rwenzori Mountains National Park p430▼; Tombs of Buganda Kings at Kasubi p663▲.

Zambia
Mosi-oa-Tunya / Victoria Falls p338.

Zimbabwe
Great Zimbabwe National Monument p240▲; Khami Ruins National Monument p257▲; Mana Pools National Park, Sapi and Chewore Safari Areas p186▼; Mosi-oa-Tunya / Victoria Falls p338; Matobo Hills p696▼.

PALAU
0

786▲

PAPUA
NEW GUINEA

127

311

699

429▼

122

AUSTRALIA

306

368▲

260

125

609

763

429▼

718

143

Only countries State Parties to the World Heritage
Convention are labelled on these maps. United Nations
countries boundaries shown as of March 2009.

World Heritage Sites
Oceania

Australia
Australian Fossil Mammal Sites (Riversleigh /
Naracoorte) p429▼; Fraser Island p393; Gondwana
Rainforests of Australia p260; Great Barrier Reef p122;
Greater Blue Mountains Area p609; Heard and
McDonald Islands p522▼ (not on map); Kakadu
National Park p127; Lord Howe Island Group p136▲;
Macquarie Island p506▼ (not on map); Purnululu
National Park p699; Royal Exhibition Building and
Carlton Gardens p718; Shark Bay, Western Australia
p368▲; Sydney Opera House p763; Tasmanian
Wilderness p143; Uluṟu-Kata Tjuṯa National Park p306;
Wet Tropics of Queensland p311; Willandra Lakes
Region p125.

New Zealand
New Zealand Sub-Antarctic Islands p556▲;
Te Wahipounamu – South West New Zealand p354;
Tongariro National Park p349.

Papua New Guinea
Kuk Early Agricultural Site p786▲.

Solomon Islands
East Rennell p558▼.

Vanuatu
Chief Roi Mata's Domain p795▼.

SOLOMON
ISLANDS

558▼

VANUATU 795▼

FIJI

788▼

New Caledonia
(France)

136▲

NEW ZEALAND

349

354
354
354
354

556▲

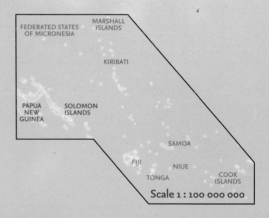

FEDERATED STATES
OF MICRONESIA

MARSHALL
ISLANDS

KIRIBATI

PAPUA
NEW
GUINEA

SOLOMON
ISLANDS

SAMOA

FIJI

NIUE

TONGA

COOK
ISLANDS

Scale 1 : 100 000 000

Henderson Island (UK)

788▼

Scale 1 : 27 000 000

RUSSIAN FEDERA

KAZAKHSTAN

601▲
GEORGIA
481▲
ARMENIA
435▼ 437
500
655▼ 698▲
630▲
AZERBAIJAN
792▼ 767▲ 607▼
215▲ 691
697▲ 746▲
767▲ 762▼
IRAQ
88▲ 38▲
ISLAMIC REP.
OF
798 IRAN
KUWAIT
714
68
746▼
SAUDI BAHRAIN
ARABIA QATAR
UNITED ARAB
EMIRATES
322 760▲
303 760▲
760▲ 760▲
245 OMAN
614
152
398▲ YEMEN

786▼
693▼
557▼
MON

750▲ 707▲ 729▲
348▼
UZBEKISTAN
778▼ 405
573 670
629 TAJIKISTAN
298▲
683 703▼
AFGHANISTAN CH
110
109▲
532▼ 124
576▲
PAKISTAN 333▲
102▲ 425
765▲ 403
216▲ 413 NEPAL 85 BHUTAN
247 161 196 676▲ 215▲
154 512▲ 50 224
710▼ 342▲ 249 688▼ 222
730 694▼ BANGLADESH 228
156 271▲ 512▼
723 160 183 MYANMAR
271▼
308▲
237▲ 252

Socotra (Yemen)
785▲

204
576▲ 276

MALDIVES
138 151▼
363 140
314 SRI LANKA
320 316

364

Jammu and Kashmir: Dotted line represents approximately the
Line of Control in Jammu and Kashmir agreed upon by India and
Pakistan. The final status of Jammu and Kashmir has not been
agreed upon by the parties.

Only countries States Parties to the World Heritage Convention
are labelled on these maps. United Nations countries boundaries
shown as of March 2009.

Inset:
236
761
761 100
569▲ 191 SYRIAN
192 205 ARAB
LEBANON 197 39 REPUBLIC
680 739
793 739
694▲ 105
ISRAEL 114
Jerusalem 221
739 676 713
740 JORDAN
212

Scale 1 : 12 000 000

SEE INSET

World Heritage Sites
Asia, Middle East and Arabian Peninsula

Afghanistan
Cultural Landscape and Archaeological Remains of the Bamiyan Valley p703▼; Minaret and Archaeological Remains of Jam p683.

Armenia
Cathedral and Churches of Echmiatsin and the Archaeological Site of Zvartnots p655▼; Monasteries of Haghpat and Sanahin p500; Monastery of Geghard and the Upper Azat Valley p630▼.

Azerbaijan
Gobustan Rock Art Cultural Landscape p767▼; Walled City of Baku with the Shirvanshah's Palace and Maiden Tower p607▼.

Bahrain
Qal'at al-Bahrain – Ancient Harbour and Capital of Dilmun p746▼.

Bangladesh
Historic Mosque City of Bagerhat p228; Ruins of the Buddhist Vihara at Paharpur p222; The Sundarbans p512▼.

Cambodia
Angkor p390; Temple of Preah Vihear p789▲.

China
Ancient Building Complex in the Wudang Mountains p429▲; Ancient City of Ping Yao p540; Ancient Villages in Southern Anhui – Xidi and Hongcun p653; Capital Cities and Tombs of the Ancient Koguryo Kingdom p722▼; Classical Gardens of Suzhou p518; Dazu Rock Carvings p584; Fujian Tulou p788▲; Historic Centre of Macao p737; Historic Ensemble of the Potala Palace, Lhasa p425; Huanglong Scenic and Historic Interest Area p395; Imperial Palaces of the Ming and Qing Dynasties in Beijing and Shenyang p277; Imperial Tombs of the Ming and Qing Dynasties p616; Jiuzhaigou Valley Scenic and Historic Interest Area p380▼; Kaiping Diaolou and Villages p774; Longmen Grottoes p654; Lushan National Park p484▼; Mausoleum of the First Qin Emperor p300; Mogao Caves p298▲; Mount Emei Scenic Area, including Leshan Giant Buddha Scenic Area p481▲; Mount Huangshan p357; Mount Qingcheng and the Dujiangyan Irrigation System p648▲; Mount Sanqingshan National Park p796▼; Mount Taishan p297; Mount Wuyi p598; Mountain Resort and its Outlying Temples, Chengde p436▲; Old Town of Lijiang p510; Peking Man Site at Zhoukoudian p298▼; Sichuan Giant Panda Sanctuaries – Wolong, Mt Siguniang and Jiajin Mountains p751; South China Karst p768; Summer Palace and Imperial Garden in Beijing p552;

continued on page 24

Asia, Middle East and Arabian Peninsula (continued)

Temple and Cemetery of Confucius and the Kong Family Mansion in Qufu p433; Temple of Heaven: an Imperial Sacrificial Altar in Beijing p559; The Great Wall p294; Three Parallel Rivers of Yunnan Protected Areas p700; Wulingyuan Scenic and Historic Interest Area p382▲; Yin Xu p759▼; Yungang Grottoes p672▼.

Georgia
Bagrati Cathedral and Gelati Monastery p435▼; Historical Monuments of Mtskheta p437; Upper Svaneti p481▼.

India
Agra Fort p161; Ajanta Caves p156; Buddhist Monuments at Sanchi p342▲; Champaner-Pavagadh Archaeological Park p730; Chhatrapati Shivaji Terminus (formerly Victoria Terminus) p723; Churches and Convents of Goa p237▲; Elephanta Caves p271▼; Ellora Caves p160; Fatehpur Sikri p247; Great Living Chola Temples p276; Group of Monuments at Hampi p252; Group of Monuments at Mahabalipuram p204; Group of Monuments at Pattadakal p308▲; Humayun's Tomb, Delhi p403; Kaziranga National Park p224; Keoladeo National Park p216; Khajuraho Group of Monuments p249; Mahabodhi Temple Complex at Bodh Gaya p688▼; Manas Wildlife Sanctuary p215▼; Mountain Railways of India p576▲; Nanda Devi and Valley of Flowers National Parks p333▲; Qutb Minar and its Monuments, Delhi p413; Red Fort Complex p765; Rock Shelters of Bhimbetka p694▼; Sun Temple, Konârak p183; Sundarbans National Park p271▲; Taj Mahal p154.

Indonesia
Borobudur Temple Compounds p376; Komodo National Park p370; Lorentz National Park p604▲; Prambanan Temple Compounds p373; Sangiran Early Man Site p503▼; Tropical Rainforest Heritage of Sumatra p712; Ujung Kulon National Park p374▼.

Iran, Islamic Republic of
Armenian Monastic Ensembles of Iran p792▼; Bam and its Cultural Landscape p710▼; Bisotun p762▼; Meidan Emam, Esfahan p38▼; Pasargadae p714; Persepolis p68; Soltaniyeh p746▲; Takht-e Soleyman p691; Tchogha Zanbil p88▲.

Iraq
Ashur (Qal'at Sherqat) p697▲; Hatra p215▲; Samarra Archaeological City p767▲.

Israel
Bahá'í Holy Places in Haifa and the Western Galilee p793; Biblical Tels – Megiddo, Hazor, Beer Sheba p739; Incense Route – Desert Cities in the Negev p740; Masada p676; Old City of Acre p680; The White City of Tel-Aviv – The Modern Movement p694▲.

Japan
Buddhist Monuments in the Horyu-ji Area p415▲; Gusuku Sites and Related Properties of the Kingdom of Ryukyu p643▲; Himeji-jo p409;

Hiroshima Peace Memorial (Genbaku Dome) p477; Historic Monuments of Ancient Kyoto (Kyoto, Uji and Otsu Cities) p438; Historic Monuments of Ancient Nara p560; Historic Villages of Shirakawa-go and Gokayama p462▲; Itsukushima Shinto Shrine p486▲; Iwami Ginzan Silver Mine and its Cultural Landscape p764▲; Sacred Sites and Pilgrimage Routes in the Kii Mountain Range p726▼; Shirakami-Sanchi p415▼; Shiretoko p748▲; Shrines and Temples of Nikko p604▼; Yakushima p400.

Jerusalem (Site proposed by Jordan)
Old City of Jerusalem and its Walls p114.

Jordan
Petra p212; Quseir Amra p221; Um er-Rasas (Kastrom Mefa'a) p713.

Kazakhstan
Mausoleum of Khoja Ahmed Yasawi p707▲; Petroglyphs within the Archaeological Landscape of Tamgaly p729▲; Saryarka – Steppe and Lakes of Northern Kazakhstan p786▼.

Korea, Democratic People's Republic of
Complex of Koguryo Tombs p715▲.

Korea, Republic of
Changdeokgung Palace Complex p521; Gochang, Hwasun and Ganghwa Dolmen Sites p626▼; Gyeongju Historic Areas p642; Haeinsa Temple Janggyeong Panjeon, the Depositories for the Tripitaka Koreana Woodblocks p462▼; Hwaseong Fortress p536; Jeju Volcanic Island and Lava Tubes p780; Jongmyo Shrine p468▲; Seokguram Grotto and Bulguksa Temple p457.

Lao People's Democratic Republic (Laos)
Town of Luang Prabang p464; Vat Phou and Associated Ancient Settlements within the Champasak Cultural Landscape p679.

Lebanon
Anjar p205; Baalbek p191; Byblos p192; Ouadi Qadisha (the Holy Valley) and the Forest of the Cedars of God (Horsh Arz el-Rab) p569▲; Tyre p197.

Malaysia
Gunung Mulu National Park p638; Kinabalu Park p631; Melaka and George Town, Historic Cities of the Straits of Malacca p790▲.

Mongolia
Orkhon Valley Cultural Landscape p719▲; Uvs Nuur Basin p693▼.

Nepal
Kathmandu Valley p50; Lumbini, the Birthplace of the Lord Buddha p512▲; Royal Chitwan National Park p196; Sagarmatha National Park p85.

Oman
Aflaj Irrigation Systems of Oman p760▲; Archaeological sites of Bat, Al-Khutm and Al-Ayn p322; Bahla Fort p303; Land of Frankincense p614.

Pakistan
Archaeological Ruins at Moenjodaro p102▲; Buddhist Ruins of Takht-i-Bahi and Neighbouring City Remains at Sahr-i-Bahlol p110; Fort and Shalamar Gardens in Lahore p124; Historic Monuments of Thatta p118▼; Rohtas Fort p532▼; Taxila p109▼.

Philippines
Baroque Churches of the Philippines p402; Historic Town of Vigan p582; Puerto-Princesa Subterranean River National Park p601▼; Rice Terraces of the Philippine Cordilleras p466; Tubbataha Reef Marine Park p412▼.

Russian Federation (see also p12)
Central Sikhote-Alin p664▲; Citadel, Ancient City and Fortress Buildings of Derbent p698▲; Golden Mountains of Altai p557▲; Lake Baikal p476; Natural System of Wrangel Island Reserve p716▼; Volcanoes of Kamchatka p490; Western Caucasus p601▲.

Saudi Arabia
Al-Hijr Archaeological Site (Madâin Sâlih) p798.

Sri Lanka
Ancient City of Polonnaruwa p140; Ancient City of Sigiriya p151▼; Golden Temple of Dambulla p363; Old Town of Galle and its Fortifications p320; Sacred City of Anuradhapura p138; Sacred City of Kandy p314; Sinharaja Forest Reserve p316.

Syrian Arab Republic (Syria)
Ancient City of Aleppo p236; Ancient City of Bosra p105; Ancient City of Damascus p39; Crac des Chevaliers and Qal'at Salah El-Din p761; Site of Palmyra p100.

Thailand
Ban Chiang Archaeological Site p388▲; Dong Phayayen-Khao Yai Forest Complex p741▲; Historic City of Ayutthaya p359; Historic Town of Sukhothai and Associated Historic Towns p375; Thungyai-Huai Kha Khaeng Wildlife Sanctuaries p364.

Turkmenistan
Kunya-Urgench p750▲; Parthian Fortresses of Nisa p778▲; State Historical and Cultural Park 'Ancient Meru' p573.

Uzbekistan
Historic Centre of Bukhara p405; Historic Centre of Shakhrisyabz p629; Itchan Kala p348▼; Samarkand – Crossroads of Cultures p670.

Vietnam
Complex of Hué Monuments p411; Ha Long Bay p440; Hoi An Ancient Town p586; My Son Sanctuary p579▲; Phong Nha-Ke Bang National Park p706.

Yemen
Historic Town of Zabid p398▲; Old City of Sana'a p245; Old Walled City of Shibam p152; Socotra Archipelago p785▲.

The World Heritage sites,
ordered by the year they were
first inscribed on the List.

World Heritage site since

1978 1979 1980 1981 1982 1983 1984 1985 1986 1987 1988 1989 1990 1991 1992 1993 1994 1995 1996 1997 1998 1999 2000 2001 2002 2003 2004 2005 2006 2007 2008

Galápagos Islands
Ecuador

Criteria – Natural phenomena or beauty; Major stages of Earth's history; Significant ecological and biological processes; Significant natural habitat for biodiversity

Situated in the Pacific Ocean approximately 1,000 km from the South American mainland, the Galápagos Archipelago of nineteen major islands and their marine reserve have been called a unique 'living museum and showcase of evolution'. Located at the confluence of three ocean currents, the Galápagos are a 'tossed salad' of marine species.

Volcanic processes formed the islands, most of which are volcanic summits, some rising over 3,000 m from the Pacific floor. They vary greatly in altitude, area and orientation and these differences, combined with their physical separation, contributed towards the species diversity and endemism on particular islands. Ongoing seismic and volcanic activity reflects the processes that formed the islands and it was these processes, together with the islands' extreme isolation, that led to the development of unusual animal life – such as the marine iguana, the giant tortoise and the flightless cormorant – that inspired Charles Darwin's theory of evolution following his visit in 1835.

The western part of the archipelago experiences intense volcanic and seismic activity. The larger islands typically comprise at least one gently sloping shield volcano, culminating in collapsed craters or calderas.

Long stretches of shoreline are only slightly eroded, but in many places faulting and marine erosion have produced steep cliffs and lava, coral or shell sand beaches.

There is coastal vegetation along beaches, salt-water lagoons and low, broken, boulder-strewn shores, and mangrove swamps dominate protected coves and lagoons. The arid zone that lies immediately inland dominates the Galápagos landscape. The humid zone emerges above the arid zone through a transition belt in which elements of the two are combined. It is very damp and is maintained in the dry season by thick, garua fogs. A fern-grass-sedge zone covers the summit areas of the larger islands where moisture is retained in temporary pools.

The endemic fauna includes invertebrate, reptile, marine and bird species. There are a few indigenous mammals. All the reptiles, except for two marine turtles, are endemic.

Marine environments are highly varied and are associated with water temperature regimes reflecting differences in nutrient and light levels. These range from warm temperate conditions brought on by vigorous upwelling (Cromwell Current) and a moderately cool, warm temperate-subtropical influence (Peru Flow).

▶ An endemic Galápagos giant tortoise. Adults in the wild can grow up to 1.2 m in length and live for 150 years. There are now only eleven subspecies remaining from the original twelve.

▶ A Sally Lightfoot (Graspus Graspus) crab which is endemic to the Galápagos Islands and lives on the rocky shore, feeding on algae and dead fish, birds and seals.

The Heritage site is situated on the Galápagos Submarine Platform and consists of about 120 islands in total. The larger islands in the group are Isabela, Santa Cruz, Fernandina, Santiago and San Cristobal.

World Heritage site since

1978 1979 1980 1981 1982 1983 1984 1985 1986 1987 1988 1989 1990 1991 1992 1993 1994 1995 1996 1997 1998 1999 2000 2001 2002 2003 2004 2005 2006 2007 2008

Mesa Verde National Park
USA

Criteria – Testimony to cultural tradition

Standing on the Mesa Verde plateau in southwest Colorado at a height of more than 2,600 m is this concentration of ancestral Pueblo Indian dwellings. Their originality derives in part from the unique local topography of mesas, or tablelands, intersected by deep canyons, that dictated their construction. The dwellings were also designed to cope with the challenging local climate: a semi-arid environment with irregular rainfall and extremes of temperature between day and night.

The dwellings were built by the Anasazi Indians, ancestors of the Pueblos, between the sixth and twelfth centuries, and the earliest habitations are found largely on the plateau. Later, villages grew in and around the cave-studded sides of the cuestas where erosion had left protective overhanging cliffs. Some of these imposing stone-built cliff dwellings comprise more than 100 rooms. In all, some 4,400 sites have been recorded.

▲
The Mesa Verde site has been protected by the Federal Antiquities Act since 1906.

The cliff-side villages had specific functions: agricultural, handicraft and religious.

The Anasazi developed irrigation techniques to cultivate the cereal crops that were central to their diet. At their civilisation's high point, they produced high-quality ceramics, weaving and wickerwork made from yucca fibre.

World Heritage site since

1978 1979 1980 1981 1982 1983 1984 1985 1986 1987 1988 1989 1990 1991 1992 1993 1994 1995 1996 1997 1998 1999 2000 2001 2002 2003 2004 2005 2006 2007 2008

Island of Gorée
Senegal
Criteria – Heritage associated with events of universal significance

The island of Gorée, which lies off the coast of Senegal opposite Dakar, is a memorial to the African diaspora. From the fifteenth to the nineteenth century, it was the largest slave-trading centre on the African coast. Ruled in succession by the Portuguese, Dutch, English and French, its architecture is characterized by the contrast between the grim slave-quarters and the elegant houses of the slave traders. An estimated 20 million Africans passed through the island between the mid-1500s and the mid 1800s. A small slave-house contained 150–200 slaves, who would often have to endure months in appalling conditions before being shipped to the Americas. Today it continues to serve as a reminder of human exploitation and as a sanctuary for reconciliation.

Slave houses contained cells, 2.6 m square, in which fifteen to twenty men would be chained to the wall by their neck and arms. Attached to the chain was a large iron ball which the men would have to carry when they were allowed, once a day, to relieve themselves.

▼

World Heritage site since

1978 1979 1980 1981 1982 1983 1984 1985 1986 1987 1988 1989 1990 1991 1992 1993 1994 1995 1996 1997 1998 1999 2000 2001 2002 2003 2004 2005 2006 2007 2008

Cracow's Historic Centre
Poland

Criteria – Significance in human history

Wawel Cathedral ▶
(the Basilica of Saints
Stanisław and Vaclav)
was for centuries the
scene of the main
events of the Polish
royal families –
coronations, weddings,
funerals and burials.

The historic centre of Cracow (Kraków), the former capital of Poland, is an outstanding example of medieval architecture. The thirteenth century merchants' town has Europe's largest market square and numerous historical houses, palaces and churches with magnificent interiors. Remnants of the fourteenth-century fortifications, the medieval site of Kazimierz with its ancient synagogues, Jagellonian University and the Gothic Wawel Cathedral all testify to the town's fascinating history.

The layout of Cracow is based on four core areas: Stare Miasto, or old town, around the market square; the Wawel, site of the imperial palace; the urban district of Kazimierz; and the Stradom quarter.

Stare Miasto is characterized by a rigid grid of perfectly orthogonal streets, a layout ordered by Boloslaw the Chaste in 1257 when he decided to unify the various peoples scattered around the hill of the Wawel. All that remains now of the medieval enclosure walls is the gate and the wall that was built in 1499 near the main city gate.

The old town is separated from the old district of Kazimierz, which was an island until 1880. Kazimierz formed the Jewish quarter of Cracow until the Second World War.

The city's university quarter is the oldest in Poland and among the oldest in Europe. Students at Jagellonian University have included Copernicus and Karol Wojtyla (Pope John Paul II).

The limestone hill of the Wawel that lies above Cracow's historic centre is the site of a complex housing some of the city's most important buildings, including the Royal Palace. Today the palace accommodates a museum displaying tapestries, the Royal Treasury, standards and antique furniture. Wawel Cathedral, where the kings of Poland were buried, also stands here.

At the entrance to the Wawel lies the start of the ancient Royal Way of monuments and remarkable historical buildings, leading to the heart of the old city: the market square (Rynek Główny). Extending 200 m along each side, this is one of the largest medieval public squares in Europe. One side is dominated by the Gothic church of the Assumption of the Virgin Mary while in the centre is the Skiennice, the ancient cloth market.

In the part of the Wawel Castle accessible on the river Wisla there is a small park at the base of the hill with the cave of the legendary Krak, prince and head of a Slav tribe.

In the Second World War the 64,000-strong Jewish community of Kazimierz was deported, eventually to the nearby concentration camps at Auschwitz; only 6,000 returned at the end of the war.

World Heritage site since

1978 1979 1980 1981 1982 1983 1984 1985 1986 1987 1988 1989 1990 1991 1992 1993 1994 1995 1996 1997 1998 1999 2000 2001 2002 2003 2004 2005 2006 2007 2008

Rock-Hewn Churches, Lalibela
Ethiopia
Criteria – Human creative genius; Interchange of values; Testimony to cultural tradition

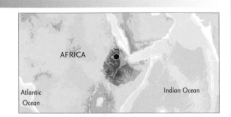

The eleven medieval monolithic churches of this thirteenth-century 'New Jerusalem' are situated in a mountainous region in the heart of Ethiopia near a traditional village with circular-shaped dwellings. The churches are carved from solid rock and are recessed below ground level. Lalibela is a high place of Ethiopian Christianity, tended by Coptic priests and still today a place of pilgrimage and devotion.

The rock churches are connected by maze-like tunnels, with a small river running between them called the Jordan. Churches on one side of the Jordan represent the earthly Jerusalem; those on the other side represent the heavenly Jerusalem.

◄ The Bete Giorgis Church.

City of Quito
Ecuador
Criteria – Interchange of values; Significance in human history

Quito, the capital of Ecuador, was founded in the sixteenth century on the ruins of an Inca city and stands at an altitude of 2,850 m. Despite the 1917 earthquake, the city has the best-preserved, least altered historic centre in Latin America. The monasteries of San Francisco and Santo Domingo, and the Church and Jesuit College of La Compañía, with their rich interiors, are pure examples of the 'Baroque school of Quito', which is a fusion of Spanish, Italian, Moorish, Flemish and indigenous art.

Quito is built on uneven land traversed by two deep ravines (quebradas). One of the ravines is arched over partly to preserve the alignment of the streets, the drainage of which escapes through a cleft in the ridge northward to the plain of Tumbaco.

◄ Santo Domingo Monastery and Church.

World Heritage site since

1978 1979 1980 1981 1982 1983 1984 1985 1986 1987 1988 1989 1990 1991 1992 1993 1994 1995 1996 1997 1998 1999 2000 2001 2002 2003 2004 2005 2006 2007 2008

Aachen Cathedral
Germany

Criteria – Human creative genius; Interchange of values; Significance in human history; Heritage associated with events of universal significance

With its columns of Greek and Italian marble, bronze doors, octagonal basilica and cupola, the Palatine Chapel of Aachen was from its inception regarded as an exceptional architectural achievement. It was the largest domed structure to be built north of the Alps since antiquity. Construction began around 790–800 under the Emperor Charlemagne, who was buried there. Throughout the Middle Ages it remained one of the prototypes of religious architecture and German emperors continued to be crowned there until 1531. Its present form has evolved over the course of more than a millennium. Two parts of the original complex have survived: the Coronation Hall, which is currently located in the Town Hall, built in the fourteenth century, and the Palatine Chapel, around which the cathedral would later be built.

When he began work on the Palatine Chapel, Charlemagne's dream was to create a 'new Rome'. Two hundred years later, when he was canonized, pilgrims began flocking to the town to see his tomb and the relics he had gathered. The town's ties with Charlemagne are reflected in numerous architectural memorials.

World Heritage site since

1978 1979 1980 1981 1982 1983 1984 1985 1986 1987 1988 1989 1990 1991 1992 1993 1994 1995 1996 1997 1998 1999 2000 2001 2002 2003 2004 2005 2006 2007 2008

Yellowstone National Park
USA

Criteria – Natural phenomena or beauty; Major stages of Earth's history; Significant ecological and biological processes; Significant natural habitat for biodiversity

Yellowstone National Park covers 9,000 km² of a vast natural forest of the southern Rocky Mountains in the North American west. The park holds half of the world's known geothermal features, with over 10,000 examples, and is equally renowned for its wildlife which includes grizzly bears, wolves, bison and wapitis.

Established as America's first national park in 1872, Yellowstone boasts an impressive array of geothermal phenomena, with geysers, lava formations, fumaroles, hot springs and waterfalls, lakes and canyons. There are more than 580 geysers – the world's largest concentration, and two-thirds of all those on the planet. The source of these phenomena lies under the ground in the area's geological origins: Yellowstone is part of the most seismically active region of the Rocky Mountains, a volcanic 'hot spot'.

Crustal uplifts 65 million years ago formed the southern Rocky Mountains and volcanic outflows were common until around 40 million years ago. A more recent period of volcanism began in the region about 2 million years ago, when thousands of cubic kilometres of magma filled immense chambers under the plateau and then erupted to the surface. Three cycles of eruption produced huge explosive outbursts of ash. The latest cycle formed a caldera 45 km wide and 75 km long when the active magma chambers erupted and collapsed. The crystallizing magma is the source of heat for hydrothermal features such as geysers, hot springs, mud pots and fumaroles.

Most of the area was glaciated during the Pleistocene (from 1.8 million to 10,000 years ago, when the last ice age ended) and many glacial features remain. The park lies at the headwaters of three major rivers – Yellowstone, Madison and Snake. Yellowstone Lake at 2,357 m is North America's largest lake at high elevation, while Lower Yellowstone Falls is the highest of more than forty named waterfalls in the park.

Great elevational differences produce a range of plant communities, from semi-arid steppe to alpine tundra. The park has seven species of coniferous tree, especially lodgepole pine, and 1,100 species of vascular plant, including an endemic grass. Thermal areas have unique assemblages of thermal algae and bacteria.

Yellowstone has six native species of ungulate, including elk. There has been intensive study and management of grizzly bear for thirty years, and regulations also protect native fish.

Castle Geyser is located in the Upper Geyser Basin in Yellowstone National Park. The geyser is thought to be approximately 5,000 years old, and will erupt hot water for 20 minutes up to a height of 27 m, once every 12 hours.

The Yellowstone Plateau, now a forested area of 6,500 km² with an average elevation of 2,000 m, was formed out of the accumulation of rhyolite magma. The plateau is flanked on the north, east and south by mountains rising to 4,000 m.

Ninety-six per cent of the Yellowstone National Park lies in Wyoming, three per cent in Montana and one per cent in Idaho.

Archaeological investigations show that human groups visited the park area for 10,000 years but none made it a permanent home.

World Heritage site since

1978 1979 1980 1981 1982 1983 1984 1985 1986 1987 1988 1989 1990 1991 1992 1993 1994 1995 1996 1997 1998 1999 2000 2001 2002 2003 2004 2005 2006 2007 2008

Wieliczka Salt Mine
Poland

Criteria – Significance in human history

The Wieliczka Salt Mine, located in southern Poland near the city of Cracow (Kraków), has been worked as a source of rock salt since the late thirteenth century. Spread over nine levels, it has 300 km of galleries, connecting more than 2,000 excavation chambers. Over the centuries, miners have established a tradition of carving sculptures out of the native rock salt. As a result, the mine contains entire underground churches, altars, bas-reliefs, and dozens of life-size or larger statues. It also houses an underground museum and has a number of special-purpose chambers such as a sanatorium for people suffering from respiratory ailments. The subterranean lake, open to tourists since the fifteenth century, completes this curious complex.

The largest of the underground chapels, the Chapel of the Blessed Kinga, is located 101 m below the surface; it is over 50 m long, 15 m wide and 12 m high, with a volume of 10,000 m³.

▼

World Heritage site since

1978 1979 1980 1981 1982 1983 1984 1985 1986 1987 1988 1989 1990 1991 1992 1993 1994 1995 1996 1997 1998 1999 2000 2001 2002 2003 2004 2005 2006 2007 2008

Simien National Park
Ethiopia

Criteria – Natural phenomena or beauty;
Significant natural habitat for biodiversity

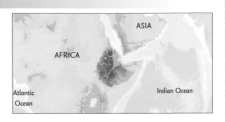

With its abundance of creviced basalt rock, Simien serves as an ideal water catchment area, replenished by two wet seasons and the Mayshasha River. Consequently the park is rich in a wide range of wildlife and vegetation.

Massive erosion over many years on the Ethiopian plateau has created one of the most spectacular landscapes in the world, with jagged mountain peaks, deep valleys and sharp precipices dropping some 1,500 m. The park is home to some extremely rare animals such as the Gelada baboon, the Simien fox and the Walia ibex, a goat found nowhere else in the world.

◀
Two of the Park's rare Gelado baboons.

L'Anse aux Meadows
National Historic Site
Canada

Criteria – Heritage associated with events of universal significance

Inscribed in 1978, L'Anse aux Meadows was the first cultural site on the list of World Heritage sites. Located at the tip of the Great Northern Peninsula on the Island of Newfoundland, L'Anse aux Meadows is the oldest known European settlement and the only authenticated Viking site on the American continent. Here, archaeologists have unearthed evidence of European exploration in North America dating back around 1000 years, well before Columbus or Cabot.

In 1960, Helge and Anne Stein Ingstad, with the help of local fisherman George

Decker, uncovered the remnants of eight wood-framed, peat-turf buildings similar to those found in Greenland and Iceland.

A forge, four workshops, and numerous artefacts have been found – a cloak pin, needle and spindle whorl indicate Norse origin and the presence of women in camp, while the metal slag found indicates the first forging of iron in the New World. Wood debris and nail fragments point to ship repair as one of the main activities by Vikings at L'Anse aux Meadows, which is likely the 'gateway to Vinland' mentioned in the sagas.

The area has been home to many different cultures. The earliest signs of human activity go back roughly 5,000 years to the Dorset and Groswater paleoeskimo. It was from here that exploration around the Gulf of St Lawrence and into Vinland by the Vikings led to the first encounters between Europeans and North American Aboriginal peoples.

World Heritage site since

1978 1979 1980 1981 1982 1983 1984 1985 1986 1987 1988 1989 1990 1991 1992 1993 1994 1995 1996 1997 1998 1999 2000 2001 2002 2003 2004 2005 2006 2007 2008

Nahanni National Park
Canada

Criteria – Natural phenomena or beauty; Major stages of Earth's history

Parks Canada is working with the Dene and Metis people to expand Nahanni National Park Reserve. The expanded park will protect more wildlife habitat and outstanding geological features.

Located in the southwest corner of the Northwest Territories, along the course of the South Nahanni and Flat rivers, the park lies in a diverse mountainous area comprising mountain ranges, rolling hills, rugged plateaus, deep canyons and spectacular waterfalls, as well as a globally significant limestone cave system. The park has many unique geological features including tufa mounds known as the Rabbitkettle Hotsprings, which rise in a succession of terraces to a height of 27 m. There are also wind-eroded sandstone landforms known as the Sand Blowouts, and large areas that have remained unglaciated for more than 100, 000 years. The park is home to animals of the boreal forest, such as bald eagles, grizzly bears and caribou.

World Heritage site since

1978 **1979** 1980 1981 1982 1983 1984 1985 1986 1987 1988 1989 1990 1991 1992 1993 1994 1995 1996 1997 1998 1999 2000 2001 2002 2003 2004 2005 2006 2007 2008

Meidan Emam, Esfahan
Islamic Republic of Iran

Criteria – Human creative genius; Traditional human settlement; Heritage associated with events of universal significance

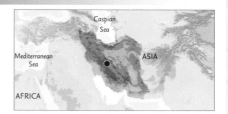

The Royal Square was the heart of the Safawid capital. Its vast sandy esplanade was used for promenades, assembling troops, playing polo, celebrations and public executions. On all sides, the arcades house shops. Above the portal of the large bazaar of Qeyssariyeh is a tribune that accommodates musicians giving public concerts.

The Royal Square (Meidan Emam) of Esfahan is a monument of Persian socio-cultural life during the Safawid period. Built by Shah Abbas the Great at the beginning of the seventeenth century, it is bordered on each side by four monumental buildings linked by a series of two-storey arcades: to the north the Portico of Qeyssariyeh (1602–19), to the south the Royal Mosque (1612–30), to the east the Mosque of Sheyx Loffollah (1602–18), and to the west the pavilion of Ali Qapu, a small fifteenth-century Timurid palace, enlarged and decorated by the shah and his successors. All of these architectural elements of the Meidan-e Shah, including the arcades, are adorned with a profusion of painted ceramic tiles.

World Heritage site since

1978 **1979** 1980 1981 1982 1983 1984 1985 1986 1987 1988 1989 1990 1991 1992 1993 1994 1995 1996 1997 1998 1999 2000 2001 2002 2003 2004 2005 2006 2007 2008

Ancient City of Damascus
Syrian Arab Republic

Criteria – Human creative genius; Interchange of values; Testimony to cultural tradition; Significance in human history; Heritage associated with events of universal significance

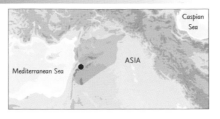

Founded in the third millennium BC, Damascus is considered to be the oldest city and oldest capital of the world. It is the cradle of historical civilizations, constituting an ancient beacon of science and art over time, and a reference for systems of architecture and town planning over several thousand years.

Damascus had many rulers, including King David of Israel and Alexander the Great,

before it became part of the Arab world as the capital of Umayyad caliph in 636.

In the Middle Ages it was the centre of a flourishing craft industry, specializing in swords and lace.

The city has some 125 monuments from the different periods of its history, one of the most spectacular being the eighth century Great Mosque of the Umayyads, built on the site of an Assyrian sanctuary.

Damascus preserves a few traces of its long history prior to the Arabic conquest, including some from the Roman period.

Over the centuries guilds of craftsmen and merchants established themselves around the city's **Great Mosque, (pictured below)** while the important Christian minority consolidated itself in the northeast quarters, around the churches and sites associated with the conversion of St Paul.

World Heritage site since

1978 **1979** 1980 1981 1982 1983 1984 1985 1986 1987 1988 1989 1990 1991 1992 1993 1994 1995 1996 1997 1998 1999 2000 2001 2002 2003 2004 2005 2006 2007 2008

Auschwitz Birkenau, German Nazi Concentration and Extermination Camp (1940–1945)
Poland

Criteria – Heritage associated with events of universal significance

Auschwitz-Birkenau was the principal and most notorious of the six concentration and extermination camps established by Nazi Germany to implement its Final Solution policy which had as its aim the mass murder of the Jewish people in Europe. It was built in Poland under Nazi German occupation initially as a concentration camp for prisoners of war. Between the years 1942–4 it became the main mass extermination camp where Jews were tortured and killed for their so-called racial origins. In addition to the mass murder of well over a million Jewish men, women and children, and tens of thousands of Polish victims, Auschwitz also served as a camp for the racial murder of thousands of Roma and Sinti and prisoners of several European nationalities. The Nazi policy of spoliation, degradation and extermination of the Jews was rooted in a racist and anti-Semitic ideology propagated by the Third Reich.

Auschwitz-Birkenau was the largest of the concentration camp complexes created by the Nazi German regime and was the one which combined extermination with forced labour. At the centre of a huge landscape of human exploitation and suffering, the remains of the two camps of Auschwitz I and Auschwitz II-Birkenau, as well as its Protective Zone were placed on the World Heritage List as evidence of this inhumane, cruel and methodical effort to deny human dignity to groups considered inferior, leading to their systematic murder. The camps are a vivid testimony to the murderous nature of the anti-Semitic and racist Nazi policy that brought about the annihilation of more than 1.2 million people in the crematoria, 90 per cent of whom were Jews.

The fortified walls, barbed wire, railway sidings, platforms, barracks, gallows, gas chambers and crematoria at Auschwitz-Birkenau show clearly how the Holocaust, as well as the Nazi German policy of mass murder and forced labour took place. The collections at the site preserve the evidence of those who were premeditatedly murdered, as well as presenting the systematic mechanism by which this was done. The personal items in the collections are testimony to the lives of the victims before they were brought to the extermination camps, as well as to the cynical use of their possessions and remains. The site and its landscape has high levels of authenticity and integrity since the original evidence has been carefully conserved without any unnecessary restoration.

Auschwitz-Birkenau, monument to the deliberate genocide of the Jews by the Nazi regime (Germany 1933–45) and to the deaths of countless others, bears irrefutable evidence to one of the greatest crimes ever perpetrated against humanity. It is also a monument to the strength of the human spirit which in appalling conditions of adversity resisted the efforts of the German Nazi regime to suppress freedom and free thought and to wipe out whole races. The site is a key place of memory for the whole of humankind for the holocaust, racist policies and barbarism; it is a place of our collective memory of this dark chapter in the history of humanity, of transmission to younger generations and a sign of warning of the many threats and tragic consequences of extreme ideologies and denial of human dignity.

World Heritage site since

1978 **1979** 1980 1981 1982 1983 1984 1985 1986 1987 1988 1989 1990 1991 1992 1993 1994 1995 1996 1997 1998 1999 2000 2001 2002 2003 2004 2005 2006 2007 2008

Ngorongoro Conservation Area

Tanzania

Criteria – Natural phenomena or beauty; Major stages of Earth's history; Significant ecological and biological processes; Significant natural habitat for biodiversity

The huge and perfect crater of Ngorongoro is one of the largest inactive, unbroken and unflooded calderas in the world. Its diverse landforms and altitudes and its variable climate have resulted in the development of several distinct habitats: scrub heath and the remains of dense montane forests on its steep slopes, with grassy plains, lakes, swamps and woodland on the crater floor. Around 25,000 large animals live in the crater, including the highest density of mammalian predators in Africa.

Ngorongoro Conservation Area also includes Empakaai Crater, filled by a deep lake, and the active volcano of Oldonyo Lenga. At nearby Olduvai Gorge, excavations uncovered the existence of one of our more distant ancestors, Homo habilis. Laitoli Site, also in the area, is one of the main localities of early hominid footprints, dating back 3.6 million years. Leopard and the endangered African elephant, mountain reedbuck, flamingo, wildebeest and buffalo are among the species that live on the crater rim.

▲
Ngorongoro crater is, on average, 16–19 km wide, its floor is 264 km² and its rim soars to 400–610 m.

One of the Park's wildebeest with flamingos in the background.
▼

World Heritage site since

1978 **1979** 1980 1981 1982 1983 1984 1985 1986 1987 1988 1989 1990 1991 1992 1993 1994 1995 1996 1997 1998 1999 2000 2001 2002 2003 2004 2005 2006 2007 2008

Bryggen
Norway
Criteria – Testimony to cultural tradition

Bryggen is the old wharf of the city of Bergen. With its traditional wooden buildings and street patterns, Bryggen is a relic of an ancient wooden urban structure once common in Northern Europe and is a reminder of the town's importance as part of the trading empire of the powerful North German and Baltic Hanseatic League from the fourteenth to the mid-sixteenth century.

Over the centuries several devastating fires, the most recent in 1955, have ravaged the traditional wooden houses. However, rebuilding has traditionally followed old patterns and methods, so leaving its main structure preserved, and the town retains its medieval appearance. Some sixty-two buildings remain of this former townscape.

Today, Bergen is the only town outside the Hanseatic League whose original structures still remain within the city limits and cityscape.

The Hanseatic League took control of Bergen in 1350 and the German colonists who arrived thereafter directly influenced the appearance of Bryggen, with buildings constructed along the narrow streets running parallel to the dock.

The urban unit revolves around a courtyard (gård) common to several three-level wooden houses whose gabled façades and lateral walls and roofs are covered with shingles.

Bryggen today. ▼

World Heritage site since

1978 **1979** 1980 1981 1982 1983 1984 1985 1986 1987 1988 1989 1990 1991 1992 1993 1994 1995 1996 1997 1998 1999 2000 2001 2002 2003 2004 2005 2006 2007 2008

Forts and Castles, Volta, Greater Accra, Central and Western Regions
Ghana

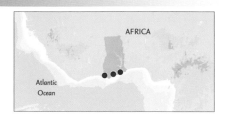

Criteria – Heritage associated with events of universal significance

The remains of fortified trading-posts, established between 1482 and 1786, can still be seen along the coast of Ghana between Keta and Beyin. They were links in the trade routes established by the Portuguese in many areas of the world during their era of great maritime exploration. Accra was first settled at the end of the sixteenth century when the Ga people migrated there. These early inhabitants engaged in farming and lagoon fishing, with sea fishing taken up during the middle of the eighteenth century. During the slave trade Accra took on greater importance owing to the nearby forts, many of which were owned and controlled by the Dutch, a prominence that lasted until the abolition of the slave trade in 1807.

The forts and castles at Accra were of great strategic value and were occupied at different times by European traders and adventurers from Portugal, Spain, Denmark, Sweden, Holland, Germany and Britain.

Prehistoric Sites and Decorated Caves of the Vézère Valley
France

Criteria – Human creative genius; Testimony to cultural tradition

The Vézère Valley contains 147 prehistoric sites dating from the Palaeolithic and twenty-five decorated caves. It is particularly interesting from an ethnological and anthropological, as well as an aesthetic point of view because of its cave paintings, especially those of the Lascaux Cave, whose discovery in 1940 was of great importance for the history of prehistoric art. The hunting scenes show some 100 animal figures, which are remarkable for their detail, rich colours and lifelike quality.

The depictions of animals involved in the hunt helped to assure or reassure the hunters of great success. The Lascaux Cave is closed to the public, but a replica has been created at Montignac, 200 m from the original cave, where two of the galleries have been reproduced: the Great Hall of the Bulls and the Painted Gallery.

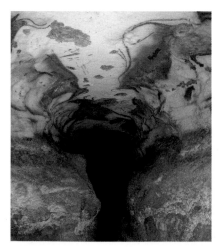

◄
Lascaux Cave.

World Heritage site since

1978 **1979** 1980 1981 1982 1983 1984 1985 1986 1987 1988 1989 1990 1991 1992 1993 1994 1995 1996 1997 1998 1999 2000 2001 2002 2003 2004 2005 2006 2007 2008

Virunga National Park
Dem. Rep. of the Congo

Criteria – Natural phenomena or beauty; Major stages of Earth's history; Significant natural habitat for biodiversity

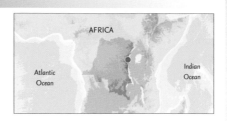

AFRICA

Atlantic Ocean

Indian Ocean

Virunga National Park lies at the border of several biogeographic zones and comprises an outstanding diversity of habitats. The range of altitudes in the park, which covers an area of 7,900 km², adds to a habitat variety ranging from swamps and savannas to the Rwenzori snowfields at a height of over 5,000 m; and from lava plains to the afromontane forests on the slopes of volcanoes. There are hot springs in the Rwindi plains, and some of the Virunga Massif volcanoes, such as Nyiragongo (3,470 m), are still active.

At the time of inscription on the World Heritage List, the rivers of Virunga had probably the largest hippopotamus concentration in Africa.

The savanna of the Rwindi area is home to elephant, buffalo, antelope, warthog, lion and various monkeys. The Semiliki Valley and the Virunga mountains are the habitat of gorilla, chimpanzee and okapi. The park also includes important wetlands, which are recognized under the Ramsar Convention on Wetlands and are important transit and wintering areas for some bird species.

The areas of lowest and highest rainfall in the Democratic Republic of the Congo are in Virunga National Park.

The park is home to mountain gorillas, and before the recent war, there were some 20,000 hippopotamuses living in its rivers. Its wetlands are a wintering ground for some bird species from Siberia.

◄

Mountain gorilla 'Bonane' holds her newborn in Virunga National Park.

World Heritage site since

1978 **1979** 1980 1981 1982 1983 1984 1985 1986 1987 1988 1989 1990 1991 1992 1993 1994 1995 1996 1997 1998 1999 2000 2001 2002 2003 2004 2005 2006 2007 2008

Site of Carthage
Tunisia

Criteria – Interchange of values; Testimony to
cultural tradition; Heritage associated with
events of universal significance

Both images show
sections of the
Antonine Baths at
Carthage, with the
top one showing
fragments of the
ruined 'caldarium'
(the hottest room)
and the steam room.

Carthage is one of the most famous historic
sites of the Roman Empire with a period of
domination that spanned several centuries,
making it one of the largest and longest-
lasting of the ancient empires.

At a time when the sea was the most
efficient means of travel and
communication, Carthage's two first-class
harbours gave it a distinct advantage over
other city states. The Carthaginians
developed superb shipbuilding and sailing
skills that were the basis of a centuries-long
naval and mercantile domination. They also
supported what was at its height a great
trading empire that spanned the
Mediterranean. The Carthaginians traded
in silver, lead, ivory and gold, beds and
bedding, simple pottery, jewellery,
glassware, African wild animals, fruit and
nuts. The huge wealth Carthage acquired
saw it become the home to a brilliant
civilization and one of the largest cities in
the pre-industrial world.

The city was founded in the ninth
century BC by Phoenician traders from
Tyre, in modern Lebanon, and by the sixth
century BC had conquered and controlled
much of the southern Mediterranean: the
North African coast from modern Morocco
to the borders of modern Egypt, Sardinia,
Malta, the Balearic Islands, and western

Sicily. This expansion inevitably brought
Carthage into conflict with the other
dominant regional powers – the empires
first of Greece and then of Rome, and the
three battled for control over territories.
Competition over Sicily sparked the war
against Greece lasting for more than 200
years; this ended in victory for the
Carthaginians. The wars against Rome,
known as the Punic Wars, were divided into
three periods between 264 BC and 146 BC,
when the Romans finally triumphed. They
razed Carthage, imposing strict controls
over further settlement; in the process
much of the evidence and artefacts of the
ancient city was destroyed.

In the first century AD the Roman Emperor
Augustus founded the city anew, as Colonia
Julia Carthago. This second, Roman
Carthage grew on the ruins of the first and
prospered quickly, becoming second only to
Rome in splendour and wealth.

The city's ultimate decline was confirmed
in AD 439 with the occupation by the Vandals
and finally in AD 637 when it was captured by
Arabs and destroyed. Carthage never
regained its importance, largely because of
the concentration of power in nearby Tunis.

The refounded first
century city of Julia
Carthago illustrated
the splendour and
wealth of Rome, and
exerted considerable
influence on the
development of
structural architecture
and of characteristic
Punic and Roman
town planning. It is
also an important
testimony to Punic
history and
constitutes an
interesting example
of the Punic city.

The Punic port is the
best place to visit, as
Carthage was a port
that was stronger on
the seas than the
Roman Empire for
many years.

World Heritage site since

1978 **1979** 1980 1981 1982 1983 1984 1985 1986 1987 1988 1989 1990 1991 1992 1993 1994 1995 1996 1997 1998 1999 2000 2001 2002 2003 2004 2005 2006 2007 2008

Dinosaur Provincial Park
Canada

Criteria – Natural phenomena or beauty; Major
stages of Earth's history

Dinosaur Provincial Park contains some of
the most important fossil discoveries ever
made from the 'Age of Reptiles' – in
particular, about thirty-nine species of
dinosaur, dating back some 75 million years.
The park stands at the heart of Alberta's
badlands, a beautiful, barren and deeply
eroded area of sparse vegetation.

During the late Cretaceous period
75 million years ago, the landscape of the
area was very different, with lush forests and
rivers flowing into a warm inland sea. Its low
swamps were home to a variety of animals,
including dinosaurs. The conditions were
also perfect for the preservation of their
bones as fossils.

The badlands now provide habitat for a
number of ecologically specialized plant
species. The mild winter microclimate
coupled with an abundant food supply,
supports native ungulates and over
165 species of bird have been recorded.

▲

Erosion has left exposed Cretaceous
shales and sandstones and created
extensive badlands, the largest in
Canada.

Geological strata of the Dinosaur
Park Formation yielded many of the
dinosaur remains for which the park
is renowned. Between 1979 and
1991, a total of 23,347 fossil
specimens were collected, including
300 dinosaur skeletons.

World Heritage site since

1978 **1979** 1980 1981 1982 1983 1984 1985 1986 1987 1988 1989 1990 1991 1992 1993 1994 1995 1996 1997 1998 1999 2000 2001 2002 2003 2004 2005 2006 2007 2008

Abu Mena
Egypt
Criteria – Significance in human history

The church, baptistry, basilicas, public buildings, streets, monasteries, houses and workshops in this early Christian holy city were built over the tomb of the martyr Menas of Alexandria, who died in AD 296. The Thermal Basilica, built in the fifth century to accommodate the increasing number of Christian pilgrims, was used to store the curative waters for the surrounding heated baths and pools.

By AD 600, the oasis at Abu Mena had become a pilgrimage destination, centred on the great basilica complex. Archaeological excavations have revealed an entire town with houses and cemeteries, which exhibit an important fusion of East and West religious influences.

◀

The ancient public baths of Abu Mena.

Thracian Tomb of Kazanlak
Bulgaria
Criteria – Human creative genius; Testimony to cultural tradition; Significance in human history

Discovered in 1944, this Hellenistic tomb is located in Bulgaria's romantic Valley of Roses. Part of a large necropolis with more than 500 burial mounds, it lies near Seuthopolis, the capital city of the Thracian king Seutes III. The art and archaeological discoveries of the site reflect a rich culture that was at its peak in the fifth to third centuries BC. The pattern of the particularly fine Kazanlak Tomb murals shows that they were not painted spontaneously: the paintings are a carefully premeditated

artistic composition executed in accordance with a precise project. The architecture and the pattern of the composition were prepared together as an integrated work of art. It is clear that both were the work of one person – an artist-architect.

The murals of the Kazanlak Tomb are particularly important because they are the only entirely preserved work of Hellenistic art that has been found in exactly the state in which it was originally designed and executed.

World Heritage site since

1978 **1979** 1980 1981 1982 1983 1984 1985 1986 1987 1988 1989 1990 1991 1992 1993 1994 1995 1996 1997 1998 1999 2000 2001 2002 2003 2004 2005 2006 2007 2008

Kathmandu Valley
Nepal

Criteria – Testimony to cultural tradition; Significance in human history; Heritage associated with events of universal significance

ASIA

Arabian
Sea

Bay of
Bengal

Kathmandu Valley has seven monumental zones with three historical palaces within their urban settings (the cities of Kathmandu, Patan and Bhaktapur), two Hindu centres (Pashupati and Changu Narayan) and two Buddhist centres (Swayambunath and Boudhanath).

Kathmandu, the Nepalese capital, is the country's political and commercial hub and is an exotic and fascinating showcase of culture, art and tradition. The valley in which the cities of Kathmandu, Patan and Bhaktapur stand is roughly an oval bowl encircled by green, terraced hills dotted with compact clusters of red tiled-roofed houses.

The world-famous Kathmandu Valley has a concentration of monuments unique and unparalleled in the world. It is the principal centre of settlement in the hill area of Nepal and is a prime cultural focus of the Himalaya. The Kathmandu Valley monumental zones represent a highly developed architectural expression of the religious, political and cultural life of the area.

The seven groups of monuments and buildings include: the Durbar Square of Hanuman Dhoka (or Hanuman Gate) in Kathmandu, the Durbar Square of Patan and the Durbar Square of Bhaktapur; the Buddhist stupas of Swayambhu and of Bauddhanath; and the Hindu temples of Pashupati and Changu Narayan.

Pashupati is Nepal's most renowned Hindu creation site. Changu Narayan Temple is an impressive double-roofed temple which is said to be the most ancient temple of the Hindu god Vishnu in the Kathmandu Valley.

The city of Kathmandu is a melting pot for the nation's population and its unique architectural heritage, palaces, temples and courtyards have inspired many writers, artists, and poets, both foreign and Nepalese. It boasts a unique symbiosis of Hinduism, Buddhism and Tantrism in its culture, which is still as alive today as it was hundreds of years ago. The religious influence can be openly seen in the city. Most of the principal monuments are in Durbar Square, the city's social, religious and urban focal point, built between the twelfth and the eighteenth centuries by the ancient Malla kings of Nepal. Some of the most interesting are the Taleju Temple, Kal Bhairab, Nautalle Durbar, Coronation Nasal Chowk, and the Gaddi Baithak, the statue of King Pratap Malla, the Big Bell, Big Drum and the Jaganath Temple.

Patan (or Lalitpur), just across the holy Bagmati River about 14 km east of Kathmandu city, is of great historic and cultural interest, and the city of Bhaktapur (or Bhadgaon) stands at an altitude of 1,401 m.

Pashupati Temple is one of the most important Hindu temples as well as Nepal's most sacred Hindu shrine and one of the subcontinent's greatest sites for the god Shiva in a sprawling collection of temples, ashrams, images and inscriptions raised over the centuries along the banks of the sacred Bagmati River.

Durbar Square of Patan. ▶

World Heritage site since

1978 **1979** 1980 1981 1982 1983 1984 1985 1986 1987 1988 1989 1990 1991 1992 1993 1994 1995 1996 1997 1998 1999 2000 2001 2002 2003 2004 2005 2006 2007 2008

Kluane / Wrangell-St Elias / Glacier Bay / Tatshenshini-Alsek
Canada and USA

Criteria – Natural phenomena or beauty; Major stages of Earth's history; Significant ecological and biological processes; Significant natural habitat for biodiversity

A unique area of mountain peaks, foothills, glacial systems, lakes, streams, valleys and coastal landscapes make up these spectacular parks stretching from British Columbia and Yukon Territory in Canada across the border to Alaska in the United States.

The Wrangell-St Elias region has the largest array of glaciers and ice fields outside the polar region. These glacial features and the high mountains of the Wrangell-St Elias, Chugach and Kluane ranges have resulted in the region being called the Mountain Kingdom of North America.

The variety of climatic zones and elevations have produced three major biomes or ecological areas: coastal coniferous; northern coniferous; and alpine tundra. The great variety of birds and wildlife, including grizzly bears, caribou and Dall's sheep, reflects this habitat diversity. There are freshwater fish and all five species of Alaskan Pacific salmon spawn in park or preserve waters.

Geologically the mountains are part of the Pacific mountain system and include the 130 km-long Bagley ice field, the second-highest peak in the USA (Mount St Elias) and the largest piedmont glacier on the North American continent (Malaspina Glacier).

Braided river.
▼

World Heritage site since

1978 **1979** 1980 1981 1982 1983 1984 1985 1986 1987 1988 1989 1990 1991 1992 1993 1994 1995 1996 1997 1998 1999 2000 2001 2002 2003 2004 2005 2006 2007 2008

Everglades National Park
USA

Criteria – Major stages of Earth's history;
Significant ecological and biological processes;
Significant natural habitat for biodiversity

NORTH AMERICA

Pacific Ocean

Atlantic Ocean

Everglades National Park, pictured below, on the southern tip of the Florida Peninsula has been called 'a river of grass flowing imperceptibly from the hinterland into the sea'. Its exceptional variety of water habitats has made it a sanctuary for large numbers of birds and reptiles, including threatened species such as the manatee.

The park lies at the interface between temperate and subtropical America, between fresh and brackish water and between shallow bays and deeper coastal waters. Consequently it hosts a complex of habitats supporting a high diversity of flora and fauna including many endemic to this area.

The Everglades protect 800 species of land and water vertebrates, over 400 bird species and sixty known species of reptile, amphibian and insect. Over 275 species of fish mostly inhabit the marine and estuarine waters and there are great numbers of economically valuable crustaceans.

The vegetation and flora of south Florida have long fascinated scientists and naturalists and were a primary reason for the establishment of the park.

Hammocks or tree islands are dominated by tropical and temperate hardwood species. The most important trees are mangroves, taxa, slash pine and cypress. Prairies can be dominated by sawgrass, muhley grass, or cordgrass in coastal areas.

World Heritage site since

1978 **1979** 1980 1981 1982 1983 1984 1985 1986 1987 1988 1989 1990 1991 1992 1993 1994 1995 1996 1997 1998 1999 2000 2001 2002 2003 2004 2005 2006 2007 2008

Grand Canyon National Park
USA

Criteria – Natural phenomena or beauty; Major stages of Earth's history; Significant ecological and biological processes; Significant natural habitat for biodiversity

NORTH AMERICA

Pacific Ocean

Atlantic Ocean

Low sunlight helps to ▶ reveal the canyon's horizontal strata.

The Grand Canyon, carved nearly 1,500 m deep into the rock by the Colorado River, is the most spectacular gorge in the world. Its horizontal strata retrace the geological history of the past two billion years as it cuts across the Grand Canyon National Park. It also contains prehistoric traces of human adaptation to a particularly harsh environment.

The Grand Canyon dominates the national park, which was created in 1919 by an act of Congress and was one of the first national parks in the United States. The steep, twisting gorge, 1.5 km deep and 445.8 km long, was formed during some 6 million years of geological activity and erosion by the Colorado River on the raised Earth's crust (2.5 km above sea level). The gorge, which ranges from 200 m–30 km wide, divides the park into the North Rim and South Rim: the buttes, spires, mesas and temples in the canyon are in fact mountains, looked down upon from the rims.

Erosion is ongoing and seasonal and permanent rivers produce impressive waterfalls and rapids of washed-down boulders along the length of the canyon and its tributaries. The horizontal geological strata that erosion has exposed span some 2,000 million years of geological history, providing evidence of the four major

geological eras: early and late Precambrian, Palaeozoic, Mesozoic and Cenozoic.

The canyon is also a vast biological museum in which there are five different life and vegetation zones. Over 1,000 plant species have so far been identified including several officially listed as threatened. The park is also home to 76 mammal, 299 bird and 41 reptile and amphibian species, and some 16 fish species inhabit the Colorado River and its tributaries.

Archaeological remains show the adaptation of human societies to the area's severe climate and landscape, with evidence of settlement. The park contains more than 2,600 documented prehistoric ruins, including evidence of Archaic cultures (the earliest known inhabitants), Cohonina Indians along the South Rim, and Anasazi Indians on both the South Rim, North Rim, and within the Inner Canyon. Hualapai and Havasupai Indians moved into the canyons at this time, remaining undisturbed until the arrival of the Anglo-Americans in 1860.

The Grand Canyon area was first protected in 1893 as a forest reserve in which mining, lumbering and hunting continued to be allowed. It was upgraded to a game reserve in 1906, giving protection to the wildlife, and redesignated a National Monument in 1908.

Altitudinal range provides a variety of climates and habitats, ranging from desert to mountain conditions.

Fossil remains found in the park include early plants, marine and terrestrial specimens, early reptiles and some mammals.

World Heritage site since

1978 **1979** 1980 1981 1982 1983 1984 1985 1986 1987 1988 1989 1990 1991 1992 1993 1994 1995 1996 1997 1998 1999 2000 2001 2002 2003 2004 2005 2006 2007 2008

Nubian Monuments from Abu Simbel to Philae

Egypt

Criteria – Human creative genius; Testimony to
cultural tradition; Heritage associated with
events of universal significance

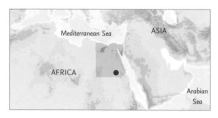

Interior of the Great Temple of Ramses II at Abu Simbel. ▲

This part of southern Egypt, extending from
Aswan to the Sudanese border, is an area of
outstanding archaeological importance
containing magnificent ancient
monuments, including the temples of the
Pharaoh Ramses II at Abu Simbel and the
Sanctuary of Isis at Philae.

Built by Ramses at Abu Simbel in Nubia to
reaffirm the Egyptian Empire's ownership of
what was then its southern neighbour, the
Great Temple has four colossal statues of the
pharaoh carved into the rock and fastened to
the cliff wall. Nearby is the Little Temple,
dedicated to the Goddess Hathor in
memory of the pharaoh's wife Queen
Nefertari.

In the 1960s the sites faced inundation by
the river Nile as a consequence of the
construction of the Aswan High Dam.
UNESCO launched a successful
international campaign to save the temples,
which led to their removal to higher ground.

▲

In addition the site includes the
temples of Amada, Derr, Ouadi Es
Sebouah, Dakka, Maharraqah,
Talmis and Beit el Ouali and the
Kiosk of ak-Kartassi.

West colonnade of the Sanctuary of
Isis at Philae.

World Heritage site since

1978 **1979** 1980 1981 1982 1983 1984 1985 1986 1987 1988 1989 1990 1991 1992 1993 1994 1995 1996 1997 1998 1999 2000 2001 2002 2003 2004 2005 2006 2007 2008

Plitvice Lakes National Park
Croatia

Criteria – Natural phenomena or beauty; Major stages of Earth's history; Significant ecological and biological processes

EUROPE

Adriatic Sea

Tyrrhenian Sea

Plitvice Lakes National Park contains a series of outstandingly beautiful lakes, caves and waterfalls. There are approximately twenty interlinked lakes between Mala Kapela Mountain and Pljesevica Mountain.

The park's features were formed over thousands of years by the waters flowing over the limestone and chalk and depositing barriers of travertine rock, in the process creating natural dams; these in turn created the lakes, caves and waterfalls. These

geological processes continue today.

Varying altitudes and soils support a variety of plants. The dense forests of the park are rich in wildlife and are home to brown bears, wolves and many rare bird species. Archaeological remains show evidence of human settlement by an Illyrian tribe dating from 1000 BC.

To preserve the natural characteristics of the lakes, the park was extended in 2000, taking in more of the area's drainage system.

The lake system is divided into the upper and lower lakes. The upper lakes lie in a dolomite valley surrounded by thick forests and interlinked by numerous waterfalls. The lower lakes, smaller and shallower, lie on the limestone bedrock and are surrounded only by sparse underbrush.

▼

World Heritage site since

1978 **1979** 1980 1981 1982 1983 1984 1985 1986 1987 1988 1989 1990 1991 1992 1993 1994 1995 1996 1997 1998 1999 2000 2001 2002 2003 2004 2005 2006 2007 2008

Mont-Saint-Michel and its Bay
France

Criteria – Human creative genius; Testimony to cultural tradition; Heritage associated with events of universal significance

Mont-Saint-Michel is one of the most important sites of medieval Christian civilization. The Gothic-style Benedictine abbey dedicated to the archangel Saint Michel and the village that grew up in the shadow of its great walls are together known as the 'Wonder of the West'. It is unequalled, as much because of the coexistence of the abbey and its fortified village within the confined limits of a small island, as for the originality of the placement of the buildings which give Mont-Saint-Michel its unforgettable silhouette. Built between the eleventh and sixteenth centuries, the abbey is a technical and artistic tour de force, having adapted to the problems posed by its unique position and site: perched on a rocky islet in the midst of vast sandbanks exposed to powerful tides between Normandy and Brittany. Mont-Saint-Michel forms an architectural complex of great originality, built by successive restructurings and additions throughout the Middle Ages.

In 966 Benedictine monks from St Wandrille founded the monastery of Saint-Michel-au-Péril-de-la-Mer on a granite tidal island in the bay of Mont-Saint-Michel. A sanctuary dedicated to Saint Michel had already existed on this location for a long time. The oldest part of the present abbey, the small pre-Romanesque church of Notre-Dame-sous-Terre, undoubtedly dates back to the tenth century. There are Romanesque influences in the nave of abbey church and in a group of convent buildings, including the chaplain's residence and the covered gallery of the monks, dating from the twelfth century.

However it was the architectural advances of the later medieval Gothic period that allowed the island's restricted area to be used to best advantage, in the high walls, soaring masses and airy pinnacles which so harmoniously crown the sharp silhouette of the rock. The new group of Romanesque convent buildings that were built from 1204, merit the name Merveille (Marvel) for the elegance of their conception. They comprise the chaplain's residence of the twelfth century; the celebrated rooms known as Salles des Hôtes and des Chevaliers, with rib vaults that spring from the central colonnades; and, on the uppermost floor beside the refectory, the cloister, which is open on one side to the sea.

This small rock outcrop, 900 m in circumference at its base and 80 m at its peak, stands about 2 km off the coast on sandy ground that is exposed at low tide.

Its exceptional situation between land and sea has determined the establishment and development of Mont-Saint-Michel, and the practical and aesthetic solutions to the challenges of its natural environment.

The abbey dominates and is controlled by a tiny village clinging to the foot of the mount.

World Heritage site since

1978 **1979** 1980 1981 1982 1983 1984 1985 1986 1987 1988 1989 1990 1991 1992 1993 1994 1995 1996 1997 1998 1999 2000 2001 2002 2003 2004 2005 2006 2007 2008

Fasil Ghebbi, Gondar Region
Ethiopia

Criteria – Interchange of values; Testimony to cultural tradition

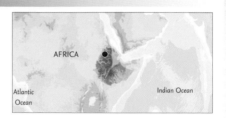

The fortress-city of Fasil Ghebbi is an outstanding testimony of the modern Ethiopian civilisation on the northern plateau of Tana. In the sixteenth and seventeenth centuries, this was the residence of the Ethiopian emperor Fasilides and his successors. Surrounded by a 900-m-long wall, the city contains palaces, churches, monasteries and unique public and private buildings marked by Hindu and Arab influences, subsequently transformed by the Baroque style brought to Gondar by the Jesuit missionaries. Beyond the confines of the city is a fine bathing palace, a two-storeyed battlemented structure situated beside a rectangular pool of water which was supplied by a canal from the nearby river. It was reached by a stone bridge, part of which could be raised for defence.

The main castle, which stands today in a grassy compound surrounded by later fortresses, was built in the late 1630s and early 1640s. With its huge towers and looming battlemented walls, it resembles a piece of medieval Europe transposed to Ethiopia. ◄

Boyana Church
Bulgaria

Criteria – Interchange of values; Testimony to cultural tradition

Located on the outskirts of Sofia, Boyana Church consists of three buildings. The eastern church was built in the tenth century, then enlarged at the beginning of the thirteenth century by Sebastocrator Kaloyan, who ordered a second two-storey building to be erected next to it. The frescoes in this second church, painted in 1259, make it one of the most important collections of medieval paintings. The ensemble is completed by a third church, built at the beginning of the nineteenth century. This site is one of the most complete and perfectly preserved monuments of east European medieval art.

Boyana's frescoes are an early example of the icon-painting style which strongly influenced the Tirnovo artistic school. The icon-style murals that became widespread in the Serbian, Russian and Mount Athel (Greece) monasteries during the fourteenth to sixteenth centuries are closely related to this innovation.

World Heritage site since

1978 **1979** 1980 1981 1982 1983 1984 1985 1986 1987 1988 1989 1990 1991 1992 1993 1994 1995 1996 1997 1998 1999 2000 2001 2002 2003 2004 2005 2006 2007 2008

Belovezhskaya Pushcha / Białowieża Forest
Belarus and Poland

Criteria – Natural phenomena or beauty

The Białowieża Primeval Forest dates back to 8000 BC and is the only remaining example of the original forests which once covered much of Europe. Located on the watershed of the Baltic and Black seas, this immense forest range consisting of evergreens and broadleaved trees is the home of some remarkable animal life, including rare mammals such as the wolf, the lynx and the otter, as well as some 300 European Bison, a species which was reintroduced into the park in 1929. The birdlife includes corncrake, white-tailed eagle, white stork, peregrine falcon and eagle owl. Białowieża National Park comprises about one-tenth of the entire forest. It is the oldest national park in Poland and one of the oldest in Europe.

Almost 90 per cent of the Białowieża National Park is covered with 'old growth' virgin stands of mixed broadleaved and conifer forests.

Over 900 vascular plant species have been recorded, including 26 tree and 138 shrub species. Almost two-thirds are indigenous.

World Heritage site since

1978 **1979** 1980 1981 1982 1983 1984 1985 1986 1987 1988 1989 1990 1991 1992 1993 1994 1995 1996 1997 1998 1999 2000 2001 2002 2003 2004 2005 2006 2007 2008

Palace and Park of Versailles
France

Criteria – Human creative genius; Interchange of values; Heritage associated with events of universal significance

The Palace of Versailles. ▶

The Palace of Versailles, built and embellished by several generations of the foremost French architects, sculptors, decorators and landscape architects, was one of the largest royal palaces in the world and provided the model of the ideal royal residence for over a century. The prestigious ensemble of the palace, the Trianons and the Park of Versailles is the result of a century and a half of works commissioned by the kings of France and entrusted to their greatest artists. Versailles was the principal residence of the French kings from Louis XIV to Louis XVI (from 1682–1789) and became both the source and symbol of absolute royal power during the Ancien Régime.

Versailles was originally a small village some 20 km southwest of Paris, set in a wooded region chosen by Louis XIII as his personal hunting preserve. The modest brick and stone château that the king ordered built here in 1623 was two storeys tall and surrounded by a moat. Enlargements followed but the strongest imprint was left by Louis XIII's son, Louis XIV. Under the direction of Louis Le Vau, the king's architect, a programme of expansion and new building began in the 1660s. The decoration of the palace interior was supervised by the painter Charles Le Brun

who, with teams of painters, decorators and craftsmen, created a remarkable complex of frescoes, marbles, stuccoes, gilded bronzes, fabrics, furniture and accessories in the palace halls.

After 1678 Versailles was considerably enlarged and radically modified by Jules Hardouin Mansart. He successfully introduced a sober and colossal architecture, homogeneous and majestic, that is inseparable even today from the memory of the 'Sun King', Louis XIV. It was during this phase of building that the palace developed the appearance that is recognizable now.

The gardens, which complete the palace, developed along with the general construction. They were designed by André Le Nôtre, who was responsible for the creation of the typology of the French garden, an open system of axial paths extending as far as the eye can see, punctuated geometrically by parterres of flowers and low hedges, little streams, large ponds and fountains.

The Orangerie and the Grand Trianon are the work of Jules Hardouin Mansart, who was assisted by Robert de Cotte in the construction of the Royal Chapel, a masterpiece of French Baroque.

The creations at Versailles during the eighteenth century are among the most perfect and most celebrated realizations of the Louis XV and Louis XVI styles: the Petit Trianon by Jacques-Ange Gabriel (1751), the decoration of the apartments of Louis XV by Verbeck and Rousseau, and the apartments and the Hameau of Marie-Antoinette by Mique.

World Heritage site since

1978 **1979** 1980 1981 1982 1983 1984 1985 1986 1987 1988 1989 1990 1991 1992 1993 1994 1995 1996 1997 1998 1999 2000 2001 2002 2003 2004 2005 2006 2007 2008

Independence Hall
USA

Criteria – *Heritage associated with events of universal significance*

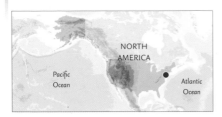

Independence Hall in Philadelphia may be considered the birthplace of the United States of America: it was here that the Declaration of Independence was signed in 1776, the Articles of Confederation uniting the thirteen colonies were ratified in 1781, and the Constitution setting out the nation's basic laws was adopted in 1787, after George Washington had presided over the debate, which ran from May to September. The universal principles of freedom and democracy set forth in these documents are of fundamental importance to American history and have also had a profound impact on law-makers around the world. The building has undergone many restorations, notably by Greek revival architect John Haviland in 1830, and by a committee from the National Park Service in 1950, returning it to its 1776 appearance.

The building, finished in 1753, is a modest brick structure with a steeple that was intended to hold a 943 kg bell. The bell has however cracked twice and stands silently on the ground in a special shelter. A reproduction now hangs in the steeple.

World Heritage site since

1978 **1979** 1980 1981 1982 1983 1984 1985 1986 1987 1988 1989 1990 1991 1992 1993 1994 1995 1996 1997 1998 1999 2000 2001 2002 2003 2004 2005 2006 2007 2008

Chartres Cathedral
France

Criteria – Human creative genius; Interchange of values; Significance in human history

Partly built, starting in 1145, and then reconstructed over a 26-year period after the fire of 1194, Chartres Cathedral marks the high point of French Gothic art. It is a place of pilgrimage, attracting throngs from all over the Christianized West. The cathedral, with its vast nave in pure ogival style, its porches adorned with fine sculptures from the mid-twelfth century, and the magnificent twelfth and thirteenth-century stained-glass windows, all in remarkable condition, combine to make it a masterpiece. Chartres Cathedral has exerted a considerable influence on the development of Gothic art both within and outside France. The fundamental design of the building was imitated at Cologne (Germany), Westminster (United Kingdom) and León (Spain).

Chartres Cathedral has retained in near totality its original stained-glass windows. The 55 lower windows are made up of large compartmentalized glass medallions in which are inscribed scenes of narrative cycles, while large figures, intended to be viewed from a distance, fill the 91 upper windows: in total there are 1,359 subjects ranged over 146 windows.

▼

World Heritage site since

1978 **1979** 1980 1981 1982 1983 1984 1985 1986 1987 1988 1989 1990 1991 1992 1993 1994 1995 1996 1997 1998 1999 2000 2001 2002 2003 2004 2005 2006 2007 2008

Vézelay, Church and Hill
France

Criteria – Human creative genius; Heritage associated with events of universal significance

Basilica of St Magdalene.

The twelfth-century monastic church of the Madeleine of Vézelay is a masterpiece of Burgundian Romanesque art and architecture. Believed to be the resting place of the relics of St Mary Magdalene, it has also been an important place of pilgrimage since medieval times. The hill of Vézelay was a site of devotion where medieval Christian spirituality produced an outflowing of expression, from prayer and the epic poem to the Crusades for which it was an important backdrop.

Established in the ninth century as a Benedictine abbey on the hill at Vézelay, the church came to prominence in the mid-eleventh century when the belief spread that it held the relics of St Mary Magdalene, the penitent. Consequently Vézelay became a popular pilgrimage destination, all the more frequented because of its situation on one of the routes leading to the great pilgrimage site of Santiago de Compostela in northwest Spain. The afflux of pilgrims profited the town, which during the twelfth century had 8,000–10,000 inhabitants, a considerable population for the period.

Vézelay became a centre of great importance in the Christian west. On Easter Sunday in 1146 it hosted the assembly called to celebrate the departure of the Second Crusade and St Bernard preached to King Louis VII, Queen Eleanor and a host of nobles, prelates and people gathered on the hill. In 1190 the kings of France and England, Philippe II and Richard I, the Lionheart, united their armies at Vézelay and left together for the Third Crusade. In 1217 Francis of Assisi chose the hill of Vézelay to found the first Franciscan establishment on French soil; and Louis IX of France, King Saint-Louis, who was especially devoted to the Madeleine of Vézelay, visited on four separate occasions.

With its sculpted capitals and portal, the superb church remains the most striking witness to this illustrious history. In particular, it was the sculpted portal between the nave and the narthex, its enclosed porch, that brought universal renown to Vézelay. The central tympanum above the doorway bears the Mission of the Apostles, an allegorical reference to the importance of the Crusades. The breadth and the complexity of the subject were well-matched by the inventiveness and passion of its sculptor who has produced one of the major monuments of Western Romanesque art.

The origins of Vézelay are linked with the memory of a knight around whom a legend had grown up. Around AD 860 Girart de Roussillon, the hero of the epic poem which bears his name, founded a monastery on the banks of the Cure together with his wife, Berthe. Ravaged some years later by the Normans, the establishment, a Benedictine abbey, was rebuilt on the summit of a neighbouring hill, the site of its present location.

World Heritage site since

1978 **1979** 1980 1981 1982 1983 1984 1985 1986 1987 1988 1989 1990 1991 1992 1993 1994 1995 1996 1997 1998 1999 2000 2001 2002 2003 2004 2005 2006 2007 2008

Persepolis
Islamic Republic of Iran

Criteria – Human creative genius; Testimony to cultural tradition; Heritage associated with events of universal significance

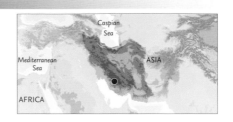

According to Plutarch, the Greeks carried away the Persepolis treasures on 20,000 mules and 5,000 camels.

The magnificent ruins of Persepolis lie at the foot of Kuh-i-Rahmat (Mountain of Mercy), about 650 km south of the present capital city of Teheran. Founded by Darius I in 518 BC, Persepolis was the capital of the Achaemenid Empire. It was built on an immense half-artificial, half-natural terrace, where an impressive palace complex was constructed, inspired by Mesopotamian models. It seems that Darius planned this impressive complex not only as the seat of government but also, and primarily, as a show place and a spectacular centre for the receptions and festivals of the Achaemenid kings and their empire. What remains today, dominating the city, is the immense stone terrace (530 m by 330 m), backed against the mountains. The importance and quality of the monumental ruins make it a unique archaeological site.

Persepolis was the example par excellence of the dynastic city, the symbol of the Achaemenid dynasty, which is why it was burned by the Greeks of Alexander the Great in 330 BC.
▼

World Heritage site since

1978 **1979** 1980 1981 1982 1983 1984 1985 1986 1987 1988 1989 1990 1991 1992 1993 1994 1995 1996 1997 1998 1999 2000 2001 2002 2003 2004 2005 2006 2007 2008

Madara Rider
Bulgaria

Criteria – Human creative genius; Testimony to
cultural tradition

The Madara Rider is a majestic horseman
carved 23 m above ground level in an almost
vertical 100 m high cliff. The horseman is
thrusting a spear into a lion lying at his
horse's feet, while a dog runs after the
horseman. Madara was the principal sacred
place of the First Bulgarian Empire before
Bulgaria's conversion to Christianity in the
ninth century. The Madara Horseman was
carved at the very beginning of the eighth
century, about three decades after the
foundation of the Bulgarian State in AD 681.
The sculpture thus marks a triumph: the
Byzantine Empire had recognised the new
state. The relief is not an abstract artistic
scene but a powerful symbolic
representation of a specific historical and
cultural transition.

In creating these
remarkable figures,
first the sculptor
outlined the images
with a 1.5 cm wide and
2 cm deep groove in
the rock. Then he
hewed out the
surrounding surface
so that the figures
project from it. Finally
he covered the figures
in red plaster to
outline them even
more clearly against
the rock.

Rock-Hewn Churches of
Ivanovo
Bulgaria

Criteria – Interchange of values; Testimony to
cultural tradition

This remarkable complex of rock-hewn
churches, chapels, monasteries and cells lies
in the valley of the Roussenski Lom River,
near the village of Ivanovo. The first hermits
dug out small cells and churches here
during the twelfth century. Then, in the
thirteenth century, Bulgaria once again
embraced Orthodox Christianity. The first
Patriarch was the monk Gioacchino, who
shared with Tsar Ivan Ansen II the plan to
expand the Bulgarian church. Before taking
over the Patriarchal throne he had lived as a
hermit in one of the Ivanovo caves, and the
monk achieved such a high level of sanctity
that the Tsar commissioned him to
construct a larger monastery at the site,
which greatly enhanced the Tsar's image
as a merciful monarch.

After 1396, the
monastery of Ivanovo
fell into ruins and was
abandoned. The solid
limestone out of
which it was carved
enabled it to resist the
weather, saving its
remarkable labyrinth
of cells, rooms,
churches and frescoes
to this day. The
fourteenth-century
murals testify to the
exceptional skill of the
artists belonging to
the Tarnovo School of
painting.

World Heritage site since

1978 **1979** 1980 1981 1982 1983 1984 1985 1986 1987 1988 1989 1990 1991 1992 1993 1994 1995 1996 1997 1998 1999 2000 2001 2002 2003 2004 2005 2006 2007 2008

Memphis and its Necropolis – the Pyramid Fields from Giza to Dahshur
Egypt

Criteria – Human creative genius; Testimony to cultural tradition; Heritage associated with events of universal significance

The capital of the Old Kingdom of Egypt in the third millennium BC, Memphis was one of the Seven Wonders of the Ancient World. The area of pyramid fields that served as the its necropolis contains a number of exceptional monuments that bear witness to the status of Ancient Egypt as one of the most brilliant civilizations of this planet.

The first sovereign of the unified Egyptian kingdom, Menes (or Narmer), was said to have ordered the construction of a new capital in the Nile Delta area around 3100 BC. It was from Menes that the name of Mennufer (City of Menes) came. The temple of Ptah built there was the most important sanctuary dedicated to this primary god of Creative Force, and its ruins are all that survive today of the grandeur of Memphis, as it was known to the ancient Greeks.

Nearby Saqqara was the necropolis of the city and the largest in the land, and it is the site of the first great stone pyramid. The pyramid was built as a mausoleum for the pharaoh Djoser, founder of the Third Dynasty, who ruled from around 2668 BC. Designed by his architect and vizier, Imhotep, it is the oldest step pyramid in the world. It stands in a funerary complex enclosed by a 10-m-high wall.

To the south lies the necropolis of Dahshur. The founder of the Fourth Dynasty, Snefru, who reigned from around 2613 BC, built here. During his twenty-nine year reign he transformed the structure of Egyptian royal tombs: he chose the now familiar pyramid shape with a square base. He built both the Red Pyramid, named after its reddish-coloured limestone, and the Rhomboid (or Bent) Pyramid, with its double-angled slope on each of the four faces; this was apparently an intermediate form. Another innovation of Snefru was the construction of an annex within the pyramid.

To the north, the great pyramids of Giza were built by Snefru's son Khufu or Cheops, and his successors Khafre (Chephren) and Menkaure (Mycerinus). The 'Horizon of Cheops' was the name given to the Pharaoh's tomb, the oldest and largest. The other two pyramids were known in antiquity as 'Great is Chepren' and 'Divine is Mycerinus' respectively. Each tomb forms part of the classic funerary complex first built by Snefru.

The Great Sphinx of Giza and the Pyramid of Khafre (Chephren). ▶

The necropolis of Saqqara dates back to the period of the formation of the pharaonic civilization.

At Giza the solar barge, one of the oldest boats preserved today, was discovered intact in the complex around the pyramid of Cheops.

The pyramid is a symbol of the Sun, the great god Ra, whose cult became pre-eminent from the Fourth Dynasty; the Pyramid Texts, found in the funerary chambers of the tombs dating from the end of the Old Kingdom, speak of the transformation of the dead king into the Sun.

World Heritage site since

1978 **1979** 1980 1981 1982 1983 1984 1985 1986 1987 1988 1989 1990 1991 1992 1993 1994 1995 1996 1997 1998 1999 2000 2001 2002 2003 2004 2005 2006 2007 2008

Medina of Tunis
Tunisia

Criteria – Interchange of values; Testimony to
cultural tradition; Traditional human settlement

The Medina of Tunis has exerted an
outstanding influence on the development
of architecture, sculpture and urban
planning. This group of buildings is rare, as
most historic Islamic centres have suffered
serious destruction and reconstruction over
the centuries, whereas Tunis still preserves
its homogeneity. The Medina of Tunis
extends over 2.7 km² and includes most of
the 700 historic monuments of the city,
including palaces, mosques, mausoleums
and fountains. It is divided between the
central core, which still bears traces from the
period of its foundation (the eighth century),
and two quarters dating back to the
thirteenth century. Today the medina is still
inhabited, but by only a small percentage of
the city's total population.

The medina is where the
main mosque of Tunis,
Zitouna, is located. Its
name means 'olive tree',
and comes from the
mosque's founder who
taught the Koran under
an olive tree. It was first
erected in the ninth
century by the Aghlabid
rulers, but its most
famous part, the
minaret is a nineteenth-
century addition. ▶

World Heritage site since

1978 **1979** 1980 1981 1982 1983 1984 1985 1986 1987 1988 1989 1990 1991 1992 1993 1994 1995 1996 1997 1998 1999 2000 2001 2002 2003 2004 2005 2006 2007 2008

Antigua Guatemala
Guatemala

Criteria – Interchange of values; Testimony to cultural tradition; Significance in human history

CENTRAL AMERICA

Caribbean Sea

Pacific Ocean

Much of Antigua's architecture today dates from the seventeenth and eighteenth centuries and provides us with a colonial jewel in the Americas.

Antigua Guatemala is an outstanding example of preserved colonial architecture. Built 1,500 m above sea level in an earthquake-prone region, Antigua, the capital of the Captaincy-General of Guatemala, was founded in the early sixteenth century as Santiago de Guatemala. It was the seat of Spanish colonial government for the Kingdom of Guatemala, which included much of present-day Central America. It was also the cultural, economic, religious and educational centre for the entire region until the capital was moved to present-day Guatemala City after the damaging earthquakes of 1773. In the space of under three centuries the city, which was built on a grid pattern inspired by the Italian Renaissance, acquired a number of superb monuments, many of which are preserved today as ruins.

The Church of our Lady of Mercy in Antigua Guatemala.
▼

World Heritage site since

1978 **1979** 1980 1981 1982 1983 1984 1985 1986 1987 1988 1989 1990 1991 1992 1993 1994 1995 1996 1997 1998 1999 2000 2001 2002 2003 2004 2005 2006 2007 2008

Ancient Thebes with its Necropolis
Egypt

Criteria – Human creative genius; Testimony to cultural tradition; Heritage associated with events of universal significance

The Great Hypostyle Hall of Columns in the Temple of Amun at the Karnak Temple Complex. There are 134 columns, resembling papyrus stalks, in 16 rows. Each column is covered in hieroglyphics or battle scenes.

Thebes was the capital of Egypt at the height of its greatest power and magnificence, the Middle and New Kingdoms that lasted over a thousand years from around 2134–1070 BC. It contains the finest relics of ancient Egyptian history, art and religion. With the temples and palaces at Karnak and Luxor and, across the river, the necropolises of the Valley of the Kings and the Valley of the Queens, Thebes is a striking testimony to Egyptian civilization at its height.

Hundreds of rulers glorified the city with architecture, obelisks and sculpture. Thebes of the Living, on the right bank of the Nile, is identifiable in the fabulous site of Luxor and Karnak, and temples built here were dedicated to the divine triad of Montu, Amon and Mut. The celebration of death took shape in the necropolis of Thebes of the Dead on the Nile's left bank.

From the Middle Kingdom to the first century AD the city was sacred to the god Amon, the supreme Sun God, and temples of incomparable splendour and size were dedicated to him. The temple of Luxor, built by Amenophis III and Ramses II, was connected to the enormous sanctuary of Karnak by a long processional avenue lined by sphinxes leading to its entrance. The monumental Karnak complex is composed of three temples: one consecrated to Mut, mother goddess of Egypt and wife of Amon; one to the warrior god Montu; and one to Amon.

On the opposite bank of the river Thebes of the Dead grew for almost fifteen centuries. Great funerary temples were built at the foot of the hills, entirely separate from their corresponding tombs which were dug into the mountains to keep them safe from violation and tomb robbers. To the north were erected the Temple of Deir el-Bahari, built by Queen Hatshepsut, dedicated to Amon-Re (one of the identities of Amon), and the temple consecrated to Hathor, the goddess of sweetness and joy who was venerated in the form of a cow.

The colossi of Memnon, massive twin statues of Amenhotep III facing eastwards over the river Nile, and the magnificent frescoes within the tombs are much admired by visitors to the site.

The tombs of the pharaohs and of their dignitaries, priests and princesses are hidden in the bowels of the mountains of Thebes of the Dead and form the great cemeteries of al-Asasif, al-Khokha, Qurnet Mura, Deir al-Medina, the Valley of the Kings and the Valley of the Queens.

Among the underground tombs of the Valley of Kings, the British explorers Howard Carter and Lord Carnarvon found in 1922 a small tomb that soon became the most renowned in Egypt, that of the young pharaoh Tutankhamun.

World Heritage site since

1978 **1979** 1980 1981 1982 1983 1984 1985 1986 1987 1988 1989 1990 1991 1992 1993 1994 1995 1996 1997 1998 1999 2000 2001 2002 2003 2004 2005 2006 2007 2008

Rock Drawings in Valcamonica
Italy

Criteria – Testimony to cultural tradition; Heritage associated with events of universal significance

Valcamonica, situated in the Lombardy plain, has one of the world's greatest collections of prehistoric petroglyphs – more than 140,000 symbols and figures carved in the rock over a period of 8,000 years, depicting themes connected with agriculture, navigation, war and magic. Several periods of carving can be identified: Upper Palaeolithic (c. 8000 BC), with scenes depicting hunting and early civilisation;

Neolithic (4000–3000 BC), in which the first depictions of a religious nature appear; and Eneolithic (3000–2000 BC), with highly detailed hunting and rural scenes and scenes depicting female initiation rituals. After 1000 BC the isolation of the area ended and battle scenes are carved into the rocks as well as drawings showing huts, wagons, harvests and weapons.

The rock engravings of Valcamonica constitute an extraordinary figurative documentation of prehistoric customs and mentality. The study of these configurations in stone has made a considerable contribution to the fields of prehistory, sociology and ethnology.

Stari Ras and Sopoćani
Serbia

Criteria – Human creative genius; Testimony to cultural tradition

On the outskirts of Stari Ras, the first capital of Serbia, there is an impressive group of medieval monuments consisting of fortresses, churches and monasteries. Situated on a hill at the border between the small kingdom of Raska and the Byzantine Empire, this city drew its strength from its strategic location and its mixture of Eastern and Western influences. Its many monuments make up a single architectural complex that testifies to the city's

architectural and cultural prime, which lasted until the early fourteenth century. The Church of St Peter, built in the ninth century on the foundations of an Illyrian cemetery and an early Christian basilica, is an example of early Christian architecture, and was the religious centre of Serbia for several centuries.

The monastery of Sopoćani, was built in 1260 by King Uros I as a family resting place. Surmounted by a cupola, it has exceptional frescoes which provide invaluable historical evidence about this ancient royal lineage.

World Heritage site since

978 **1979** 1980 1981 1982 1983 1984 1985 1986 1987 1988 1989 1990 1991 1992 1993 1994 1995 1996 1997 1998 1999 2000 2001 2002 2003 2004 2005 2006 2007 2008

Natural and Cultural Heritage of the Ohrid region
The Former Yugoslav Republic of Macedonia

Criteria – Human creative genius; Testimony to cultural tradition; Significance in human history; Natural phenomena or beauty

Ohrid is one of the most ancient human settlements in Europe. More than 250 archaeological sites with material remains dating from between the Neolithic period and the late Middle Ages have been excavated.

Writing, education and Slavic culture – all spread out from Ohrid in the seventh to nineteenth centuries. It is a cultural centre of great importance for the history not only of this part of the Balkan Peninsula, but for world history and literature in general. The city and its historic-cultural region are located in a natural setting of exceptional beauty, while its architecture represents the best preserved and most complete ensemble of ancient urban architecture of the Slavic lands. It also has the oldest Slav monastery, St Pantelejmon, and more than 800 Byzantine-style icons dating from the eleventh to the fourteenth century. After those of the Tretiakov Gallery in Moscow, this is considered to be the most important collection of icons in the world.

St Pantelejmon monastery, fully restored in 2002.
▼

World Heritage site since

1978 **1979** 1980 1981 1982 1983 1984 1985 1986 1987 1988 1989 1990 1991 1992 1993 1994 1995 1996 1997 1998 1999 2000 2001 2002 2003 2004 2005 2006 2007 200▪

Historic Cairo
Egypt

Criteria – Human creative genius; Traditional
human settlement; Heritage associated with
events of universal significance

The Mosque and
Madrasa of Sultan
Hassan.

Lying at the heart of modern, urban Cairo
lies one of the world's oldest Islamic cities.
Cairo, which was founded in the tenth
century, was the new capital and centre of
the Islamic world, and its wealth of
architectural treasures – mosques, madrasas,
palaces and caravanserais, hammams and
fountains – reflect its importance at that
time.

In the seventh century AD, after the death
of the Prophet Muhammad, the founder of
Islam, Arab armies marched with speed to
conquer neighbouring lands. In 640 Caliph
Omar with his army reached the Nile and
founded his own capital of al-Fustat; this
was the first capital in Egypt under Arab rule.

Over the following centuries of factional
battles for supremacy the capital was
moved, first in 870 to al Qatai to the north
east, where the Great Mosque of Ibn-Tulun
was built. With its large courtyard
surrounded by porticoes and punctuated by
elegantly decorated arches, this mosque was
the only building of the period to survive
and remains one of the finest monuments
in historic Cairo.

In AD 969, during the Fatimid era, Cairo (al-
Qahira) was established as the capital. This
was the start of a great period of city
splendour. The present-day quarter of al-
Azhar has contemporary monuments,

including the three large gates and the
huge, square towers of the city's walls and
five mosques. Among these, the Mosque of
al-Hakim, compact and severe, is the last
example of a military mosque. The Mosque
of al-Azhar was built between 970 and 972
under the Caliph Muizz to serve as a
sanctuary and as a meeting place; it also
housed a university which became the most
important centre for Islamic studies.

In the following centuries Egypt had
successive changes of dynasties and rulers,
including Saladin, the successful leader of
the Islamic forces in the Third Crusade of
1189–1192, who constructed the Citadel
which remained the heart of Egyptian
government until the nineteenth century.
Cairo thrived under the dynasty of the
Mamelukes, rulers from 1250–1517. In
addition to religious structures the sultans
of that time built splendid mausoleums in
the City of the Dead, the huge cemetery to
the east of the city proper, and one of a
number of cemeteries within this World
Heritage site.

In 1517, the Ottomans defeated the
Mamelukes and during their rule from the
sixteenth to the nineteenth centuries many
impressive buildings, such as the
magnificent mosque of Muhammad Ali,
were erected.

Few other cities in the
world are as rich in old
buildings: the historic
centre on the east
bank of the river Nile
includes 600 classified
monuments that date
from the seventh
century through
Cairo's golden age in
the fourteenth
century, right up to
the twentieth century.
Collectively they
reflect the strategic,
political, intellectual
and commercial
importance of the city.

The ancient capitals of
al-Fustat, al Qatai and
al-Qahira are included
in Historic Cairo.

World Heritage site since

1978 **1979** 1980 1981 1982 1983 1984 1985 1986 1987 1988 1989 1990 1991 1992 1993 1994 1995 1996 1997 1998 1999 2000 2001 2002 2003 2004 2005 2006 2007 2008

Amphitheatre of El Jem
Tunisia

Criteria – Significance in human history; Heritage associated with events of universal significance

EUROPE

Mediterranean Sea

AFRICA

The impressive ruins of the largest colosseum in North Africa are found in the small village of El Jem, known in classical times as Thysdrus, located 60 km south of Sousse. It is one of the most accomplished examples of Roman amphitheatre construction, approaching the status of the Colosseum in Rome. Built during the first half of the third century, it may have accommodated up to 60,000 spectators.

Elliptical in form, it is constructed from large stone blocks and probably comprised four floors. The theatre was never completed, because of political rivalries and lack of funds within the Empire. Later it served as a stronghold: it was the last Berber bastion against Arab invaders. Following the Roman period, the amphitheatre was used at various times as a citadel and was twice attacked by cannon fire.

The amphitheatre is well preserved and little altered, one of the last surviving monuments of this type from the Roman world. Underneath it run two passageways, in which animals, prisoners and gladiators were kept until the moment when they were brought up into the bright daylight to perform what was in most cases the last show of their lives.

▼

World Heritage site since

1978 **1979** 1980 1981 1982 1983 1984 1985 1986 1987 1988 1989 1990 1991 1992 1993 1994 1995 1996 1997 1998 1999 2000 2001 2002 2003 2004 2005 2006 2007 2008

Urnes Stave Church
Norway

Criteria – Human creative genius; Interchange of values; Testimony to cultural tradition

The wooden stave church of Urnes, built in the twelfth and thirteenth centuries, is an outstanding example of traditional Scandinavian wooden architecture. It brings together traces of Celtic art, Viking traditions and Romanesque architectural design. The stave churches constitute one of the most elaborate types of wood construction in northern Europe. They are constructed just from wood – even the roof is covered with wooden shingles. At Urnes, the structure uses wood for the columns, capitals and semicircular arches to replicate the style of stone Romanesque architecture. On the outside, there are outstanding panels of eleventh-century carved strapwork taken from an earlier stave church built here. Inside is an amazing series of twelfth-century figurative capitals and a wealth of medieval liturgical objects. The embellishments of the seventeenth century and the restoration of 1906–1910 preserved its authenticity.

Christianity was introduced into Norway by St Olav in the eleventh century. The stave church at Urnes stands in a glacial valley on the north bank of Sognefjord, and is the finest example of a medieval stave church from around thirty that remain in Norway.

World Heritage site since

1978 **1979** 1980 1981 1982 1983 1984 1985 1986 1987 1988 1989 1990 1991 1992 1993 1994 1995 1996 1997 1998 1999 2000 2001 2002 2003 2004 2005 2006 2007 2008

Historical Complex of Split with the Palace of Diocletian
Croatia

Criteria – Interchange of values; Testimony to cultural tradition; Significance in human history

A view of the historic ▶ complex of Split with St Dominus' Bell Tower in the foreground.

The ancient city of Split is renowned for the massive palace built at the turn of the fourth century AD by the Roman Emperor Diocletian. Unique in scale and completeness, it is integrated into an array of important medieval buildings including the cathedral, twelfth- and thirteenth-century Gothic palaces, and other palaces in Renaissance and Baroque style.

In the early Middle Ages the town was built within the palace until commercial prosperity inspired its growth and spread outside in the thirteenth and the fourteenth centuries. A new centre developed along the western walls of the palace, which was fortified in the fourteenth century. Later, in the seventeenth century, a new defence system with projecting bastions was erected.

Much of the Roman and medieval town exists today. On the eastern side of the palace lies the Porta Argentea (Silver Gate) with the church of St Dominic on the opposite side. The Silver Gate gives access to the Plain of King Tomislav and thence to the peristil (peristyle), the central open-air courtyard area of the palace.

In the eastern part of the peristyle is the Mausoleum of Diocletian (today's Cathedral of St Dominus dedicated to St Mary); its original octagonal form is complete and a dome, once covered by mosaics, forms the roof. The monumental wooden gateposts and the stone pulpit from the thirteenth century represent the oldest monuments in the cathedral.

A small temple opposite the mausoleum, probably dedicated to Jupiter, became the baptistry in the early Middle Ages. Only the closed part of the temple (the cella), with its richly decorated portal, has been preserved.

Diocletian Street runs north from the peristyle to the Porta Aurea (Golden Gate); Agubio Palace, with a Gothic portal and inner yard is to the left. To the right, in Papaliceva Street, is the Papalic Palace (fifteenth century), the most important example of Gothic architecture in Split.

Kresimir Street leads from the peristyle to the Porta Ferrea (Iron Gate) in the west; Cindro Palace (seventeenth century), the city's most beautiful Baroque palace, lies on the right. Beyond the Iron Gate is the square Narodni Trg (Piaca), centre of the medieval commune and the liveliest square of modern Split. Of the original Gothic houses at the northern end of the square, only the Town Hall which opened in 1443 survives.

Diocletian's Palace is internationally important not only for its level of preservation but also for the buildings of succeeding historical periods within and around it that together form the fabric of old Split.

Diocletian spent his later years after his retiral in the palace that he built between AD 293 and 305 near his Dalmatian birthplace. The most valuable example of Roman architecture on the eastern Adriatic, its form and the arrangement of the buildings within, represent a transitional style of imperial villa, Hellenistic town and Roman camp.

World Heritage site since

1978 **1979** 1980 1981 1982 1983 1984 1985 1986 1987 1988 1989 1990 1991 1992 1993 1994 1995 1996 1997 1998 1999 2000 2001 2002 2003 2004 2005 2006 2007 2008

Natural and Culturo-Historical Region of Kotor
Montenegro

Criteria – Human creative genius; Interchange of values; Testimony to cultural tradition; Significance in human history

This natural harbour on the Adriatic coast in Montenegro was an autonomous city of the Byzantine Empire until the late twelfth century, when it became a free city of medieval Serbia. From 1420 it was under Venetian control and was then occupied by the French from 1807–1914.

The region of Kotor has played an exceptionally important role in the diffusion of Mediterranean culture in the Balkans along the southern Adriatic coast. Its art, goldsmithing and architecture schools have had a long-lasting influence on the region. Founded by the Romans, Kotor developed in the Middle Ages into an important commercial and artistic centre and many empires battled for control of the city. Its most impressive building is St Tryphon Cathedral, built in 1166, damaged during the 1667 earthquake and then restored. Most of Kotor's palaces and houses, many Romanesque churches and all of neighbouring Dobrota's palaces and Perast's main buildings have suffered from earthquakes. Kotor was evacuated by all its inhabitants after the most recent, on 15 April 1979. An intensive restoration programme has now been completed and the city is flourishing again.

St Tryphon Cathedral and part of the old town of Kotor, viewed from above.
▼

World Heritage site since

1978 **1979** 1980 1981 1982 1983 1984 1985 1986 1987 1988 1989 1990 1991 1992 1993 1994 1995 1996 1997 1998 1999 2000 2001 2002 2003 2004 2005 2006 2007 2008

Sagarmatha National Park
Nepal

Criteria – Natural phenomena or beauty

This is an exceptional area with dramatic mountains, glaciers and deep valleys, dominated by Mount Everest (Sagarmatha), the highest peak in the world (8,848 m), and is the homeland of the Sherpa, with their unique culture. The park is fan-shaped and enclosed on all sides by high, geologically young mountain ranges. The deeply incised valleys cut through sedimentary rocks and underlying granites to drain southwards into the Dudh Kosi and its tributaries, which form part of the Ganges River system. Most of the park (69 per cent) comprises barren land above 5,000 m, 28 per cent is grazing land and about 3 per cent is forested. The low number of mammals is almost certainly the result of human activities. Several rare species, such as the snow leopard and the lesser panda, are found in the park.

There are approximately 2,500 Sherpa people living within the park. They belong to the Nyingmapa sect of Tibetan Buddhism, which was founded by the revered Guru Rimpoche who was, according to legend, born of a lotus in the middle of a lake. There are several monasteries in the park, the most important being Tengpoche.

▲
Namche Bazaar village.

World Heritage site since

1978 **1979** 1980 1981 1982 1983 1984 1985 1986 1987 1988 1989 1990 1991 1992 1993 1994 1995 1996 1997 1998 1999 2000 2001 2002 2003 2004 2005 2006 2007 2008

Old City of Dubrovnik
Croatia

Criteria – Human creative genius; Testimony to
cultural tradition; Significance in human history

The distinctive ▶
rooftops of Dubrovnik,
many of which were
reconstructed after the
1990s' armed conflict.

The 'Pearl of the Adriatic', Dubrovnik was an
important Mediterranean power from the
thirteenth century. The city's self-
confidence, wealth and culture are reflected
in its beautiful Gothic, Renaissance and
Baroque churches, monasteries, palaces and
fountains, many of which are survivors of a
massive earthquake in 1667. UNESCO is
currently coordinating a major restoration
programme to restore Dubrovnik's
architecture, damaged in armed conflict in
the 1990s after the break-up of the
communist state of Yugoslavia, of which it
had been part.

From the city's founding in the seventh
century, it was under the control of the
Byzantine Empire. Sovereignty passed to
Venice in 1205 after the crusaders sacked
Constantinople, until 1358 when Venice
ceded control and Dubrovnik became part
of the Hungarian-Croatian kingdom. It
operated effectively as a republican free
state, reaching its peak in the fifteenth and
sixteenth centuries. But an economic crisis
in Mediterranean shipping and, particularly,
a catastrophic earthquake in April 1667
destroyed the well-being of the republic.
It proved a turning point for the previously
powerful city.

Dubrovnik is a remarkably well-preserved
example of a late-medieval walled city, with a
regular street layout. Among the
outstanding medieval, Renaissance and
Baroque monuments within the magnificent
fortifications and the monumental gates to
the city are the eleventh century Town Hall
(now the Rector's Palace); the Franciscan
Monastery (completed in the fourteenth
century, but now largely Baroque in
appearance) with its imposing church; the
Dominican Monastery; the cathedral (rebuilt
after the earthquake); the customs house
(Sponza); and a number of other Baroque
churches, such as that of St Blaise, the patron
saint of the city.

The original World Heritage site consisted
solely of the defences and the intra-mural
city. It was later extended to include the Pile
medieval industrial suburb, a planned
development of the fifteenth century, and
the Lovrijenac Fortress, set on a cliff, which
was probably begun as early as the eleventh
century but owes its present appearance to
building work in the fifteenth and sixteenth
centuries. Also included were the Lazarets,
built in the early seventeenth century to
house potential plague-carriers from abroad;
the late fifteenth-century Kase moles, built to
protect the port against southeasterly gales;
and the Revelin Fortress, dating from 1449,
which was built to command the town moat
on its northern side.

Dubrovnik was
founded on an island
in the seventh century
by refugees from the
Latin city of
Epidaurum, who
named their
settlement Laus (from
the Latin lausa, 'rock').
Dubrovnik's old name
of Ragusa (or Rausa),
originated from the
name Laus. Across the
river, at the foot of Srđ
Mountain, the Slavs
later developed their
own settlement
named Dubrovnik
(from the Croatian
word dubrava, 'oak
woods'). The two
settlements were
united in the twelfth
century when the
channel separating
them was filled in.

World Heritage site since

1978 **1979** 1980 1981 1982 1983 1984 1985 1986 1987 1988 1989 1990 1991 1992 1993 1994 1995 1996 1997 1998 1999 2000 2001 2002 2003 2004 2005 2006 2007 2008

Tchogha Zanbil
Islamic Republic of Iran

Criteria – Testimony to cultural tradition;
Significance in human history

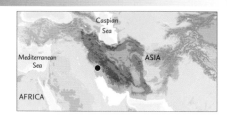

The ruins of the holy city of the Kingdom of
Elam, surrounded by three huge concentric
walls, are found at Tchogha Zanbil. Founded
c. 1250 BC, the city remained unfinished after it
was invaded by Ashurbanipal, as shown by the
thousands of unused bricks left at the site.

◄
The site contains the
best preserved and the
largest of all
the ziggurats of
Mesopotamia. Today
the ziggurat is over
25 m high, but with its
last two stages (since
destroyed) it would
originally have risen to
a height of 60 m.

Tikal National Park
Guatemala

Criteria – Human creative genius; Testimony to
cultural tradition; Significance in human history;
Significant ecological and biological processes;
Significant natural habitat for biodiversity

Tikal National Park protects some 221 km²
of rainforest and possesses a great wealth of
animal life. In the heart of this jungle lies
one of the major sites of Mayan civilization,
inhabited from the sixth century BC to the
tenth century AD. The ceremonial centre
contains superb temples and palaces, and
public squares accessed by means of ramps.
Remains of dwellings are scattered
throughout the surrounding countryside.

At its height from AD 700–800, the city
supported a population of 90,000 Mayan
Indians. There are over 3,000 separate
buildings, including religious monuments
decorated with hieroglyphic inscriptions,
and tombs. The ruined city reflects the
cultural evolution of Mayan society from
hunter-gathering to farming, with an
elaborate religious, artistic and scientific
culture.

The reserve contains
the largest area of
tropical rainforest in
Guatemala and
Central America, with
a wide range of
unspoilt natural
habitats. Tikal is a
well-known site for
viewing jungle
animals, especially the
howler monkey and
the spider monkey,
large numbers of
which often
congregate around
the Mayan site.

World Heritage site since

1978 1979 **1980** 1981 1982 1983 1984 1985 1986 1987 1988 1989 1990 1991 1992 1993 1994 1995 1996 1997 1998 1999 2000 2001 2002 2003 2004 2005 2006 2007 2008

Historic Centre of Warsaw
Poland

Criteria – Interchange of values; Heritage
associated with events of universal significance

The historic centre of the Polish capital
Warsaw was almost completely rebuilt after
fighting during the Second World War
destroyed 85 per cent of the Old Town. The
city is an outstanding example of a near-
total reconstruction of a span of history
covering the thirteenth to the twentieth
centuries.

Poland was occupied by German Nazi
troops in 1939 and its capital was the scene
of two large-scale insurrections: the 1943
Warsaw Ghetto Uprising of the Jewish

population; and the general Warsaw
Uprising of August 1944. After the 1944
Uprising, occupying forces systematically
destroyed the city in reprisal. From these
total ruins, between 1945 and 1966, Warsaw's
historic centre was reconstructed.

Almost every building in Warsaw Old Town
is of a unique architectural style, from
Gothic to Baroque. All have been restored:
among the more famous are the market
square, city walls and Barbican as well as the
Royal Castle and numerous churches.

The reconstruction
of the city was
meticulously planned
in a five-year
programme that
began after the war.
Among the references
utilized to reconstruct
the Old Town, it is
believed that paintings
of Warsaw streetscapes
by the Italian artist
Canaletto were used.

The precise
reconstruction of
the historic centre
showcases the
restoration techniques
of the later twentieth
century.

◄

Kanonia Square is
next to Warsaw's main
square. It is triangular
in shape and contains
Warsaw's narrowest
house (the yellow
house in the corner).
This was an attempt
by the seventeenth/
eighteenth century
owners to pay less
property tax (which
depended on the width
of the property's street-
facing wall and the
number of windows
that wall had).

World Heritage site since

1978 1979 **1980** 1981 1982 1983 1984 1985 1986 1987 1988 1989 1990 1991 1992 1993 1994 1995 1996 1997 1998 1999 2000 2001 2002 2003 2004 2005 2006 2007 2008

Maya Site of Cópan
Honduras

Criteria – Significance in human history;
Heritage associated with events of universal
significance

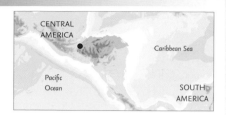

CENTRAL
AMERICA

Caribbean Sea

Pacific
Ocean

SOUTH
AMERICA

Mayan civilization
developed from
around 2000 BC and
at the height of its
power in Central
America (around AD
250–900), Cópan was
the largest and most
influential city in the
southeastern area.

Cópan is under threat
from river erosion,
agricultural activity
and earthquakes,
while its natural
surroundings are
being threatened by
the infringement of
the neighbouring
town of Cópan Ruins.

The ruins of Cópan are among the most
important sites of the indigenous American
Mayan civilization, with temples, plazas and
terraces that are among the most
characteristic of any complex of Mayan
architecture and building.

There is evidence that Cópan was
inhabited from 2000 BC onwards, but its full
flourishing, along with other Mayan cities,
came in the period AD 250–900, with
important advances in mathematics,
astronomy and hieroglyphic writing, as well
as major cultural developments.

Cópan's ruined citadel and imposing
public squares reveal the three main stages
of city development, during which the
temples, plazas, altar complexes and ball
courts that can be seen today, evolved. For
unknown reasons the site was abandoned
shortly after AD 900.

Discovered in 1570 by the Spanish
conquistador Diego García de Palacio, the
ruins of Cópan were not excavated until the
nineteenth century.

◄ ►
A Mayan temple, left,
and a carved stone
pillar, right, in the
ruins of Cópan.

World Heritage site since

1978 1979 **1980** 1981 1982 1983 1984 1985 1986 1987 1988 1989 1990 1991 1992 1993 1994 1995 1996 1997 1998 1999 2000 2001 2002 2003 2004 2005 2006 2007 2008

Historic Centre of Rome, the Properties of the Holy See in that City Enjoying Extraterritorial Rights and San Paolo Fuori le Mura
Holy See and Italy

Criteria – Human creative genius; Interchange of values; Testimony to cultural tradition; Significance in human history; Heritage associated with events of universal significance

Founded, according to legend, by Romulus and Remus in 753 BC, Rome was first the centre of the Roman Republic, then of the Roman Empire, and from the fourth century the capital of the Christian world.

In 1990 the World Heritage site in Rome was extended out to the seventeenth-century city walls built by Pope Urban VIII.

The World Heritage site in the historic centre of Rome comprises outstanding buildings and monuments from ancient and medieval Roman history. These include the forums, the Mausoleum of Augustus, the Mausoleum of Hadrian, the Pantheon (pictured below), Trajan's Column and the Column of Marcus Aurelius, and the religious and public buildings of papal Rome.

The 1929 Lateran Treaty between Italy and the Holy See that established Vatican City State also confirmed that some properties termed 'extraterritorial' and situated on Italian soil would remain the property of the Holy See. Those extraterritorial properties comprise a series of unique artistic achievements: Santa Maria Maggiore, St John Lateran and St Paul Outside the Walls. All were architecturally and artistically influential in the Christian world for centuries. Also included are several remarkable palaces: the Cancelleria, the Palazzo Maffei, the Palazzo di San Callisto and the Palazzo di Propaganda Fide.

◄

The Pantheon is one of the best preserved ancient Roman buildings. The original Pantheon was built in c. 27 BC by Agrippa but was destroyed in AD 80 by fire. The current building dates from c. AD 125 during Emperor Hadrian's reign.

World Heritage site since

978 1979 **1980** 1981 1982 1983 1984 1985 1986 1987 1988 1989 1990 1991 1992 1993 1994 1995 1996 1997 1998 1999 2000 2001 2002 2003 2004 2005 2006 2007 2008

City of Valletta
Malta

Criteria – Human creative genius; Heritage
associated with events of universal significance

EUROPE

AFRICA

Mediterranean Sea

The three islands of
Malta were long
fought over for their
perceived strategic
importance in the
Mediterranean and
were held successively
by Phoenicians,
Greeks, Carthaginians,
Romans, Byzantines,
Arabs, and the Order
of the Knights of St
John, who ruled for
over two centuries.

Valletta is inextricably linked to the military,
charitable and religious Order of St John of
Jerusalem, the Knights Hospitaller. The
Knights were based in Malta from 1530 to
1798 and left a strong mark, bequeathing the
capital with a grid of broad, straight streets
lined with hundreds of well-planned
monuments. The result is that at a total size
of just 0.5 km², Valletta is one of the most
concentrated historic areas in the world.

The city's design is late Renaissance with
a grid-based street layout, fortified and
bastioned walls modelled around its
peninsular site, and the planned building of
great monuments in well-chosen locations.
Churches and palaces, museums and
theatres, gardens and piazzas retain their
original features almost completely.
Remarkably, the city has undergone no
important modifications since 1798, when
the Knights left.

Valletta, Malta. ▼

World Heritage site since

1978 1979 **1980** 1981 1982 1983 1984 1985 1986 1987 1988 1989 1990 1991 1992 1993 1994 1995 1996 1997 1998 1999 2000 2001 2002 2003 2004 2005 2006 2007 2008

Asante Traditional Buildings
Ghana
Criteria – Traditional human settlement

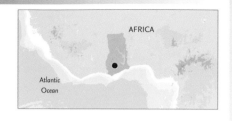

AFRICA

Atlantic
Ocean

The majority of the Asante villages were destroyed during the nineteenth century in the wars undertaken by these people against British domination between 1806 and 1901. The royal mausoleum (Barem) was burned by Baden-Powell in 1895.

To the northeast of Kumasi, these buildings are the last material remains of the great Asante civilization, which reached its high point in the eighteenth century. Since the dwellings are made of earth, wood and straw, they are vulnerable to the onslaught of time and weather. There exist today only a few of the traditional homes and temples, of which the majority are less than 100 years old. These are constructed with a framework of poles and wooden imposts linked by bamboo slats which support the thatched roof. The floor is of puddled clay. The main façade is built of earth over a core of wood and has a balustrade, imposts and sometimes windows. The decoration consists of geometric, floral, animal or anthropomorphic motifs.

Garamba National Park
Dem. Rep. of the Congo
Criteria – Natural phenomena or beauty; Significant natural habitat for biodiversity

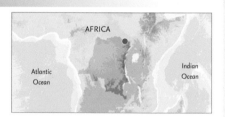

AFRICA

Atlantic
Ocean

Indian
Ocean

Lying on the watershed between the river Nile and the river Congo the park is a vast undulating plateau broken up by inselbergs (generally of granitic formation) and sizeable marshland depressions.

The park's immense savannas, grasslands and woodlands, interspersed with gallery forests along the river banks and the swampy depressions, are home to flagship species such as the Northern white rhinoceros and the Congo giraffe, which are found nowhere else. High concentrations of elephants are also found here.

◄

Hippos in the 'maternité' pool in Garamba National Park.

World Heritage site since

1978 1979 **1980** 1981 1982 1983 1984 1985 1986 1987 1988 1989 1990 1991 1992 1993 1994 1995 1996 1997 1998 1999 2000 2001 2002 2003 2004 2005 2006 2007 2008

Church and Dominican Convent of Santa Maria delle Grazie with 'The Last Supper' by Leonardo da Vinci

Italy

Criteria – Human creative genius; Interchange of values

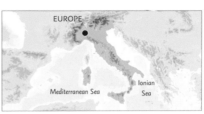

The fifteenth-century Renaissance Church and Convent of Santa Maria delle Grazie in Milan, pictured below, is a striking monument in itself, made all the more remarkable by bearing on the fabric of a wall one of the supreme art masterpieces: 'The Last Supper' by Leonardo da Vinci.

Da Vinci was commissioned in 1495 to paint a mural on the north wall of the refectory and finished work in 1497. The painting depicts the moment in the Gospel of John 13:21 immediately after Jesus says, 'One of you will betray me'. In composition, perspective and execution the painting broke with tradition and heralded a new era in the history of art.

It was not technically a fresco in that Leonardo worked on dry plaster, sealing the wall and painting in tempera. Within years the painting began to deteriorate and repeated conservation has been carried out.

Within the building, the genius of Leonardo da Vinci is seen especially in the use of light and strong perspective. The three windows and the landscape beyond create a luminosity that, set against the backlight, illuminates the characters from the side as well. The result is a combination of particularly classical Florentine and chiaroscuro perspectives.

World Heritage site since

1978 1979 **1980** 1981 1982 1983 1984 1985 1986 1987 1988 1989 1990 1991 1992 1993 1994 1995 1996 1997 1998 1999 2000 2001 2002 2003 2004 2005 2006 2007 2008

Megalithic Temples of Malta
Malta

Criteria – Significance in human history

The seven megalithic temples that make up this World Heritage site on Malta and Gozo are outstanding examples of structures that represent a major development in culture, art and technology. All date from the third millennium BC and each is the result of an individual development, differing from the others in plan, execution and construction techniques.

The two temples of Ggantija on the island of Gozo are notable for their gigantic

Bronze Age structures. On Malta, the temples of Hagar Qim, Mnajdra and Tarxien are unique architectural masterpieces, given the limited resources available to their builders. The Ta'Hagrat and Skorba complexes show how the tradition of temple-building was handed down in Malta. Each one is remarkable for diversity of form and decoration.

Professor Lord Renfrew, a leading prehistorian, described the megalithic temples of Malta and Gozo as 'the oldest freestanding monuments in the world'. They are also remarkable for their diversity of form and decoration.

The elaborate rituals to which the temples are testimony are a remarkable manifestation of the human spirit, especially on a remote island at such an early date.

Section of the Temple of Hagar Qim. ▼

World Heritage site since

1978 1979 **1980** 1981 1982 1983 1984 1985 1986 1987 1988 1989 1990 1991 1992 1993 1994 1995 1996 1997 1998 1999 2000 2001 2002 2003 2004 2005 2006 2007 2008

Kahuzi-Biega National Park
Dem. Rep. of the Congo

Criteria – Significant natural habitat for
biodiversity

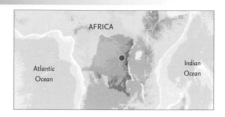

A vast area of primary tropical forest
dominated by two spectacular mountains,
Kahuzi (3,308 m) and Biega (2,790 m), the
park has a diverse and abundant fauna. It
consists of a smaller sector on the eastern
side, covering part of the Mitumba
Mountains, and a larger western sector in
the central Congo basin. The two zones are
connected by a narrow corridor. The western
zone is forested by equatorial rainforest,
with transition forest between 1,200 m and
1,500 m. In the eastern zone, six different
primary vegetation types have been
distinguished. The park was established to
protect 200–300 eastern lowland (Grauer's)
gorilla occurring mainly in the forests at
2,100–2,400 m, but also in the lower
rainforest.

The park's mosaic of
biotypes makes it an
excellent gorilla
habitat, but it is also
home to other
primates including
eastern chimpanzee,
and numerous
Cercopithecinae and
Colobinae. Other
mammals include
elephant, forest hog
and many antelope
and duiker.

Ichkeul National Park
Tunisia

Criteria – Significant natural habitat for
biodiversity

The Ichkeul lake and wetland are a major
stopover point for hundreds of thousands of
migrating birds, such as ducks, geese, storks
and pink flamingos, who come to feed and
nest there. Ichkeul is the last remaining lake
in a chain that once extended across North
Africa. Due to increased salinity in the lake
and marshes, conservation efforts in recent
years, including a high-quality scientific
monitoring programme, have resulted in
improved water quality, leading toward the
restoration of vegetation critical to the
functioning of the ecosystem, the gradual
return of wintering and breeding birds, and
the recovery of fish populations.

Ichkeul National Park
consists of an isolated
and wooded massif,
Djebel Ichkeul, and a
brackish permanent
lake, Lake Ichkeul, the
area of which varies
with the seasons.
Ichkeul is recognized
as being extremely
diverse largely due to
the wide variety of
habitats.

◄

World Heritage site since

1978 1979 **1980** 1981 1982 1983 1984 1985 1986 1987 1988 1989 1990 1991 1992 1993 1994 1995 1996 1997 1998 1999 2000 2001 2002 2003 2004 2005 2006 2007 2008

Hal Saflieni Hypogeum
Malta

Criteria – Testimony to cultural tradition

The Hypogeum is an enormous subterranean structure excavated c. 2500 BC. The megalithic walls are constructed of cyclopean masonry – large irregular blocks of chalky coralline stone without mortar. The workmanship is all the more impressive when it is considered that the chambers were meticulously carved using only flint and stone tools. The principal rooms distinguish themselves by their domed vaulting and by the elaborate structure of false bays inspired by the doorways and windows of contemporary terrestrial constructions. Curvilinear and spiral paintings in red ochre are still visible in some areas. The carved façade is magnificent and the quality of its architecture is in a remarkable state of preservation. Perhaps originally a sanctuary, the Hypogeum became a necropolis in prehistoric times.

Hal Saflieni Hypogeum is a cultural property of exceptional prehistoric value. The only known example of a subterranean structure of the Bronze Age, it was only discovered accidentally in 1902 by a stonemason laying housing foundations.

Aksum
Ethiopia

Criteria – Human creative genius; Significance in human history

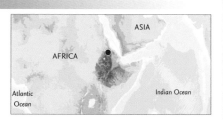

The ruins of the ancient city of Aksum are found close to Ethiopia's northern border. They mark the location of the heart of ancient Ethiopia, when the Kingdom of Aksum was the most powerful state between the Eastern Roman Empire and Persia. The massive ruins, dating from between the first and the thirteenth centuries AD, include giant monolithic stelae, royal tombs and the ruins of ancient castles. Long after its political decline in the tenth century, Ethiopian emperors continued to be crowned in Aksum.

◀

Transported to Rome by the troops of Mussolini in 1937, the stela 2 was returned by the Italian Government in April 2005. Weighing 150 tons and 24 m high, it is around 1,700 years old and a symbol of the Ethiopian people's identity.

World Heritage site since

1978 1979 **1980** 1981 1982 1983 1984 1985 1986 1987 1988 1989 1990 1991 1992 1993 1994 1995 1996 1997 1998 1999 2000 2001 2002 2003 2004 2005 2006 2007 2008

Fortifications on the Caribbean Side of Panama: Portobelo – San Lorenzo

Panama

Criteria – Human creative genius; Significance in human history

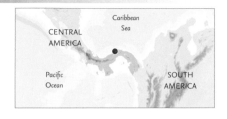

The group of seventeenth- and eighteenth-century fortifications of the Caribbean coast of Panama are magnificent examples of contemporary Spanish colonial military architecture, located in a natural setting of great beauty. The forts of Portobelo and San Lorenzo formed part of the defence system of the Spanish Crown to guard the access to the Isthmus of Panama that was crucial to Europe's trade with its colonies.

Portobelo's forts, castles, barracks and batteries created a defensive line around its bay and protected the harbour, while the works at San Lorenzo guarded the mouth of the Chagres River.

The forts are now in a poor state of preservation. The Pan American Institute of Geography and History is among the international organizations acknowledging the importance of the Portobelo and San Lorenzo sites as an essential link in the understanding of American history.

The forts came under regular attack and were rebuilt three times: after their capture by the privateer Henry Morgan in 1668 and by British Admiral Vernon in 1739, and again in 1761. However, trade routes had changed and the fort suffered no new attacks.

Fort Geronimo, Portobelo.
▼

World Heritage site since

1978 1979 **1980** 1981 1982 1983 1984 1985 1986 1987 1988 1989 1990 1991 1992 1993 1994 1995 1996 1997 1998 1999 2000 2001 2002 2003 2004 2005 2006 2007 2008

Site of Palmyra
Syrian Arab Republic

Criteria – *Human creative genius; Interchange of values; Significance in human history*

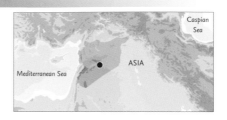

The grand colonnade, 1,100 m in length, which links the temple of Bel with the so-called Camp of Diocletian, is the monumental axis of the city, with its open central street flanked by covered lateral passages.

An oasis in the Syrian desert, northeast of Damascus, Palmyra contains the monumental ruins of a great city that was one of the most important cultural centres of the ancient world. From the first to the second century, the art and architecture of Palmyra, standing at the crossroads of several civilizations, married Graeco-Roman techniques with local traditions and Persian influences. The city offers the consummate example of an ancient urbanized complex, for the most part protected, with its large public monuments such as the Agora, the Theatre and the temples. Alongside these, the inhabited quarters are preserved, and there are immense cemeteries outside the fortified enceinte. Palmyra exerted a decisive influence on the evolution of neoclassical architecture and modern urbanization.

Part of the ruined city of Palmyra.
▼

World Heritage site since

1978 1979 **1980** 1981 1982 1983 1984 1985 1986 1987 1988 1989 1990 1991 1992 1993 1994 1995 1996 1997 1998 1999 2000 2001 2002 2003 2004 2005 2006 2007 2008

Paphos
Cyprus

Criteria – Testimony to cultural tradition;
Heritage associated with events of universal
significance

The Tombs of the
Kings, in Kato Paphos,
is a necropolis carved
out of solid rock with
some tombs decorated
with Doric pillars.
Spread over a vast area,
these impressive
underground tombs
date back to the fourth
century BC. High
officials rather than
kings were buried here.

Paphos has been inhabited since the
Neolithic period. It was a centre of the cult
of Aphrodite and of pre-Hellenic fertility
deities. According to legend, Aphrodite's
birthplace was on this island, and her temple
was erected here by the Mycenaeans in the
twelfth century BC. The remains of villas,
palaces, theatres, fortresses and tombs
mean that the site is of exceptional
architectural and historic value. The mosaics
of Nea Paphos are among the most
beautiful in the world. Excavations have also
unearthed the spectacular third- to fifth-
century mosaics of the Houses of Dionysus,
Orpheus and Aion, and the Villa of Theseus,
buried for sixteen centuries and yet
remarkably intact. The mosaic floors of
these noblemen's villas are considered
among the finest in the Eastern
Mediterranean.

◄
Mosaic floor from the
Villa of Theseus,
depicting Theseus' fight
with the minotaur in
the labyrinth. The villa
dates from the second
half of the second
century AD and
contained 100 rooms.

World Heritage site since

1978 1979 **1980** 1981 1982 1983 1984 1985 1986 1987 1988 1989 1990 1991 1992 1993 1994 1995 1996 1997 1998 1999 2000 2001 2002 2003 2004 2005 2006 2007 2008

Archaeological Ruins at Moenjodaro
Pakistan
Criteria – Interchange of values; Testimony to cultural tradition

The ruins of the huge city of Moenjodaro – built entirely of unbaked brick in the third millennium BC – lie in the Indus valley. The acropolis, set on high embankments, the ramparts, and the lower town, which is laid out according to strict rules, provide evidence of an early system of town planning.

◀
Moenjodaro is the most ancient and best-preserved urban ruin on the Indian subcontinent. It has exercised a considerable influence on the subsequent development of urbanization on the Indian peninsula.

Lower Valley of the Awash
Ethiopia
Criteria – Interchange of values; Testimony to cultural tradition; Significance in human history

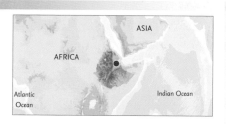

The Awash valley contains one of the most important groupings of palaeontological sites on the African continent. The remains found at the site, the oldest of which date back at least four million years, provide evidence of human evolution which has modified our conception of the history of humankind. The most spectacular discovery came in 1974, when fifty-two fragments of a skeleton enabled the famous Lucy to be reconstructed.

◀
In life, Lucy would have stood about 1 m tall and weighed 27–30 kg. By dating the deposits in which her fossilised remains were found, she is estimated to have lived 3.18 million years ago.

World Heritage site since

1978 1979 **1980** 1981 1982 1983 1984 1985 1986 1987 1988 1989 1990 1991 1992 1993 1994 1995 1996 1997 1998 1999 2000 2001 2002 2003 2004 2005 2006 2007 2008

Redwood National Park
USA

Criteria – Natural phenomena or beauty;
Significant ecological and biological processes

Redwood National Park comprises a region of coastal mountains bordering the Pacific Ocean north of San Francisco. It is covered with a magnificent forest of coastal redwood trees, the tallest and most impressive trees in the world. The park was established specifically to protect these trees, because it is only here and in Oregon that they now survive. Descendants of the giant evergreens that grew during the age of the dinosaurs, redwoods thrived in moist temperate regions of the world. They take 400 years to mature and some of the survivors are more than 2,000 years old. The park's marine and land life are equally remarkable, in particular the sea lions, the bald eagle and the endangered California brown pelican.

Archaeological surveys, test excavations, research and consultations conducted over the past twenty years have resulted in the recording of fifty prehistoric archaeological sites, nineteen historic sites and at least twenty-one places of significance to local Indian communities. The archaeological sites span 4,500 years and represent changing settlement and subsistence systems.

▼ Redwood trees can grow to around 100 m in only a few hundred years.

World Heritage site since

1978 1979 **1980** 1981 1982 1983 1984 1985 1986 1987 1988 1989 1990 1991 1992 1993 1994 1995 1996 1997 1998 1999 2000 2001 2002 2003 2004 2005 2006 2007 2008

Røros Mining Town
Norway

Criteria – Testimony to cultural tradition;
Significance in human history; Traditional human
settlement

Røros is in a
remarkably complete
state of preservation.
An engraving of the
town as seen from the
slag heaps in the
1860s is virtually the
same as a 1970s
photograph taken
from the same
viewpoint.

Røros is an extensive mining settlement
dating from 1644, when the development of
the copper works began. Completely rebuilt
after its destruction by Swedish troops in
1679, the town has some eighty wooden
houses, most of them standing around
courtyards. Many retain their dark pitch-log
façades, giving the town a medieval
appearance. The buildings reflect the dual
occupations of the inhabitants – mining and
farming – the domestic groups being
arranged as compact farmyards. These
groups are disposed on a regular urban
pattern adapted to the mountain terrain,
reflecting the particular kind of industrial
planning introduced by the Danish kings of
Norway in the sixteenth and seventeenth
centuries. Mining continued in the town
until as recently as 1977.

*A modern day view
of Røros.*
▼

World Heritage site since

1978 1979 **1980** 1981 1982 1983 1984 1985 1986 1987 1988 1989 1990 1991 1992 1993 1994 1995 1996 1997 1998 1999 2000 2001 2002 2003 2004 2005 2006 2007 2008

Ancient City of Bosra
Syrian Arab Republic

Criteria – Human creative genius; Testimony to cultural tradition; Heritage associated with events of universal significance

Bosra, once the capital of the Roman province of Arabia, was an important stopover on the ancient caravan route to Mecca. A magnificent second-century Roman theatre, early Christian ruins and several mosques and madrasas are found within its great walls. The Roman theatre, pictured below, probably built under Trajan, is enclosed by the walls and towers of a citadel fortified between 481 and 1231. From outside it could be an Arab fortress, with great square towers built from enormous blocks of stone, but right at its heart lies this ancient theatre, with room for 15,000 spectators. The cathedral of Bosra, completed in 513 by Archbishop Julianus, has influenced Christian and, to a lesser extent, Islamic architectural forms. The Mosque of Omar, restored in 1950, is one of the rare buildings of the first century of the Hegira preserved in Syria.

Bosra is associated with important events in the history of ideas and beliefs: according to tradition its bishop took part in the Council of Antioch, while the Prophet Muhammad came there twice and, at the time of his first visit, is said to have learned the precepts of Christianity from a Nestorian monk named Bahira.

The Roman theatre of Bosra.
▼

World Heritage site since

1978 1979 **1980** 1981 1982 1983 1984 1985 1986 1987 1988 1989 1990 1991 1992 1993 1994 1995 1996 1997 1998 1999 2000 2001 2002 2003 2004 2005 2006 2007 2008

Tiya
Ethiopia
Criteria – Human creative genius; Significance in human history

Tiya is among the most important of approximately 160 archaeological sites discovered so far in the Soddo region, south of Addis Ababa. The site contains thirty-six monuments, including thirty-two carved stelae covered with symbols, most of which are difficult to decipher. They are the remains of an ancient Ethiopian culture whose age has not yet been precisely determined.

◀
Tiya contains several representational configurations including low relief carvings of sword designs and one of a human figure. The stelae could also have had a funerary significance, as there are tombs scattered around the carvings.

Lower Valley of the Omo
Ethiopia
Criteria – Testimony to cultural tradition; Significance in human history

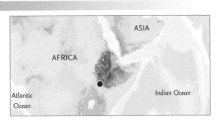

Evidence of the oldest-known humanoid technological activity has been found in this region, as well as stone objects attesting to an encampment of prehistoric human beings that is among the oldest known today.

A prehistoric site near Lake Turkana, the lower valley of the Omo is renowned the world over. The discovery of many fossils there, especially Homo gracilis, has been of fundamental importance in the study of human evolution. The area is unlike any other place on Earth in that so many different types of people have inhabited such a small area of land over many millennia. It is believed that it was the crossroads of a wide assortment of cultures where early humans of many different ethnicities passed as they migrated to and from lands in every direction.

◀
Drinking mixture being prepared for coming of age 'Bull ceremony'.

World Heritage site since

1978 1979 **1980** 1981 1982 1983 1984 1985 1986 1987 1988 1989 1990 1991 1992 1993 1994 1995 1996 1997 1998 1999 2000 2001 2002 2003 2004 2005 2006 2007 2008

Durmitor National Park
Montenegro

Criteria – Natural phenomena or beauty; Major stages of Earth's history; Significant natural habitat for biodiversity

EUROPE

Adriatic
Sea

Black
Sea

This breathtaking national park was formed by glaciers and is traversed by rivers and underground streams. Along the Tara River canyon, which has the deepest gorges in Europe (1,300 m), the dense pine forests are interspersed with clear lakes. The park comprises the Mount Durmitor plateau and the valley formed by the canyon of the Tara River, one of the last wild rivers in Europe. There are numerous examples of weathering processes, rock shapes and land features characteristic of karstic, fluvial and glacial

erosion. Because of its location and range in altitude, the park is influenced by both Mediterranean and alpine microclimates, resulting in an exceptional range of flora and fauna. The park contains one of the last virgin black pine forests in Europe. The Tara and its tributaries, as well as the lakes, contain a large number of salmon.

The park supports a rich karstic flora with many rare and endemic plants. There are thirty-seven species endemic to the area and six specific to Durmitor. Forest fauna includes the brown bear, wolf, wild boar, wild cat, chamois, various species of eagle, capercaillie, black grouse and rock partridge.

Black Lake, below the slopes of Mount Durmitor, is the largest and deepest lake in the park.

▼

World Heritage site since

1978 1979 **1980** 1981 1982 1983 1984 1985 1986 1987 1988 1989 1990 1991 1992 1993 1994 1995 1996 1997 1998 1999 2000 2001 2002 2003 2004 2005 2006 2007 2008

Historic Town of Ouro Preto
Brazil

Criteria – Human creative genius; Testimony to cultural tradition

Located 513 km north of Rio de Janeiro, Ouro Preto (Black Gold), pictured below, played a leading role in Brazil's Golden Age in the eighteenth century. It was created by thousands of soldiers of fortune eager to exploit local gold deposits; they were followed by many artists who produced works of outstanding quality, such as the Church of São Francisco de Assis by Antônio Francisco Lisboa (Aleijadinho), a masterpiece of Brazilian architecture.

A 'Mining Baroque' style developed which successfully fused Brazilian influences with European Baroque and Rococo. At its centre is Tiradentes Square containing imposing public and private buildings, such as the old Parliament House (1784) and the Palace of the Governors. In the closing years of the eighteenth century it became a centre of the movement for the emancipation of Brazil from colonial rule known as Inconfidência Mineira.

Ouro Preto was shaped by the grouping together of small settlements (arriais) in a hilly landscape, forming an irregular urban layout that follows the contours of the landscape. With the exhaustion of the gold mines in the nineteenth century, the city's influence declined but many churches, bridges and fountains remain as a testimony to its past prosperity.

World Heritage site since

1978 1979 **1980** 1981 1982 1983 1984 1985 1986 1987 1988 1989 1990 1991 1992 1993 1994 1995 1996 1997 1998 1999 2000 2001 2002 2003 2004 2005 2006 2007 2008

Al Qal'a of Beni Hammad
Algeria
Criteria – Testimony to cultural tradition

The ensemble of preserved ruins known as the Al Qal'a of Beni Hammad is situated on the southern flank of the Jebel Maâdid 1,000 m above sea level in a setting of striking beauty. It bears exceptional witness to a cultural tradition: it is one of the most interesting and most precisely dated monumental complexes of the Islamic civilisation and provides an authentic picture of a fortified Muslim city. The first capital of the Hammadid emirs, it was established in 1007 by Hammad, son of Bologhine, the founder of Algiers. The city was abandoned in 1090 when it was menaced by a Hilalian invasion, and finally destroyed in 1152 by the Almohads. It enjoyed particular splendour during the eleventh century.

The Al Qal'a encompasses a large number of monumental remains, including the Great Mosque. This is one of the largest in Algeria: its minaret is 25 m high and its prayer room has thirteen aisles with eight bays.

Taxila
Pakistan
Criteria – Testimony to cultural tradition; Heritage associated with events of universal significance

From the ancient Neolithic tumulus of Saraikala to the ramparts of Sirkap second century BC and the city of Sirsukh first century AD, Taxila illustrates the different stages in the development of a city on the Indus that was alternately influenced by Persia, Greece and Central Asia and which, from the fifth century BC to the second century AD, was an important Buddhist centre of learning.

Situated strategically on a branch of the Silk Road that linked China to the West, Taxila flourished economically and culturally. Buddhist monuments in the valley became destinations for pilgrims from as far afield as Central Asia and China.

◄

Archaeologists excavating a site in Taxila.

World Heritage site since

1978 1979 **1980** 1981 1982 1983 1984 1985 1986 1987 1988 1989 1990 1991 1992 1993 1994 1995 1996 1997 1998 1999 2000 2001 2002 2003 2004 2005 2006 2007 2008

Buddhist Ruins at Takht-i-Bahi and Neighbouring City Remains at Sahr-i-Bahlol
Pakistan

Criteria – Significance in human history

The complex is the most impressive and complete Buddhist monastery in Pakistan. In 1871, many sculptures were found at Takht-i-Bahi. Some depicted stories from the life of the Buddha while others, more devotional in nature, included the Buddha and Bodhisattava.

The Buddhist monastic complex of Takht-i-Bahi (Throne of Origins) was founded in the early first century. Owing to its location on the crest of a high hill, it escaped successive invasions and is still exceptionally well preserved. Nearby are the ruins of Sahr-i-Bahlol, a small fortified city dating from the same period. The complex consists of four main groups: the Court of Stupas, embellished with a series of tall niches enshrining Buddhist statues; the early monastic complex with residential cells around an open court; the temple complex with a main stupa in the middle of a courtyard; and the tantric monastic complex with an open courtyard in front of a series of dark cells with low openings for mystical meditation, in keeping with tantric practice.

Hill-top ruins of Takht-i-Bahi.
▼

World Heritage site since

1978 1979 1980 **1981** 1982 1983 1984 1985 1986 1987 1988 1989 1990 1991 1992 1993 1994 1995 1996 1997 1998 1999 2000 2001 2002 2003 2004 2005 2006 2007 2008

Los Glaciares
Argentina

Criteria – Natural phenomena or beauty; Major
stages of Earth's history

The Los Glaciares National Park is an area of
exceptional natural beauty, with rugged,
towering mountains and numerous glacial
lakes.

   This vast alpine area includes the
Patagonian ice field: at over 14,000 km² it
is the largest ice mantle outside Antarctica
and occupies about half of the park. It has
a total of forty-seven glaciers while a further
200 smaller glaciers are independent of the
main ice field.

   Los Glaciares is the best place in South
America to see glaciers in action. Glacial
activity is concentrated around two main
lakes, Argentino and Viedma. Lake
Argentino, 160 km long, is particularly
spectacular, with three glaciers dumping
their massive blue icebergs into the lake
with thunderous splashes.

   The most impressive wildlife in the park
are the birds: the waterfowl include swans,
ducks, geese and flamingos, while overhead
glides the huge Andean condor.

In Los Glaciares, the
effects of retreating
and advancing
glaciers can be clearly
seen. The Perito
Mereno glacier,
pictured below, often
advances so far that
its snout cuts off the
normal escape stream
of Lake Rico, forming
a natural dam which
inundates vast areas.
When the glacier
retreats in the heat
of summer a wall of
water roars down
the valley.

▼

World Heritage site since

1978 1979 1980 **1981** 1982 1983 1984 1985 1986 1987 1988 1989 1990 1991 1992 1993 1994 1995 1996 1997 1998 1999 2000 2001 2002 2003 2004 2005 2006 2007 2008

Serengeti National Park
Tanzania

Criteria – Natural phenomena or beauty;
Significant natural habitat for biodiversity

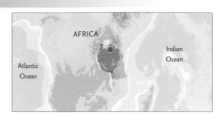

The vast plains of the Serengeti comprise 15,000 km² of savanna and open woodland. They contain the largest herds of grazing animals in the world and the carnivores that prey on them, providing a wildlife spectacle that is second to none.

The annual migration is dominated by wildebeest, gazelles and zebras, each harvesting the grass most suited to it. The herds are followed by prides of lion numbering up to 3,000 individuals, hyenas and jackals.

The great herds are continuously moving through the entire ecosystem but the sight is most impressive in May and June, when millions of animals travel en masse from the central plains to the permanent water holes on the western side of the park.

Serengeti is contiguous with Ngorongoro Conservation Unit, an area of 5,280 km² that was declared a World Heritage site in 1979.

Characteristic larger mammals of the Serengeti include leopard, cheetah, African elephant, black rhino, hippopotamus and giraffe, pictured below. Smaller mammals include numerous species of bat, bushbaby, monkey and baboon, aardvark, hare, porcupine, fox, mongoose, otter, wildcat, bushpig and rodent. Reptiles include Nile crocodile, Nile monitor lizard, python, cobra and puff adder. Over 500 bird species include raptors, vultures and over 20,000 water birds.

World Heritage site since

1978 1979 1980 **1981** 1982 1983 1984 1985 1986 1987 1988 1989 1990 1991 1992 1993 1994 1995 1996 1997 1998 1999 2000 2001 2002 2003 2004 2005 2006 2007 2008

Palace and Park of Fontainebleau
France

Criteria – Interchange of values; Heritage
associated with events of universal significance

Standing at the heart of a vast forest in the Île-de-France, Fontainebleau was transformed from a medieval royal hunting lodge into a dazzling Italianate palace that became one of the most important and prestigious sites of the French court. Surrounded by an immense and beautiful park, Fontainebleau combines Renaissance and French artistic influences. Its architecture and decor were influential on the artistic evolution not only of France but of Europe as a whole.

Fontainebleau was first enlarged and embellished by François I who commissioned artists from Renaissance Italy to begin work in 1528. The fashion of painting, stucco work, sculpture and statuary at Fontainebleau gave its name to a style that became influential throughout Europe in the sixteenth and seventeenth centuries. Successive modifications continued until in the nineteenth century the palace complex reached its present layout with five courtyards, arranged irregularly and surrounded by wings of buildings and gardens.

Royal domicile, 'house of the centuries' – Fontainebleau retains the imprint of every reign and style. Henri IV, Louis XIII, Louis XV and Louis XVI carried on the embellishment of this beautiful palace, which Napoleon I preferred above all others.

▼

World Heritage site since

1978 1979 1980 **1981** 1982 1983 1984 1985 1986 1987 1988 1989 1990 1991 1992 1993 1994 1995 1996 1997 1998 1999 2000 2001 2002 2003 2004 2005 2006 2007 2008

Old City of Jerusalem and its Walls
Jerusalem (Site proposed by Jordan)

Criteria – Interchange of values; Testimony to cultural tradition; Heritage associated with events of universal significance

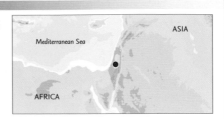

The golden-domed building is the Dome of the Rock and is one of the earliest surviving Islamic monuments.

As a holy city for Judaism, Christianity and Islam, Jerusalem has always been of enormous symbolic importance. Among its hundreds of historic monuments are key sites for each religious community: the Temple Mount (known as the Haram ash-Sharif for the Muslims) and the Western Wall for Jews; the Church of the Holy Sepulchre and the Via Dolorosa for Christians and for Muslims, the Dome of the Rock and Al-Aqsa Mosque on the Haram ash-Sharif. Jerusalem's symbolic importance for three of the world's religions has prompted power struggles that have lasted for centuries.

It is considered that the First Temple was built on Mount Moriah, or Temple Mount, by King Solomon and was completed in 957 BC. It was destroyed in 586 BC by Nebuchadnezzar II of Babylon. In 515 BC the Second Temple was completed and during the Roman rule, Herod the Great (73–4 BC) enlarged it, the famous Western Wall being part of the supporting structure for the levelled platform on which the temple stood. The Romans destroyed the Second Temple in AD 70 and established a citadel in the area, the Aelia Capitolina.

Jerusalem's period of Christian rule, from the fourth century, was one of its most peaceful and prosperous epochs. Among the Christian shrines was the Church of the Holy Sepulchre, completed in 335, marking the site of the crucifixion, tomb and resurrection of Jesus and the most sacred place in the Christian world.

The Arabs captured Jerusalem in 638 and the new rulers commissioned the Dome of the Rock. Intended as a shrine rather than a mosque, it was completed around 691 and is one of the most ancient Islamic buildings in existence. Close by, the Al-Aqsa Mosque was built between the late seventh and early eighth centuries.

This era of peaceful coexistence ended in 969 when control of the city passed to the Egyptian Arab Fatimids who systematically destroyed all synagogues and churches. Their prohibition on Christian pilgrimage became a contributing cause of the Crusades that culminated in the Christians' capture of Jerusalem in 1099. During the Christian Kingdom, the Dome of the Rock was converted to a Christian shrine (Templum Domini), the Church of the Holy Sepulchre was rebuilt, and hospices and monasteries were founded.

The Tower of David, the Citadel, Jerusalem.

World Heritage site since

1978 1979 1980 **1981** 1982 1983 1984 1985 1986 1987 1988 1989 1990 1991 1992 1993 1994 1995 1996 1997 1998 1999 2000 2001 2002 2003 2004 2005 2006 2007 2008

Ruins of Kilwa Kisiwani and Ruins of Songo Mnara
Tanzania
Criteria – Testimony to cultural tradition

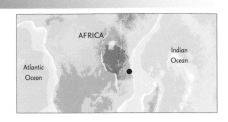

These sites are of prime importance to the understanding of the Swahili culture and the Islamization of the east coast of Africa. They include the remarkable Great Mosque, constructed in the twelfth century of coral tiles imbedded in a core of puddled clay.

The remains of two great East African ports admired by early European explorers are situated on two small islands near the coast. From the thirteenth to the sixteenth century, the merchants of Kilwa dealt in gold, silver, pearls, perfumes, Arabian crockery, Persian earthenware and Chinese porcelain; much of the trade in the Indian Ocean thus passed through their hands.

◄
Ruins of Kilwa's fort or gezira (prison).

Niokolo-Koba National Park
Senegal
Criteria – Significant natural habitat for biodiversity

Niokolo-Koba National Park covers 9,130 km² of the Guinea savanna of Senegal, with significant areas of bush land and gallery forest along both banks of the upper Gambia River. The area is rich in wildlife, including lions that are reputed to be Africa's largest.

Niokolo-Koba National Park covers almost 10,000 km² of the Guinea savanna of Senegal, with significant areas of bush land and gallery forest along both banks of the upper Gambia River. This varied landscape is home to an incredible range of wildlife, with over seventy species of mammal, 329 bird, thirty-six reptile, twenty amphibian and vast numbers of invertebrates. Derby's eland, an endangered species, is the world's largest antelope. Other important species include chimpanzee, leopard, dwarf crocodile and hippopotamus. The park is the last refuge in Senegal for giraffe and elephant, of which there is a large population. Niokolo-Koba is large enough to demonstrate the key aspects of the functioning Guinea savanna ecosystem, and to ensure the survival of its many endangered species.

World Heritage site since

1978 1979 1980 **1981** 1982 1983 1984 1985 1986 1987 1988 1989 1990 1991 1992 1993 1994 1995 1996 1997 1998 1999 2000 2001 2002 2003 2004 2005 2006 2007 2008

Amiens Cathedral
France

Criteria – Human creative genius, Interchange of values

Amiens Cathedral, in the heart of Picardy, is one of the largest classic thirteenth-century Gothic churches. It is notable for the coherence of its plan, the beauty of its three-tier interior elevation, and the particularly fine display of sculptures on the principal façade and in the south transept.

Building of Notre Dame d'Amiens began in 1220, two years after its Romanesque predecessor was destroyed by fire. Its builders used the technical knowledge and experience of other church builders to construct their cathedral quickly, resulting in a uniform style that is very rare: the nave, the largest part of the church, was completed in 1245.

The great height of the cathedral and its design allow in an exceptional amount of light. Despite its vast amount of stained glass, the building survived the First World War virtually unscathed.

The western façade, which is flanked by two square towers without spires, has three portals decorated with the elaborate statuary for which the cathedral is famous (**the central one is pictured above**).

From 1292–1375 the cathedral was enhanced by a series of chapels built between the buttresses of the side aisles. This style of the seven radiating chapels became a model for other cathedrals.

World Heritage site since

1978 1979 1980 **1981** 1982 1983 1984 1985 1986 1987 1988 1989 1990 1991 1992 1993 1994 1995 1996 1997 1998 1999 2000 2001 2002 2003 2004 2005 2006 2007 2008

Head-Smashed-In Buffalo Jump
Canada

Criteria – Heritage associated with events of universal significance

Among the many buffalo hunting techniques of the Plains Peoples, the Buffalo Jump was perhaps the most effective. When a bison herd was in proper position, young men well-instructed in animal behaviour, the "buffalo runners," enticed the animals towards the kill. They did so by dressing in the skins of calves and imitating the sound of a calf separated from its mother. Behind and upwind of the herd, fellow runners in wolf or coyote pelts moved the herd onward with their calls. Moving closer to the cliff edge, the herd was funnelled into converging lines of stone cairns where other band members shouted and waved large robes to frighten the buffalo into a headlong stampede. Confronted with the plunge over the cliff, the lead buffalo were forced over the precipice by the pressure of the herd racing behind.

In a typical hunt, dozens of animals were killed in the fall from the cliff or by the weapons of the hunters waiting below. For days after, the spoils of the kill were processed in the butchering camp, ensuring survival of the hunting group for another season.

Historic Monuments of Thatta
Pakistan

Criteria – Testimony to cultural tradition

The capital of three successive dynasties and later ruled by the Mughal emperors of Delhi, Thatta was constantly embellished from the fourteenth to the eighteenth century. Covering a distance of about 12 km, the site preserves an imposing monumental complex with the remains of the city itself in the valley and those of the necropolis at the edge of the Makli plateau. The remains of the city and its necropolis provide a unique view of civilization in Sind.

Within the broad family of Islamic monuments, those of Thatta represent a particular type, notable for the fusion of diverse influences into a local style. The effect of the Grand Mosque of Shah Jahan with its complex of blue and white buildings capped by ninety-three domes is unique.

◄

The tomb of Prince Sultan Ibrahim bin Norza Mohammad Isa Tarkhan has stood here since the eleventh century.

World Heritage site since

1978 1979 1980 **1981** 1982 1983 1984 1985 1986 1987 1988 1989 1990 1991 1992 1993 1994 1995 1996 1997 1998 1999 2000 2001 2002 2003 2004 2005 2006 2007 2008

Medina of Fez
Morocco

Criteria – Interchange of values; Traditional human settlement

Fez is a jewel of Spanish-Arabic civilization, an outstanding and well-preserved example of an ancient city. The Medina of Fez is densely packed with monuments – madrasas (schools), fondouks (shops), palaces, residences, mosques and fountains.

Founded in the ninth century and home to the oldest university in the world, Fez el Bali (the name of the old city) has two distinct centres on the right and left banks of the river Fez, settled by Arab refugees from Córdoba in Spain and from Kairouan (in modern-day Tunisia). In the fourteenth century a Jewish quarter, the Mellah, was joined to the newly founded city. The urban fabric and principal monuments in the Medina date from this period.

Previously the capital of Morocco, Fez lost that status to Rabat in 1912.

A medina typically forms a distinct quarter in many North African cities. It is generally the oldest part of a city, with walls and labyrinthine streets.

The Medina of Fez is thought to be the largest car-less urban area in the world.

▲
The minaret and city walls in Fez Medina.

World Heritage site since

1978 1979 1980 **1981** 1982 1983 1984 1985 1986 1987 1988 1989 1990 1991 1992 1993 1994 1995 1996 1997 1998 1999 2000 2001 2002 2003 2004 2005 2006 2007 2008

Roman Theatre and its Surroundings and the 'Triumphal Arch' of Orange
France

Criteria – Testimony to cultural tradition;
Heritage associated with events of universal
significance

Situated in the Rhone valley, the ancient theatre of Orange, with its 103 m-long façade, is one of the best preserved of all the great Roman theatres. The theatre was closed by imperial command in 391: by this time Christianity had become the *de facto* state religion and the Church opposed all pagan spectacles. The theatre was abandoned and later sacked and pillaged by barbarians. It was only in the nineteenth century that it slowly recovered its original splendour, thanks to the restoration work begun in 1825. The Triumphal Arch, pictured below, located to the north of the town, is one of the most beautiful and interesting surviving examples of a provincial triumphal arch from the reign of Augustus. It is decorated with low reliefs commemorating the establishment of the Pax Romana.

The most striking feature of the theatre is the stage wall 'scenae frons'. This would originally have been covered with marble slabs and decorated with bas-reliefs, carved friezes, statues in niches and columns. Their purpose was not purely decorative because the projections and cavities in the wall would have helped to eliminate the problem of echoes.

The Triumphal Arch.
▼

World Heritage site since

1978 1979 1980 **1981** 1982 1983 1984 1985 1986 1987 1988 1989 1990 1991 1992 1993 1994 1995 1996 1997 1998 1999 2000 2001 2002 2003 2004 2005 2006 2007 2008

Cistercian Abbey of Fontenay
France
Criteria – Significance in human history

EUROPE

Atlantic Ocean

Fontenay Abbey is located in northern Burgundy, 80 km north of Dijon in a small marshy valley a few kilometres from Montbard. It was founded by St Bernard in 1119 and built from 1130 onwards, making it one of the oldest Cistercian monasteries in Europe. With its church, cloister, refectory, sleeping quarters, bakery, and ironworks, it is an excellent illustration of the ideal of self-sufficiency practised by the earliest communities of Cistercian monks.

The forge, dated to the end of the twelfth century, recalls the part which the Cistercians played in technological progress during the Middle Ages: this is one of the oldest industrial buildings in France. Despite the transformations undergone in the thirteenth, fifteenth, and sixteenth centuries, the Abbey of Fontenay, restored after 1906, has the appearance today of a generally authentic and well-preserved whole.

The Fontenay gardens were completely redesigned in the 1970s by the garden architect Peter Holmes, who created a magnificent landscape of greenery and plants to bring out the abbey buildings. The beauty of the gardens adds an extra charm to the old Romanesque stone buildings of Fontenay.

Djoudj National Bird Sanctuary
Senegal
Criteria – Natural phenomena or beauty; Significant natural habitat for biodiversity

AFRICA

Atlantic Ocean

Situated in the Senegal River delta, the Djoudj Sanctuary is a wetland of 160 km², comprising a large lake surrounded by streams, ponds and backwaters. It forms a living but fragile sanctuary for some 1.5 million birds, such as the white pelican, the purple heron, the African spoonbill, the great egret and the cormorant.

The park is one of the first sources of fresh water for more than three million migrant birds after crossing 200 km of the Sahara. The waters also hold populations of crocodile and African manatee.

◀ Great Egret.

World Heritage site since

1978 1979 1980 **1981** 1982 1983 1984 1985 1986 1987 1988 1989 1990 1991 1992 1993 1994 1995 1996 1997 1998 1999 2000 2001 2002 2003 2004 2005 2006 2007 200

Great Barrier Reef
Australia

Criteria – Natural phenomena or beauty; Major stages of Earth's history; Significant ecological and biological processes; Significant natural habitat for biodiversity

The Great Barrier Reef viewed from above.

The Great Barrier Reef is a site of remarkable variety and beauty on the northeast coast of Australia. It is the world's most extensive stretch of coral reefs. The great diversity of its fauna reflects the maturity of an ecosystem that has evolved over millions of years on the northeast continental shelf of Australia.

The site contains a huge range of species including over 1,500 species of fish, about 360 species of hard coral and 5,000 species of mollusc, plus a great diversity of sponges, sea anemones, marine worms and crustaceans. About 215 species of birds are found in its islands and cays. Extending to Papua New Guinea, the reef system comprises some 2,900 individual reefs covering more than 20,000 km², including 760 fringing reefs. The reefs range in size from under 0.01 km² to over 100 km² and vary in shape to provide the most spectacular marine scenery on Earth. There are approximately 600 continental islands including many with towering forests and freshwater streams, and some 300 coral cays and unvegetated sand cays. A rich variety of landscapes and seascapes, including rugged mountains with dense and diverse vegetation, provide spectacular scenery.

The form and structure of the individual reefs show great variety. There are two main classes: platform or patch reefs, resulting from radial growth; and wall reefs, resulting from elongated growth, often in areas of strong currents. There are also the many fringing reefs where growth is established on subtidal rock of the mainland coast or continental islands.

The site includes major feeding grounds for the endangered dugong and nesting grounds of world significance for four species of marine turtle including the endangered loggerhead turtle. Given the severe pressures on these species elsewhere, the Great Barrier Reef may be a last stronghold. It is also an important breeding area for humpback and other whale species.

A wide range of fleshy algae occurs, often small and inconspicuous but highly productive and heavily grazed by turtles, fish, molluscs and sea urchins. In addition, calcareous algae are an important component of reef building processes. Fifteen species of seagrass grow throughout the area, forming over 3,000 km² of seagrass meadows and providing an important food source for grazing animals, such as dugongs and turtles.

The Great Barrier Reef is important in the historic and contemporary culture of the Aboriginal and Torres Strait Islander groups of the coastal areas of northeast Australia. The contemporary use of and association with the Marine Park plays an important role in the maintenance of their cultures and there is a strong spiritual connection with the ocean and its inhabitants. New species continue to be discovered throughout the Great Barrier Reef. A new species of dolphin, the Australian snub-nose dolphin, was discovered in 2005 within inshore areas, but only in low numbers. A recent survey of the inter-reef areas has found at least four new species of fish and one of the sponges that commonly occurs is likely to be a new genus.

A clown fish. ▶

World Heritage site since

1978 1979 1980 **1981** 1982 1983 1984 1985 1986 1987 1988 1989 1990 1991 1992 1993 1994 1995 1996 1997 1998 1999 2000 2001 2002 2003 2004 2005 2006 2007 200

Fort and Shalamar Gardens in Lahore
Pakistan

Criteria – Human creative genius; Interchange of values; Testimony to cultural tradition

The Fort and Shalamar Gardens in Lahore are a unique artistic realization which, while bearing exceptional testimony to the Mughal civilization, have exercised a considerable influence long after their creation in the Punjab and throughout the Indian subcontinent. The first historic references to Lahore Fort, situated northwest of the city, date from before the eleventh century. Destroyed and rebuilt several times by the Mughals from the thirteenth to the fifteenth centuries, it was definitively rebuilt and reorganized starting with the reign of Emperor Akbar (1542–1605). Based on the twenty-one monuments preserved within its boundaries, it comprises the most beautiful repertory of the forms of Mughal architecture, whose evolution may be followed over more than two centuries. The elegance of the splendid gardens, built on three terraces with lodges, waterfalls and large ornamental ponds, is unequalled.

The complex of fairytale-like buildings surrounding the Court of Shah Jahan, and especially the Shah Burj or Shish Mahal, make it one of the most beautiful palaces in the world. Built in 1631–2, it sparkles with mosaics of glass, gilt, semi-precious stones and marble screening.

Shalamar Gardens.
▼

World Heritage site since

1978 1979 1980 **1981** 1982 1983 1984 1985 1986 1987 1988 1989 1990 1991 1992 1993 1994 1995 1996 1997 1998 1999 2000 2001 2002 2003 2004 2005 2006 2007 2008

Willandra Lakes Region
Australia

Criteria – Testimony to cultural tradition; Major stages of Earth's history

The Willandra Lakes Region is a remarkable example of a site where the economic and cultural life of Homo sapiens can be partly reconstructed, showing a fascinating interaction between Aboriginal people and the changing natural environment. The fossil remains of a series of lakes and sand formations that date from the Pleistocene (2.5 million to 5,000 years ago) can be found in this region, together with archaeological evidence of human occupation dating from

45–60,000 years ago. It is a unique landmark in the study of human evolution on the Australian continent. Several well-preserved fossils of giant marsupials and other animals, some of which are now extinct, have also been found here. The site includes the entire lake and river system from Lake Mulurulu, the latest to hold water, to the Prungle Lakes, dry for more than 15,000 years.

When the Willandra Billabong Creek ceased to flow, the lakes dried in series from the Prungle Lakes in the south to Lake Mulurulu in the north over several thousand years. As each lake evaporated, it became an independent system undergoing a basic transformation from fresh water to saline water to dry lake bed.

Lake Mungo Lunette. Erosion of this dune has exposed extensive evidence of Aboriginal occupation over the millennia. ▼

World Heritage site since

1978 1979 1980 **1981** 1982 1983 1984 1985 1986 1987 1988 1989 1990 1991 1992 1993 1994 1995 1996 1997 1998 1999 2000 2001 2002 2003 2004 2005 2006 2007 200

Mammoth Cave National Park
USA

Criteria – Natural phenomena or beauty; Major stages of Earth's history; Significant natural habitat for biodiversity

Mammoth Cave National Park, located in the state of Kentucky, has the world's largest network of natural limestone caves and underground passageways, carved by twenty-five million years of cave-forming action by the Green River and its tributaries. The park and its underground network of more than 560 km of surveyed passageways is the most extensive and diverse cave ecosystem in the world, with over 200 species indigenous to the network of caves, and forty-two species adapted to life in total darkness. Surface features are also important and Big Woods, a temperate deciduous oak-hickory dominated forest, is reputed to be one of the largest and best remaining examples of the ancient forests of eastern North America that once covered most of Kentucky.

Almost every type of cave formation is known within the site and the geological processes involved in cave formation are continuing. Mammoth boasts superlative examples of chambers, shafts, stalagmites, stalactites, as well as gypsum 'flowers' and 'needles'.

Mount Nimba Strict Nature Reserve
Côte d'Ivoire and Guinea

Criteria – Significant ecological and biological processes; Significant natural habitat for biodiversity

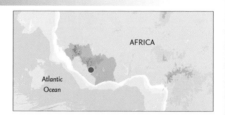

Located on the borders of Guinea, Liberia and Côte d'Ivoire, Mount Nimba rises above the surrounding savanna. Its slopes are covered by dense forest at the foot of grassy mountain pastures. They harbour an especially rich flora and fauna, with endemic species such as the viviparous toad, and chimpanzees that use stones as tools.

The park includes significant portions of Mount Nimba, a geographically unique area with more than 200 endemic species. These species include multiple types of duikers, big cats and civets.

◄

The critically endangered Mount Nimba toad.

World Heritage site since

1978 1979 1980 **1981** 1982 1983 1984 1985 1986 1987 1988 1989 1990 1991 1992 1993 1994 1995 1996 1997 1998 1999 2000 2001 2002 2003 2004 2005 2006 2007 2008

Kakadu National Park
Australia

Criteria – Human creative genius; Heritage associated with events of universal significance; Natural phenomena or beauty; Significant ecological and biological processes; Significant natural habitat for biodiversity

This spectacular living cultural landscape has been cared for continuously by Aboriginal people – called Bininj/Mungguy - for more than 50,000 years. Their deep spiritual connection to the land dates back to the creation time and Kakadu is inscribed on the World Heritage List for both its cultural and natural values.

Kakadu's rock art, pictured right, and archaeological sites record the way of life of the region's inhabitants, from the hunter-gatherers of prehistoric times to the Aboriginal people still living there today.

Bininj/Mungguy believe that during the creation time, ancestral beings known as the first people, or Nayahunggi, journeyed across the landscape. The ancestors created the landforms, plants, animals and Aboriginal people we see today, and they left language, ceremonies, kinship, and rules to live by.

The park covers a huge 20,000 km² and moving through the park, the landscape changes dramatically, from the soaring sandstone escarpment to floodplains, monsoon forests, savanna woodlands and tidal flats. These environments provide habitat for a wide range of rare plant and animal species, some of which are found nowhere else in the world.

Kakadu's landscapes undergo dramatic seasonal changes. Wet season rains create a vast shallow wetland sea for hundreds of square kilometres, where ducks, geese and wading birds abound. As the floodplains start to dry, huge numbers of these waterbirds congregate around the permanent rivers and billabongs.

Because of its diversity, from marine and coastal habitats (which are home to turtles and dugongs) through to the stone country, Kakadu offers a fantastic cultural and wildlife experience.

One-third of Australia's bird species and almost one-fifth of its mammal species are found in Kakadu. Millions of waterbirds make seasonal use of the floodplains. Kakadu's great diversity of invertebrates includes 55 species of termite, at least 350 species of ant and more than 160 species of grasshopper.

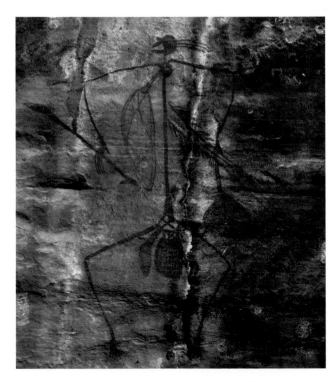

World Heritage site since

1978 1979 1980 **1981** 1982 1983 1984 1985 1986 1987 1988 1989 1990 1991 1992 1993 1994 1995 1996 1997 1998 1999 2000 2001 2002 2003 2004 2005 2006 2007 200

Olympic National Park
USA

Criteria – Natural phenomena or beauty;
Significant ecological and biological processes

Renowned for the diversity of its ecosystems, Olympic National Park contains glacier-clad peaks interspersed with alpine meadows surrounded by an extensive forest, the best example of temperate rainforest in the Pacific Northwest. The park is divided into two segments: a mountainous core and a separate coastal strip. The mountains contain about sixty active glaciers; the area is unique because it is the lowest latitude in the world at which glaciers begin, at an elevation lower than 2,000 m. The coastal strip stretches along 80 km of wilderness beach, characterized by rocky headlands and a wealth of intertidal life, and the arches, caves and buttresses are evidence of the continuous battering of the waves. The coniferous forest of Olympic is of prime commercial interest and practically all the original forest outside the park has been harvested.

Olympic National Park has an extensive temperate rainforest dominated by conifers.

The main danger to the integrity of the site is, oddly, one of its attractions: the mountain goats, introduced in the 1920s. They have increased erosion, reduced and altered plant cover, such that more resistant or less palatable species have become dominant. As a result three of the endemic plants may now be endangered.

Speyer Cathedral
Germany

Criteria – Interchange of values

Speyer Cathedral exerted a considerable influence on the development of Romanesque architecture in the eleventh and twelfth centuries, and also on the evolution of the principles of building restoration in Europe and in the world from the eighteenth century to the present. A basilica with four towers and two domes, it was founded by Conrad II in 1030 and remodelled at the end of the eleventh century. It is one of the most important Romanesque monuments from the time of the Holy Roman Empire. The cathedral was the burial place of the German emperors for almost 300 years. A huge stone font, with a capacity of 1,560 litres, stands on the square in front of the main portal of the cathedral. This font once symbolized the borderline between the diocese and the city.

In 1689 the cathedral was seriously damaged by fire, and it was reconstructed in the Romanesque style. The Bavarian King Ludwig I commissioned the painting of the interior. A new western block was added in 1854–8, a Romanesque pastiche. Starting in 1957, nineteenth-century paintings and the layers of painted plaster were removed to restore its eleventh-century form.

World Heritage site since

1978 1979 1980 **1981** 1982 1983 1984 1985 1986 1987 1988 1989 1990 1991 1992 1993 1994 1995 1996 1997 1998 1999 2000 2001 2002 2003 2004 2005 2006 2007 200£

Darien National Park
Panama

Criteria – Natural phenomena or beauty;
Significant ecological and biological processes;
Significant natural habitat for biodiversity

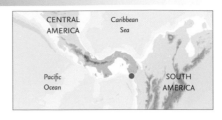

Forming a land-bridge between the Central and South American continents, Darien National Park contains an exceptional variety of habitats – sandy beaches, rocky coasts, mangroves, swamps, and lowland and upland tropical forests – containing remarkable wildlife. The area has been under protection since 1972, with the establishment of Alto Darién Protection Forest, and was declared a national park in

1980. It extends along about 80 per cent of the Colombian border and includes part of the Pacific coast. Access is by river and heavy truck. The area is both anthropologically and historically rich, with two major indigenous groups – Chocó and Kuna Indians – and a number of smaller groups still living by traditional practices. Today, conservation of Indian culture is included as a management objective.

The park's remarkably varied fauna is largely unstudied and includes bush dog, giant anteater, jaguar, ocelot, capybara and howler monkey. Harpy eagles are also found in the park, as are cayman and American crocodile.

Archaeological Park and Ruins of Quirigua
Guatemala

Criteria – Human creative genius; Interchange of values; Significance in human history

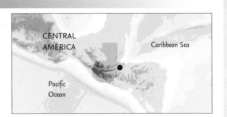

Quirigua's monumental complexes are remarkable for their elaborate system of pyramids, terraces, and staircases. The ruins contain some outstanding eighth-century monuments and an impressive series of carved stelae. These are the principal written chronicles of this lost civilisation, as well as the key to their highly advanced calendar system. Like most Mayan monuments, they were erected to commemorate the passage of time, and significant historic events.

Inhabited since the second century AD, by the reign of Cauac Sky (723–84), Quirigua had become the capital of an autonomous and prosperous state. The extraction of jade and obsidian in the upper valley of the Rio Motagua, which was tightly controlled, gave rise to a profitable goods trade with the coastal ports of the Caribbean.

Quirigua's huge stone monolithic sculptures were artfully carved without the benefit of metal tools; stone chisels, driven by other stones or wooden mallets, were the only tools available. They include the largest known quarried stone in the Maya world, which stands 10.6 m tall and weighs over 59 tons.

World Heritage site since

1978 1979 1980 **1981** 1982 1983 1984 1985 1986 1987 1988 1989 1990 1991 1992 1993 1994 1995 1996 1997 1998 1999 2000 2001 2002 2003 2004 2005 2006 2007 2008

Arles, Roman and Romanesque Monuments
France

Criteria – Interchange of values; Significance in human history

In Roman times Arles was surrounded by graveyards, including one known as Les Alyscamps. This cemetery became important when the Christian martyr St Genest and the first bishops of Arles were buried there. In 1040 the site became the St-Honorat priory, one of the required stops on the pilgrimage route to Santiago de Compostela in Spain.

Church of St-Trophime.
▼

Arles is an especially significant example of the appropriation of a classical Roman city by a medieval European civilization. It has some impressive Roman monuments, of which the earliest – the arena, the theatre and the cryptoporticus (subterranean galleries) – date back to the first century BC. The Roman theatre could hold 10,000 people in thirty-three rows of seats, and the arena 20,000 spectators. Gladiator fights and animal hunts took place here until the end of the fifth century. During the fourth century Arles experienced a second golden age, as attested by the Baths of Constantine and the necropolis of Les Alyscamps. In the eleventh and twelfth centuries, Arles once again became one of the most attractive cities in the Mediterranean. Within the city walls, the church of St-Trophime, pictured below, with its cloister, is one of Provence's major Romanesque monuments.

World Heritage site since

1978 1979 1980 **1981** 1982 1983 1984 1985 1986 1987 1988 1989 1990 1991 1992 1993 1994 1995 1996 1997 1998 1999 2000 2001 2002 2003 2004 2005 2006 2007 2008

Würzburg Residence with the Court Gardens and Residence Square
Germany

Criteria – Human creative genius; Significance in human history

This magnificent Baroque palace – one of the largest and most beautiful in Germany and surrounded by wonderful gardens – was created under the patronage of the Prince-Bishops of Schönborn and was home to one of the most brilliant courts of Europe during the eighteenth century. The main part of the Residence was built between 1720 and 1744 and decorated internally between 1740 and 1770. The most renowned architects of the time – the Viennese Lukas von de Hildebrandt and the Parisians Robert de Cotte and Germain Boffrand – drew up plans that were supervised by the official architect of the Prince Bishop, Balthasar Neumann. Sculptors and stucco-workers came from Italy, Flanders and Munich. The Venetian painter Giovanni Battista Tiepolo decorated the staircase and the walls of the Imperial Hall with frescoes.

Tiepolo painted the staircase vault with a fresco that depicts Apollo and the Four Continents. In the portion of the fresco representing Europa, he included portraits of those responsible for the design of the Residence: Neumann (portrayed as an artillery colonel), Tiepolo with his son Giandomenico and the Prince-Bishop supported by Fame and crowned by Virtue.

World Heritage site since

1978 1979 1980 **1981** 1982 1983 1984 1985 1986 1987 1988 1989 1990 1991 1992 1993 1994 1995 1996 1997 1998 1999 2000 2001 2002 2003 2004 2005 2006 2007 2008

SGang Gwaay
Canada

Criteria – Testimony to cultural tradition

Gwaii Haanas is on the spring migration route of the grey whales which spend their summers in feeding grounds in the Bering Sea. Killer whales (orcas), humpback and minke whales are also seen in the waters surrounding Gwaii Haanas, along with Pacific white-sided dolphins, steller sea lions and harbour seals.

SGang Gwaay Llnagaay (Ninstints) is a village site of the Haida people and part of the Gwaii Haanas National Park Reserve and Haida Heritage Site in the Queen Charlotte Islands (Haida Gwaii) of British Columbia. The Haida people lived on SGang Gwaay (Anthony Island) for thousands of years, but the population was decimated by disease by the 1880s. Most of the village has been taken by nature and returned to the forest. What is left is unique in the world: remains of a nineteenth-century Haida village where the remnants of twenty-three houses and thirty-two memorial or mortuary poles depict a rich and flamboyant society. The protected area epitomises the rugged beauty and ecological character of the Pacific coast, and commemorates the living culture of the Haida people and their relationship to the land and sea, offering a tangible key to their oral traditions.

World Heritage site since

1978 1979 1980 1981 **1982** 1983 1984 1985 1986 1987 1988 1989 1990 1991 1992 1993 1994 1995 1996 1997 1998 1999 2000 2001 2002 2003 2004 2005 2006 2007 2008

Aldabra Atoll
Seychelles

Criteria – Natural phenomena or beauty; Significant ecological and biological processes; Significant natural habitat for biodiversity

Aldabra is the least-disturbed large island in the Indian Ocean and the only place in the world where a reptile is the dominant herbivore. The only endemic mammal is a flying fox.

The atoll is comprised of four large coral islands which enclose a shallow lagoon; the group of islands is itself surrounded by a coral reef. Due to difficulties of access and the atoll's isolation, Aldabra has been protected from human influence and thus retains some 152,000 giant tortoises, the world's largest population of this reptile.

◄ Giant Tortoise.

World Heritage site since

1978 1979 1980 1981 **1982** 1983 1984 1985 1986 1987 1988 1989 1990 1991 1992 1993 1994 1995 1996 1997 1998 1999 2000 2001 2002 2003 2004 2005 2006 2007 2008

Historic Centre of Florence
Italy

Criteria – Human creative genius; Interchange of values; Testimony to cultural tradition; Significance in human history; Heritage associated with events of universal significance

EUROPE

Mediterranean Sea

Ionian Sea

The Duomo dominates the Florence skyline.

The Medicis, Florence's great patrons and benefactors, ruled the Grand Duchy of Tuscany until the family died out in 1737. Florence became part of the Kingdom of Italy in 1859 and was the country's political capital for a short time, between 1865 and 1870.

The historic centre of Florence may be viewed from the surrounding hills, especially Piazzale Michelangelo (just under the Romanesque Basilica of San Miniato), or Fiesole; both offer some of the most spectacular views in the Arno valley.

Florence is the symbol and cradle of the Renaissance and its beautiful historic centre may best be described as a treasure chest of works of art and architecture. Its 600 years of extraordinary artistic activity can be seen in a wealth of buildings and artefacts, especially the cathedral Santa Maria del Fiore (the Duomo), the church of Santa Croce, the Uffizi Gallery and the Medicis' Pitti Palace, now a gallery with paintings by great masters such as Giotto, Brunelleschi, Botticelli and Michelangelo.

Founded in 59 BC as a Roman colony known as Florentia, the free commune of Florence gradually gained supremacy over rival towns in Tuscany until, in the fifteenth century, the city reached the apex of its splendour. Defined by the fourteenth-century walls, and built up thanks to the enormous business and economic power Florence achieved, the two succeeding centuries were its golden age. Under the powerful Medici family, its rulers in the fifteenth and sixteenth centuries, Florence exerted a major influence on the development of architecture and the monumental arts, first in Italy and then throughout Europe.

The spiritual focus of the city is the Cathedral Piazza of Santa Maria del Fiore;

Giotto's campanile is on one side and the Baptistry of St John in front, with the Gates of Paradise by Lorenzo Ghiberti.

To the north lies the Palazzo Medici-Riccardi by Michelozzo and St Lawrence's Basilica by Brunelleschi, their sacristies designed by Donatello and Michelangelo. Further on are the Museum of St Mark's, with Fra Angelico's masterpieces, the Galleria dell'Accademia with Michelangelo's David (1501–4) and the Santissima Annunziata Piazza with the Lodge of the Holy Innocents by Brunelleschi.

On the south side of the cathedral is the political and cultural centre of Florence, with the Palazzo Vecchio, and the Galleria degli Uffizi nearby. Close to these are the Museo del Bargello and the Basilica of the Holy Cross. Across the Ponte Vecchio, over the river Arno, is the Oltrarno quarter, with the Pitti Palace and Boboli Gardens. Also in the Oltrarno, is the Holy Ghost Basilica by Filippo Brunelleschi and the Carmelite Church, with its frescoes by Masolino, Masaccio and Filippino Lippi. To the west of the cathedral stand the imposing Strozzi Palace and the Basilica of Santa Maria Novella, its façade designed by Leon Battista Alberti.

World Heritage site since

1978 1979 1980 1981 **1982** 1983 1984 1985 1986 1987 1988 1989 1990 1991 1992 1993 1994 1995 1996 1997 1998 1999 2000 2001 2002 2003 2004 2005 2006 2007 2008

Lord Howe Island Group
Australia

Criteria – Natural phenomena or beauty;
Significant natural habitat for biodiversity

A remarkable example of isolated oceanic
islands, born of volcanic activity more than
2,000 m under the sea, the Lord Howe
Island Group boasts an exceptional diversity
of spectacular and scenic landscapes within
a small area. The sheer slopes of its volcanic
mountains and the dramatic rock formation
Ball's Pyramid rise out of an underwater
world that is one of the most beautiful in the

world. The isolation of this special place at
the junction of tropical and temperate
latitudes has led to tremendous biodiversity.
This group of islands is one of the major
breeding sites for seabirds in the southwest
Pacific and is home to numerous endemic
species of flora and fauna.

The Lord Howe Island
Group supports the
southernmost true
coral reef in the world,
which differs
considerably from
more northerly warm
water reefs. It is
unique in being a
transition between
the algal and coral
reef, due to
fluctuations of hot
and cold water around
the island.

Archaeological Site of Cyrene
Libyan Arab Jamahiriya

Criteria – Interchange of values; Testimony to
cultural tradition; Heritage associated with
events of universal significance

A colony of the Greeks of Thera, Cyrene was
one of the principal cities in the Hellenic
world. It was Romanized and remained a
great capital until the earthquake of AD 365.
The site contains the three monumental
complexes of the sanctuary of Apollo, the
Acropolis and the Agora, and preserves a
necropolis complex which is numbered
among the most extensive and varied of the
ancient world. A thousand years of history is
written into its ruins.

Once given by Mark
Anthony to Cleopatra,
Cyrene is not only one
of the cities of the
Mediterranean world
around which myths,
legends and stories
have been woven over
more than 1,000
years, but it is also
one of the most
impressive complexes
of ruins in the entire
world.

◀

Temple of Zeus,
Cyrene.

World Heritage site since

1978 1979 1980 1981 **1982** 1983 1984 1985 1986 1987 1988 1989 1990 1991 1992 1993 1994 1995 1996 1997 1998 1999 2000 2001 2002 2003 2004 2005 2006 2007 2008

Selous Game Reserve
Tanzania

Criteria – Significant ecological and biological processes; Significant natural habitat for biodiversity

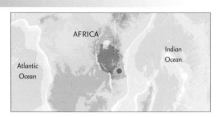

Some 400 species of animal are known, and in 1986 approximately 750,000 large animals of 57 species were recorded.

The Selous Game Reserve is a 50,000 km² ecosystem and one of the largest wildlife reserves in the world. It is noted particularly for the range of its wildlife – large numbers of elephant, black rhinoceros, cheetah, giraffe, impala, hippopotamus and crocodile live in this immense sanctuary – and for its remoteness, being relatively undisturbed by humans.

The reserve, which includes Mikumi National Park and Kilombero Game Controlled Area, has a variety of vegetation zones, ranging from dense thickets to open wooded grasslands. The deciduous miombo woodland is dominant, providing the world's best example of this vegetation type. Soils are relatively poor and infertile and winters bring drought, but despite these facts, the reserve has a higher density and species diversity than any other miombo woodland area, thanks to its size, the diversity of its habitats, the availability of food and water, and the lack of human settlement.

In 1994, in the reserve and surrounding buffer area, there were 52,000 of the endangered African elephant, 50 per cent of the country's total, which is now growing again after years of decline due to ivory poaching.

Impalas in the reserve.
▼

World Heritage site since

1978 1979 1980 1981 **1982** 1983 1984 1985 1986 1987 1988 1989 1990 1991 1992 1993 1994 1995 1996 1997 1998 1999 2000 2001 2002 2003 2004 2005 2006 2007 2008

Sacred City of Anuradhapura
Sri Lanka

Criteria – Interchange of values; Testimony to cultural tradition; Heritage associated with events of universal significance

Anuradhapura, a political and religious capital that flourished for 1,300 years, is one of the principal shrines of Buddhism. Abandoned and hidden away in dense jungle for many years, the splendid site, with its palaces, monasteries and monuments, is now accessible again.

Founded in the fourth century BC, Anuradhapura quickly became both the capital of Ceylon, as the country was then called, and the sacred city of Buddhism on the island. It has remarkable monuments, particularly the huge dagabas (relic chambers), set on circular foundations and surrounded by a ring of monolithic columns, much like Sinhalese stupas.

The city was attacked by waves of invaders from southern India and was finally abandoned in 993. It stands as a permanent manifesto of the culture of Sri Lanka, impervious to outside influences.

The religious significance of Anuradhapura was confirmed when a cutting from the 'tree of enlightenment', under which the Buddha meditated and gained enlightenment, was brought to the city in the third century BC. The cutting flourished and today, the bodhi tree spreads out over the centre of the site from a sanctuary near the Brazen Palace.

The dome-shaped roof of a large dagaba.
▼

World Heritage site since

1978 1979 1980 1981 **1982** 1983 1984 1985 1986 1987 1988 1989 1990 1991 1992 1993 1994 1995 1996 1997 1998 1999 2000 2001 2002 2003 2004 2005 2006 2007 2008

National History Park – Citadel, Sans Souci, Ramiers
Haiti

Criteria – Significance in human history; Heritage associated with events of universal significance

Situated within the National History Park created by presidential decree in 1978, these buildings enjoy splendid natural setting of mountainous peaks covered with luxuriant vegetation.

These Haitian monuments date from the beginning of the nineteenth century, when Haiti proclaimed its independence. The Palace of Sans Souci, the buildings at Ramiers and, especially, the Citadelle Henry serve as universal symbols of liberty, as they were the first to be built by black slaves who had gained their freedom. The Citadelle, constructed at an altitude of 970 m and covering a surface area of about 10,000 m², is one of the best examples of the art of military engineering of the early nineteenth century. It was designed specifically to allow an integrated use of artillery capabilities; an elaborate system of cisterns supplied water; and colossal defensive walls rendered this citadel impregnable. It can shelter a garrison of up to 5,000 men.

Taï National Park
Côte d'Ivoire

Criteria – Natural phenomena or beauty; Significant natural habitat for biodiversity

The park contains some 1,300 species of higher plants, of which 54 per cent occur only in the Guinea zone. Vegetation is predominantly dense evergreen forest of a Guinean type characterized by tall trees (40–60 m) with massive trunks. Plants thought to be extinct, such as Amorphophallus staudtii, have been discovered in the area.

The park is one of the last remaining portions of the vast primary forest that once stretched across present-day Ghana, Côte d'Ivoire, Liberia and Sierra Leone, and is the largest island of forest remaining in West Africa. Its rich natural flora, and threatened mammal species such as the pygmy hippopotamus and eleven species of monkeys, are of great scientific interest. The park lies in southwest Côte d'Ivoire about 200 km south of Man and 100 km from the coast, between the Cavally River (which marks the western border with Liberia) and the Sassandra River on the east. There is a gradation from north to south, with the southern third of the park being the moistest and richest area, especially of leguminous trees.

World Heritage site since

1978 1979 1980 1981 **1982** 1983 1984 1985 1986 1987 1988 1989 1990 1991 1992 1993 1994 1995 1996 1997 1998 1999 2000 2001 2002 2003 2004 2005 2006 2007 2008

Ancient City of Polonnaruwa
Sri Lanka

Criteria – Human creative genius; Testimony to cultural tradition; Heritage associated with events of universal significance

A carving of Buddha in Polonnaruwa.

Polonnaruwa was the second capital of Ceylon (now Sri Lanka) after the destruction of Anuradhapura in 993. This immense new capital created by the megalomaniac sovereign, Parakramabahu I, in the twelfth century, is one of history's most astonishing urban creations, both because of its unusual dimensions and because of the very special relationship of its buildings with their natural setting. In addition to the monumental ruins of Parakramabahu's fabulous garden city, Polonnaruwa contains the Brahmanic monuments built in the eleventh century by the Chola invaders from southern India.

After the destruction of Anuradhapura by Rajaraja, Polonnaruwa, a temporary royal residence during the eighth century, became the capital. The conquering Cholas constructed monuments to their religion (Brahmanism), and especially temples to Shiva where fine bronze statues, today in the Museum of Colombo, were found. The reconquest of Ceylon by Vijayabahu I did not put an end to the city's role as capital: it became covered, after 1070, with Buddhist sanctuaries, of which the Atadage, the Temple of the Tooth Relic which held the tooth of the Buddha, is the most renowned.

The apogee of Polonnaruwa occurred in the twelfth century AD when two sovereigns endowed it with monuments. Parakramabahu I (1153–1186) created within the boundary walls a garden city, where palaces and sanctuaries prolonged the enchantment of the countryside. This required the construction of a series of sophisticated irrigation systems; these systems are still used today. In addition, other notable monuments were built during his reign: the Lankatilaka, an enormous brick structure which has preserved a colossal image of Buddha; the Gal Vihara, with its gigantic rock sculptures which is among the masterpieces of Sinhalese art; and the Tivanka Pilimage, where wall paintings of the thirteenth century illustrate the jataka (narratives of the previous lives of Buddha).

The successor to Parakramabahu, Nissamkamalla I, ruler until 1196, built monuments that were less refined than those of his predecessor but nonetheless splendid: the Rankot Vihara, an enormous stupa or relic chamber 175 m in diameter and 55 m high, is one of the most impressive; its plan and its dimensions are reminiscent of the dagabas at Anuradhapura.

After this golden age the city fell into a long decline until government finally moved to Kurunegala at the end of the thirteenth century.

Polonnaruwa bears witness to several civilizations, notably that of the conquering Cholas, disciples of Brahmanism, and of the Sinhalese sovereigns of the twelfth and thirteenth centuries.

The city is a Buddhist shrine. The tooth of Buddha, a remarkable relic placed in the Atadage, or Temple of the Tooth Relic, by King Vijayabahu I, was considered the talisman of the Sinhalese monarchy. Its removal by King Bhuvanaikabahu II to Kurunegala at the end of the thirteenth century only confirmed the decline of Polonnaruwa.

World Heritage site since

1978 1979 1980 1981 **1982** 1983 1984 1985 1986 1987 1988 1989 1990 1991 1992 1993 1994 1995 1996 1997 1998 1999 2000 2001 2002 2003 2004 2005 2006 2007 2008

Tipasa
Algeria

Criteria – Testimony to cultural tradition;
Significance in human history

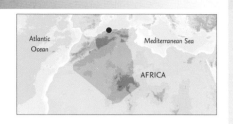

The extraordinary archaeological complexes of Tipasa, on the Mediterranean coast 70 km west of Algiers, are perhaps the most significant to the study of the contacts between the indigenous civilizations and various waves of colonization from the sixth century BC to the sixth century AD.

On the shores of the Mediterranean, Tipasa was an ancient Punic trading post conquered by Rome and turned into a strategic base for the conquest of the kingdoms of Mauritania. It comprises a unique group of Phoenician, Roman, palaeochristian and Byzantine ruins alongside indigenous monuments such as the Kbor er Roumia, the great royal mausoleum of Mauritania. The oldest Roman settlement is in the centre of the city on a steep slope protected by cliffs. The impressive ruins of the civic buildings are set in the heart of a dense network of private houses (many decorated with paintings and mosaics), commercial warehouses, and industrial establishments of the second and third centuries. Of the numerous Christian religious buildings, the immense fourth-century seven-aisled basilica is particularly striking.

Río Plátano Biosphere Reserve
Honduras

Criteria – Natural phenomena or beauty; Major stages of Earth's history; Significant ecological and biological processes; Significant natural habitat for biodiversity

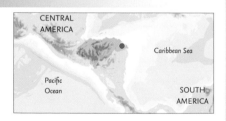

The site of Ciudad Blanca (White City) within the protected area constitutes one of the most important archaeological sites of Mayan civilization. The reserve also contains the site where Christopher Columbus first landed in the Americas in 1492. There are some 200 sites of archaeological importance.

Situated in the Mosquita region of northeast Honduras, the site comprises a belt of approximately 15 km by 150 km extending inland from the Caribbean coast in a southwesterly direction. The reserve protects virtually the entire watershed of the 100 km-long Plátano River, as well as major portions of the Paulaya, Guampu and Sicre rivers. These three waterways, together with the Caribbean, form the boundaries of the reserve, which is the largest surviving area of virgin tropical rainforest in Honduras. Its varied topography ranges from mountains with spectacular rock formations (such as Dama Peak) and waterfalls to placid coastal lagoons. Thirty-nine species of mammal, 377 species of bird and 126 reptiles and amphibians have been recorded.

World Heritage site since

1978 1979 1980 1981 **1982** 1983 1984 1985 1986 1987 1988 1989 1990 1991 1992 1993 1994 1995 1996 1997 1998 1999 2000 2001 2002 2003 2004 2005 2006 2007 2008

Tasmanian Wilderness
Australia

Criteria – Testimony to cultural tradition;
Significance in human history; Heritage
associated with events of universal significance;
Natural phenomena or beauty; Major stages of
Earth's history; Significant ecological and
biological processes; Significant natural habitat
for biodiversity

Covering an area of 13,800 km², the
Tasmanian Wilderness contains some of the
last expanses of temperate rainforest and is
one of the three largest temperate
wilderness regions in the southern
hemisphere. It comprises a vast network of
reserved lands that extends over much of
southwestern Tasmania, including several
coastal islands. Glacial erosion and other
geomorphological processes have
contributed to spectacular landforms and
the area contains rocks from almost every
geological period. The flora and fauna is
correspondingly diverse, with living
evidence of its ancient Gondwanan heritage.
The isolation of the Tasmanian Wilderness
has contributed to the uniqueness of its flora
and fauna, which includes some of the
world's longest-lived trees and largest
carnivorous marsupials. Remains found in
limestone caves attest to human occupation
of the area for at least 35,000 years through
periods of great climatic variation.

Liffey falls. ▲

Tasmania was cut off from mainland
Australia by rising sea levels
approximately 8,000 years ago,
thereby isolating Tasmanian
Aborigines for some
500 generations.

World Heritage site since

1978 1979 1980 1981 **1982** 1983 1984 1985 1986 1987 1988 1989 1990 1991 1992 1993 1994 1995 1996 1997 1998 1999 2000 2001 2002 2003 2004 2005 2006 2007 2008

Archaeological Site of Leptis Magna
Libyan Arab Jamahiriya

Criteria – Human creative genius; Interchange of values; Testimony to cultural tradition

Leptis Magna was one of the most beautiful cities of the Roman Empire, with its imposing public monuments, harbour, marketplace, storehouses, shops and residential districts. It was enlarged and embellished by Septimius Severus, who was born there and later became emperor. It is still one of the best examples of Severan urban planning. Thereafter, Leptis fell prey to the same vicissitudes of fortune as the majority of the coastal cities of Africa. Pillaged from the fourth century and reconquered by the Byzantines who transformed it into a stronghold, it definitively succumbed to the second wave of Arab invasion, that of the Hilians in the eleventh century. Buried under drifting sands, the city has been disengaged, piece by piece, over the course of a long archaeological exploration.

The city, which was constructed during the reign of Augustus and Tiberius but which was entirely remodelled along very ambitious lines under the Severan emperors, incorporates major monumental elements of that period. The forum, basilica and Severan arch rank among the foremost examples of a new Roman art, strongly influenced by African and Eastern traditions.

Leptis Magna Theatre with the ruins of the market, the forum and the harbour in the background.
▼

World Heritage site since

978 1979 1980 1981 **1982** 1983 1984 1985 1986 1987 1988 1989 1990 1991 1992 1993 1994 1995 1996 1997 1998 1999 2000 2001 2002 2003 2004 2005 2006 2007 2008

Tassili n'Ajjer
Algeria

Criteria – Human creative genius; Testimony to cultural tradition; Natural phenomena or beauty; Major stages of Earth's history

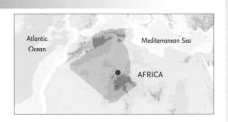

Tassili, a mountainous region in the centre of the Sahara, is a strange lunar landscape of deep gorges, dry river beds and 'stone forests'. During the prehistoric period Tassili had a very different climate, with abundant game, regular rainfall and fertile land. In 1933 one of the most important groupings of prehistoric cave art in the world was discovered here. More than 15,000 drawings and engravings record the climatic changes, the animal migrations and the evolution of

human life on the edge of the Sahara from 6000 BC to the first centuries AD. This art covers several periods, each of which corresponds to a particular fauna, yet each can equally be characterized by stylistic differences, without reference to an ecosystem.

At the end of the Upper Pleistocene period there were huge lakes in this region, fed by rivers whose dry beds can still be seen. The plants and animals found on the plateaus bear witness to former wetter periods. Relict species surviving in wet microclimates include fish and shrimp and, until the 1940s, the dwarf Saharan crocodile, many thousands of kilometres from the nearest population in Egypt.

M'Zab Valley
Algeria

Criteria – Interchange of values; Testimony to cultural tradition; Traditional human settlement

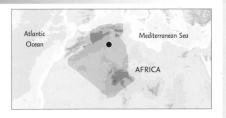

The five ksour (fortified cities) in the M'Zab Valley preserve a traditional human habitat of the tenth century. Simple, functional and perfectly adapted to the environment, the architecture of M'Zab was designed for community living, while respecting the structure of the family. The pattern of the life in the M'Zab Valley included a seasonal migration. Each summer the population moved to palm groves, where the 'summer

cities' were marked by a looser organization, the highly defensive nature of the houses, the presence of watchtowers, and a mosque without a minaret, comparable with those in the cemeteries. The settlement of the M'Zab Valley has exerted considerable influence on architects and city planners of the twentieth century, from Le Corbusier to Pouillon.

Each of the five M'Zab miniature citadels is encircled by walls and dominated by a mosque, whose minaret functioned as a watchtower. The mosque, with its arsenal and grain stores, was conceived as a fortress, the last bastion of resistance in the event of a siege.

World Heritage site since

1978 1979 1980 1981 **1982** 1983 1984 1985 1986 1987 1988 1989 1990 1991 1992 1993 1994 1995 1996 1997 1998 1999 2000 2001 2002 2003 2004 2005 2006 2007 200

Old Havana and its Fortifications
Cuba

Criteria – Significance in human history; Traditional human settlement

The Castillo de la Real Fuerza is the oldest extant colonial fortress in the Americas: its west tower is crowned by a bronze weathervane dating back to 1632.

The Plaza de Armas has been the seat of authority and power in Cuba for 400 years. The imposing Palacio de los Capitanes Generales on the west side of the square is one of Cuba's most majestic buildings and is now the City Museum. Calle Obispo, which runs off the Plaza de Armas, was a haunt of Ernest Hemingway.

Havana was the last city the Spanish conquistadors founded in Cuba and is the finest surviving Spanish complex in the Americas. The city's situation made it a perfect gathering point for the annual treasure fleets bound for Spain from Mexico and Peru and it became the front door to the vast Spanish colonial empire. The modern-day city is a large metropolis but its old centre retains an interesting mix of Baroque and neoclassical monuments, and a homogeneous ensemble of private houses with arcades, balconies, wrought-iron gates and internal courtyards.

Havana was established at its present location in 1519 and by 1550 it had become the most important city on the island, a position it has held ever since. It became one of the Caribbean's main centres for shipbuilding and has been Cuba's capital since 1607.

Spain fortified the city in the 1760s when Europe's Seven Years' War spilled over into the Americas; those fortifications can still be seen today. Havana was allowed to trade freely, growing in wealth through the eighteenth and nineteenth centuries, remaining physically untouched by the wars of independence in Central and South America in the first half of the nineteenth century. After political agitation throughout the nineteenth century, Cuba finally gained its independence from Spain in 1902. Havana suffered little damage during periods of unrest, including the Cuban Revolution of the 1950s, and the old city stands today much as it was 100 years ago or more.

The Cuban government has been involved in efforts to restore the character of a colonial city to the historic centre. Many of Old Havana's finest buildings have been converted into museums, and there are churches, palaces, castles, revolutionary monuments and markets to visit. The fortress of La Fuerza has been restored, as have the palaces of the Segundo Cabo and of Los Capitanes Generales. Restoration is slowly extending to residential areas.

The pattern of early urban planning still exists with the city's four large squares: Plaza de La Cathedral, Plaza de San Francisco, Plaza Vieja and Plaza de Armas. There is also a notable complex of seventeenth–nineteenth-century buildings. The Plaza de la Catedral, with the towers of the Catedral de San Cristóbal de La Habana dominating the square, is one of the city's most outstanding sights.

World Heritage site since

1978 1979 1980 1981 **1982** 1983 1984 1985 1986 1987 1988 1989 1990 1991 1992 1993 1994 1995 1996 1997 1998 1999 2000 2001 2002 2003 2004 2005 2006 2007 2008

Djémila
Algeria

Criteria – Testimony to cultural tradition;
Significance in human history

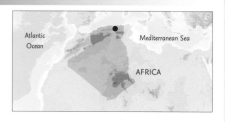

Situated 900 m above sea level, Djémila, or Cuicul, is one of the world's most beautiful Roman ruins. The classic formula of Roman urban planning was adapted to the physical constraints of the site: at both ends of the cardo maximus, the backbone of the city, are two gates. In the centre is the forum, an enclosed square surrounded by buildings essential to the functioning of civil life. Aristocratic dwellings set with rich mosaics multiplied during the course of the second century in this central quarter. However, this cramped defensive situation, hemmed in by walls, hindered the development of the city. In the mid-second century AD the city therefore expanded to the south, where a new quarter, rich in both public buildings and private dwellings, was established.

Christianity was implanted in the southern quarter at an early date. The remains of a group of episcopal buildings have been located there: two basilicas, a baptistry, a chapel and several houses, the residence of the bishop and the priest.

Timgad
Algeria

Criteria – Interchange of values; Testimony to cultural tradition; Significance in human history

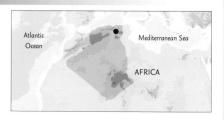

Timgad, which lies on the northern slopes of the Aurès mountains in a site of great natural beauty, is a consummate example of a Roman military colony created ex nihilo by the Emperor Trajan in AD 100. With its square enclosure measuring 355 m on each side and its precise orthogonal design based on the cardo and decumanus (the two perpendicular routes running through the city), it is an excellent example of Roman town planning at its height. Large public buildings in the south of the city include the forum and its annexes (basilica and curia), temples, a theatre with a seating capacity of 3,500, a market and baths. In the northeast sector, there are other baths and a public library.

Timgad's development was meticulously planned. The streets were paved with large rectangular slabs of limestone, and particular attention was paid to the disposition of public conveniences. The houses, including many immense private residences, sparkled under a décor of sumptuous mosaics.

World Heritage site since

1978 1979 1980 1981 **1982** 1983 1984 1985 1986 1987 1988 1989 1990 1991 1992 1993 1994 1995 1996 1997 1998 1999 2000 2001 2002 2003 2004 2005 2006 2007 2008

Archaeological Site of Sabratha
Libyan Arab Jamahiriya
Criteria – Testimony to cultural tradition

A Phoenician trading post that served as an outlet for the products of the African hinterland, Sabratha was part of the short-lived Numidian kingdom of Massinissa before being Romanized and rebuilt in the second and third centuries AD. The spectacular ruins highlight its wealth in Roman times, which saw the construction of grandiose monuments, of which the most renowned is the theatre (pictured below), probably built during the reign of the Emperor Commodus (AD 161–192), with a capacity of 5,000 seats. The best conserved part is the frons scena, at the back of the stage, which has been reconstructed with original fragments and is divided into three levels with overlapping marble columns. Near the theatre stands the amphitheatre, and other monuments include the temples of Liber Pater, Serapis, Hercules and Isis, the Basilica of Justinian and the Capitolium.

The decline of Sabratha began in the fourth century: commerce with Africa was less active, it was wracked by religious quarrels, and much of the city was destroyed by earthquakes, particularly that of AD 365. The Vandals invaded Sabratha in 455 and the city was definitively abandoned after the Arab invasions of the seventh and eleventh centuries.

Roman Theatre at Sabratha.
▼

World Heritage site since

1978 1979 1980 1981 **1982** 1983 1984 1985 1986 1987 1988 1989 1990 1991 1992 1993 1994 1995 1996 1997 1998 1999 2000 2001 2002 2003 2004 2005 2006 2007 2008

Historic Centre of the Town of Olinda
Brazil

Criteria – Interchange of values; Significance in human history

Founded in 1537 by the Portuguese, the town's history is linked to the sugar cane industry. Rebuilt after being looted by the Dutch, its basic urban fabric dates from the eighteenth century. The harmonious balance between the buildings, gardens, twenty Baroque churches, convents and numerous small passos (chapels) all contribute to Olinda's particular charm. From the sixteenth century, few of the churches and convents built by religious missions survive.

Among the more important of the eighteenth-century buildings are the Episcopal Church, the Jesuit College and Church (now the Church of Graça), the Franciscan, Carmelite, Benedictine and other monasteries, convents and churches. The studied refinements of the décor of these Baroque architectural structures contrasts with the charming simplicity of the houses, which are painted in vivid colours or faced with ceramic tiles.

Over recent decades, Olinda – a city much appreciated by artists – has been the object of numerous preservation measures. Outstanding buildings such as the Church of Graça, with the former Jesuit College, the **Convent do Carmo (pictured below)**, and the Episcopal Palace have all been restored.

World Heritage site since

1978 1979 1980 1981 **1982** 1983 1984 1985 1986 1987 1988 1989 1990 1991 1992 1993 1994 1995 1996 1997 1998 1999 2000 2001 2002 2003 2004 2005 2006 2007 2008

Cahokia Mounds State Historic Site
USA

Criteria – Testimony to cultural tradition; Significance in human history

Cahokia Mounds, some 13 km northeast of St Louis, Missouri, is the largest pre-Columbian settlement north of Mexico. It was occupied primarily during the Mississippian period 800–1400, when it covered nearly 16 km² and included some 120 mounds (earthworks). It is a striking example of a complex chiefdom society, with many satellite mound centres and numerous outlying hamlets and villages.

This agricultural society may have had a population of 10–20,000 at its peak between 1050 and 1150. Primary features at the site include Monks Mound, the largest prehistoric earthwork in the Americas, covering over 50,000 m² and standing 30 m high.

Cahokia has numerous large oval-shaped pits arranged in arcs of circles. Archaeologists believe that posts set in these pits lined up with the rising sun at certain times of the year, serving as a calendar, called Woodhenge.

Ancient City of Sigiriya
Sri Lanka

Criteria – Interchange of values; Testimony to cultural tradition; Significance in human history

The ruins of the capital built by the parricidal King Kassapa I (477–495) lie on the steep slopes and at the summit of a granite peak standing some 370 m high, the 'Lion's Rock', which dominates the jungle from all sides. A series of galleries and staircases emerging from the mouth of a gigantic lion constructed of bricks and plaster provide access to the site.

The 'Sigiri graffiti' are poems inscribed on the rock, and are among the most ancient texts in the Sinhalese language. They show the considerable influence exerted by the city on the literature and thought of ancient Ceylon.

◄ Sigiriya or Lion's Rock.

World Heritage site since

1978 1979 1980 1981 **1982** 1983 1984 1985 1986 1987 1988 1989 1990 1991 1992 1993 1994 1995 1996 1997 1998 1999 2000 2001 2002 2003 2004 2005 2006 2007 200

Old Walled City of Shibam
Yemen

Criteria – Testimony to cultural tradition;
Significance in human history; Traditional
human settlement

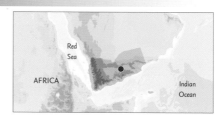

Surrounded by a fortified wall, the sixteenth-century city of Shibam is one of the oldest and best examples of urban planning based on the principle of vertical construction, giving the city the nickname of 'the Manhattan of the desert'. The city is built on a hillock, which has allowed it to escape the devastating floods of Wadi Hadramaut. Its plan is almost rectangular; and it is enclosed by earthen walls within which blocks of dwellings, also built from earth, have been laid out on a grid. The tallest house is eight storeys and the average is five, and the buildings for the most part date from the sixteenth century. However, some older houses and large buildings still remain from the first centuries of Islam, such as the Friday Mosque, built in 904, and the castle, built in 1220.

In Shibam there are some mosques, two ancient sultan's palaces and more than 500 buildings all made uniform by the material of which they are constructed: unfired clay blocks which continue to be made locally.

▼

World Heritage site since

1978 1979 1980 1981 **1982** 1983 1984 1985 1986 1987 1988 1989 1990 1991 1992 1993 1994 1995 1996 1997 1998 1999 2000 2001 2002 2003 2004 2005 2006 2007 2008

Royal Saltworks of Arc-et-Senans
France

Criteria – Human creative genius; Interchange of values; Significance in human history

The Royal Saltworks of Arc-et-Senans is a factory designed with the same sense of architectural quality as that of a palace. It is a temple to labour, perfectly illustrating the cultural changes that came about with the birth of industrial society. Located near Besançon, the Royal Saltworks was designed by Claude-Nicolas Ledoux. Its construction, begun in 1775 and completed in 1779 during the reign of Louis XVI, was the first major achievement of industrial architecture, reflecting the ideal of progress of the Enlightenment. This vast semicircular

complex was designed to permit a rational and hierarchical organization of work and was to have been followed by the building of an ideal city, the city of Choux, a project that was never realized.

The Saltworks' exceptional architecture included a system of double canals composed of wooden cylinders which drew saline water from Salons, 21 km away. At Arc-et-Senans the water was evaporated in immense vats heated night and day by a wood fire.

◄

The guardhouse, the entrance to the Royal Saltworks.

World Heritage site since

1978 1979 1980 1981 1982 **1983** 1984 1985 1986 1987 1988 1989 1990 1991 1992 1993 1994 1995 1996 1997 1998 1999 2000 2001 2002 2003 2004 2005 2006 2007 2008

Comoé National Park
Côte d'Ivoire

Criteria – Significant ecological and biological processes; Significant natural habitat for biodiversity

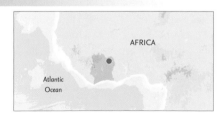

One of the largest protected areas in West Africa, this park is characterized by its great plant diversity. It is one of the few remaining natural areas in the region that is large enough to ensure the ecological integrity of the species contained within the site. The park comprises the land between the Comoé and Volta rivers, with a mean

altitude of 250–300 m and a series of ridges rising to 600 m. The Comoé and its tributaries form the principal drainage; the river runs through the park for 230 km. Due to the presence of the Comoé, the park contains plants which are normally found only much farther south, such as shrub savannas and patches of thick rainforest.

The park provides an outstanding example of an area of transitional habitat from forest to savanna. Its remarkable variety of habitats is home to a high number of mammal species, including eleven species of monkey.

World Heritage site since

1978 1979 1980 1981 1982 **1983** 1984 1985 1986 1987 1988 1989 1990 1991 1992 1993 1994 1995 1996 1997 1998 1999 2000 2001 2002 2003 2004 2005 2006 2007 2008

Taj Mahal
India

Criteria – Human creative genius

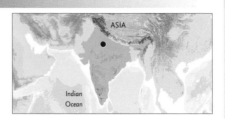

Situated on the right bank of the Yamuna River, the Taj Mahal stands in a vast Mughal garden of some 0.17 km^2 bounded by four minarets. The unique Mughal style combines elements and styles of Persian, Central Asian and Islamic architecture. Its octagonal structure is capped by a bulbous dome. The architectural precision is balanced by the delicate white marble decoration of floral arabesques, decorative bands and calligraphic inscriptions.

The Taj Mahal is the jewel of Muslim art in India and one of the universally admired masterpieces of the world's heritage. An immense mausoleum of white marble, it was built in Agra between 1631 and 1648 by order of the Mughal emperor Shah Jahan in memory of his third and favourite wife.

The Taj Mahal partially owes its renown to the moving circumstances of its construction. To perpetuate the memory of Mumtaz Mahal, who died in 1631, Shah Jahan had this funerary mosque built.

The monument, begun in 1632, was finished in 1648. Unverified legends attribute its construction to an international team of several thousands of masons, marble workers, mosaicists and decorators working under the orders of the architect of the emperor, Ustad Ahmad Lahori using building materials from all over India and central Asia.

The Darwaza, or gateway, was completed in 1648. It is a three-storey red sandstone structure with a lofty central arch and two-storeyed wings on either side. Its walls are inscribed with verses from the Koran in Arabic in black calligraphy. The small domed pavilions on top are Hindu in style and signify royalty. The gate was originally lined with silver, now replaced with copper, and decorated with 1,000 nails whose heads were contemporary silver coins.

The Bageecha, or gardens, are planned along classical Mughal char-bagh style. Two cypress-tree-lined canals quarter the garden into equal squares. The garden is laid out in such a way as to maintain perfect symmetry.

The Taj Mahal stands in the north end of the garden on two bases, one of sandstone and above it a square platform in a chequerboard design topped by a huge white marble terrace. On the corners are four minarets.

Two identical red sandstone buildings stand on either side; the western one is the masjid (mosque), which sanctifies the area and provides a place of worship. The rauza, the mausoleum itself, is square with bevelled corners. Each corner has a small dome while in the centre the main double dome is topped by a brass finial. The main chamber inside is octagonal with a high domed ceiling. This chamber contains false tombs of Mumtaz and Shah Jahan, both inlaid with precious stones.

World Heritage site since

1978 1979 1980 1981 1982 **1983** 1984 1985 1986 1987 1988 1989 1990 1991 1992 1993 1994 1995 1996 1997 1998 1999 2000 2001 2002 2003 2004 2005 2006 2007 2008

Ellora Caves
India

Criteria – Human creative genius; Testimony to
cultural tradition; Heritage associated with
events of universal significance

The Ellora Caves not only bear witness to the
three great religions of ancient India
(Buddhism, Brahminism and Jainism), they
also illustrate the spirit of tolerance which
permitted these three religions to establish
their sanctuaries and their communities in a
single place. The thirty-four monasteries and
temples of Ellora, extending over more than
2 km, were dug side by side in the wall of a
high basalt cliff, not far from Aurangabad, in

Maharashtra. The caves, with their
uninterrupted sequence of remarkable
reliefs, sculptures and architecture dating
from AD 600–1000, bring the civilization of
ancient India to life. The caves of the
Brahmin group are no doubt the best known
of Ellora with the 'Cavern of the Ten Avatars'
and especially the Kailasha Temple, an
enormous complex most likely undertaken
during the reign of Krishna I.

Progressing from
south to north along
the cliff, one discovers
successively the twelve
caves of the Buddhist
group (between c. AD
600 and 800),
comprising
monasteries and a
single large temple,
then the caves of the
Brahmin group
(c. 600–900), and
finally the Jain group
(c. 800–1000).

Rock carved images of
Buddha in meditation
in cave number twelve.
▼

World Heritage site since

1978 1979 1980 1981 1982 **1983** 1984 1985 1986 1987 1988 1989 1990 1991 1992 1993 1994 1995 1996 1997 1998 1999 2000 2001 2002 2003 2004 2005 2006 2007 2008

Agra Fort
India

Criteria – Testimony to cultural tradition

ASIA

Indian
Ocean

Near the gardens of the Taj Mahal stands the sixteenth-century Mughal monument known as the Red Fort of Agra. With its walls of red sandstone rising above a moat and interrupted by graceful curves and lofty bastions, the fortress encompasses within its enclosure walls of 2.5 km the imperial city of the Mogul rulers. It comprises many fairy-tale palaces, such as the Jahangir Palace and the Khas Mahal; audience halls, such as the Diwan-i-Khas; and two very beautiful mosques. Like the Delhi Fort, that of Agra is one of the most obvious symbols of the Mogul grandeur which asserted itself under Akbar, Jahangir and Shah Jahan. Several of the buildings are made from pure marble with beautiful carvings; all of these monuments mark the high point of an Indo-Muslim art strongly marked by influences from Persia.

The wall has two gates, the Delhi Gate and the Amar Singh Gate. The original and grandest entrance was through the Delhi Gate, which leads to the inner portal called the Hathi Pol or Elephant Gate. The entrance to the fort is now only through the Amar Singh Gate.

World Heritage site since

1978 1979 1980 1981 1982 **1983** 1984 1985 1986 1987 1988 1989 1990 1991 1992 1993 1994 1995 1996 1997 1998 1999 2000 2001 2002 2003 2004 2005 2006 2007 2008

Talamanca Range-La Amistad Reserves / La Amistad National Park
Costa Rica and Panama

Criteria – Natural phenomena or beauty; Major stages of Earth's history; Significant ecological and biological processes; Significant natural habitat for biodiversity

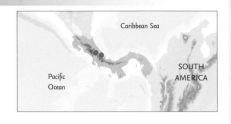

The Cordillera de Talamanca is the highest and wildest non-volcanic mountain range in Central America. The mountains contain the largest tracts of virgin forest in Costa Rica.

Tropical rainforests have covered this area since the last glaciations, about 25,000 years ago. The park includes lowland tropical rainforest and cloudforest, and four communities not found elsewhere in Central America: subalpine paramo forests, pure oak stands, glacial lakes and high-altitude bogs. Species diversity is perhaps unequalled in any other reserve of equivalent size in the world. Signs of tapir, a species as yet unrecorded in Costa Rica, are abundant on the Panama side of the border. All Central American wild cats are found including puma, ocelot, jaguarundi, tiger cat and jaguar, as well as the Central American squirrel monkey. A green and black high-altitude viper, that has rarely been seen or collected, is present. Four different Indian tribes inhabit this property, which benefits from close cooperation between Costa Rica and Panama.

Central Zone of the Town of Angra do Heroismo in the Azores
Portugal

Criteria – Significance in human history; Heritage associated with events of universal significance

Angra has preserved the better part of its monumental heritage and a homogenous urban ensemble. Its original vernacular architecture is perfectly adapted to unique climatic conditions.

Angra do Heroismo played a highly significant role in maritime exploration from the fifteenth to the nineteenth centuries. Situated on the island of Terceira in the Azores archipelago, the site was protected from the prevailing winds by a series of hills and had two natural basins: that of the Beacon and that of the Anchorage (Angra) from which the village took its name. The port was an obligatory port of call for fleets from Africa and the Indies, and is an eminent example of the maritime framework that allowed exchanges between the world's great civilizations. The 400-year-old San Sebastião and San João Baptista fortifications are unique examples of military architecture. Damaged by an earthquake in 1980, Angra is now being restored.

World Heritage site since

1978 1979 1980 1981 1982 **1983** 1984 1985 1986 1987 1988 1989 1990 1991 1992 1993 1994 1995 1996 1997 1998 1999 2000 2001 2002 2003 2004 2005 2006 2007 2008

Monastery of the Hieronymites and Tower of Belém in Lisbon
Portugal

Criteria – Testimony to cultural tradition; Heritage associated with events of universal significance

The interior of the church of Belém incorporates three naves of equal height. The ribs of the vaulting spring from their piers, all of which are covered with sculptures where the luxurious Gothic flora is muted with decorative elements of the Renaissance.

Standing at the entrance to Lisbon harbour, the Monastery of the Hieronymites, construction of which began in 1502, exemplifies Portuguese art at its best. The very rich ornamentation carved on the columns and walls – animals, vegetables, and twining ropes – earns this monument its fame.

The nearby Tower of Belém, built around 1514 to commemorate Vasco da Gama's expedition, is a reminder of the great maritime discoveries that laid the foundations of the modern world. The cross of the Knights of Christ is repeated on the parapets of this fortress, while the watchtowers that flank it are capped with ribbed cupolas inspired by Islamic architecture.

Created by the dynasty of Avis at its height, the complex of Belém is one of the most representative monuments to Portuguese power during the time of the Great Discoveries.

The two-storey cloisters of the Monastery of the ▶ Hieronymites.

World Heritage site since

1978 1979 1980 1981 1982 **1983** 1984 1985 1986 1987 1988 1989 1990 1991 1992 1993 1994 1995 1996 1997 1998 1999 2000 2001 2002 2003 2004 2005 2006 2007 2008

Great Smoky Mountains National Park
USA

Criteria – Natural phenomena or beauty; Major stages of Earth's history; Significant ecological and biological processes; Significant natural habitat for biodiversity

NORTH AMERICA

Pacific Ocean

Atlantic Ocean

Great Smoky Mountains National Park is the most important natural area in the eastern United States and is of world importance as an example of temperate deciduous hardwood forest. Stretching over more than 2,000 km², this exceptionally beautiful park is home to more than 3,500 plant species, including almost as many species of tree, 130, as in all of Europe.

The name 'Smoky' comes from the natural fog that often forms in the morning or after rainfall.
▼

Many endangered animal species are also found there, including probably the greatest variety of salamanders in the world. The dominant topographic feature of the park is the range of the Great Smoky Mountains with sixteen peaks over 1,829 m. Since the park is relatively untouched, it gives an idea of temperate flora before the influence of humankind.

The park contains evidence of four pre-Columbian Indian cultures: Mississippian, Woodland, Archaic and palaeo-Indian. The early Woodland culture period is of special archaeological importance because it shows the first evidence of organized horticulture in North America, with primitive agriculture on river floodplains.

World Heritage site since

1978 1979 1980 1981 1982 **1983** 1984 1985 1986 1987 1988 1989 1990 1991 1992 1993 1994 1995 1996 1997 1998 1999 2000 2001 2002 2003 2004 2005 2006 2007 2008

Ancient City of Nessebar
Bulgaria

Criteria – Testimony to cultural tradition;
Significance in human history

EUROPE

Black Sea

Situated on a rocky peninsula on the Black
Sea, Nessebar has been occupied for over
3,000 years. Originally a Thracian
settlement, at the beginning of the sixth
century BC, it became a Greek colony. The
city's remains from the Hellenistic period
include the acropolis, a temple of Apollo,
an agora and a wall from the Thracian
fortifications. Until its capture by the Turks
in 1453, Nessebar was one of the most
important Byzantine towns on the west
coast of the Black Sea. From this period
come monuments of exceptional quality:
for example, the Stara Mitropolia, a large
basilica without transept rebuilt in the ninth
century, the Church of the Virgin and the
Nova Mitropolia, founded in the eleventh
century and continually embellished until
the eighteenth century. Nessebar was
enriched in the nineteenth century by
numerous houses in the 'Plovdiv style', with
stone foundations and broad wooden eaves.

▲
The ruins of an ancient bastion and walls in Nessebar.

Nessebar, one of the oldest towns in
Europe, still exudes the spirit of
numerous ages and peoples –
Thracians, Hellenes, Romans, Slavs,
Byzantines and Bulgarians. Its
cobbled streets, well-kept medieval
churches, and timbered houses from
the nineteenth century illustrate its
chequered past. Nessebar's churches
can be best described as a cross
between Slav and Greek Orthodox
architecture.

World Heritage site since

1978 1979 1980 1981 1982 **1983** 1984 1985 1986 1987 1988 1989 1990 1991 1992 1993 1994 1995 1996 1997 1998 1999 2000 2001 2002 2003 2004 2005 2006 2007 2008

La Fortaleza and San Juan National Historic Site in Puerto Rico
USA

The walls of Fort San Felipe del Moro (El Morro) in San Juan.

Criteria – Heritage associated with events of universal significance

Founded in 1521, San Juan was the second city established by the Spanish in the Americas. It stands at a strategically important point in the Caribbean, and between the fifteenth and nineteenth centuries a series of defensive structures was built to protect the city and the Bay of San Juan. The city defences represent a fine display of European military architecture adapted to harbour sites on the American continent.

For the explorers and the colonists of the New World who came from the east, Puerto Rico was an obligatory stopping-place in the Caribbean. From this evolved its primordial strategic role at the beginning of the Spanish colonization. The island was for centuries a stake disputed by the Spanish, French, English and Dutch. The fortifications of the Bay of San Juan, the magnificent port to which Puerto Rico owes its name, bear witness to its long military history.

La Fortaleza, completed in 1540, was the first of the defensive fortifications to be built to protect Spain's interests against Carib Indians, pirates and enemy warships. It is a vast system with ramparts, fortlets and fortresses, attesting formerly to its effectiveness and today to its historic significance. The main components of this defensive system are La Fortaleza, El Morro (see photograph on the right) and San Cristóbal.

La Fortaleza is an exemplary monument of Hispano-American colonial architecture. It served as an arsenal, prison, and residence for the Governor-General of the island. El Morro, also built to protect San Juan Bay, stands on a rocky peak of land on the western extremity of the island. This triangular fort developed into a masterpiece of military engineering with stout walls, carefully planned steps and ramps for moving men and artillery. By the end of the eighteenth century, more than 400 cannon defended the fort, making it almost impregnable. San Cristóbal, with its dependencies, is another accomplished example of the military architecture of the second half of the eighteenth century.

San Juan National Historic Site includes forts, bastions, powder houses, walls and El Cañuelo Fort, also called San Juan de la Cruz – defensive fortifications that once surrounded the old colonial portion of San Juan, Puerto Rico. El Cañuelo Fort is located on the Isla de Cabras at the western end of the entrance to San Juan Bay.

By the nineteenth century, the old city had become a charming residential and commercial district. The city itself, with its institutional buildings, museums, houses, churches, plazas and commercial buildings, is part of the San Juan Historic Zone which is administered by municipal, state and federal agencies.

The entire historic site of San Juan with its different monumental components maintains, at present, a balance between constructed and non-constructed zones.

World Heritage site since

1978 1979 1980 1981 1982 **1983** 1984 1985 1986 1987 1988 1989 1990 1991 1992 1993 1994 1995 1996 1997 1998 1999 2000 2001 2002 2003 2004 2005 2006 2007 2008

Srebarna Nature Reserve
Bulgaria

Criteria – Significant natural habitat for biodiversity

Until 1949, the lake was connected to the river Danube. Its disconnection prevented annual flooding and the level of the lake fell one metre per year. However, the lake was reconnected with the Danube by canal in 1978 to prevent water levels from becoming too low and to restore the lake's fish population.

The Srebarna Nature Reserve is a freshwater lake adjacent to the Danube and extending over 6 km². The reserve was set up primarily to protect the rich diversity of wildfowl, which represent half of all Bulgarian bird species. It is the breeding ground of almost a hundred species of bird, many of which are rare or endangered. Some eighty other bird species migrate and seek refuge there every winter. Sixty-seven plant species can be found in Srebarna Nature Reserve, including water lily and a number of rare marsh plants. Reeds occupy two-thirds of the reserve and form a thick barrier around the lake. They form reed-mace islands which birds use for nesting.

Abbey Church of Saint-Savin sur Gartempe
France

Criteria – Human creative genius; Testimony to cultural tradition

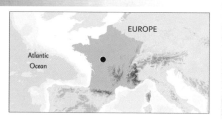

Saint-Savin sur Gartempe is the second-oldest church still standing in France. It boasts the largest Romanesque frescoes in Europe.

Known as the 'Romanesque Sistine Chapel', the Abbey Church of Saint-Savin contains many beautiful eleventh- and twelfth-century murals which are still in a remarkable state of preservation. During the reign of Charlemagne the bodies of two martyrs, Savin and Cyprian, who had been persecuted in the fifth century, were discovered under miraculous circumstances by Baidilius, Abbot of Marmoutier, who ordered a church built to shelter the holy remains. By coincidence, Charlemagne decided to have a castle erected next to the sanctuary. Because the abbey was sheltered by the castle it escaped pillage from the Viking raids and survives to this day. The beautifully proportioned building dates to the end of the eleventh century, but includes some older parts, such as the crypts of the saints.

World Heritage site since

1978 1979 1980 1981 1982 **1983** 1984 1985 1986 1987 1988 1989 1990 1991 1992 1993 1994 1995 1996 1997 1998 1999 2000 2001 2002 2003 2004 2005 2006 2007 2008

Place Stanislas, Place de la Carrière and Place d'Alliance in Nancy
France
Criteria – Human creative genius; Significance in human history

These three squares in Nancy represent the oldest and most typical example of a modern capital where an enlightened monarch proved to be sensitive to the needs of the public. Built between 1752 and 1756 by a brilliant team led by the architect Heré, this was a carefully conceived project that succeeded in creating a capital that not only enhanced the sovereign's prestige but was also functional. The foundation stone of the first building in the square was officially laid in March 1752 and the royal square solemnly inaugurated in November 1755. In addition to some prestigious architecture conceived to exalt a sovereign with triumphal arches, statues, and fountains, the project provided the public with three squares that gave access to the town hall, the courts of law, and the Palais des Fermes as well as to other public buildings.

▲
Fountain of Amphitrite in Place Stanislas.

The project was carried out under the patronage of Stanislas Leszczynski, unhappy pretender to the Polish throne and father-in-law of Louis XV, King of France. Stanislas received, as a recompense for his abdication, the Dukedom of Lorraine for life. He 'reigned' (peacefully) from Nancy from 1737 to 1766.

World Heritage site since

1978 1979 1980 1981 1982 **1983** 1984 1985 1986 1987 1988 1989 1990 1991 1992 1993 1994 1995 1996 1997 1998 1999 2000 2001 2002 2003 2004 2005 2006 2007 200

Wood Buffalo National Park
Canada

*Criteria – Natural phenomena or beauty;
Significant ecological and biological processes;
Significant natural habitat for biodiversity*

Pacific
Ocean

NORTH
AMERICA

Atlantic
Ocean

Situated in a vast wilderness area, Wood Buffalo was created specifically to protect North American bison, one of the largest free-roaming, self-regulating herds in existence. The park is also the natural nesting place of the whooping crane and contains the world's largest inland delta, located at the mouth of the Peace and Athabasca rivers. The park has four main landscape features: a glacially eroded plateau; glaciated plains; a major freshwater delta; and alluvial river lowlands (including a unique salt mudflat in dry weather). Vegetation is typical of the boreal forest zone with white spruce, black spruce, jack pine and tamarack predominant. The upper surface of the plateau is about 1,500 m above the rest of the park and supports a spruce-willow-birch upland tundra community. Some areas of prairie occur. The park contains the largest undisturbed grass and sedge meadows in North America.

The forty-seven mammal species in the park include **North American Bison (pictured below)**, black bear, woodland caribou, Arctic fox, moose, grey wolf, lynx snowshoe hare, muskrat, beaver and mink, while 227 bird species have been recorded, including great grey owl and snowy owl, willow ptarmigan, redpoll, crossbill and boreal chickadee.

World Heritage site since

1978 1979 1980 1981 1982 **1983** 1984 1985 1986 1987 1988 1989 1990 1991 1992 1993 1994 1995 1996 1997 1998 1999 2000 2001 2002 2003 2004 2005 2006 2007 2008

Sangay National Park
Ecuador

Criteria – Natural phenomena or beauty; Major
stages of Earth's history; Significant ecological
and biological processes; Significant natural
habitat for biodiversity

The site is situated in the Cordillera Oriental
region of the Andes in central Ecuador.
With its outstanding natural beauty and two
active volcanoes, Tungurahua (5,016 m) and
Sangay (5,230 m), the park illustrates the
entire spectrum of ecosystems, ranging
from tropical rainforests to glaciers, with
striking contrasts between the snowcapped
peaks and the forests of the plains. A third
volcano, El Altar (5,139 m), is considered

extinct. Major rivers drain eastwards into
the Amazon Basin and are characterized by
dramatic variations in level. Numerous
waterfalls occur, especially in the hanging
valleys of the glaciated zone and along the
eastern edge of the Cordillera. The area's
isolation has encouraged the survival of
indigenous species such as the mountain
tapir and the Andean condor.

The park's varied
vegetation zones offer
homes to many rare
and vulnerable fauna
species. At the highest
altitudes mountain
tapir, puma, guinea
pig and Andean fox
occur. Elsewhere
spectacled bear,
jaguar, ocelot, margay,
pudu and giant otter
are found.

Vallée de Mai Nature Reserve
Seychelles

Criteria – Natural phenomena or beauty; Major
stages of Earth's history; Significant ecological
and biological processes; Significant natural
habitat for biodiversity

Vallée de Mai is a valley in the heart of
Praslin National Park, an area which was
untouched until the 1930s and still retains
primeval palm forest in a near-natural state.
This palm forest includes the endemic
species coco de mer, the bearer of the
largest nut in the world. The other five
species of endemic palm are also found
here, the only place in the Seychelles where
they all occur together. Notable bird

species include the endemic black parrot,
which is totally dependent on the Vallée de
Mai and the surrounding palm forest. In a
densely populated island, the survival of
the Vallée de Mai in itself is a remarkable
achievement; too small to be self-
sustaining, its present status is due to some
replanting of coco de mer.

A rich body of legend
has developed around
the coco de mer, often
with religious
significance. In the
nineteenth century
the British General
Gordon produced
detailed 'proof' that
the Vallée de Mai was
the Garden of Eden
and that coco de mer
was the tree of
knowledge.

World Heritage site since

1978 1979 1980 1981 1982 **1983** 1984 1985 1986 1987 1988 1989 1990 1991 1992 1993 1994 1995 1996 1997 1998 1999 2000 2001 2002 2003 2004 2005 2006 2007 200

Historic Sanctuary of Machu Picchu
Peru

Criteria – Human creative genius; Testimony to cultural tradition; Natural phenomena or beauty; Significant ecological and biological processes

Machu Picchu stands 2,430 m above sea level in an extraordinarily beautiful setting in the middle of a tropical mountain forest. The city was probably the greatest urban achievement of the Inca Empire at its height: its giant walls, terraces and ramps seem as if they have been cut naturally in the continuous rock escarpments. The natural setting, on the eastern slopes of the Andes, encompasses the upper Amazon basin with its rich diversity of flora and fauna.

Set on the vertiginous site of a granite mountain sculpted by erosion and dominating a meander in the Rio Urubamba, Machu Picchu is a world-renowned archaeological site. Its construction, set out according to a very rigorous plan, comprises one of the most spectacular creations of the Incas, the largest civilization in the Americas before the arrival of Europeans. It appears to date from the period of the two great Inca rulers, Pachacutec Inca Yupanqui (1438–71) and Tupac Inca Yupanqui (1472–93). The function of this city, which is over 100 km from the Inca capital, Cuzco, is still unknown. Without making a judgement as to their purpose, several individual quarters may be noted in the ruins: a 'farmers' quarter near the colossal terraces whose slopes were cultivated and transformed into hanging gardens; an 'industrial' quarter; a 'royal' quarter and a 'religious' quarter.

The Historic Sanctuary of Machu Picchu covers 325 km² in some of the scenically most attractive mountainous territory of the Peruvian Andes. It was the last stronghold of the Incas, is of superb architectural and archaeological importance, and remains one of the most important cultural sites in Latin America. The site's stonework is a first-class example of the use of a natural raw material to create outstanding architecture totally appropriate to the surroundings.

The surrounding valleys have been cultivated continuously for well over 1,000 years, providing one of the world's greatest long-term examples of a productive man–land relationship. The people living around Machu Picchu continue a way of life closely resembling that of their Inca ancestors, being based on potatoes, maize and llamas. Machu Picchu also provides a secure habitat for several endangered species, notably the spectacled bear.

Machu Picchu bears, with Cuzco and the other archaeological sites of the valley of the Urubamba (Ollantautaybo, Runcuracay, Sayacmarca, Phuyupamarca, Huiñay Huayna, Intipucu and others), a unique testimony to the Inca civilization. Machu Picchu in particular is an outstanding example of man's interaction with his natural environment.

World Heritage site since

1978 1979 1980 1981 1982 **1983** 1984 1985 1986 1987 1988 1989 1990 1991 1992 1993 1994 1995 1996 1997 1998 1999 2000 2001 2002 2003 2004 2005 2006 2007 20C

Gulf of Porto: Calanche of Piana, Gulf of Girolata, Scandola Reserve
France

Criteria – Natural phenomena or beauty; Major stages of Earth's history; Significant natural habitat for biodiversity

This reserve, on the central western coast of Corsica, includes a coastline of astonishing beauty studded with offshore islets and sea pillars rising out of translucent waters. On the shore there are hidden coves and long beaches of fine sand, sea grottoes and high cliffs of blood-red porphyry. Seagulls, cormorants and sea eagles can be found there. The clear waters, with their islets and inaccessible coves, host a rich marine life. The reserve is divided into two sectors: the

Elpa Nera inlet (between Pointe Bianca and Pointe Validori) and the Scandola peninsula. The vegetation is an outstanding example of typical Mediterranean maquis (shrubland). This is replaced by arborescent plants at an altitude of 200 m and oaks replace this in certain areas. This area conserves traditional agriculture and grazing activities, and contains complete systems of architecturally interesting fortifications.

Scandola Nature Reserve contains a rich sedentary and migrant fauna including the peregrine falcon, osprey and Eleonora falcon, with Cory's shearwater and Audouin's gull occurring in the littoral zone. The marine environment contains considerable numbers of spiny lobster and a wide range of littoral and sublittoral invertebrates and fish

World Heritage site since

~~1978~~ 1978 1979 1980 1981 1982 **1983** 1984 1985 1986 1987 1988 1989 1990 1991 1992 1993 1994 1995 1996 1997 1998 1999 2000 2001 2002 2003 2004 2005 2006 2007 2008

Convent of Christ in Tomar
Portugal

Criteria – Human creative genius; Heritage associated with events of universal significance

EUROPE

Atlantic Ocean

Mediterranean Sea

AFRICA

The Convent of Christ, originally conceived as a symbolic monument to the Reconquista, the retaking of Portuguese territory from Islamic rule, became the most spectacular example of highly decorated Manueline architecture. During the second half of the twelfth century, the Knights Templar came to Tomar to assist in the Reconquista. Their original church, built at the end of the twelfth century by the first great Master of the Templars, Gualdim Pais, was based on a polygonal ground plan of sixteen bays with a central octagonal choir and an ambulatory, typical of Templar architecture. It was under Manuel I that the choir was lavishly decorated and Diego de Arruda created the enormous nave built above the chapter room. The prodigious exterior decoration combines with stupefying ease Gothic and Moorish influences.

When, in the fourteenth century, the Order of the ▶ Knights Templar was abolished and replaced by the Knights of the Order of Christ, Tomar lost none of its importance. Successive embellishments rendered it one of the most prestigious monuments in Portugal.

World Heritage site since

1978 1979 1980 1981 1982 **1983** 1984 1985 1986 1987 1988 1989 1990 1991 1992 1993 1994 1995 1996 1997 1998 1999 2000 2001 2002 2003 2004 2005 2006 2007 200

Convent of St Gall
Switzerland
Criteria – Interchange of values; Significance in human history

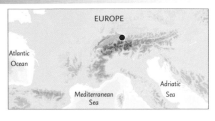

In 747, the abbot Othmar established a community of Benedictine monks in the place made famous by St Gall. By the ninth century, the abbey of St Gall was one of the most renowned centres of Western culture and science.

The Convent of St Gall is a perfect example of a great Carolingian monastery and was, from the eighth century to its secularization in 1805, one of the most important in Europe. Its library is one of the richest and oldest in the world with more than 160,000 books and many precious manuscripts. It has 400 volumes more than 1,000 years old and the earliest-known architectural plan drawn on parchment. An abbey has existed on this site since 719 and successive restructurings of its buildings attest to its ongoing religious and cultural function. From 1755–68, the conventual area was rebuilt in Baroque style. The cathedral and the library are the main features of this remarkable architectural complex, reflecting twelve centuries of continuous activity.

Pilgrimage Church of Wies
Germany
Criteria – Human creative genius; Testimony to cultural tradition

The prodigious stucco decoration is the work of Dominikus Zimmermann, assisted by his brother Johann Baptist. The lively colours of the paintings bring out the sculpted detail, while the frescoes and stuccowork interpenetrate to produce a décor of unprecedented richness. The ceilings, painted as trompe-l'œil, appear to open on to an iridescent sky, across which angels fly.

The sanctuary of Wies, a pilgrimage church in the beautiful setting of an Alpine valley, is a masterpiece of Rococo art. The hamlet of Wies, near Steingaden in Bavaria, was the setting of a miracle in 1738: a simple wooden image of Christ, mounted on a column, appeared to some of the faithful to be in tears. A wooden chapel was initially built to house the miraculous statue. However, pilgrims from Germany and abroad became so numerous that the Abbot of Steingaden decided to construct a splendid sanctuary. Accordingly, work began in 1745 under the direction of the celebrated architect, Dominikus Zimmermann. The choir was consecrated in 1749 and the remainder of the church finished by 1754.

World Heritage site since

78 1979 1980 1981 1982 **1983** 1984 1985 1986 1987 1988 1989 1990 1991 1992 1993 1994 1995 1996 1997 1998 1999 2000 2001 2002 2003 2004 2005 2006 2007 2008

Monastery of Batalha
Portugal

Criteria – Human creative genius; Interchange of values

EUROPE

Atlantic Ocean

Mediterranean Sea

AFRICA

The Monastery of the Dominicans of Batalha was built to commemorate the victory of the Portuguese over the Castilians at the battle of Aljubarrota in 1385. It was to be the Portuguese monarchy's main building project for the next two centuries, out of which a highly original, national Gothic style evolved. The majority of the complex dates from the reign of João I, when the church (finished in 1416), the royal cloister, the chapterhouse, and the funeral

chapel of the founder were constructed. The last great period of Batalha coincided with the reign of Manuel I, who built the monumental vestibule and the principal portal, and restored the royal cloister, creating a masterpiece of Manueline art. The arcades were embellished with finely carved tracery displaying the emblems of Manuel I, the Cross of the Order of Christ and the armillary sphere.

The church's interior maintains a sober Gothic majesty that has remained undisturbed by later additions. The nave and aisles are separated by thick pillars crowned by capitals with plant motifs. The chancel windows, decorated with beautiful sixteenth-century stained glass, project a diffused light that gives a feeling of great spirituality.

World Heritage site since

1978 1979 1980 1981 1982 **1983** 1984 1985 1986 1987 1988 1989 1990 1991 1992 1993 1994 1995 1996 1997 1998 1999 2000 2001 2002 2003 2004 2005 2006 2007 200

Rila Monastery
Bulgaria

Criteria – Heritage associated with events of
universal significance

Rila Monastery was founded in the tenth century by St John of Rila, a hermit whose ascetic dwelling and tomb became a holy site. It was transformed into a monastic complex that played an important role in the spiritual and social life of medieval Bulgaria. Destroyed by fire at the beginning of the nineteenth century, the monastery was rebuilt between 1834 and 1862. Only the Hrelyu Tower, built in 1355 by Stefan Hrelyu,

a local prince, survives from an earlier period. The remaining nineteenth-century structures occupy a vast irregular square, with two entrances, both decorated with frescoes. In the centre is the Cathedral of Our Lady of the Assumption, built in 1833. The buildings that surround it contain four chapels, a refectory and some 300 cells for the monks, a library and rooms for the guests of the monastery.

During the Ottoman Turkish domination of Bulgaria, the monaste became a bulwark of national identity. It was a destination for pilgrimages from all over the Balkan region especially after 1469, when the relics of St John of Rila were brought here. Its reconstruction in the nineteenth century symbolized an awakening of a Slavic cultural identity.

Rila Monastery fresco.
▼

World Heritage site since

1978 1979 1980 1981 1982 **1983** 1984 1985 1986 1987 1988 1989 1990 1991 1992 1993 1994 1995 1996 1997 1998 1999 2000 2001 2002 2003 2004 2005 2006 2007 2008

Pirin National Park
Bulgaria

Criteria – Natural phenomena or beauty; Major stages of Earth's history; Significant ecological and biological processes

Extending over an area of 274 km² in southwest Bulgaria, Pirin National Park has a limestone landscape, with lakes, waterfalls, caves and pine forests. The rugged mountains, with some seventy glacial lakes scattered throughout them, are home to hundreds of endemic and rare species, many of which are representative of the Balkan Pleistocene flora. There are many rivers and waterfalls. Generally, the timberline has developed as a result of

human interference over a long period and descends as low as 2,000 m, but in some places reaches 2,200–2,300 m. In the subalpine zone there are thickets of dwarf mountain pine and Juniperus sibirica. Above 2,400–2,600 m is a layer of alpine meadows, stony slopes, screes and rocks. There is a wide variety of animal species including many endemic species and glacial relicts among the invertebrate fauna.

The presence of limestone rocks, the southerly position of the range and close proximity to the Aegean, coupled with its relative isolation, have made Pirin National Park an important refuge for many species. Winter in the upper parts is cold and long with snow cover remaining for five to eight months. Summer is cool and short.

World Heritage site since

1978 1979 1980 1981 1982 1983 **1984** 1985 1986 1987 1988 1989 1990 1991 1992 1993 1994 1995 1996 1997 1998 1999 2000 2001 2002 2003 2004 2005 2006 2007 200

Vatican City
Holy See

Criteria – Human creative genius; Interchange of values; Significance in human history; Heritage associated with events of universal significance

EUROPE

Adriatic
Sea

Tyrrhenian Sea

St Peter's was founded as a longitudinal basilica with five aisles, with a transept, apse, and large atrium with quadriporticus.

The basilica was reconstructed in the sixteenth century under the guidance of the most brilliant architects of the Renaissance. The remains of Constantine's original basilica still exist, with fragments of the circus of Caligula and Nero, and an entire first century AD Roman necropolis where Christian sepulchres are placed side-by-side with pagan tombs.

The Vatican City, one of the most sacred places in Christendom, attests to a great history and a formidable spiritual venture. A unique collection of artistic and architectural masterpieces lies within the boundaries of this small state. As the site of the tomb of St Peter and a pilgrimage centre, the Vatican is directly and materially linked with the history of Christianity. Furthermore, it is both an ideal and an exemplary creation of the Renaissance and of Baroque art. It exerted an underlying influence on the development of art from the sixteenth century.

The independent Vatican City State defined by the 1929 Lateran Treaty between the Holy See and Italy, has a territorial sovereignty of less than 0.5 km². However, this tiny enclave within Rome has, in the heritage of mankind, an importance which is inversely proportional to its area. The centre of Christianity since the time of Constantine in the fourth century and the seat of papal power, the Vatican is at once an important archaeological site of the Roman world, the pre-eminently holy city of Catholics and one of the major cultural reference points of both Christians and non-Christians.

Its prestigious past explains the development of an architectural and artistic ensemble of exceptional value. The churches and palaces rest on a substratum impregnated with history. At its centre is the Basilica of St Peter, with its double colonnade and a circular piazza in front and bordered by palaces and gardens. The basilica was first built in 315 over the tomb of St Peter the Apostle but its present-day appearance dates from the sixteenth century, when Pope Julius II inaugurated a massive artistic project for the refoundation of the entire basilica, along with the decoration of the Stanze Vaticane (the papal apartments) and the Sistine Chapel, and the construction of his own tomb. The result is the fruit of the combined genius of Bramante, Raphael, Michelangelo, Bernini and Maderna.

The Apostolic Palace or Vatican Palace is the result of a long series of construction campaigns in which successive popes, from the Middle Ages onwards rivalled each other in their munificence. It is the official residence of the Pope and houses the Vatican Museum, the Vatican Library and various chapels including the Sistine Chapel.

World Heritage site since

1978 1979 1980 1981 1982 1983 **1984** 1985 1986 1987 1988 1989 1990 1991 1992 1993 1994 1995 1996 1997 1998 1999 2000 2001 2002 2003 2004 2005 2006 2007 2008

Works of Antoni Gaudí
Spain

Criteria – Human creative genius; Interchange of values; Significance in human history

The works of the architect Antoni Gaudí (1852–1926) represent outstanding examples of early twentieth century residential and public architecture. In particular, seven of his properties in and around Barcelona testify to his exceptional creative contribution to the development of architecture and building technology in the late nineteenth and early twentieth centuries.

The seven buildings in the World Heritage Site are Casa Vicens, Gaudí's work on the Nativity façade and Crypt of La Sagrada Familia, Casa Batlló, Casa Milà, Park Güell, Palacio Güell and the crypt in Colonia Güell.

These monuments represent an eclectic and very personal style which was given free reign in the design of gardens, sculpture and all decorative arts, as well as architecture.

Construction began on La Sagrada Familia in 1882 and continues to this day. It is expected to be complete by 2026.

Gaudí's work exhibits an important interchange of values closely associated with the cultural and artistic currents of his time, as represented in El Modernisme movement of Catalonia. It also anticipated and influenced many of the forms and techniques that were relevant to the development of modern construction in the twentieth century.

World Heritage site since

1978 1979 1980 1981 1982 1983 **1984** 1985 1986 1987 1988 1989 1990 1991 1992 1993 1994 1995 1996 1997 1998 1999 2000 2001 2002 2003 2004 2005 2006 2007 2008

Sun Temple, Konârak
India

Criteria – Human creative genius; Testimony to cultural tradition; Heritage associated with events of universal significance

ASIA

Indian
Ocean

On the shores of the Bay of Bengal, bathed in the rays of the rising sun, the temple at Konârak was one of the earliest centres of sun worship in India. Built around 1250, the entire temple was conceived as a representation of the Sun God Surya's chariot, with a set of spokes and elaborate carvings. Its twenty-four wheels are decorated with symbolic designs referring to the cycle of the seasons and the months, and it is led by a team of seven horses, six of which still exist. The temple is carefully oriented so as to permit the first rays of the Sun to strike its principal entry. It is one of the most famous Brahmin sanctuaries of Asia.

Legend has it that the temple was constructed by Samba, the son of Lord Krishna. Samba was afflicted by leprosy and after twelve years of penance he was cured by Surya, the Sun God, in whose honour he built this temple.

One of the twenty-four ornately carved stone wheels on the Sun Temple at Konârak.
▼

World Heritage site since

1978 1979 1980 1981 1982 1983 **1984** 1985 1986 1987 1988 1989 1990 1991 1992 1993 1994 1995 1996 1997 1998 1999 2000 2001 2002 2003 2004 2005 2006 2007 200

Yosemite National Park
USA

Criteria – Natural phenomena or beauty; Major stages of Earth's history

El Capitan is a 910 m rock formation predominantly carved out during the Sherwin Glaciation around one million years ago.

Yosemite National Park lies in the heart of California in an area of outstanding wilderness and great scenic beauty. With its hanging valleys, many waterfalls, cirque lakes, polished domes, moraines and U-shaped valleys, it provides an excellent overview of all kinds of granite relief fashioned by glaciation.

At 600–4,000 m, a great variety of flora and fauna can also be found here. The National Park, on the west slope of the central Sierra Nevada Mountains, represents practically all the different environments found within the Sierra Nevada, including sequoia groves, historic resources, evidence of Indian habitation, and domes, valleys, polished granites and other geological features that illustrate the formation of the mountain range.

Yosemite's natural beauty was the impetus for the implementation of the concept of the national park. Its archaeological features add to the area's cultural importance.

The variety of flora in the park is reflected in the existence of six distinct vegetation zones which are governed by altitude. There are 1,200 species of flowering plant along with various other ferns, bryophytes and lichens.

Port, Fortresses and Group of Monuments, Cartagena
Colombia

Criteria – Significance in human history; Heritage associated with events of universal significance

The quarter of San Pedro, where the nobles and the notables resided, still preserves monuments of high quality such as the cathedral, the Church and Convent of San Pedro Claver, the Church of Santo Domingo, and the building that once was the Monastery of San Diego.

Cartagena, together with La Habana and San Juan de Puerto Rico, was one of the three most important ports in the West Indies. It is an outstanding example of the military architecture of the sixteenth–eighteenth centuries, the most extensive in the New World and one of the most complete. The fortifications, finally completed in the seventeenth century, made Cartagena an impregnable stronghold, which successfully resisted attack until 1697. Within the shelter of the formidable defences, the city continued to grow. The plan, characteristic of colonial foundations of the sixteenth century, illustrates a rigorous zoning system, divided into three quarters corresponding to the major social categories: San Pedro, with the cathedral and many Andalusian-style palaces; San Diego, where merchants and the middle class lived; and Gethsemani, the 'popular quarter'.

World Heritage site since

1978 1979 1980 1981 1982 1983 **1984** 1985 1986 1987 1988 1989 1990 1991 1992 1993 1994 1995 1996 1997 1998 1999 2000 2001 2002 2003 2004 2005 2006 2007 2008

Salonga National Park
Dem. Rep. of the Congo

Criteria – Natural phenomena or beauty;
Significant ecological and biological processes

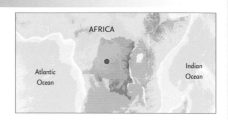

Salonga is the largest tropical forest national park in the world. Situated at the heart of the central basin of the Congo river, the park is very isolated and accessible mainly by water. Equatorial forest covers most of the area, varying in composition according to the geomorphology. There are three types of landscape: low plateaus, river terraces and high plateaus. The total area of grassland is under 0.5 per cent of the park area. The park is the habitat of many endangered species, the most important of which is the endemic dwarf chimpanzee, or bonobo. Other species include colobus monkeys, hippopotamus and leopard.

Rivers in the west of the north sector are large and meandering with marshy banks. In the higher east, valleys are deeper with cliffs up to 80 m high. The south sector includes the watershed between the basin of the Luilaka to the north and east, Likoro to the west, and Lukenje to the south.

Mana Pools National Park, Sapi and Chewore Safari Areas
Zimbabwe

Criteria – Natural phenomena or beauty;
Significant ecological and biological processes;
Significant natural habitat for biodiversity

On the banks of the Zambezi, great cliffs overhang the river and the floodplains. The Mana Pools are former channels of the Zambezi. With the Sapi and Chewore areas, they total some 6,766 km². The area contains the last remaining natural stretch of the Middle Zambezi and there is virtually no permanent human habitation. The area is home to a remarkable concentration of wild animals, including elephants, buffalo, leopards and cheetahs. An important concentration of Nile crocodiles is also found there. Much of the Chewore is heavily dissected and the Mupata Gorge (some 30 km long) occurs along the northern border of this area. Above the Mupata Gorges the river is broad and sandy, flowing through numerous channels, sandbanks and islands.

Well-grassed miombo woodland dominates the mountainous escarpment and higher Chewore areas with small but significant riparian communities along the numerous streams. The valley floor is dominated by mopane woodlands or dry, highly deciduous thickets known as Jesse.

World Heritage site since

1978 1979 1980 1981 1982 1983 **1984** 1985 1986 1987 1988 1989 1990 1991 1992 1993 1994 1995 1996 1997 1998 1999 2000 2001 2002 2003 2004 2005 2006 2007 2008

Statue of Liberty
USA

Criteria – Human creative genius; Heritage
associated with events of universal significance

The Statue of Liberty is a masterpiece of the
human creative spirit. Its construction in the
studios of Bartholdi in Paris represents one
of the greatest technical exploits of the
nineteenth century.

It welcomed immigrants at the entrance
to New York harbour, and so is directly
associated with an event of outstanding
historical significance: the populating of the
USA, the melting pot of disparate peoples in
the later nineteenth century. That the statue
was paid for by international subscription
and built in Europe by a French sculptor
only strengthens its symbolic significance.

Liberty was dedicated on 28 October 1886
and features a woman holding a book and a
torch. It stands on Liberty Island and was
designated a National Monument in 1924.
Ellis Island, the immigrants' former landing
place, became part of the Statue of Liberty
National Monument in 1965.

▲

This towering monument (46 m tall)
to liberty was a gift from France in
1886 to celebrate the centenary of
American independence – ten years
late. 'Liberty Enlightening the World',
symbolizing the ideals of Washington
and Lincoln, was extensively restored
for its centenary and the annual
celebration of American independence
on 4 July 1986.

World Heritage site since

1978 1979 1980 1981 1982 1983 **1984** 1985 1986 1987 1988 1989 1990 1991 1992 1993 1994 1995 1996 1997 1998 1999 2000 2001 2002 2003 2004 2005 2006 2007 2008

Iguazu National Park
Argentina

Criteria – Natural phenomena or beauty;
Significant natural habitat for biodiversity

Iguazu Falls.

One of the world's most spectacular waterfalls, the Iguazu Falls, lies at the heart of this vast, rich and diverse national park. The waterfall is semicircular, some 80 m high and 2,700 m in diameter and stands on a basaltic line that spans the border between Argentina and Brazil.

The site consists of the national park and national reserves in Misiones Province, northeastern Argentina. The Iguazu River forms the northern boundary of both the reserves and park, and also the southern boundary of Iguaçu National Park World Heritage site in Brazil.

The Iguazu Falls lie on the Argentina-Brazil border and are made up of many cascades that generate vast sprays of water and produce one of the most magnificent waterfalls in the world. The vegetation is mostly subtropical wet forest rich in lianas and epiphytes, although the forests have less species diversity when compared with others in Brazil and parts of Paraguay. Nonetheless, over 2,000 species of vascular plant have been identified.

Vegetation around the falls is particularly luxuriant due to the constant spray. The site is particularly rich in bird life, with almost half of Argentina's bird species found there. The fauna are typical of the region and include tapir, coatimundi, and tamandua.

Threatened mammals such as the jaguar, ocelot and tiger-cat number among the carnivores, and the giant anteater and Brazilian otter are also found. Primates include the black-capped capuchin and black howler monkey. There are also small populations of the endangered broad-nosed cayman and the threatened Brazilian merganser (sawbill duck).

The first inhabitants in the area were the Caingangues Indians. This tribe was dislodged by the Tupi-Guaranies who coined the name Iguazu (Big Water). The first European to reach the Iguazu falls was the Spanish explorer Don Alvar Nuñes Cabeza de Vaca in 1541; some ten years later Spanish and Portuguese colonization began. There are at least two sites of particular archaeological interest within the national park.

World Heritage site since

1978 1979 1980 1981 1982 1983 **1984** 1985 1986 1987 1988 1989 1990 1991 1992 1993 1994 1995 1996 1997 1998 1999 2000 2001 2002 2003 2004 2005 2006 2007 2008

Canadian Rocky Mountain Parks
Canada

Criteria – Natural phenomena or beauty; Major stages of Earth's history

Pacific Ocean — NORTH AMERICA — Atlantic Ocean

Athabasca Glacier.

Covering 325 km², the massive Columbia Ice Field spans the east-west continental divide and straddles the boundary between Jasper and Banff National Parks. The Columbia Ice Fields of Jasper National Park are regarded as the hydrographic apex of North America and are the headwaters to three major river systems: the North Saskatchewan, the Athabasca and the Columbia.

Studded with mountain peaks, glaciers, lakes, waterfalls, canyons and limestone caves, the Canadian Rocky Mountain Parks form a spectacular mountain landscape. The Burgess Shale fossil site, well known for its fossil remains of soft-bodied marine animals, is also found there.

The Canadian Rocky Mountain Parks comprise seven parks in total: the contiguous national parks of Banff, Jasper, Kootenay and Yoho, and the Mount Robson, Mount Assiniboine and Hamber provincial parks.

The Canadian Rocky Mountains are oriented in a southeastern to northwestern direction along the continental divide which separates the orientation of rivers that flow to the east and to the west. It consists of the Western Ranges, the Main Ranges, the Front Ranges and the Foothills, all of which are represented within the parks. Active glaciers and ice fields still exist throughout the region, particularly in the Main Ranges. The most significant is the Columbia Ice Field, the largest in North America's subarctic interior.

The Rockies have been divided into three life zones or ecoregions: montane, subalpine and alpine. Montane vegetation occurs in major valley bottoms and on the foothills and lower sun-exposed mountain slopes, especially in the Front Ranges. Montane wetlands and meadows occupy areas by major rivers. Forest is generally found between 1,200 m and 1,800 m and typical species include Douglas fir, white spruce, aspen and poplar. The subalpine ecoregion occupies mountainsides between 1,800 m and 2,100 m, and valley bottoms of high elevations and is the most extensive ecoregion in the Rockies. The alpine ecoregion occurs above the timberline and is characterized by diminutive and hardy vegetation such as low-growing willow and dwarf birch, heath and sedge.

Fifty-six mammal species have been recorded. Characteristic species of alpine meadows include Rocky Mountain goat, bighorn sheep, northern pika and hoary marmot. Forest mammals include moose, mule deer, white-tailed deer, caribou, elk and red squirrel, while carnivores include grey wolf, grizzly bear, black bear, wolverine, lynx and cougar. Some 280 bird species have been noted, including northern three-toed woodpecker, white-tailed ptarmigan, grey jay, mountain bluebird, Clark's nutcracker, golden eagle, mountain chickadee and rock pipit. Other recorded fauna includes toad, frog, salamander and snake.

Emerald Lake, Yoho National Park.

World Heritage site since

1978 1979 1980 1981 1982 1983 **1984** 1985 1986 1987 1988 1989 1990 1991 1992 1993 1994 1995 1996 1997 1998 1999 2000 2001 2002 2003 2004 2005 2006 2007 200

Historic Centre of Córdoba
Spain

Criteria – Human creative genius; Interchange of values; Testimony to cultural tradition; Significance in human history

EUROPE

Atlantic
Ocean

Mediterranean Sea

AFRICA

Mezquita, mosque-cathedral.

Córdoba's historic centre, clustering round the mosque-cathedral, the Mezquita, preserves much of its medieval urban fabric with narrow, winding streets. The city's Roman past can be seen in the sixteen-span bridge that crosses the river Guadalquivir, the fine mosaics in the Alcázar and sections of the Roman wall. The gardens of the Alcázar are good examples of Moorish Andalusian garden design. The remains of the monumental Caliphal Baths are nearby. La Judería, the old Jewish quarter, best preserves the medieval street pattern.

Córdoba has one of the finest historical heritages of any city in the world. It was the capital of the extensive Moorish Caliphate of Córdoba, and one of the largest cities in the world by the tenth century, with innumerable palaces, public buildings and mosques that rivalled the splendours of Constantinople, Damascus and Baghdad. After the city's recapture by the Christian Spanish, a series of new defensive structures was built. Córdoba's extent and plan, its historical significance as a living expression of the different cultures that existed there, and its relationship with its river, the Guadalquivir, is a historical ensemble of extraordinary value.

Córdoba was a flourishing Carthaginian township when the Romans captured it in 206 BC. Recognising its strategic and commercial importance, they made it the capital of Hispania Inferior, fortified it and adorned it with fine buildings.

The city fell to successive invasions: first to the Visigoths in AD 550, then to the Moors from North Africa in 716. In 756 the Caliph of Damascus set up court at Córdoba and laid the foundations for an illustrious period of the city's history. He began building the Great Mosque on the site of a church that was itself originally a Roman temple. Córdoba became the centre of a caliphate renowned for its artistic and intellectual predominance, but the kingdom collapsed after an eleventh-century civil war.

In 1236 Ferdinand III of Castile and Leon, captured the city. The mosque became the cathedral and new defensive structures were raised as befitted the status of a frontier town under constant threat from the Moors.

The Alcázar de los Reyes Cristianos dates from the early fourteenth century and was built as a royal residence. The Torre de la Calahorra formed part of a medieval fortress. Churches include San Jacinto (now the Palace of Congresses and Exhibitions); the Chapel of San Bartolomeo, Moorish in origin and now clearly Christian; and San Francisco and San Nicolás. There are important buildings from the sixteenth century: the Seminary of San Pelagio, Puerta del Puente, Casa Solariega de los Pàez de Castillo and Casa del Marqués de la Fuensanta del Valle, which illustrate the religious, military and architectural styles. From the eighteenth century come the civic buildings the Triunfos de San Rafael and Hospital del Cardenal Salazar.

World Heritage site since

1978 1979 1980 1981 1982 1983 **1984** 1985 1986 1987 1988 1989 1990 1991 1992 1993 1994 1995 1996 1997 1998 1999 2000 2001 2002 2003 2004 2005 2006 2007 2008

Burgos Cathedral
Spain

Criteria – Interchange of values; Significance in human history; Heritage associated with events of universal significance

Begun in 1221 at the same time as the great cathedrals of the Île-de-France and completed in 1567, Santa María de Burgos is a striking summary of the evolution of Gothic architecture. The plan of the cathedral is based on a Latin cross of pleasing proportions. Initial work on the cathedral was completed in 1293. After a hiatus of nearly 200 years, work was resumed in the mid-fifteenth century and continued for more than 100 years. These were embellishments of a profuse splendour – including paintings, choir stalls, reredos, tombs, cupolas (pictured far right), and stained-glass windows – which have ensured the status of this magnificent building. The two-storeyed cloister that was completed towards 1280 still fits within the framework of 'French' High Gothic.

The three-storey elevation, the vaulting, and the tracery of the windows are closely related to contemporary models of the north of France. The portals of the transept (the Puerta del Sarmental to the south and the Puerta de la Coronería to the north) may also be compared with the great sculpted ensembles of the French royal domain.

World Heritage site since

1978 1979 1980 1981 1982 1983 **1984** 1985 1986 1987 1988 1989 1990 1991 1992 1993 1994 1995 1996 1997 1998 1999 2000 2001 2002 2003 2004 2005 2006 2007 2008

Alhambra, Generalife and Albayzín, Granada
Spain

Criteria – Human creative genius; Testimony to cultural tradition; Significance in human history

A view across the basin in the Patio de los Arrayanes (Myrtle Courtyard) looking towards the Hall of the Ambassadors, Alhambra Palace.

Unique artistic creations set on a hill above Granada, the Alhambra palace and fortress and the Generalife residence and gardens bear exceptional testimony to Muslim Spain and are exceptional examples of royal Arab residences of the medieval period. Standing on an adjacent hill is the residential district of the Albayzín, a rich repository of Moorish vernacular architecture into which traditional Andalusian architecture blends harmoniously. It survives as a remarkable example of a Spanish-Moorish town. During the early period of Moorish rule in Spain, power initially rested with the Caliphate of Córdoba until civil war ended its dominance in 1031. From then on, the Emirate of Granada grew steadily in importance and prosperity but the town did not become an important centre of Muslim Spain until 1238, when Muhammad ibn al Ahmar founded the present Alhambra.

The palace was completed in the fourteenth century. Organized around two rectangular courts, the Patio de los Arrayanes and the Patio de los Leones, it has many, highly-decorated rooms with marble columns, stalactite cupolas, stucco work, azulejos (ceramic tiles), precious wood, and paintings on leather, all competing with the delicacy of the natural decor: water, still and flowing in immense basins, narrow canals

and fountains, including the spectacular fountain of the Patio de los Leones (Court of Lions).

The Alhambra (in Arabic 'the Red') incorporates palaces, guard rooms, patios and gardens as well as workshops, shops, baths and mosque (independently of the church of Santa María built in the sixteenth century on the site of the royal mosque). It is enclosed by a massive fortified wall with towers.

At a short distance to the east of the Alhambra, the enchantment is extended to the gardens of the Generalife, rural residence of the emirs who ruled this part of Spain in the thirteenth and fourteenth centuries. Here the relationship between the architectural and the natural has been reversed: gardens and water predominate over the pavilions, summerhouses and living quarters. The massive boxwood trees, rose, carnation and gillyflower bushes, and shrubs ranging from willow to cypress, comprise an absolute masterpiece of the art of horticulture by restoring the Koranic image of paradise to the believers.

Much of the Albayzín's significance lies in its medieval town plan, with narrow streets and small squares and the relatively modest Moorish and Andalusian houses that line them.

After the Reconquista, the Christian reconquest of Spain in 1492, the influx of a substantial Christian population into Granada brought new developments. Late Gothic or early Plateresque churches and monasteries harmonized with the existing architecture. However, the Albayzín still bears witness to its medieval Moorish origins as its urban fabric, architecture and main characteristics adapted successfully to new ways of living.

World Heritage site since

1978 1979 1980 1981 1982 1983 **1984** 1985 1986 1987 1988 1989 1990 1991 1992 1993 1994 1995 1996 1997 1998 1999 2000 2001 2002 2003 2004 2005 2006 2007 2008

Group of Monuments at Mahabalipuram
India

Criteria – Human creative genius; Interchange of values; Testimony to cultural tradition; Heritage associated with events of universal significance

These exceptional sanctuaries on the Coromandel coast, founded by the Pallava kings, were carved out of rock in the seventh and eighth centuries. The monuments may be subdivided into five categories: ratha temples in the form of processional chariots, cut into the rocks which emerge from the sand; mandapa, or rock sanctuaries, modelled as rooms covered with bas-reliefs; rock reliefs in the open air that illustrate a popular episode in the iconography of Siva; temples built from cut stone, like the Temple of Rivage, with its high-stepped pyramidal tower and thousands of sculptures dedicated to the glory of Siva; and monolithic rathas, of one to three storeys, in a variety of architectural forms. In addition the Shore Temple represents the peak of Pallava architecture, although it has been eroded by seawater and air and the sculptures have become indistinct.

The rock reliefs tell the story of the Descent of the Ganges. Begged by King Baghirata, Siva ordered the river Ganges to descend to Earth and to nourish the world. The sculptors used the natural fissure in the rock to suggest this cosmic event to which a swarming crowd of gods, goddesses, mythical beings, wild animals and domestic animals bear witness.

▼

World Heritage site since

1978 1979 1980 1981 1982 1983 **1984** 1985 1986 1987 1988 1989 1990 1991 1992 1993 1994 1995 1996 1997 1998 1999 2000 2001 2002 2003 2004 2005 2006 2007 2008

Anjar
Lebanon

Criteria – Testimony to cultural tradition;
Significance in human history

The ruins of Anjar reveal a very regular layout, reminiscent of the palace-cities of ancient times, and are a unique testimony to city planning under the Umayyads. It was discovered when archaeological explorations began in 1949 in the Beqaa. On a site that had long been occupied, the city of Anjar was founded at the beginning of the eighth century by Caliph Walid I (705–715). Re-used parts of earlier Greek,

Roman and early Christian buildings are frequently found in the masonry of its walls. It flourished for only 20–30 years before the Abbasids overran the city and it fell into disuse. At its peak, it housed more than 600 shops, Roman-style baths, two palaces and a mosque. It takes its name from the Arabic term ayn al-jaar (water from the rock), referring to the streams that flow from the nearby mountains.

Public and private buildings are laid out according to a strict plan: the principal palace and mosque in the southeast quadrant; the secondary palace and baths in the northeast and northwest quadrants; and the densely inhabited southwest quadrant, criss-crossed by a network of streets.

World Heritage site since

1978 1979 1980 1981 1982 1983 1984 **1985** 1986 1987 1988 1989 1990 1991 1992 1993 1994 1995 1996 1997 1998 1999 2000 2001 2002 2003 2004 2005 2006 2007 2008

Historic District of Old Québec
Canada

Criteria – Significance in human history; Heritage associated with events of universal significance

Château Frontenac grand hotel overlooks Québec's Lower City.

The city is an early example of urban heritage conservation as a result of action in the 1870s by the Governor General Lord Dufferin, who took a stand against the demolition of the fortifications which, from a strategic standpoint, had become useless. He simply had new gates to the city cut into them. From the beginning of the twentieth century, long before its listing as a historic monument in 1957, the fortified walls of Québec were maintained by Canadian government funds.

Québec is one of the finest examples of a fortified colonial city. It is the only North American city to have preserved its ramparts, together with the numerous bastions, gates and defensive works that surround the historic district.

The Upper City, built on a cliff and defended by walls with bastions, has remained the religious and administrative centre, with its churches, convents and other monuments like the Dauphine Redoubt, the Citadel and Château Frontenac (which can be seen at the top of the photo on the right). Together with the Lower City and its harbour and old quarters, it forms a coherent urban ensemble which is by far the most complete fortified colonial town in North America.

Québec illustrates one of the major stages in the population and growth of the Americas during the modern and contemporary period. When Samuel de Champlain founded Québec, the capital of New France in 1608, he chose the natural site of a steep plateau overlooking the St Laurent River. The old heart of the city was established on this promontory, Cap-aux-Diamants, which is protected by Fort St Louis.

Québec had an urban organization early on and a zoning system which stemmed from its various functions as a town, a fortified city and a harbour for trade from the North and Europe. Its cliff divided the city into two districts: business and naval district in the Lower City, and the administrative and religious centre in the Upper City.

The construction of a citadel at the far southeast end of Cap-aux-Diamants by the engineer Elias Durnford from 1819–31 and the expansion of the system of fortifications to cover the city's entire perimeter were in keeping with the original spatial organization of the city and gave Québec its current topographical features.

The oldest quarters are located in the Lower City around the Place Royale which, along with the Rue Notre Dame, is lined with old seventeenth- and eighteenth-century houses. In the Upper City the seventeenth-century convents and seminary still have some original elements. Of 700 old civil or religious buildings remaining, 2 per cent date to the seventeenth century, 9 per cent to the eighteenth and 43 per cent to the first half of the nineteenth century. The city took on its present aspect under the influence of the Baillairgés, a dynasty of architects who, for generations, imposed an interpretation of neoclassical style.

World Heritage site since

1978 1979 1980 1981 1982 1983 1984 **1985** 1986 1987 1988 1989 1990 1991 1992 1993 1994 1995 1996 1997 1998 1999 2000 2001 2002 2003 2004 2005 2006 2007 2008

Historic Centre of Salvador de Bahia
Brazil

Criteria – Significance in human history; Heritage associated with events of universal significance

Salvador de Bahia, with Ouro Preto, is the colonial city par excellence in the Brazilian northeast. An eminent example of Renaissance town planning, the city has many outstanding monuments and Renaissance buildings. It was the first slave market in the New World and became a major point of convergence of European, African and American Indian cultures from the sixteenth to the eighteenth century.

Founded by the Portuguese, the city was the first colonial capital of Brazil from 1549 until the administration was transferred to Rio de Janeiro in 1763.

Commercial activity revolved around the port while the upper city, in the area of Bahia de Todos los Santos, was hilly and picturesque, with residential and administrative buildings. Its situation on a ridge parallel to the Atlantic coast also made it defensible against attack in the sixteenth and seventeenth centuries.

In addition to a number of major buildings of the seventeenth and eighteenth centuries, the historic centre of Salvador retains a host of sixteenth-century open spaces and Baroque palaces. There are streets characteristic of the colonial city, lined with bright, multicoloured houses, often decorated with high-quality stucco.

▼

World Heritage site since

1978 1979 1980 1981 1982 1983 1984 **1985** 1986 1987 1988 1989 1990 1991 1992 1993 1994 1995 1996 1997 1998 1999 2000 2001 2002 2003 2004 2005 2006 2007 2008

Old Town of Segovia and its Aqueduct
Spain

Criteria – Human creative genius; Testimony to cultural tradition; Significance in human history

The Roman Aqueduct of Segovia (pictured below), the symbol of the city, is a magnificent, double-tiered construction. Probably built around AD 50, it is the best known of Spain's remaining Roman aqueducts owing to its monumentality, its excellent state of conservation and its location in one of the most beautiful historic cities in the world.

The Roman hydraulic engineers brought the waters of the Río Frío in the Sierra de Guadarrama 18 km to Segovia via a canal with an average gradient of 1 per cent. The biggest natural obstacle they encountered was the valley of the Río Clamores at the end of the route. In order to reach the city they built an enormous stone aqueduct 813 m long, with four straight segments and two superimposed arcades borne by 128 pillars.

At the valley's lowest point, the aqueduct stands 28.5 m above the ground.

Other important monuments in the old city include the Alcázar, the castle begun around the eleventh century, and the stunning sixteenth-century Gothic cathedral, one of the last in Europe of the style.

World Heritage site since

1978 1979 1980 1981 1982 1983 1984 **1985** 1986 1987 1988 1989 1990 1991 1992 1993 1994 1995 1996 1997 1998 1999 2000 2001 2002 2003 2004 2005 2006 2007 200

Chavín (Archaeological site)
Peru

Criteria – Testimony to cultural tradition

Chavín's location in a high valley of the Peruvian Andes at an altitude of 3,177 m didn't stop its architects and artists producing some remarkable structures These include immense sculpted megaliths, one of which is more than 4 m in height.

The archaeological site of Chavín gave its name to the culture that developed between 1500 and 300 BC in this high valley of the Peruvian Andes. This former place of worship is one of the earliest and best-known pre-Columbian sites. Its appearance is striking, with the complex of terraces and squares, surrounded by structures of dressed stone, and the mainly zoomorphic ornamentation.

◄

A feline head, one of the zoomorphic carvings at Chavín.

Painted Churches in the
Troodos Region
Cyprus

Criteria – Interchange of values; Testimony to cultural tradition; Significance in human history

The Church of the Transfiguration of the Saviour (Ayia Sotira) in Palaichori has a steep-pitched wooden roof with flat hooked tiles. This type of roofing over a Byzantine church is not found elsewhere, making the wooden-roofed churches of Cyprus a unique group example of religious architecture.

In the Troodos Mountains, in the heart of Cyprus, can be found one of the greatest concentrations of churches and monasteries of the Byzantine Empire, by which the island was annexed during the conquest of 965. The complex of ten monuments included on the World Heritage List, all richly decorated with murals, provides an overview of Byzantine and post-Byzantine painting in Cyprus. They range from small churches, whose rural architectural style is in stark contrast to their highly refined decoration, to monasteries such as that of St John Lampadistis. Among the most significant cycles is that of Panagia Phorbiotissa of Nikitari, which was painted in 1105–6, and that of Panagia tou Arakou in Lagoudera, which was executed during the last six months of 1192.

World Heritage site since

1978 1979 1980 1981 1982 1983 1984 **1985** 1986 1987 1988 1989 1990 1991 1992 1993 1994 1995 1996 1997 1998 1999 2000 2001 2002 2003 2004 2005 2006 2007 2008

Santiago de Compostela (Old Town)
Spain

Criteria – Human creative genius; Interchange of values; Heritage associated with events of universal significance

The Old Town of Santiago de Compostela, with Romanesque, Gothic and Baroque buildings, is one of the world's most beautiful urban areas and one of the most famous pilgrimage sites in the Christian world. It became a symbol in the Spanish Christians' struggle against the Islamic Moors after its destruction by the Muslims at the end of the tenth century.

Santiago conserves a valuable historic centre worthy of one of Christianity's greatest holy cities. The oldest monuments are grouped around the tomb of St James and the cathedral, which is a masterpiece of Romanesque art. Civil and religious architectural elements from the Middle Ages and the Renaissance are also integrated into a high-quality urban fabric.

During the Romanesque and Baroque periods Santiago exerted a decisive influence on the development of architecture and art across the northern Iberian peninsula.

Santiago de Compostela is associated with pilgrimage, one of the major themes of the medieval centuries. From the shores of the North Sea and the Baltic, thousands carrying the pilgrims' scallop shell and staff walked to the Galician sanctuary along the paths of Santiago (St James) to the tomb of St James the Great.

The Cathedral at Santiago de Compostela.
▼

World Heritage site since

1978 1979 1980 1981 1982 1983 1984 **1985** 1986 1987 1988 1989 1990 1991 1992 1993 1994 1995 1996 1997 1998 1999 2000 2001 2002 2003 2004 2005 2006 2007 20C

Petra
Jordan

Criteria – Human creative genius; Testimony to cultural tradition; Significance in human history

Petra is one of the world's most famous and spectacular archaeological sites, where ancient Eastern traditions blend with Hellenistic architecture. Situated between the Red Sea and the Dead Sea and half-built, half-carved in the rock, Petra was an important crossroads between Arabia, Egypt and Syria-Phoenicia.

Petra was first established around the sixth century BC by the Nabataean Arabs, a Semitic people who laid the foundations of a commercial empire that extended into Syria. In AD 106 the Roman Emperor Trajan annexed the Nabataean Kingdom as part of the province of Arabia. The many earthquakes that hit Petra triggered a slow decline for the city, which was not halted by its designation as an archepiscopal see. Muslim Arabs conquered the city in 636 but it remained distant from the pilgrim road to Mecca. The Crusaders constructed a fort there in the twelfth century and Petra returned to its ancient splendour, but soon they withdrew, leaving it to the local people until the early nineteenth century, when it was visited by the Swiss explorer Burckhardt.

Petra lies south of modern Amman on the edge of the mountainous desert of Wadi Araba, surrounded by towering hills of sandstone which gave the city some natural protection against invaders. It was for centuries the meeting point of the main routes used by camel caravans transporting spices between the Mediterranean and the Near East, Africa and India.

The Khazneh el Faroun, or the Treasury of the Pharaoh, is one of Petra's most famous sights: an imposing façade some 40 m tall, carved into the mountain rock like a half-finished sculpture. The Khazneh is the only rock-cut building in Petra with no Nabataean element and links exclusively with the Alexandrian world and Hellenistic artistic traditions. Behind the impressive façade, a large square room has been carved out of the rock of the cliff. Typically for the tombs in Petra, its interior is as plain as its exterior is intricate.

North from the Khazneh lies the massif of Jebel Khubtha. Three large structures (Royal Tombs) are carved into the rock face, which is known as the King's Wall: the Urn Tomb, the Corinthian Tomb and the Palace Tomb (Silk Tomb), named for the extraordinary chromatic effect of the rock.

The main entrance to Petra is through the Siq, a gorge formed by the Musa, which the Nabataeans blocked with a dam and channelled to carry drinking water to the city. The Siq narrows to little more than 5 m in width, twisting and turning through sheer walls towering hundreds of metres on either side. The Khazneh, or Treasury, stands at the end of the gorge.

Petra has notable Roman relics, including its first century AD theatre: carved almost entirely in the rock, it could hold more than 8,000 spectators.

World Heritage site since

1978 1979 1980 1981 1982 1983 1984 **1985** 1986 1987 1988 1989 1990 1991 1992 1993 1994 1995 1996 1997 1998 1999 2000 2001 2002 2003 2004 2005 2006 2007 200

Huascarán National Park
Peru

Criteria – Natural phenomena or beauty; Major stages of Earth's history

Huascarán National Park is located in the Cordillera Blanca Range in the Peruvian Andes. The park covers most of the Cordillera Blanca, the highest tropical mountain range in the world. It has twenty-seven snow-capped peaks 6,000 m above sea level, of which El Huascarán (6,768 m) is the highest. The deep ravines watered by numerous torrents, the glacial lakes and the variety of the vegetation make it a site of spectacular beauty. It is the home of such species as the spectacled bear and the Andean condor. Puma, mountain cat, white-tailed deer and vicuna are other important indigenous species, but all have been heavily hunted in the past. The national park is uninhabited, although there is some grazing in the lowlands by native livestock (llama and alpaca) under an agreement with the local people.

The Cordillera region has for centuries been a site for the settlement of ethnic groups, as witnessed by ruins at Gekosh and Chuchumpunta and elsewhere. These represent the largest collection of such remains in the world.

The peak of El Huascarán.
▼

World Heritage site since

1978 1979 1980 1981 1982 1983 1984 **1985** 1986 1987 1988 1989 1990 1991 1992 1993 1994 1995 1996 1997 1998 1999 2000 2001 2002 2003 2004 2005 2006 2007 2008

Hatra
Iraq

Criteria – Interchange of values; Testimony to cultural tradition; Significance in human history; Heritage associated with events of universal significance

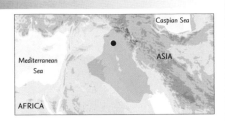

Hatra was a large fortified city under the influence of the Parthian Empire and capital of the first Arab Kingdom. It flourished as a major staging-post on the famous oriental Silk Road to become another of the great Arab cities like Palmyra in Syria, Petra in Jordan, and Baalbek in Lebanon. This Eastern monarchy was a source of concern for the Romans who sought unsuccessfully to destroy it. Hatra withstood invasions by

the Romans in AD 116 and 198 thanks to its high, thick walls reinforced by towers. The remains of the city, especially the temples where Hellenistic and Roman architecture blend with Eastern decorative features, attest to the greatness of an entire facet of Assyro-Babylonian civilisation influenced by Greeks, Parthians, Romans and Arabs.

Hatra is a circular city, almost 2 km in diameter, with four fortified gates set in immense double walls. The external wall is an earthen bank; a wide ditch separates this from a stone wall. The perfect condition of the double wall in an untouched environment sets it aside as an outstanding example of this type of fortification.

Manas Wildlife Sanctuary
India

Criteria – Natural phenomena or beauty; Significant ecological and biological processes; Significant natural habitat for biodiversity

On a gentle slope in the foothills of the Himalaya, where wooded hills give way to alluvial grasslands and tropical forests, the Manas sanctuary is home to a great variety of wildlife. Named after the Goddess Manasa, the site is noted for its spectacular scenery, with a variety of habitat types that support a diverse fauna, making it the richest of all Indian wildlife areas. The park represents the core of an extensive tiger reserve that protects an important

migratory wildlife resource along the West Bengal to Arunachal Pradesh and Bhutan borders. Its wetlands are of international importance. It is also the single most important site for the survival of pygmy hog, hispid hare and golden langur.

Over 450 species of bird have been recorded, including the threatened Bengal florican, great pied hornbill, wreathed hornbill and other hornbills. Uncommon waterfowl species include spot-billed pelican, lesser adjutant and greater adjutant.

Reptiles include the Assam roofed turtle, a variety of snakes, gharial and monitor lizard.

World Heritage site since

1978 1979 1980 1981 1982 1983 1984 **1985** 1986 1987 1988 1989 1990 1991 1992 1993 1994 1995 1996 1997 1998 1999 2000 2001 2002 2003 2004 2005 2006 2007 2008

Keoladeo National Park
India

Criteria – Significant natural habitat for biodiversity

This former duck-hunting reserve of the Maharajas is one of the major wintering areas for large numbers of aquatic birds from Afghanistan, Turkmenistan, China and Siberia. Some 364 species of bird, including the rare Siberian crane, have been recorded in the park, giving it a unique assemblage of species. The park's location in the Gangetic Plain makes it an unrivalled breeding site for herons, storks and cormorants and an important wintering ground for large numbers of migrant ducks. An estimated sixty-five million fish fry are carried into the park's water impoundments by river flooding every year during the monsoon season, providing the food base for large numbers of wading and fish-eating birds. There are also many birds of prey including the osprey, peregrine and various species of eagle.

The area consists of a flat patchwork of marshes, artificially created in the 1850s and maintained ever since by a system of canals, sluices and dykes. Large predators are absent, leopard having been deliberately exterminated by 1964, but many species of small carnivores inhabit the park along with primates such as rhesus macaque.

World Heritage site since

1978 1979 1980 1981 1982 1983 1984 **1985** 1986 1987 1988 1989 1990 1991 1992 1993 1994 1995 1996 1997 1998 1999 2000 2001 2002 2003 2004 2005 2006 2007 2008

Monuments of Oviedo and the Kingdom of the Asturias

Spain

Criteria – Human creative genius; Interchange of values; Significance in human history

In the ninth century the flame of Christianity was kept alive in the Iberian peninsula in the tiny Kingdom of the Asturias. Here an innovative pre-Romanesque architectural style was created that was to play a significant role in the development of the religious architecture of the peninsula. Its highest achievements can be seen in the churches of Santa María del Naranco, San Miguel de Lillo, Santa Cristina de Lena, the Cámara Santa and San Julián de los Prados, in and around the ancient capital city of Oviedo. These churches, which are basilical in layout and entirely vaulted, and which make use of columns instead of piers, have very rich decors combining influences from a variety of sources. Associated with them is the remarkable contemporary hydraulic engineering structure known as La Foncalada.

Santa María del Naranco is a former royal residence built on two levels. Excavations in 1930–4 revealed the existence of baths in one of the lower rooms. This rectangular palace was converted into a church between 905 and 1065.

▼

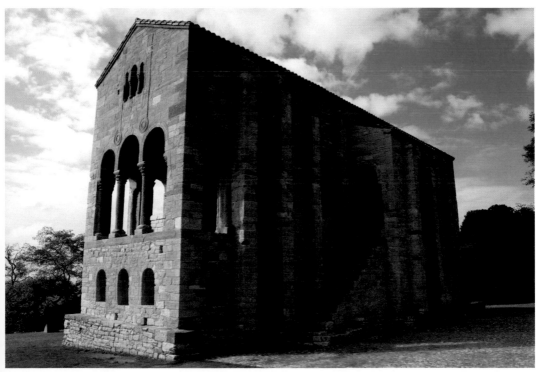

World Heritage site since

1978 1979 1980 1981 1982 1983 1984 **1985** 1986 1987 1988 1989 1990 1991 1992 1993 1994 1995 1996 1997 1998 1999 2000 2001 2002 2003 2004 2005 2006 2007 2008

Medina of Marrakesh
Morocco

Criteria – Human creative genius; Interchange of values; Significance in human history; Traditional human settlement

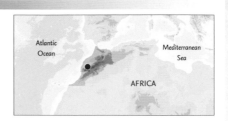

The minaret of the Koutoubia Mosque.

The legacy of Marrakesh's Saadian rulers can be seen in the ruins of the El Badi Palace and especially in the magnificent Saadian tombs. Elaborately decorated with delicate carvings, coloured tiles and Arabic script, these tombs of the city's rulers and wealthy were sealed by the Alawites and remained closed until 1917. Materials for their building came from great distances, such as the fine marble from Carrara which the writer Montaigne observed being cut in Tuscany 'for the king of Morocco in Berberia'.

Marrakesh, which gave its name to the Moroccan Empire, is the textbook example of a large Islamic capital in the Western world. With its maze of narrow streets, houses, souks (markets), traditional crafts and trade activities, and its medina, this ancient settlement is an outstanding example of a vibrant historic city.

Founded in 1070–2 by the Almoravids, Marrakesh became the true capital of these conquering nomads and a political, economic and cultural centre that was influential throughout the western Muslim world from North Africa to Andalusia. For the next two centuries, Marrakesh played a decisive role in the development of urban planning.

In Marrakesh, as in other North African cities, the medina is the oldest, walled part of the city. The original layout of Marrakesh's medina dates back to the Almoravid period from which there still remain various monumental vestiges. The medina walls were built in 1126–7 and it is believed the Almoravids planted the palm groves that still grow there, now covering a 1.3 km² area east of the city.

The Berber Almohads took Marrakesh in 1147, destroying many of their predecessors' buildings. The city nevertheless remained the capital and for two centuries under its Almohad rulers, Marrakesh experienced new and unprecedented prosperity.

The Almohads built the Koutoubia Mosque on Almoravid foundations. Its incomparable minaret (pictured on the right), a key monument of Muslim architecture, is one of the major features of the cityscape and the symbol of the city. They built new quarters, extended the city wall, fortified the Kasbah (fortress) which was an extension of the city to the south with its own ramparts and gates (Bab Agnaou, Bab Robb), its mosque, palace, market, hospital, parade ground and gardens. These leaders strengthened their control over their domains by planting crops and by civil engineering achievements such as the Tensift Bridge.

Although Marrakesh went into decline after the Almohads lost power in 1269, notable buildings, such as the Ben Salih Mosque and minaret, were still built. The rebirth of the capital under the Saadian rulers from 1510–1669 brought a new blossoming of the arts, and from the seventeenth century onwards Marrakesh's Alawite rulers also added to the town fabric, building a new mosque, madrasas, palaces and residences, all harmoniously integrated into the old town.

World Heritage site since

1978 1979 1980 1981 1982 1983 1984 **1985** 1986 1987 1988 1989 1990 1991 1992 1993 1994 1995 1996 1997 1998 1999 2000 2001 2002 2003 2004 2005 2006 2007 2008

Cave of Altamira and Paleolithic Cave Art of Northern Spain
Spain

Criteria – Human creative genius; Testimony to cultural tradition

Altamira, the 'Sistine Chapel of Prehistory', is a unique artistic achievement and bears exceptional testimony to the Magdalenian cultures of southern Europe around 15,500 years ago. Discovered by chance in 1869, the Cave of Altamira revealed the existence of two very rich archaeological levels. It was inhabited as early as the Aurignacian, the date of the first figurative representations depicted on its walls, around 18,500 years ago. It was at the start of the Magdalenian period that the largest room in the cavern was decorated. Under its vault there are

superb paintings of bison, horses, deer and boars using a palette of colours consisting of only a small number of shades (ochre, reds and blacks). In 2008, seventeen decorated caves, representing the apogee of Paleolithic cave art that developed across Europe, from the Urals to the Iberian Peninsula, from 35,000 to 11,000 BC, were inscribed as an extension to the Altamira Cave site. With deep galleries, and isolated from external climatic influences, these caves are particularly well preserved.

The large animal images (the doe is 2.2 m long) are striking in their naturalism, and their specific features are scrupulously reproduced. They are also outstanding in the variety of fur and mane textures, which are extremely well rendered, and the variety of poses which make masterly use of the cave's surface and crevices that provide surprising trompe-l'œil effects.

Punic Town of Kerkuane and its Necropolis
Tunisia

Criteria – Testimony to cultural tradition

This Phoenician city was active for over 400 years and then abandoned during the First Punic War c. 250 BC. It was not rebuilt by the Romans and its remains constitute the only example of a Phoenicio-Punic city to have survived, offering a unique snapshot of Phoenician city life in the third century BC. Its port, ramparts, residential districts, shops, workshops, streets, squares, temples and necropolis remain as they were when the city was abandoned. The city was carefully

laid out, with advanced use of hydraulics and high standards of hygiene. Houses were built to a standard plan: a single entrance and a corridor gave access to an interior courtyard with a well, a washbasin and a bath; around the courtyard were the reception rooms.

The necropolis of Arg el Ghazouani, located on a rocky hill less than 1 km northwest of the city, is the best conserved portion of the great Kerkuane necropolis whose tombs are scattered throughout the coastal hills at the extreme end of Cap Bon. In the protected area there are approximately 200 tombs, including 50 that have not been excavated.

World Heritage site since

1978 1979 1980 1981 1982 1983 1984 **1985** 1986 1987 1988 1989 1990 1991 1992 1993 1994 1995 1996 1997 1998 1999 2000 2001 2002 2003 2004 2005 2006 2007 2008

Quseir Amra
Jordan

Criteria – Human creative genius; Testimony to
cultural tradition; Significance in human history

Built in the early eighth century, this
exceptionally well-preserved desert castle
was both a fortress with a garrison and a
residence of the Umayyad caliphs. The most
outstanding features of this small pleasure-
palace are the reception hall and the
hammam (steam bath), whose walls and
vaults are both richly decorated with
figurative murals that reflect the secular art
of the time. Quseir Amra was probably built

under Walid I (705–15), although a more
recent theory suggests the reign of Walid II
(743–4). Of great interest is the remarkable
architectural structure of the reception hall
and also the existence of a very extensive
bath complex, fed by an aqueduct, that
resembles Roman baths with its three
rooms: the changing-room (apodyterium), the
warm bath (tepidarium) and the hot bath
(caldarium), in addition to the service room.

The murals consist of
historical themes (royal
figures who were
defeated by the
Umayyad caliphs),
mythological
representations (the
muses of Poetry,
Philosophy and
History, with their
names in Greek), a
zodiac, hunting scenes
and hammam scenes as
well as some imaginary
themes (such as animal
musicians, and a
hunter being chased by
a lion).

World Heritage site since

1978 1979 1980 1981 1982 1983 1984 **1985** 1986 1987 1988 1989 1990 1991 1992 1993 1994 1995 1996 1997 1998 1999 2000 2001 2002 2003 2004 2005 2006 2007 2008

Ruins of the Buddhist Vihara at Paharpur
Bangladesh

Criteria – Human creative genius; Interchange of values; Heritage associated with events of universal significance

ASIA

Bay of Bengal

Enclosed by thick walls up to 5 m high, this remarkable monastery is quadrangular, and has a colossal temple with a cross-shaped floor plan in the centre of the courtyard and with an elaborate gateway complex on the north.

Somapura Mahavira (Great Monastery) was a renowned intellectual centre until the twelfth century, epitomising the rise of Mahayana Buddhism in Bengal from the seventh century onwards. With its simple, harmonious lines and its profusion of carved decoration, this monastery-city represents a unique artistic as well as religious achievement. The monastery had cells for 177 monks, ranged around its central quadrangle. Its layout was perfectly adapted to its function and it influenced Buddhist architecture as far away as Cambodia. The wide range of excavated finds include terracotta plaques, images of different gods and goddesses, pottery, coins, inscriptions and ornamental bricks. The site is the most important and largest known monastery south of the Himalaya to have been excavated.

World Heritage site since

.978 1979 1980 1981 1982 1983 1984 **1985** 1986 1987 1988 1989 1990 1991 1992 1993 1994 1995 1996 1997 1998 1999 2000 2001 2002 2003 2004 2005 2006 2007 2008

Sanctuary of Bom Jesus do Congonhas
Brazil

Criteria – Human creative genius; Significance in human history

In the eighteenth century, the region of Minas Gerais in Brazil was in its heyday – there were more than 30,000 gold prospectors in 1712. Moreover, the devotion of these pioneers was responsible for a remarkable blossoming of religious art, full of Baroque touches and influenced by Rococo invention.

The wish of a Portuguese immigrant who had been miraculously cured of a crippling infirmity was the impetus for the construction of one of Christian art's most amazing groupings of monuments. Buried in the still luxuriant Brazilian highlands, the Sanctuary was completed in the 1770s after little more than sixty years. The Church of Bom Jesus is a simple construction in the tradition of the area. However, after the death of its founder, Feliciano Mendes, in 1765 it was given a sumptuous Rococo interior. Externally, the flight of steps, begun in 1770, was later decorated with twelve soapstone statues of the Prophets by Antônio Francisco Lisboa (Aleijadinho). In seven small chapels in the grounds are housed 'The Passos', seven Stations of the Cross, sculpted in wood by Aleijadinho. Christian art in Latin America reached its unquestioned zenith with these multicoloured groups.

World Heritage site since

1978 1979 1980 1981 1982 1983 1984 **1985** 1986 1987 1988 1989 1990 1991 1992 1993 1994 1995 1996 1997 1998 1999 2000 2001 2002 2003 2004 2005 2006 2007 2008

Kaziranga National Park
India

Criteria – Significant ecological and biological processes; Significant natural habitat for biodiversity

ASIA

Indian Ocean

In the heart of Assam, this park is one of the last areas in eastern India undisturbed by human activity. It is inhabited by the world's largest populations of one-horned rhinoceroses and Indian elephants (pictured below) as well as many other mammals, including tigers, panthers and bears, and thousands of birds. The site is in the flood plains on the southern bank of the Brahmaputra River at the foot of the Mikir Hills. The riverine habitat consists primarily of tall, dense grasslands interspersed with open forests, streams and numerous small lakes (bheels). Over three-quarters of the area is submerged annually by the flood waters of the Brahmaputra. The park contains about fifteen species of India's threatened mammals. The numerous areas of water are rich reservoirs of food (including fish) and thousands of migratory birds, representing over 100 species, visit the park seasonally.

Other mammals include the capped langur, a small population of hoolock gibbon, Ganges dolphin, otter, swamp deer, hog deer and Indian muntjac. Birds of interest include grey pelican, black-necked stork, Pallas's fish eagle, Bengal florican, swamp partridge, grey peacock-pheasant, great pied hornbill and green imperial pigeon. The reptilian fauna include the water monitor, Indian python, common cobra and king cobra.

World Heritage site since

1978 1979 1980 1981 1982 1983 1984 **1985** 1986 1987 1988 1989 1990 1991 1992 1993 1994 1995 1996 1997 1998 1999 2000 2001 2002 2003 2004 2005 2006 2007 2008

Great Mosque and Hospital of Divriği
Turkey

Criteria – Human creative genius; Significance in human history

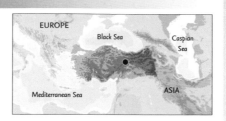

The Divriği mosque was founded in 1228-9 by Emir Ahmet Shah, along with a marestan (hospital for the insane) endowed by his wife. These two complementary monuments were built simultaneously by the same architect. Divriği is an outstanding example of Selçuk mosques in Anatolia, having neither a courtyard, colonnades, nor an uncovered ablutions basin, but which (due to the harshness of the climate) organises all religious functions in an enclosed area. The mosque has a single prayer room and is crowned by two cupolas. The highly sophisticated technique of vault construction, and a creative, exuberant type of decorative sculpture – particularly on the three doorways, in contrast to the unadorned walls of the interior – are the unique features of this masterpiece of Islamic architecture.

The Divriği ensemble presents a gripping contrast between the low, blind walls of its rectangular enclosure and the three immense gates which afford access to the hospital at the west and to the mosque at the north and west.

St Mary's Cathedral and St Michael's Church at Hildesheim
Germany

Criteria – Human creative genius; Interchange of values; Testimony to cultural tradition

The ancient Benedictine abbey church of St Michael, built between 1010 and 1022 by Bernward, Bishop of Hildesheim, is one of the key monuments of mediaeval art. It has a symmetrical plan with two apses that was characteristic of Ottonian Romanesque art in Old Saxony. In the nave, square impost pillars alternate in an original rhythm with cubic-capital columns to create a distinctive type of elevation. St Mary's Cathedral, rebuilt after the fire of 1046, still retains its original crypt, and was modelled after St Michael's. It has a similar nave arrangement, but its proportions are more slender. Notable treasures of the churches include the corona of light of Bishop Hezilon and the baptismal fonts of gold-plated bronze of Bishop Conrad.

The Church of St Michael and St Mary's Cathedral have an exemplary selection of interior decorative elements which beautifully express the spirit of Romanesque architecture. St Michael's famous bronze doors (c. 1015) retrace the events from the book of Genesis and the life of Christ, and the Bernward bronze column (c. 1020) depicts scenes from the New Testament.

World Heritage site since

1978 1979 1980 1981 1982 1983 1984 **1985** 1986 1987 1988 1989 1990 1991 1992 1993 1994 1995 1996 1997 1998 1999 2000 2001 2002 2003 2004 2005 2006 2007 200

Historic Areas of Istanbul Turkey

Criteria – Human creative genius; Interchange of values; Testimony to cultural tradition; Significance in human history

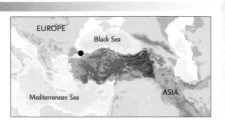

EUROPE

Black Sea

Mediterranean Sea

ASIA

The Blue Mosque, Istanbul.

The monuments of the city have greatly influenced the development of architecture, monumental arts and the organization of space, in both Europe and Asia for many centuries. Thus, the 6,650 m terrestrial wall of Byzantine Emperor Theodosius II, created in AD 447, was a leading reference for military architecture; St Sophia became a model for an entire family of churches and later mosques; and the mosaics of Constantinople's palaces and churches influenced Eastern and Western Christian art.

Istanbul was built at the crossroads of two continents and was successively the capital of three empires, the Eastern Roman empire, the Byzantine Empire and the Ottoman Empire. First as Byzantium, then as Constantinople and currently as Istanbul, the city has constantly been associated with major events in political, religious and artistic history in both Europe and Asia for two millennia. Masterpieces of the city's built heritage include the ancient Hippodrome of Constantine, the sixth-century Hagia Sophia and the sixteenth-century Süleymaniye Mosque, all now under threat from population pressure, industrial pollution and uncontrolled urbanization.

The World Heritage site covers four zones, illustrating the major phases of the city's history using its most prestigious monuments: the Archaeological Park, at the tip of the peninsula on which the western city stands; the Süleymaniye quarter; the Zeyrek quarter; and the zone of the ramparts.

The ancient city and the capital of the Eastern Roman Empire are both represented, by the Hippodrome of Constantine (dating from 324) in the Archaeological Park, by the aqueduct of Valens (378) in the Süleymaniye quarter, and

by the ramparts (begun in 413), located in the fourth zone.

The capital of the Byzantine Empire is highlighted by several major monuments. In the Archaeological Park are the churches of St Sophia and St Irene, built in the reign of Justinian (527–65). In the Zeyrek quarter is the ancient Pantocrator Monastery, founded under John II Comnenus (1118–43); in the zone of the ramparts there is the old church of the Holy Saviour in Chora (now the Kariye Camii) with its marvellous mosaics and fourteenth- and fifteenth-century paintings. Moreover, the current layout of the walls results from modifications performed in the seventh and twelfth centuries to include the quarter and the Palace of the Blachernes.

The capital of the Ottoman Empire is represented by its most important monuments architecturally: Topkapı Saray and the Blue Mosque in the archaeological zone; the Sehzade and Süleymaniye mosques in the Süleymaniye quarter; and the vernacular settlement vestiges of this quarter, in 525 listed and protected wooden houses.

World Heritage site since

1978 1979 1980 1981 1982 1983 1984 **1985** 1986 1987 1988 1989 1990 1991 1992 1993 1994 1995 1996 1997 1998 1999 2000 2001 2002 2003 2004 2005 2006 2007 2008

Historic Mosque City of Bagerhat
Bangladesh
Criteria – Significance in human history

ASIA

Bay of
Bengal

This remarkable city had no sooner been created than it was swallowed up by the jungle, after its founder died in 1459. The density of Islamic religious monuments is explained by the piety of Khan Jahan, which is shown by the engraved inscription on his tomb. The lack of fortifications is attributable to the possibilities of retreat into the impenetrable swamps of the Sundarbans.

Formely known as Khalifatabad this historic city is an outstanding example of an architectural ensemble which illustrates a significant stage in human history. Situated in the suburbs of Bagerhat, at the meeting point of the Ganges and Brahmaputra rivers, this ancient city was founded by the Turkish general Ulugh Khan Jahan in the fifteenth century. The city had 360 mosques, public buildings, mausoleums, bridges, roads, water tanks and other public buildings, all built from baked brick. Shait Gumbad Mosque and Khan Jahan's Mausoleum are just two examples of these historic buildings. The quality of the infrastructures – the supply and evacuation of water, the cisterns and reservoirs, the roads and bridges – reveal a perfect mastery of the techniques of planning and a will towards spatial organisation.

World Heritage site since

1978 1979 1980 1981 1982 1983 1984 **1985** 1986 1987 1988 1989 1990 1991 1992 1993 1994 1995 1996 1997 1998 1999 2000 2001 2002 2003 2004 2005 2006 2007 2008

Pont du Gard (Roman Aqueduct)
France

Criteria – Human creative genius; Testimony to cultural tradition; Significance in human history

The Pont du Gard was built shortly before the Christian era to carry an aqueduct across the Gardon River. The Roman architects and hydraulic engineers who designed this bridge, which stands almost 50 m high and is on three levels – the longest measuring 275 m – created a technical and artistic masterpiece. The first level, which was used as a bridge in the Middle Ages, consists of six semi-circular arches, the second level eleven, and the third level, at the top of which runs the water channel, has thirty-five. On the two lower levels large stone blocks, which can weigh up to 6 tonnes, were used while the upper level was constructed of small stone rubble. It is one of the most significant monuments of the early Imperial period, most likely built upon the initiative of Agrippa about 20 BC.

In order to meet the needs of the colony of Nemausus (Nîmes), springs were tapped near Uzès. To achieve an average gradient of 34 cm per 1,000 m, the 50 km aqueduct compensates for the topography by following a winding path. Near Remoulins it had to cross the deep valley of the Gardon, requiring the construction of this exceptional bridge. ▼

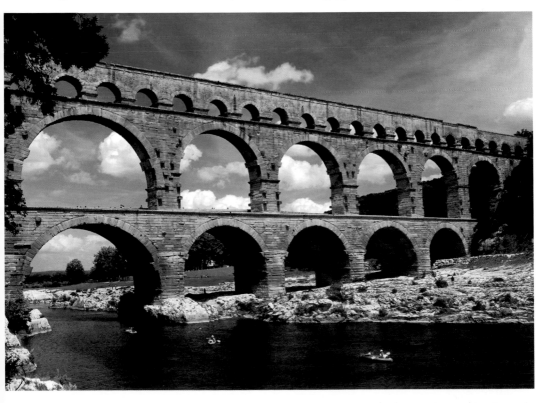

World Heritage site since

1978 1979 1980 1981 1982 1983 1984 **1985** 1986 1987 1988 1989 1990 1991 1992 1993 1994 1995 1996 1997 1998 1999 2000 2001 2002 2003 2004 2005 2006 2007 2008

Göreme National Park and the Rock Sites of Cappadocia
Turkey

Criteria – Human creative genius; Testimony to cultural tradition; Traditional human settlement; Natural phenomena or beauty

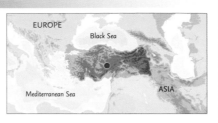

EUROPE

Black Sea

Mediterranean Sea

ASIA

The Göreme Valley is a spectacular example of the effects of wind and water erosion on soft volcanic tuff. Pillars, columns, towers, obelisks and needles reaching heights of 40 m are typical features of the landscape. The nearby Erciyas volcano is still active, with occasional minor eruptions.

Within these rock formations people have excavated a network of caves which served as refuges, residences, storage and places of worship dating from the fourth century. The surrounding landscape is agricultural with a number of small, scattered rural villages.

In a spectacular landscape, entirely sculptured by erosion, the Göreme valley and its surroundings contain rock-hewn sanctuaries that provide unique evidence of Byzantine art in the post-iconoclastic period of the ninth century AD onwards. Dwellings, troglodyte villages and underground towns – the remains of a traditional human habitat dating back to the fourth century – can also be seen there.

In the ruin-like landscape of the Cappadocia plateau where natural erosion sculpted the rock into shapes eerily reminiscent of towers, spires, domes and pyramids, man has added to the workmanship of the elements by digging cells, churches and veritable subterranean cities which together make up one of the world's largest cave-dwelling complexes. The historical setting, the rock-hewn churches and the unusual eroded landforms combine to produce a mixed cultural and natural landscape of unusual appearance.

Although the area has been extensively used and modified by man for centuries, the resulting landscape is one of harmony and consideration of the intrinsic values of the natural landforms. Architectural styles are based on the local stone and the valley has changed little over time.

Although interesting from a geological and ethnological perspective, this phenomenal site excels especially for the incomparable beauty of the rupestral decor in the Christian sanctuaries whose features make Cappadocia one of the leading examples of post-iconoclast-period Byzantine art.

The first signs of monastic activity in Cappadocia are thought to date from the fourth century when, on the instructions of Basil the Great, Bishop of Caesarea (Kayseri), small, hermit communities began inhabiting the cells dug into the rock. Later, to resist Arab forays they began banding together into troglodyte villages or subterranean towns such as Kaymakli or Derinkuyu, which served as places of refuge.

Cappadocian monasticism was already well established in the iconoclast period (725–842), as illustrated by the decoration in many of the sanctuaries: a minimum of symbols, most often sculpted or tempera-painted crosses. But after 842 many rupestral churches were dug in Cappadocia and these were were richly decorated with brightly coloured figurative painting. Among them in the Göreme valley were Tokali Kilise and El Nazar Kilise in the tenth century, Barbara Kilise and Sakli Kilise in the eleventh century, and El Mali Kilise and Karanlik Kilise, created in the late twelfth and early thirteenth centuries.

World Heritage site since

1978 1979 1980 1981 1982 1983 1984 **1985** 1986 1987 1988 1989 1990 1991 1992 1993 1994 1995 1996 1997 1998 1999 2000 2001 2002 2003 2004 2005 2006 2007 2008

Royal Palaces of Abomey
Benin

Criteria – Testimony to cultural tradition;
Significance in human history

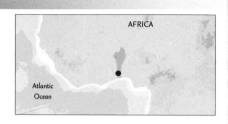

The West African Kingdom of Abomey (later Dahomey), founded in 1625 by the Fon people, developed into a powerful military and commercial empire. Under the twelve kings who succeeded one another from 1695 to 1900, the kingdom became one of the most powerful on the west coast of Africa. Until the late nineteenth century its primary source of wealth was from selling prisoners of war as slaves to European slave traders

for transport across the Atlantic to the New World. Each of the twelve kings built a lavish palace on the royal grounds in Abomey, the capital city, all within the same cob-walled area. Still used regularly for rituals and royal ceremonies, the palace buildings both represent the past and continue its traditions.

The bas-relief decorations provide a unique history of this society that had no written records. They glorify each king and document the myths, customs and rituals of the Fon people whose kingdom this was. The walls show that the military might of the Abomey kingdom was based, in part, on companies of female warriors who matched their male counterparts in fierceness and courage.

Rock Art of Alta
Norway

Criteria – Testimony to cultural tradition

The petroglyphs of the Alta fjord in the province of Tromsø are among the leading rock art sites in the world. Close to the Arctic Circle, they are a valuable illustration of human activity between 4200 and 500 BC in the far North. The thousands of paintings and engravings are located at forty-five sites scattered over seven localities and show a definite chronological sequence. The position of the paintings and engravings

with respect to sea level at different postglacial periods constitutes a relative dating element, which is corroborated by objective iconography data. According to the principle of reverse stratigraphy, the most ancient works are generally the highest, the most recent being close to the present sea level (the height difference is roughly 26 m).

The drawings represent northern wildlife, including reindeer, elk and bear, geese, cormorants and salmon. There are depictions of hunting, trapping and fishing, dancing and ceremonial acts. The climate warmed as time passed and the latest drawings show agricultural activities.

World Heritage site since

1978 1979 1980 1981 1982 1983 1984 **1985** 1986 1987 1988 1989 1990 1991 1992 1993 1994 1995 1996 1997 1998 1999 2000 2001 2002 2003 2004 2005 2006 2007 2008

Old Town of Ávila with its Extra-Muros Churches
Spain

Criteria – Testimony to cultural tradition;
Significance in human history

Ávila is an outstanding example of a fortified city of the Middle Ages, the surrounding walls of which are fully intact (pictured below). The density of religious and secular monuments, both inside (intra) and outside (extra) the city walls (muros), makes it an urban ensemble of exceptional value. Founded in the eleventh century to protect the Spanish territories from the Moors, this 'City of Saints and Stones', the birthplace of St Teresa and

the burial place of the Grand Inquisitor Torquemada, has kept its medieval austerity. The walls, which date back to 1090 (but mostly rebuilt during the twelfth century), have an average thickness of 3 m and are flanked by eighty-two semicircular towers, with nine gates of different periods. The cathedral, with its crenellated Romanesque choir linked to the curtain wall, is part of the system of fortifications.

The church of San Pedro Extra-Muros was begun in 1100. In its beautiful open atrium, royal receptions took place, and it was here that the Catholic kings in June 1475 and Emperor Charles I in May 1534 vowed to respect the city's charters.

World Heritage site since

1978 1979 1980 1981 1982 1983 1984 **1985** 1986 1987 1988 1989 1990 1991 1992 1993 1994 1995 1996 1997 1998 1999 2000 2001 2002 2003 2004 2005 2006 2007 2008

Thracian Tomb of Sveshtari
Bulgaria

Criteria – Human creative genius; Testimony to cultural tradition

EUROPE

Black Sea

The tomb is an exceptional testimony to the culture of the Getae, a Thracian people living in the north of Hemus, who were in contact with the Greek and Hyperborean worlds, according to ancient geographers.

The discovery in 1982 of the Thracian tomb of Sveshtari was one of the most spectacular archaeological events of the twentieth century. This third-century BC tomb reflects the fundamental structural principles of Thracian cult buildings. The tomb has a unique architectural décor, with polychrome half-human, half-plant caryatids and painted murals. The ten female figures carved in high relief on the walls of the central chamber and the decoration of the lunette in its vault are the only examples of this type found so far in the Thracian lands. The tomb consists of a corridor (dromos) and three square chambers: antechamber, lateral chamber, and main burial chamber covered by a semi-cylindrical vault.

Rock-Art Sites of Tadrart
Acacus
Libyan Arab Jamahiriya

Criteria – Testimony to cultural tradition

Mediterranean Sea

ASIA

AFRICA

As well as its remarkable cave paintings, Tadrart Acacus boasts some of the most extraordinary scenery in the world. Its unique natural wonders include isolated towers emerging from the sand and eroded into bizarre shapes, petrified arches, vast dunes, and canyons carved by ancient rivers.

On the borders of Tassili N'Ajjer in Algeria, also a World Heritage site, this rocky massif has thousands of cave paintings reflecting marked changes in fauna and flora, and the different ways of life of the populations that succeeded one another in this region of the Sahara. The oldest images (12,000 BC) are outline engravings of large savanna mammals such as elephant and rhinoceros. Paintings of magic religious scenes and representations of a humid landscape appear alongside engravings around 8000 BC. The domesticated horse features about 1500 BC in a semi-arid climate, while the first centuries BC saw the intensification of a desert environment. The dromedary settled in the region at this time and became the main subject of the last rock art paintings.

World Heritage site since

1978 1979 1980 1981 1982 1983 1984 1985 **1986** 1987 1988 1989 1990 1991 1992 1993 1994 1995 1996 1997 1998 1999 2000 2001 2002 2003 2004 2005 2006 2007 2008

Iguaçu National Park
Brazil

Criteria – Natural phenomena or beauty;
Significant natural habitat for biodiversity

The park shares with Iguazú National Park in Argentina one of the world's largest and most impressive waterfalls. The Iguaçu Falls span the border between Argentina and Brazil. At 80 m high and 2,700 m wide, the Falls produce vast clouds of spray that encourage the growth of lush vegetation of which around 90 per cent is subtropical rainforest.

The park is home to many rare and endangered species of flora and fauna, among them the giant otter and the giant anteater.

Among the many other species recorded are La Plata otter, ocelot, jaguar, puma, margay, brocket deer, pampas deer, American tapir, black howler monkey, collared peccary, white-lipped peccary and urutu viper.

Noteworthy birds include great dusky swift, solitary tinamou, ornate hawk eagle, red-breasted toucan, glaucous macaw, vinaceous-breasted and red-spectacled parrots, white-tailed trogon and harpy eagle.

Vegetation is lush and abundant. The lower park has subtropical rainforest with tree ferns, lianas and epiphytes. The upper part is mainly humid subtropical deciduous forest and has stands of the Brazilian pine with two palms, the Assai palm and wild coconut palm, and the imbuya. These stands are limited to a small section in the north-east of the peak.

World Heritage site since

1978 1979 1980 1981 1982 1983 1984 1985 **1986** 1987 1988 1989 1990 1991 1992 1993 1994 1995 1996 1997 1998 1999 2000 2001 2002 2003 2004 2005 2006 2007 200

Ancient City of Aleppo
Syrian Arab Republic

Criteria – Testimony to cultural tradition;
Significance in human history

The souks or bazaars of Aleppo's old town are among the most famous in the Middle East. Enclosed by stone vaulted roofs, they twist for 7 km along narrow streets.

The Great Mosque was founded around AD 715 but little from that time remains: it was largely rebuilt around 1250 although its highest minaret dates from 1090.

The ancient city of Aleppo is one of the oldest continuously inhabited cities in the world. Located at the crossroads of trade routes between the Mediterranean and the East, it prospered from the third millennium BC. Successive occupiers – including Byzantines, Romans, Greeks, Crusaders and Arabs – have left their mark in the architecture and the city plan.

Aleppo is dominated by its huge medieval castle, the Citadel (pictured below), which stands on a partly artificial mound towering 50 m above the city. Although the structure dates from around the twelfth century, the hill on which it stands has been used for defensive purposes from around the third millennium BC.

The old city was surrounded by a defensive wall of which parts still exist along with seven of its fortified gates; Bab Qinnesrin is the most impressive remaining.

◄ The Citadel of Aleppo.

World Heritage site since

978 1979 1980 1981 1982 1983 1984 1985 **1986** 1987 1988 1989 1990 1991 1992 1993 1994 1995 1996 1997 1998 1999 2000 2001 2002 2003 2004 2005 2006 2007 2008

Churches and Convents of Goa
India

Criteria – Interchange of values; Significance in human history; Heritage associated with events of universal significance

The churches and convents of Goa, the former capital of the Portuguese Indies – particularly the Church of Bom Jesus, which contains the tomb of St Francis-Xavier – illustrate the evangelization of Asia. These monuments were influential in spreading forms of Manueline, Mannerist and Baroque art in all the countries of Asia where missions were established.

The Sé Cathedral is a fine example of Renaissance architecture, with a Tuscan exterior, Corinthian columns, raised platform with steps leading to the entrance, and barrel-vault.

◄

The church of Bom Jesus.

Hattusha: the Hittite Capital
Turkey

Criteria – Human creative genius; Interchange of values; Testimony to cultural tradition; Significance in human history

Hattusha exerted a dominating influence upon the civilizations in the thirteenth century BC in Anatolia and northern Syria. The palaces, temples, trading quarters and necropolis of this political and religious metropolis provide a comprehensive picture of a capital city and bear unique testimony to the vanished Hittite civilization. The city's fortifications, along with the Lion Gate, the Royal Gate and the Yazılıkaya rupestral ensemble with its sculptured friezes,

represent unique artistic achievements as monuments. Inside the walls, the city is built on two levels. To the northwest is the lower town with its great temple, dedicated to the god of storms and the goddess of the Sun, Arinna. Thousands of cuneiform tablets were found in this area. To the south is the upper city with the royal residence of Büyükkale.

At its largest, the city spread over a sloping, uneven plateau, covering 2.1 km from north to south and 1.3 km from east to west. In the thirteenth century, the city was surrounded by a system of double walls forming a perimeter of roughly 8 km.

World Heritage site since

1978 1979 1980 1981 1982 1983 1984 1985 **1986** 1987 1988 1989 1990 1991 1992 1993 1994 1995 1996 1997 1998 1999 2000 2001 2002 2003 2004 2005 2006 2007 200

Chan Chan Archaeological Zone
Peru

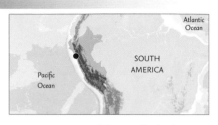

Criteria – Human creative genius; Testimony to cultural tradition

The planning of Chan Chan, largest city of pre-Hispanic America and a unique testimony to the disappeared Chimú kingdom, is a masterpiece of inhabited space and hierarchical construction. Chan Chan was the capital of the kingdom of Chimor which reached its zenith in the fifteenth century, shortly before its fall to the Chimús' great rivals, the Incas.

The planning of this huge, 6 km² city reflects a strict political and social strategy, illustrated by its division into nine 'palaces' –

autonomous units delineated by high, earthen walls. Each 'palace' was grouped around one or more squares and their buildings include temples, dwellings, storehouses, kitchens, orchards, gardens and funeral platforms.

Nearby lie four industrial sectors where the main activities appear to have been woodworking, weaving and precious-metal working. The remains of an irrigation system suggest another area used for farming.

An irrigation system channeled water from the Moche River to Chan Chan. It is difficult to imagine the fertility of this now-desert region during the height of the Chimú civilization.

Hatred of the Incas led the Chimú to welcome the Spanish conquistadores. In 1535 the Spanish founded a new capital at Trujillo, 5 km from Chan Chan which was abandoned.

World Heritage site since

1978 1979 1980 1981 1982 1983 1984 1985 **1986** 1987 1988 1989 1990 1991 1992 1993 1994 1995 1996 1997 1998 1999 2000 2001 2002 2003 2004 2005 2006 2007 2008

Ironbridge Gorge
United Kingdom

Criteria – Human creative genius; Interchange of values; Significance in human history; Heritage associated with events of universal significance

Ironbridge, the site of the world's first cast-iron bridge, is known throughout the world as the symbol of the Industrial Revolution. In the blast furnace at nearby Coalbrookdale were developed the coke-based iron-production techniques that drove the eighteenth-century steel revolution. Coalbrookdale and Ironbridge exerted great influence on the development of industrial techniques and architecture, and the area provides a fascinating summary of the progress of the industrial age.

In the area there are five major areas of interest: Coalbrookdale, with Abraham Darby's coke-fired blast furnace; Ironbridge, which takes its name from the bridge built in 1779; Hay Brook valley with its open-air museum featuring extraction galleries and preserved blast furnaces; Jackfield, a former mining town; and Coalport with a porcelain museum in its former porcelain-production factory.

Ironbridge Gorge is a concentration of mining areas, foundries, factories and warehouses, coexisting with a network of paths, roads, canals and railways, as well as substantial remains of traditional landscape and housing, public buildings and infrastructure of the eighteenth and nineteenth centuries. These are well preserved and together comprise an excellent educational environment.

▼

World Heritage site since

1978 1979 1980 1981 1982 1983 1984 1985 **1986** 1987 1988 1989 1990 1991 1992 1993 1994 1995 1996 1997 1998 1999 2000 2001 2002 2003 2004 2005 2006 2007 200

Great Zimbabwe National Monument
Zimbabwe

Criteria – Human creative genius; Testimony to cultural tradition; Heritage associated with events of universal significance

The ruins of Great Zimbabwe – the capital of the Queen of Sheba, according to legend – are a unique testimony to the civilisation of the Shona people of the eleventh to fifteenth centuries. There are three main areas of the site. The Hill Ruins are the remains of a royal city, perched atop a large granite spur. The Great Enclosure below the hill dates from the fourteenth century and holds a series of brick-built living quarters. The Valley Ruins are a series of living ensembles surrounded by drystone masonry walls. Finds made in the inhabited areas have revealed vital information about the farming and pastoral activities of the inhabitants at the time of Great Zimbabwe's heyday and about earthenware and smithing craft activities.

The city of Great Zimbabwe has captured the imagination of travellers since the Middle Ages, and remains a powerful symbol of national identity today. It covers an area of nearly 0.8 km² and was an important trading centre with both European and Asian cultures. In the fourteenth century, it was the principal city of a major state extending over the gold-rich plateaus, with a population of more than 10,000.

Temple of Apollo Epicurius at Bassae
Greece

Criteria – Human creative genius; Interchange of values; Testimony to cultural tradition

This famous temple to the god of healing and the sun was built towards the middle of the fifth century BC in the lonely heights of the Arcadian mountains. The temple, which has the oldest Corinthian capital yet found, combines the Archaic style and the serenity of the Doric style with some daring architectural features. The decoration is notable, particularly on account of the different materials used: the walls and the bases of the columns are limestone, and the Ionic capitals and the Corinthian capital are in Doliana marble. The discovery of the Ionic frieze's twenty-two sculptured plates ultimately divested the site of these remarkable sculptures, which were transferred to the British Museum along with the Corinthian capital.

The temple was dedicated by the inhabitants of Phigalia to Apollo Epicurius, the god-healer who came to their aid when they were beset by plague. Its remoteness kept it undiscovered until 1765 when a French architect happened upon it.

World Heritage site since

1978 1979 1980 1981 1982 1983 1984 1985 **1986** 1987 1988 1989 1990 1991 1992 1993 1994 1995 1996 1997 1998 1999 2000 2001 2002 2003 2004 2005 2006 2007 2008

St Kilda
United Kingdom

Criteria – Testimony to cultural tradition; Traditional human settlement; Natural phenomena or beauty; Significant ecological and biological processes; Significant natural habitat or biodiversity

The archipelago comprises the islands of Hirta, Dun, Soay and Boreray and lies in the Atlantic 66 km west of the island of Benbecula.

St Kilda is a remote volcanic archipelago of exceptional natural beauty which supports significant natural habitats. It is unique in the very high bird densities that occur in a relatively small area, conditioned by the site's complex and different ecological niches. Its cliffs are among the highest in Europe and are home to large colonies of endangered birds, especially puffins and gannets. There is also a complex ecological dynamic in the site's three marine zones that is essential to the maintenance of both marine and terrestrial biodiversity.

St Kilda bears the evidence of more than 2,000 years of human occupation in extreme conditions. Human vestiges include built structures and field systems, cleits (dry-stone huts) and traditional Highland stone houses (both pictured below) – the vulnerable remains of a subsistence economy based on the products of birds, agriculture and sheep farming and reflecting age old traditions.

The islanders of St Kilda were evacuated at their own request by the British government in 1930 and the only inhabitants now are British government Ministry of Defence staff and a seasonal nature warden.

World Heritage site since

1978 1979 1980 1981 1982 1983 1984 1985 **1986** 1987 1988 1989 1990 1991 1992 1993 1994 1995 1996 1997 1998 1999 2000 2001 2002 2003 2004 2005 2006 2007 20C

Historic City of Toledo
Spain

Criteria – Human creative genius; Interchange of values; Testimony to cultural tradition; Significance in human history

Toledo is the repository of more than 2,000 years of history with architectural and artistic masterpieces that are the product of several civilizations and their three religions – Christianity, Judaism and Islam.

A centre of power for centuries, the city was successively a self-governing Roman municipium, the capital of the Visigothic Kingdom, a fortress of the Emirate of Córdoba, an outpost of the Christian kingdoms fighting the Moors and, in the sixteenth century, the temporary seat of supreme power under Holy Roman Emperor and King of Spain Charles V, who endowed it with the status of imperial and crowned city. During the Renaissance the city became one of the most important artistic centres in Spain.

Ironically, the irreversible economic and political decadence of Toledo after 1561, when Phillip II of Spain chose Madrid as his capital, miraculously spared this museum-city.

Each set of inhabitants left their mark on Toledo: Rome, with vestiges of the circus, aqueduct and sewer; the remains of Visigoth walls; the Islamic monuments of the Emirate of Córdoba; and remarkable churches and synagogues built after the reconquest of 1085. Furthermore, Toledo possesses a broad range of medieval structures, walls and fortified buildings, bridges, streets and houses.

The cathedral, pictured right, and the bridge, gates and Alcázar of Toledo, pictured below. ▼

World Heritage site since

1978 1979 1980 1981 1982 1983 1984 1985 **1986** 1987 1988 1989 1990 1991 1992 1993 1994 1995 1996 1997 1998 1999 2000 2001 2002 2003 2004 2005 2006 2007 2008

Giant's Causeway and Causeway Coast
United Kingdom

Criteria – Natural phenomena or beauty; Major stages of Earth's history

The site lies on the north coast of the County of Antrim, Northern Ireland, and includes the Causeway Coast, a 6 km stretch of extraordinary geological formations representing volcanic activity during the early Tertiary period some 50–60 million years ago. The most characteristic and unique feature of the site is the Giant's Causeway, a sea-level promontory of around 40,000 polygonal columns of basalt in perfect horizontal sections forming a pavement. The dramatic sight has inspired legends of giants striding over the sea to Scotland. The Giant's Causeway featured in the eighteenth-century geological controversies on the origins of basalts, and geological studies of these formations over the last 300 years have greatly contributed to the development of the earth sciences.

Other features of the Causeway Coast include the Giant's Organ (about sixty regular columns, 12 m high), the Chimney Tops (a number of columns separated from the cliffs by erosion), and Hamilton's Seat (a viewpoint).

World Heritage site since

1978 1979 1980 1981 1982 1983 1984 1985 **1986** 1987 1988 1989 1990 1991 1992 1993 1994 1995 1996 1997 1998 1999 2000 2001 2002 2003 2004 2005 2006 2007 2008

Old City of Sana'a
Yemen

Criteria – Significance in human history;
Traditional human settlement; Heritage
associated with events of universal significance

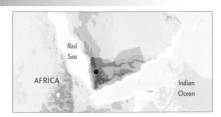

Situated in a mountain valley at an altitude
of 2,200 m, Sana'a has been inhabited for
more than 2,500 years and has been the
capital of the Yemen since 1962. The Great
Mosque is said to have been constructed
while the Prophet Muhammad was still
living, and in the seventh and eighth
centuries the city became a major centre
for the propagation of Islam. This religious
and political heritage can be seen in the
103 mosques, 14 hammams (steam baths) and
over 6,000 houses, all built before the
eleventh century. Sana'a's many-storeyed
tower-houses, built of rammed earth, add to
the beauty of the site. The successive
reconstructions of Sana'a under Ottoman
domination, beginning in the sixteenth
century, respected the proportions and
balance of the medieval city while changing
its appearance.

Given official status in
the second century BC
when it was an
outpost of the
Yemenite kingdoms,
Sana'a (Arabic for
'fortified place') was
associated with all the
major historical
events that took place
in Arabia Felix.

World Heritage site since

1978 1979 1980 1981 1982 1983 1984 1985 **1986** 1987 1988 1989 1990 1991 1992 1993 1994 1995 1996 1997 1998 1999 2000 2001 2002 2003 2004 2005 2006 2007 2008

Durham Castle and Cathedral
United Kingdom

Criteria – Interchange of values; Significance in human history; Heritage associated with events of universal significance

The present castle is a veritable labyrinth of halls and galleries of different periods, and in its north wing it houses various vestiges of the Romanesque epoch, including the castral chapel, built in 1080.

Located on a rocky promontory overlooking a bend in the Wear River, the monumental array constituted by the cathedral and its outbuildings to the south and by the castle to the north makes up one of the best-known cityscapes of medieval Europe. Durham Cathedral was built in the late-eleventh and early twelfth centuries to house the relics of St Cuthbert, evangelizer of Northumbria, and the Venerable Bede. It attests to the importance of the early Benedictine monastic community and is the largest and finest example of Norman architecture in England. The innovative audacity of its vaulting foreshadowed Gothic architecture. Behind the cathedral stands the castle, an ancient Norman fortress which regularly faced the onslaught of Scottish troops. It later became the residence of the prince-bishops of Durham and in the nineteenth century was incorporated into Durham University.

World Heritage site since

1978 1979 1980 1981 1982 1983 1984 1985 **1986** 1987 1988 1989 1990 1991 1992 1993 1994 1995 1996 1997 1998 1999 2000 2001 2002 2003 2004 2005 2006 2007 2008

Fatehpur Sikri
India

Criteria – Interchange of values; Testimony to cultural tradition; Significance in human history

ASIA

Indian Ocean

Built during the second half of the sixteenth century by the Emperor Akbar, Fatehpur Sikri – the City of Victory – was the capital of the Mughal Empire for a mere fourteen years. Constructed between 1571 and 1573, it was abandoned in 1585 when Akbar went to fight against the Afghan tribes and chose a new capital, Lahore. Briefly, for three months in 1619, the city resumed its role as the seat of the great Mughal court when Jahangir sought refuge there from the plague that devastated Agra. The site was then finally abandoned, until its archaeological exploration in 1892. The complex of monuments and temples in the city, all in a uniform architectural style, includes one of the largest mosques in India, the Jama Masjid, able to accommodate some 10,000 of the faithful.

This capital without a future was considerably more than the fancy of a sovereign during the fourteen years of its existence. Described by the English traveller Ralph Fitch in 1585 as 'considerably larger than London and more populous', it comprised a series of palaces, public buildings and mosques, as well as living areas for the court, the army, servants of the king and an entire population.

Walls of the fort at Fatehpur Sikri. ▼

World Heritage site since

1978 1979 1980 1981 1982 1983 1984 1985 **1986** 1987 1988 1989 1990 1991 1992 1993 1994 1995 1996 1997 1998 1999 2000 2001 2002 2003 2004 2005 2006 2007 200

Studley Royal Park including the Ruins of Fountains Abbey
United Kingdom

Criteria – Human creative genius; Significance in human history

The Fountains site in Yorkshire owes its originality to the fact that a strikingly beautiful landscape was constituted around the largest medieval ruins of the United Kingdom. Fountains Abbey was founded in 1132 by thirteen Cistercian monks from York, who were searching for an ideal of life in closer keeping with St Benedict's teachings. But by the time the monastic community was abolished by Henry VIII in 1530 it had become the richest abbey in the kingdom, owing to massive donations. These four centuries of prosperity are reflected in the utter magnitude of the ruins of the buildings. Essential to the ruins are the small Fountains Hall Castle, the landscaping, the gardens, the canal created by John Aislabie in the eighteenth century, the plantations and vistas of the nineteenth century, and Studley Royal Church.

▲
Vaulted ceilings in the Cellarium at Fountains Abbey.

The nave of the abbey church is close to the pristine ideal of Cistercian austerity. However, the rich array of monastic buildings grouped together to the south testifies to the deep-seated changes occurring in a community which rapidly grew away from the pristine ideal owing to its land wealth and its spiritual influence.

World Heritage site since

1978 1979 1980 1981 1982 1983 1984 1985 **1986** 1987 1988 1989 1990 1991 1992 1993 1994 1995 1996 1997 1998 1999 2000 2001 2002 2003 2004 2005 2006 2007 2008

Khajuraho Group of Monuments
India

Criteria – Human creative genius; Testimony to cultural tradition

The temples at Khajuraho were built during the Chandela dynasty, which reached its apogee between 950 and 1050. The Temple of Kandariya is decorated with a profusion of sculptures that are among the greatest masterpieces of Indian art. Of the eighty-five temples built, only about twenty remain; they fall into three distinct groups and belong to two different religions – Hinduism and Jainism. They strike a perfect balance between architecture and sculpture. Yasovarman (AD 954) built the temple of

Vishnu, now famous as Lakshmana temple; this ornate example proclaims the prestige of the Chandelas. The largest and grandest temple is the Kandariya Mahadeva, attributed to Ganda (1017–29). Greatly influenced by the Tantric school of thought, the Chandela kings promoted various Tantric doctrines through royal monuments, including temples. The sculptors of Khajuraho depicted all aspects of life, secular, spiritual and sexual.

The temples of Khajuraho comprise an elevated platform, on which rises the richly decorated building, the 'jangha', covered with sculpted panels. This is crowned by a series of bundled towers with curvilinear contours, the 'sikharas'. Each of these towers, which is characteristic of the temples in the Nagera style, symbolizes the 'cosmic mountain', Mount Kailasha.

◄
Vishwanath Temple at Khajuraho.

World Heritage site since

1978 1979 1980 1981 1982 1983 1984 1985 **1986** 1987 1988 1989 1990 1991 1992 1993 1994 1995 1996 1997 1998 1999 2000 2001 2002 2003 2004 2005 2006 2007 200

Group of Monuments at Hampi
India

Criteria – Human creative genius; Testimony to cultural tradition; Significance in human history

The city of Hampi bears exceptional testimony to the vanished civilization of the Hindu kingdom of Vijayanagar, which reached its peak under Krishna Deva Raya (1509–30). Its fabulously rich princes built Dravidian temples and palaces which won the admiration of travellers between the fourteenth and sixteenth centuries. Conquered by the Deccan Muslim confederacy in 1565, the city was pillaged over a period of six months before being abandoned. Hampi, enriched by the cotton and the spice trade, was one of the most beautiful cities of the medieval world.

The temples of Ramachandra (1513) and Hazara Rama (1520), with their sophisticated structures, may be counted among the most extraordinary buildings in India. Besides the temples, the impressive complex of civil, princely and public buildings (elephant stables, Queen's Bath, Lotus Mahal, bazaars, markets) are enclosed in the massive fortifications.

In one of the interior courtyards of the temple of Vitthala can be found a stone monument of a chariot pulled by two small elephants, a favourite of tourists today as it was of travellers of the past.

Ancient water pool and temple (background) at Krishna market, Hampi. ▼

World Heritage site since

78 1979 1980 1981 1982 1983 1984 1985 **1986** 1987 1988 1989 1990 1991 1992 1993 1994 1995 1996 1997 1998 1999 2000 2001 2002 2003 2004 2005 2006 2007 2008

Old Town of Cáceres
Spain

Criteria – Testimony to cultural tradition;
significance in human history

Cáceres is an outstanding example of a feudal city that developed after the conflicts between the Christians and Moors. Its history is reflected in its architecture, which is a blend of Roman, Islamic, Northern Gothic and Italian Renaissance styles. Originally a Roman city, the Islamic Almohads built remarkable fortifications which completely changed the appearance of the Roman walls. Of the thirty or so towers from the Muslim period, the Torre

del Bujaco is the most famous. The street pattern, with winding backstreets that open on to tiny squares, also dates from this period. After the Reconquista, the city became the stage for power struggles between rival clans and fortified houses dotted the landscape. In the fifteenth and sixteenth centuries, noble pride was demonstrated by adding richly decorated coats of arms to the front of houses and by building many defensive towers.

When Spanish adventurers returned from America in the sixteenth century, new palaces were constructed: Palacio Godoy, built by a newly-rich conquistador and Palacio de los Toledo-Moctezuma, built in the second half of the sixteenth century for the grandson of Moctezuma II, the Aztec ruler when Cortes reached Mexico.

World Heritage site since

1978 1979 1980 1981 1982 1983 1984 1985 **1986** 1987 1988 1989 1990 1991 1992 1993 1994 1995 1996 1997 1998 1999 2000 2001 2002 2003 2004 2005 2006 2007 200

Stonehenge, Avebury and Associated Sites
United Kingdom
Criteria – Human creative genius; Interchange of values; Testimony to cultural tradition

Two different materials were used for the Stonehenge constructions: irregular sandstone blocks, known as sarsens, which were quarried in a plain near Salisbury; and bluestones, quarried about 200 km away in Pembrokeshire in Wales.

Stonehenge and Avebury are among the most famous groups of megaliths in the world. Together with their associated sites, they represent a masterpiece of human creative genius of the Neolithic age.

The megalithic sites of Stonehenge and Avebury consist of circles of menhirs arranged in a pattern of obvious astronomical significance which is still being explored. However, a number of satellite sites make it possible to better understand the more famous sites by examining them in a broader context.

Stonehenge was built in several distinct phases from 3100–1100 BC and its size, height and perfection make it one of the most impressive megalithic monuments in the world. Its plan was based on a series of concentric circles and the menhirs used are huge: from the third phase of construction onwards, large lintels were placed upon the vertical blocks, thereby creating a type of bonded entablature. The Avenue, an earthwork cut into the chalk soil, runs straight into the northeast corner of Stonehenge.

Avebury lies about 30 km to the north of Stonehenge, and although less well known, it is nevertheless Europe's largest circular megalithic ensemble: its exterior circle

comprises some 100 menhirs. In all, 180 standing stones were put into place here before the beginning of the third millennium BC, as demonstrated by abundant ceramic samples found on the site.

There are four avenues at Avebury of which only the southern one, West Kennet Avenue, is still lined with megaliths; the avenues lead to the four cardinal points of the circle. West Kennet Avenue leads to the site of The Sanctuary at Overton Hill 2.5 km away. The Sanctuary was a series of concentric timber and stone circles; their purpose remains unknown.

There are several other Neolithic satellite sites around Avebury, including Silbury Hill, the largest known man-made earthen mound in Europe. As with The Sanctuary, its purpose is not known. Windmill Hill, a Neolithic causewayed enclosure, is 2 km northwest of Avebury and West Kennet chambered long barrow lies to the south.

Although the ritual function of Stonehenge is not known in detail, the cosmic references of its structure appear essential. An old theory is that the site was a sanctuary for worship of the sun. Although there is no unanimous agreement among prehistorians on the subject, Stonehenge nevertheless attracts a folkloric gathering at dawn each Midsummer Day.

World Heritage site since

1978 1979 1980 1981 1982 1983 1984 1985 **1986** 1987 1988 1989 1990 1991 1992 1993 1994 1995 1996 1997 1998 1999 2000 2001 2002 2003 2004 2005 2006 2007 2008

Mudéjar Architecture of Aragon
Spain

Criteria – Significance in human history

Tower in Mudéjar architectural style, Teruel.
▼

Mudéjar art represents the fusion of two artistic traditions, Islamic and Christian, in the region of Aragon, after the Reconquista of the twelfth century. Present until the early seventeenth century, it is characterized by an extremely refined and inventive use of brick and glazed tiles, especially in the belfries, the most visible element of Mudéjar architecture. The towers of Teruel together form a coherent ensemble which is truly characteristic of Mudéjar art after the Reconquista. The architects of the Christian churches copied the structure and decoration of Almohad minarets, although giving them new functions right from the start. Another typical feature of Mudéjar architecture is found in the painted and decorated wooden ceilings (e.g. Santa María de Mediavilla) of Teruel. Mudéjar architecture is also found in monasteries, castles, and residential buildings.

Mudéjar art continued to predominate over Gothic, except in some minor areas in the south until, in the sixteenth seventeenth centuries the Mudéjars were forced to convert to Christianity, becoming 'new Christians' (Moriscos). This was followed by a period of intolerance, resulting in their expulsion in 1609–10, and the extinction of Mudéjar art.

World Heritage site since

1978 1979 1980 1981 1982 1983 1984 1985 **1986** 1987 1988 1989 1990 1991 1992 1993 1994 1995 1996 1997 1998 1999 2000 2001 2002 2003 2004 2005 2006 2007 2008

Khami Ruins National Monument
Zimbabwe

Criteria – Testimony to cultural tradition; Significance in human history

Khami, which developed after the capital of Great Zimbabwe had been abandoned in the mid-sixteenth century, is of great archaeological interest. It is scattered over more than 2 km, from Passage Ruin to North Ruin. Although located in an area where a human presence can be traced back roughly 100,000 years, the city grew between 1450 and 1650. As is the case in Great Zimbabwe, several sectors can be clearly differentiated in terms of use.

The chief's residence (mambo) is towards the north, on the Hill Ruins site, which is a hill, created largely of earth used to level the terraces, contained by bearing walls. Some highly significant imported goods have been found here: sixteenth century Rhineland stoneware, Ming porcelain pieces which date back to the seventeenth century, Portuguese imitations of Chinese porcelain and seventeenth century Spanish silverware.

The people of Khami lived in huts made from cob (a mixture of earth, sand and straw) surrounded by granite walls. The fences and walls are similar to later constructions at Great Zimbabwe. Worthy of note are the many decorative friezes, with chevron and chequered patterns, and the great number of narrow passageways and galleries.

Castles and Town Walls of King Edward in Gwynedd
United Kingdom

Criteria – Human creative genius; Testimony to cultural tradition; Significance in human history

The castles of Beaumaris and Harlech, largely the work of the greatest military engineer of the time, James of St George, and the fortified complexes of Caernarfon and Conwy are located in the former principality of Gwynedd, in north Wales. These extremely well-preserved monuments are examples of the colonization and defence works carried out throughout the reign of Edward I (1272–1307) and the military architecture of the time.

From 1283, Edward I undertook a castle-building programme of unprecedented scale. In twenty years, ten fortresses were built, not to mention those restored after being wrested from the enemy, creating a strategic and symbolic expression of English power.

◄

Caernarfon castle and battlements along the river Seiont in north Wales.

World Heritage site since

1978 1979 1980 1981 1982 1983 1984 1985 **1986** 1987 1988 1989 1990 1991 1992 1993 1994 1995 1996 1997 1998 1999 2000 2001 2002 2003 2004 2005 2006 2007 200

Historic Centre of Évora

Portugal

Criteria – Interchange of values; Significance in human history

University of The Holy Spirit in Évora.

Évora's unique character derives from the coherence of the minor architecture of the sixteenth, seventeenth and eighteenth centuries. This finds its expression in the myriad low, whitewashed, tile-roofed houses, and in the terraces which line the narrow medieval streets of the old city centre and other areas. Wrought iron and azulejo decoration splendid in the convents and palaces and charming in the most humble dwellings, strengthens the fundamental unity of a type of architecture which is perfectly adapted to the climate and the site.

Évora is the finest example of a city of the golden age of Portugal after the destruction of Lisbon by the earthquake of 1755. A museum-city whose roots go back to Roman times, it reached its apogee in the fifteenth century when it became the residence of the Portuguese kings. Its unique quality stems from the whitewashed houses decorated with azulejos (ceramic tiles) and wrought-iron balconies dating from the sixteenth to the eighteenth century. The cityscape of Évora demonstrates the influence exerted by Portuguese architecture in Brazil, in sites such as Salvador de Bahia.

Évora has been shaped by some twenty centuries of history, going back to pre-Roman Celtic times. The city fell under Roman domination, when it was called Liberalitas Julia; among other ruins, the Temple of Diana still stands in the town. During the Visigothic period, the Christian city occupied the area surrounded by the Roman wall, which was reworked. Under Moorish domination, which ended in 1165, further improvements were made to the original defensive system as seen in a fortified gate and the remains of the ancient Kasbah. Moreover, place names are indicative of the Maghreb population, which remained after the Reconquest in the La Mouraria quarter of the northeast.

There are a number of buildings from the medieval period, the best known of which is unquestionably the cathedral, begun in 1186 and essentially completed in the thirteenth to fourteenth centuries. It was in the fifteenth century, however, when the Portuguese kings began living in the city on an increasingly regular basis, that Évora's golden age began. Convents and royal palaces sprang up across the city. These splendid monuments, which were either entirely new buildings or else constructed within already existing establishments, are characterized by the Manueline style which survived in the major creations of the sixteenth century, such as the Palace of the Counts of Basto, built on the site of the Alcazar, and the Church of the Knights of Calatrava.

The sixteenth century was a time of major urban planning: the Agua da Prata aqueduct was built in 1537 and many fountains remain from that time. Évora's intellectual and religious influence was also strong: the Jesuits taught at the University of the Holy Spirit from 1553 until their expulsion in 1759, when the city's rapid decline began.

World Heritage site since

1978 1979 1980 1981 1982 1983 1984 1985 **1986** 1987 1988 1989 1990 1991 1992 1993 1994 1995 1996 1997 1998 1999 2000 2001 2002 2003 2004 2005 2006 2007 200

Gondwana Rainforests of Australia
Australia

Criteria – Major stages of Earth's history;
Significant ecological and biological processes;
Significant natural habitat for biodiversity

This site, comprising approximately forty separate reserves, is situated predominantly along the Great Escarpment on Australia's east coast. The outstanding geological features displayed around shield volcanic craters and the 200 rare and threatened rainforest species are of international significance for science and conservation. The evolution of new species is encouraged by the natural separation and isolation of rainforest stands. The site includes the most extensive areas of subtropical rainforest in the world, large areas of warm temperate rainforest and almost all of the Antarctic beech cool temperate rainforest. Although rainforests cover only about 0.3 per cent of Australia, they contain about half of all Australian plant families and about a third of Australia's mammal and bird species. Many plants and animals found here are locally restricted to a few sites or occur in widely separated populations.

Rainforest near Protestors Falls, Nightcap National Park, ▲ New South Wales.

Few places on Earth contain so many plants and animals whose ancestors can be traced through the fossil record to Gondwanan origins that today remain relatively unchanged. There is a concentration of primitive plant families that shows a direct link with the origin of flowering plants over 100 million years ago, as well as some of the oldest of the world's ferns and conifers.

World Heritage site since

1978 1979 1980 1981 1982 1983 1984 1985 **1986** 1987 1988 1989 1990 1991 1992 1993 1994 1995 1996 1997 1998 1999 2000 2001 2002 2003 2004 2005 2006 2007 2008

Škocjan Caves
Slovenia

Criteria – Natural phenomena or beauty; Major stages of Earth's history

This system of subterranean passages, fashioned by the Reka river, constitutes a dramatic example of large-scale karst drainage. An underground system of passages runs from the Reka's source to Timavo on the Gulf of Trieste in Italy. In places the surfaces of the galleries at several levels have collapsed and give the appearance of deep chasms. The river enters the Škocjan grotto through an underground passage 350 m long, reappearing at the

bottom of a chasm 150 m deep and 300 m long, before disappearing into a passage 2 km long. There are five galleries and a canal. A gallery of stalactites and stalagmites leads to the surface. In total there are twenty-five cascades along the river.

The site, located in the Kras region (literally meaning 'karst'), is one of the most famous in the world for the study of karstic phenomena. The protected area extends over 2 km² and includes four deep and picturesque chasms, Sokolak in the south, Globocak in the west, and Lisicina and Sapen dol in the north.

Old Town of Ghadamès
Libyan Arab Jamahiriya

Criteria – Traditional human settlement

The historic city of Ghadamès, known as 'the pearl of the desert,' is today a small oasis city situated next to a palm grove. It is one of the oldest pre-Saharan cities and an outstanding example of a traditional settlement. Its domestic architecture is characterized by a vertical division of functions: the ground floor was used to store supplies; then came another floor for

the family, overhanging covered alleys that create what is almost an underground network of passageways; and at the top were open-air terraces reserved for the women. Roughly circular in layout, the city comprises a cluster of houses. The reinforced outer walls of the houses on the edge of the city form a fortified wall.

The unique layout of this unusual city cannot be perceived as a whole. Ground level passageways form arcades rather than actual streets, and are principally the reserve of the men. At the upper level, women socialize on connected terraces.

World Heritage site since

1978 1979 1980 1981 1982 1983 1984 1985 1986 **1987** 1988 1989 1990 1991 1992 1993 1994 1995 1996 1997 1998 1999 2000 2001 2002 2003 2004 2005 2006 2007 200

Archaeological Site of Delphi
Greece

Criteria – Human creative genius; Interchange of values; Testimony to cultural tradition; Significance in human history; Heritage associated with events of universal significance

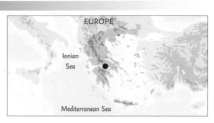

EUROPE

Ionian
Sea

Mediterranean Sea

Delphi Tholos.

The development of the sanctuary and oracle began in the eighth century BC when the cult of Apollo was established. The sanctuary had strong religious and political influence across Greece.

The Pan-Hellenic sanctuary of Delphi, standing in a magnificent natural setting on Mount Parnassus, was the site of the Delphic Oracle through which the god Apollo spoke. Blending harmoniously with the superb landscape and charged with sacred meaning, Delphi in the sixth century BC was both the religious centre and symbol of unity of the ancient Greek world.

The modular elements of the Dephi site – terraces, temples and treasuries – combine to form a strong expression of its physical and moral values. There are several monuments; the following are among the most important.

Temple of Apollo: Dated to the fourth century BC, the temple was erected on the remains of an earlier, sixth-century temple. Inside was the seat of the Pythia, the priestess who presided over the Oracle and delivered the prophesies inspired by Apollo. Delphi was the most important oracle in the Greek world.

Altar of the Chians: The large altar of the sanctuary, in front of the temple, erected in the fifth century BC.

Treasury of the Athenians: A small building in Doric order, built by the Athenians at the end of the sixth century BC to house their offerings to Apollo.

Stoa of the Athenians: Built in the Ionic order, has seven fluted columns, each made from a single stone. According to an inscription, it was built by the Athenians after 478 BC to house the trophies taken in their naval victories over the Persians.

Theatre: Originally built in the fourth century BC, its visible ruins actually date from the Roman imperial period. The theatre was used mostly for the performances during the great festivals.

Stadium: Constructed in the fifth century BC and remodelled in the second century AD at the expense of Herodes Atticus. The Pan-Hellenic Pythian Games, one of the forerunners of the Olympic Games, took place in this stadium.

Tholos: A circular building in Doric order, built around 380 BC (pictured on the right). Although its function is unknown it must have been important, judging from the fine workmanship.

Polygonal Wall: Built after the destruction of the old temple of Apollo in 548 BC to support the terrace on which the new temple was to be erected. Many inscriptions, mostly manumissions of slaves, are carved into its stones.

The Pythian Games were reorganized and held here from the sixth century BC. At that time the sanctuary was enlarged and enriched with fine buildings, statues and other offerings.

During the Roman period Delphi's fortunes were mixed. As Christianity spread it lost its religious significance and was finally closed down by Emperor Theodosius I in AD 395.

World Heritage site since

1978 1979 1980 1981 1982 1983 1984 1985 1986 **1987** 1988 1989 1990 1991 1992 1993 1994 1995 1996 1997 1998 1999 2000 2001 2002 2003 2004 2005 2006 2007 2008

Hawaii Volcanoes National Park
USA

Criteria – Major stages of Earth's history

The park extends from the volcanic sea-cliff headlands of the southern coast to the summit calderas of Kilauea, the world's most active volcano, and Mauna Loa. The latter is a massive, flat-domed shield volcano built by lava flow layers and is considered to be the best example of its type in the world, extending from 5,581 m below sea level to 4,170 m above.

Two of the most active volcanoes in the world, Mauna Loa (4,170 m) and Kilauea (1,250 m), tower over the Pacific Ocean at this site where volcanic eruptions have created a constantly changing landscape.

Climate varies with altitude from tropical humid to alpine desert and the park contains a high diversity of plant communities with striking life-form and physiognomic differences. There are twenty-three distinct vegetation types, grouped into five major ecosystems: subalpine, montane

seasonal, montane rainforest, submontane seasonal and coastal lowlands.

Except for a single species of Hawaiian hoary bat, the park has no native mammals and most endemic birds are rare or endangered. Ranching and the introduction of species such as the pig, goat and mongoose have had serious biological consequences, including destruction of native ecosystems and widespread extinction of endemic species.

◄

Part of the volcanic shelf collapsed into the ocean five days after this photograph was taken in November 2005.

World Heritage site since

1978 1979 1980 1981 1982 1983 1984 1985 1986 **1987** 1988 1989 1990 1991 1992 1993 1994 1995 1996 1997 1998 1999 2000 2001 2002 2003 2004 2005 2006 2007 2008

Piazza del Duomo, Pisa
Italy

Criteria – Human creative genius; Interchange of values; Significance in human history; Heritage associated with events of universal significance

The world-famous monuments of the Piazza del Duomo are masterpieces of medieval architecture. The cathedral, its baptistry, its campanile the Leaning Tower and its walled cemetery, had a huge influence on monumental art in medieval Italy.

The huge marble-and-stone cathedral has been identified as the best example of Pisan Romanesque style. Work began in 1064 and was completed in the twelfth century. The cathedral is in the shape of a Latin cross, with five naves. Its Romanesque baptistry, with a pulpit by the sculptor Nicola Pisano, was completed in 1363. The cathedral's famous leaning bell tower began subsiding soon after construction began in 1173. War meant it remained unfinished until almost 200 years later.

The monumental cemetery was virtually destroyed by Allied bombing during the Second World War but has since been restored to its original state.

Cathedral and its campanile (the Leaning Tower). ▲

Pisa was an ancient port and one of Italy's four wealthy Maritime Republics, powerful trading city-states. Its enormous wealth meant the city flourished until commercial rivalry and war precipitated a decline in the late thirteenth century.

In the cathedral hangs the bronze lamp of Possenti da Pietrasanta; its swinging was said to have inspired Galileo's discovery of the principle of isochronism.

World Heritage site since

1978 1979 1980 1981 1982 1983 1984 1985 1986 **1987** 1988 1989 1990 1991 1992 1993 1994 1995 1996 1997 1998 1999 2000 2001 2002 2003 2004 2005 2006 2007 2008

Cathedral, Alcázar and Archivo de Indias in Seville
Spain

Criteria – Human creative genius; Interchange of values; Testimony to cultural tradition; Heritage associated with events of universal significance

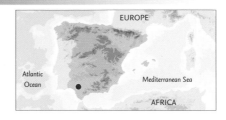

These three buildings in the heart of Seville together form a remarkable monumental complex. The cathedral and Alcázar date from the Reconquista, the recapture from the Moors of the Iberian peninsula, and reflect elements of the civilizations of the Berber Almohads and of Christian Andalusia. Begun in the late sixteenth century, the Casa Lonja was intended as a Trades Hall but by the 1790s it had become the Archivo General de Indias, housing collections of documents of the Spanish overseas empire.

The cathedral, the largest Gothic cathedral in the world (pictured below), houses the tomb of Christopher Columbus. The Alcázar is a palatial fortress first built in 712 by the conquering Arabs. After the Spanish retook Seville in 1248 it became a Spanish royal residence.

The cathedral reflects a chequered history in its complex structure. The capitals of several columns in one area date from the time of the Visigoths; these represent the last vestiges of the original cathedral which Seville's Arab conquerors destroyed in 712. The Great Mosque they built in the twelfth century was itself built over, although the spectacular Giralda minaret remains.

View of the cathedral from the Giralda tower. ▼

World Heritage site since

1978 1979 1980 1981 1982 1983 1984 1985 1986 **1987** 1988 1989 1990 1991 1992 1993 1994 1995 1996 1997 1998 1999 2000 2001 2002 2003 2004 2005 2006 2007 2008

Sian Ka'an
Mexico

Criteria – Natural phenomena or beauty;
Significant natural habitat for biodiversity

The area has 103 types of mammal including jaguar, puma, ocelot, margay and jaguarondi; Central American tapir and red brocket; spider monkey and howler monkey; kinkajou, collared anteater and Caribbean manatee. Some 339 bird and 42 amphibian and reptile species have been recorded as well as over 52 types of fish, which are abundant.

This biosphere reserve on the east coast of the Yucatán peninsula contains tropical forests, mangroves, marshes and a large marine section intersected by a barrier reef. The complex hydrological system provides a habitat for a rich flora and a fauna comprising more than 300 species of birds and a large number of the region's characteristic terrestrial vertebrates.

Sian Ka'an lies on a partially emerged coastal limestone plain which forms part of the extensive barrier-reef system along

Central America's east coast. The hydrological cycle is complex and the water table is permanently close to the surface although shallow limestone soils mean that there is little surface running water within the reserve.

There are an estimated 1,200 plant species. In the medium and low semi-deciduous forest, abundance of palm is a characteristic feature. Coastal dunes stretch along 64 km of the coast.

World Heritage site since

1978 1979 1980 1981 1982 1983 1984 1985 1986 **1987** 1988 1989 1990 1991 1992 1993 1994 1995 1996 1997 1998 1999 2000 2001 2002 2003 2004 2005 2006 2007 2008

Venice and its Lagoon
Italy

Criteria – Human creative genius; Interchange of values; Testimony to cultural tradition; Significance in human history; Traditional human settlement; Heritage associated with events of universal significance

Part of the Grand Canal and one of the numerous side canals that dissect the city.

Venice is a unique artistic achievement: built on 118 small islands, the city seems to float on the waters of the Venetian Lagoon. Venice possesses an incomparable series of architectural ensembles that collectively illustrate the longevity of its splendour.

Standing in this inland sea on a tiny archipelago at the very edge of the waves, Venice is one of the most extraordinary built-up areas of the Middle Ages. From Torcello in the north to Chioggia in the south, almost every small island had its own settlement. In the centre of the 50,000-km² lagoon, Venice stood as one of the greatest capitals in the medieval world.

In the fifth century AD, Venetian populations fleeing hostile invaders first found refuge on the sandy islands of Torcello, Iesolo and Malamocco. What had begun as temporary settlements gradually became permanent and the one-time refuge of land-dwelling peasants and fishermen grew into one of the world's foremost maritime powers.

Venetian power grew steadily from the twelfth to the fifteenth century. The independent city-state was one of the four Italian Maritime Republics and from its strategically crucial position at the head of the Adriatic Sea, Venice controlled the length of the eastern Mediterranean to the Ionian Sea. This control allowed Venice to dominate trade between Europe and the Byzantine Empire and the Near East, including the Crusader States. In 1204, led by its ruler, the Doge Enrico Dàndolo, Venice allied with the Crusaders to loot Constantinople. Among the abundant spoils brought back were the bronze horses that still stand over the entrance to St Mark's Cathedral.

With the development of overseas empires in the fifteenth century and the expansion of the Turkish Ottoman Empire into Byzantium, the balances of power on which Venice's wealth was built began to change, and the city's long decline began. It finally lost its independence to Napoleon in 1797, after which many of its palaces and notable buildings fell into disrepair.

The influence of Venice on the development of architecture and monumental arts has been considerable. The fabulous wealth that its trading empire brought is reflected in the scale and grandeur of its beautiful buildings and monuments. The city has been inspiration to some of the world's greatest artists, including Canaletto, Giorgione, Titian, Tintoretto and Veronese.

The islands of the Comune of Venice are grouped around the historic city at its centre. Transport and communication in this necessarily car-free area is by water. Canals – such as the Giudecca Canal, St Mark's Canal and the Grand Canal – and the network of rii, or rivers, are the arteries of the city. In this unreal space where there is no notion of the concept of terra firma, masterpieces of one of the most extraordinary architectural museums on Earth have been accumulated for over 1,000 years.

World Heritage site since

1978 1979 1980 1981 1982 1983 1984 1985 1986 **1987** 1988 1989 1990 1991 1992 1993 1994 1995 1996 1997 1998 1999 2000 2001 2002 2003 2004 2005 2006 2007 2008

Blenheim Palace
United Kingdom

Criteria – Interchange of values; Significance in human history

Atlantic Ocean

North Sea

EUROPE

Blenheim Palace and park illustrate the beginnings of the English Romantic movement, characterized by the eclecticism of its inspiration, its return to national sources and its love of nature. The influence of Blenheim on the architecture and the organization of space in the eighteenth and nineteenth centuries was greatly felt both in England and abroad.

Situated near Oxford, Blenheim Palace stands in an 8.5-km² romantic park created by the famous landscape gardener Lancelot 'Capability' Brown. It was presented by the English nation to John Churchill, first Duke of Marlborough, in recognition of his victory in 1704 over French and Bavarian troops at the Battle of Blenheim. Built by John Vanbrugh and Nicholas Hawksmoor between 1705 and 1722, it is a perfect example of an eighteenth-century princely dwelling.

Between 1764 and 1774 'Capability' Brown, one of the most famous English landscape gardeners, turned the classical park laid out by Vanbrugh into a wonderful artificial landscape by the creation of two lakes. In the later eighteenth century, Gothic or neo-Gothic style buildings were added.

The British Prime Minister Winston Churchill was born in the palace in 1874.

World Heritage site since

1978 1979 1980 1981 1982 1983 1984 1985 1986 **1987** 1988 1989 1990 1991 1992 1993 1994 1995 1996 1997 1998 1999 2000 2001 2002 2003 2004 2005 2006 2007 2008

Sundarbans National Park
India

Criteria – Significant ecological and biological processes; Significant natural habitat for biodiversity

The Sundarbans covers 10,000 km² of land and water in the Ganges delta, more than half of it in India, the rest in Bangladesh . It contains the world's largest area of mangrove forests. A number of rare or endangered species live in the park, including tigers, aquatic mammals, birds and reptiles.

The Sundarban waterways now carry little freshwater as the outflow of the Ganges has shifted progressively eastwards since the seventeenth century. This is due to subsidence of the Bengal Basin and a gradual eastward tilting of the overlying crust. Waterways in the tiger reserve are maintained largely by the diurnal tidal flow.

Elephanta Caves
India

Criteria – Human creative genius; Testimony to cultural tradition

The 'City of Caves', consists of seven caves on an island in the Arabian Sea close to Mumbai. With their decorated temples and images from Hindu mythology, they bear a unique testimony to a civilisation that has disappeared and are one of the most striking collections of rock art in India. There are two groups of caves, dating from the sixth – eighth centuries AD. To the east, Stupa Hill (named after the small brick Buddhist monument at the top) contains two caves, one of which is unfinished, and several cisterns. To the west, the larger group consists of five rock-cut Hindu shrines. The main cave is universally famous for its huge high reliefs to the glory of Shiva, who is exalted in various forms and actions.

The island of Elephanta owes its name to the enormous stone elephant found there by Portuguese navigators. This elephant was cut into pieces, removed to Mumbai and somehow put together again. It is today the melancholy guardian of Victoria Gardens Zoo in Mumbai.

◄
Entrance to the secondary cave on Elephanta Island.

World Heritage site since

1978 1979 1980 1981 1982 1983 1984 1985 1986 **1987** 1988 1989 1990 1991 1992 1993 1994 1995 1996 1997 1998 1999 2000 2001 2002 2003 2004 2005 2006 2007 200

Budapest, including the Banks of the Danube, the Buda Castle Quarter and Andrássy Avenue
Hungary

Criteria – Interchange of values; Significance in human history

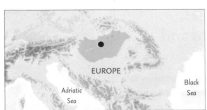

A panoramic view of Budapest with the Danube River and the Hungarian Houses of Parliament on the far riverbank.

With no attempts made at organized urban development since the Middle Ages, the Hungarian capital caught up in one great leap in respect of public services, transportation and city planning. The route of Andrássy Avenue cut straight through an unregulated suburban area, thereby radically transforming its urban structure. The Siemens and Halske companies built the first underground railway on the European continent there between 1893 and 1896, which in turn led to the construction of more city monuments.

Budapest is one of the world's outstanding urban landscapes and illustrates in its architecture the great periods in the history of the Hungarian capital. It has the remains of monuments from various periods, such as the Roman city of Aquincum and the Gothic castle of Buda, which have had a considerable influence on the architecture.

Within the unified perspective of an immense urban panorama, the Danube is the dividing line between what were originally two cities: Buda on the spur on the right bank, and Pest in the plain on the left bank. Human occupation can be traced back to the Palaeolithic period, but the city's historic importance dates from the Roman period when it was part of Lower Pannonia, a border province of the Empire in the second century AD.

After the Hungarian invasion in the ninth century, Pest became the first medieval urban centre, only to be devastated by Mongol raids in 1241–2. A few years later Bela IV built the castle of Buda on the right bank and the inhabitants of Pest found shelter within its fortified outer walls. Buda Castle played an essential role in the diffusion of Gothic art in the Magyar region from the fourteenth century.

Buda's history became closely identified with that of the Hungarian monarchy and the city followed its changing fortunes. Two centuries of ascendancy from 1308 to the 1490s and beyond were ended when the Turks ransacked the city in 1526, precipitating its final fall in 1541. Recovery did not really begin again until the eighteenth century.

In the nineteenth century the city's role as capital was enhanced by the foundation of the Hungarian Academy (1830) and especially by the construction of the imposing neo-Gothic Parliament building (1884–1904). The parliament (pictured on the right) is an outstanding example of a great official building on a par with those of London, Munich, Vienna and Athens. It exemplifies the eclectic architecture of the nineteenth century while symbolizing the political function of the second capital of the Austro-Hungarian monarchy. Since 1849, W. T. Clark's suspension bridge over the Danube has symbolized the reunification of Buda and Pest, which did not become official until 1873. With this union, Budapest truly became the nation's capital.

World Heritage site since

1978 1979 1980 1981 1982 1983 1984 1985 1986 **1987** 1988 1989 1990 1991 1992 1993 1994 1995 1996 1997 1998 1999 2000 2001 2002 2003 2004 2005 2006 2007 2008

Kilimanjaro National Park
Tanzania
Criteria – Natural phenomena or beauty

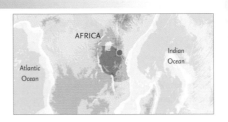

At 5,895 m, the volcanic massif of Mount Kilimanjaro is the highest point in Africa. It stands in splendid isolation above the surrounding plains, encircled by mountain forest, its snowy peak looming over the savanna. Numerous mammals, many of them endangered, live in the park.

The national park and forest reserve occupy the upper part of Kilimanjaro adjacent to the Kenyan border. The national park comprises all the mountain above the timberline and six forest corridors stretching down through the montane forest belt. Kilimanjaro, one of the largest volcanoes in the world, last showed signs of major activity in the Pleistocene period (between 1.8 million and 10,000 years ago). It stands alone but is the largest of an east-west belt of volcanoes stretching across northern Tanzania. It has three main volcanic peaks of varying ages, Shira, Mawenzi and Kibo, and a number of smaller parasitic cones.

Kilimanjaro has five main vegetation zones: savanna bushland at 700–1,000 m (on the south slopes) 1,400–1,600 m (on the north slopes) and densely populated submontane agroforest on southern and southeastern slopes, the montane forest belt, subalpine moorland and alpine bogs. Above this is alpine desert.

The montane forest belt circles the mountain between 1,300 m (about 1,600 m on the drier north slopes) to 2,800 m. Forests above 2,700 m are within the national park. According to a 2001 study there are 2,500 plant species on the mountain, 1,600 of them on the southern slopes and 900 within the forest belt. There are 130 species of tree, with the greatest diversity being found between 1,800 and 2,000 m.

The whole mountain including the montane forest belt, part of which extends into the national park, is very rich in animal life: there are 140 mammals (eighty-seven forest species), including seven primates, twenty-five carnivores, twenty-five antelopes and twenty-four species of bat. Above the timberline at least seven of the larger mammal species have been recorded, although it is likely that many of these also use the lower montane forest habitat.

Although 179 highland bird species have been recorded on the mountain, species recorded in the upper zones are few in number. The white-necked raven is the most conspicuous bird species at higher altitudes.

The area around the mountain is quite heavily populated, principally by the Chagga people, and the northern and western slopes of the forest reserve surrounding the national park have eighteen medium-to-large forest villages. Although it is illegal, these people use the forest for many household and medicinal products, for fuelwood, small-scale farming, beekeeping, hunting, charcoal production and logging.

Some 12 per cent of the forest is plantation, some of which almost reaches down to the moorland.

World Heritage site since

1978 1979 1980 1981 1982 1983 1984 1985 1986 **1987** 1988 1989 1990 1991 1992 1993 1994 1995 1996 1997 1998 1999 2000 2001 2002 2003 2004 2005 2006 2007 2008

Acropolis, Athens
Greece

Criteria – Human creative genius; Interchange of values; Testimony to cultural tradition; Significance in human history; Heritage associated with events of universal significance

Parthenon at the Acropolis, Athens.

The Acropolis of Athens and its monuments are universal symbols of the classical spirit and civilization, and form the greatest architectural and artistic complex of Greek antiquity.

The Acropolis stands on a rocky promontory 156 m above the Ilissos valley and covers an area of less than 30,000 m². From the third millennium BC it was a fortress protecting places of worship and royal palaces.

The Acropolis is now a testing ground for innovative open-air conservation techniques aimed at safeguarding the marble sections which are being affected by pollution.

The Athenian Acropolis is the supreme expression of the adaptation of architecture to a natural site, a unique series of public monuments built and conserved in one of the densest spaces of the Mediterranean. This grand composition of perfectly balanced massive structures creates a monumental landscape of unique beauty consisting of a complete series of masterpieces of the fifth century BC. The monuments of the Acropolis have exerted an exceptional influence, not only in Graeco-Roman antiquity, a time in the Mediterranean world when they were considered exemplary models, but also in contemporary times.

In the later fifth century BC, Athens followed its victory against the Persians and the establishment of democracy by taking a leading position among the other city-states of the ancient world. In the age that followed, as philosophy and art flourished, an exceptional group of artists put into effect the ambitious plans of Athenian statesman Pericles and, under the inspired guidance of the sculptor Pheidias, transformed the rocky hill of the Acropolis into a unique monument of thought and the arts.

The years from 447–406 BC saw the successive building of the Parthenon (pictured on the right), the main temple dedicated to Athena; the Propylaea, the monumental entrance to the Acropolis built on the site of one of the entrances to the citadel of the ancient kings; the temple of Athena Nike; and the Erechtheion – the four masterpieces of classical Greek art.

The sacred hill of Athens was protected throughout the period of Roman domination until the Herulian raid in AD 267. Since then and despite long periods of relative calm, the monuments and site have been damaged many times.

The Byzantines converted the temples into churches and removed their art treasures to Constantinople. After the Byzantine Empire fell in 1204, Athens was put into the hands of Frankish lords who had little respect for its ruins. When the Turks took the city in 1456, it became a mosque and the Erechtheion was the occasional harem of the Turkish governor. In 1687 the siege of the Acropolis by Venetian armies resulted in the explosion of the Parthenon, used as the Turks' powder magazine. Finally in the nineteenth century, the British ambassador Lord Elgin pillaged the marble sections which since 1815 have been in the British Museum.

World Heritage site since

1978 1979 1980 1981 1982 1983 1984 1985 1986 **1987** 1988 1989 1990 1991 1992 1993 1994 1995 1996 1997 1998 1999 2000 2001 2002 2003 2004 2005 2006 2007 2008

Historic Centre of Mexico City and Xochimilco
Mexico

Criteria – Interchange of values; Testimony to cultural tradition; Significance in human history; Traditional human settlement

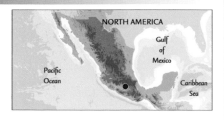

Built in the sixteenth century by the Spanish on the ruins of Tenochtitlan, the old Aztec capital, Mexico City is now one of the world's largest and most densely populated cities. It has five Aztec temples, the ruins of which have been identified, the largest cathedral on the continent and some fine nineteenth- and twentieth-century public buildings such as the Palacio de las Bellas Artes.

Xochimilco lies 28 km south of Mexico City. With its network of canals and artificial islands, it testifies to the efforts of the Aztec people to build a habitat in the midst of an unfavourable environment. Its characteristic urban and rural structures, built since the sixteenth century and during the colonial period, have been preserved in an exceptional manner.

The value of these two properties is unequalled. The historic centre of Mexico city includes the archaeological site of the Templo Mayor with its remarkable array of colonial monuments and famous cathedral. The lakeside area of Xochimilco still features some chinampas, the floating gardens that the Spanish so admired.

Historic Centre of Puebla
Mexico

Criteria – Interchange of values; Significance in human history

Puebla, which was founded ex nihilo in 1531, is situated about 100 km east of Mexico City, at the foot of the Popocatepetl volcano. It was the first city in central Mexico founded by the Spanish conquerors that was not built upon the ruins of a conquered Amerindian settlement. On 5 May 1862, it was in Puebla that General Zaragoza won the first significant victory over the French expeditionary corps. The city was subsequently renamed Puebla de Zaragoza in memory of this event of national

importance. It has preserved its great religious structures such as the sixteenth–seventeenth-century cathedral and fine buildings such as the old archbishop's palace, as well as a host of houses with walls covered in tiles (azulejos). The new aesthetic concepts resulting from the fusion of European and American styles were adopted locally and are peculiar to the Baroque district of Puebla.

The Historic Centre of Puebla comprises major religious buildings such as the Cathedral Santo Domingo and the Jesuit Church, as well as superb palaces. Nineteenth-century transformations of the urban landscape have further endowed Puebla with high-quality public and private architecture.

World Heritage site since

978 1979 1980 1981 1982 1983 1984 1985 1986 **1987** 1988 1989 1990 1991 1992 1993 1994 1995 1996 1997 1998 1999 2000 2001 2002 2003 2004 2005 2006 2007 2008

Chaco Culture
USA

Criteria – Testimony to cultural tradition

For over 2,000 years, Pueblo peoples occupied a vast region of the southwestern United States. Chaco Canyon is the area with the highest concentration of archaeological sites of the whole zone. Chaco society is characterized by an elaborate ground occupation system, which includes a constellation of towns surrounded by satellite villages and linked by a road network. The Chaco people combined pre-planned architectural designs, astronomical alignments, geometry, landscaping and engineering to create an ancient urban centre of spectacular public architecture. It became a major centre of ancestral Pueblo culture between 850 and 1250, and was a focus for ceremonials, trade and political activity for the prehistoric Four Corners area. In addition to the Chaco Culture National Historical Park, the World Heritage property includes the Aztec Ruins National Monument and several smaller Chaco sites.

Between the twelfth and thirteenth centuries the Chaco population died out and the pueblos were abandoned. After 1250, the people migrated from the area, moving south, east and west. The region remained practically uninhabited until the seventeenth century, when it was taken over by Navajo Indians.

◄
Pueblo village ruins, including a 'kiva' – an underground/partially underground chamber used by the men especially for ceremonies or councils.

World Heritage site since

1978 1979 1980 1981 1982 1983 1984 1985 1986 **1987** 1988 1989 1990 1991 1992 1993 1994 1995 1996 1997 1998 1999 2000 2001 2002 2003 2004 2005 2006 2007 2008

Historic Centre of Oaxaca and Archaeological Site of Monte Albán
Mexico

Criteria – Human creative genius; Interchange of values; Testimony to cultural tradition; Significance in human history

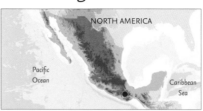

Three distinct cultural properties stand in the Oaxaca valley: the historic centre of the city founded in 1529 by the Spanish; the pre-Hispanic archaeological site of Monte Albán; and the village of Cuilapan, where the Dominicans built a vast monastery in the mid-sixteenth century.

Monte Albán is an outstanding example of a pre-Columbian ceremonial centre. It stands in the middle zone of present-day Mexico which was subjected to influences from the north – first from Teotihuacan and later the Aztecs – and from the Maya from the south. With its pelota court, magnificent temples, tombs and bas-reliefs with hieroglyphic inscriptions, Monte Albán bears unique testimony to the successive civilizations occupying the region during the pre-Classic and Classic periods that stretched from around 1800 BC to AD 900. For more than a millennium it exerted considerable influence on the whole cultural area.

Among some 200 pre-Hispanic archaeological sites inventoried in the valley of Oaxaca, the Monte Albán complex best represents the singular evolution of a region inhabited by a succession of peoples: the Olmecs, Zapotecs and Mixtecs.

Monte Albán was literally carved out from a solid mountain, in various stages spanning 1,500 years. Man-made terraces and esplanades replaced the natural unevenness of the site with a whole new sacred topography of pyramids, and artificial knolls and mounds. The ensemble began to decline around 800 when the Mixtecs, descending from the mountains, threatened the Zapotecs living in the valley.

A short time before the arrival of the Spanish conquistadores, the Aztecs took control of the valley and founded the stronghold of Huaxyacac. This place name survived, when in 1521 the Spanish erected the fort of Antequera de Oaxaca on the same site.

Oaxaca is built on a grid pattern and is a good example of Spanish colonial town planning. Its monumental heritage is one of the richest and most coherent in the area that was known as New Spain. The solidity and volume of the city's buildings show that these architectural gems were adapted to the earthquake-prone region in which they were constructed.

The Cuilapan Convent, 10 km to the south, was founded in 1555. Its roofless, open-air church was never finished.

Pre-Hispanic archaeological site of Monte Albán.

The modern city of Oaxaca has retained its historic centre. The major religious monuments, the superb patrician town houses (including the house of conquistador Hernán Cortés) and whole streets lined with other dwellings combine to create a harmonious cityscape and reconstitute the image of a former colonial city whose monumental aspect has been kept intact.

Near Oaxaca is the birthplace of Benito Juárez, the first indigenous president of Mexico and a national hero. On his death in 1872, the city formally took the name of Oaxaca de Juárez.

World Heritage site since

1978 1979 1980 1981 1982 1983 1984 1985 1986 **1987** 1988 1989 1990 1991 1992 1993 1994 1995 1996 1997 1998 1999 2000 2001 2002 2003 2004 2005 2006 2007 2008

Pre-Hispanic City of Teotihuacan
Mexico

Criteria – *Human creative genius; Interchange of values; Testimony to cultural tradition; Significance in human history; Heritage associated with events of universal significance*

Located 48 km northeast of Mexico City, the holy city of Teotihuacan ('the place where the gods were created') is one of the oldest known archaeological sites in Mexico. Built between the first and seventh centuries AD, it is characterized by the vast size of its sacred monuments, which were laid out on geometric and symbolic principles. Lining the immense Avenue of the Dead, this unique group of monuments and places of worship (Pyramids of the Sun, the Moon and Quetzalcoatl, and Palaces of Quetzalmariposa, Jaguars, Yayahuala and others) constitutes an outstanding example of a pre-Columbian ceremonial centre. As one of the most powerful cultural centres in Mesoamerica, Teotihuacan extended its cultural and artistic influence throughout the region and beyond.

The location of the first sanctuary, the Pyramid of the Sun (built on a cave discovered in 1971), was calculated on the position of the sun at its zenith, and astronomical calculations determined the organization of the space: the Avenue of the Dead was drawn out perpendicularly to the principal axis of the solar temple.

World Heritage site since

78 1979 1980 1981 1982 1983 1984 1985 1986 **1987** 1988 1989 1990 1991 1992 1993 1994 1995 1996 1997 1998 1999 2000 2001 2002 2003 2004 2005 2006 2007 2008

City of Potosí
Bolivia

Criteria – Interchange of values; Significance in human history; Heritage associated with events of universal significance

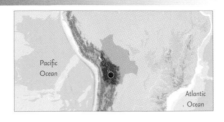

Potosí owes its importance to the discovery, between 1542 and 1545, of the New World's biggest silver lodes in the Cerro de Potosí, the mountain south of the city and which overlooks it. As a result it quickly became the world's largest industrial complex, with the extraction of silver ore relying on a series of hydraulic mills. The site consists of the industrial monuments of the Cerro Rico, where water is provided by an intricate system of aqueducts and artificial lakes; the colonial town with the Casa de la Moneda; the Church of San Lorenzo; several patrician houses; and the barrios mitayos, the areas where the workers lived. Production continued until the nineteenth century, slowing down only after the country's independence in 1825.

The 'Imperial City' of Potosí, which it became following the visit of Francisco de Toledo in 1572, exerted lasting influence on the development of architecture and monumental arts in the central region of the Andes by spreading the forms of a Baroque style incorporating Indian influence.

World Heritage site since

1978 1979 1980 1981 1982 1983 1984 1985 1986 **1987** 1988 1989 1990 1991 1992 1993 1994 1995 1996 1997 1998 1999 2000 2001 2002 2003 2004 2005 2006 2007 200

City of Bath
United Kingdom

Criteria – Human creative genius; Interchange
of values; Significance in human history

Founded by the Romans as a thermal spa,
Bath has an exceptionally rich architectural
heritage which reflects the city's history
through several periods from its time as an
important Roman spa town.

The town grew around a temple the
Romans built between AD 60 and 70 on the
site of a geothermal spring. The mineral
spring was a shrine to the British goddess
Sulis, a local divinity whom the Romans
associated with Minerva, goddess of
wisdom and medicine. The Romans kept the
old reference in the name they gave the
town, Aquae Sulis (the waters of Sulis).

The spring yielded over 1,200,000 litres of
water daily at more than 46°C, and between
the first and fourth centuries the Romans
built a hot bath (calidarium), warm bath
(tepidarium) and cold bath (frigidarium)
together with all the standard equipment of
tepidaria, frigidaria and hypocausts, the
underfloor central-heating systems that
maintained the water temperature.

The town retained its importance after the
departure of the Romans in the fifth
century. In 1090 Bath succeeded Wells as
the seat of the see of Somerset and
consequently the church that had been Bath
Abbey became a cathedral. To match Bath's
new status, a new cathedral was begun and
new baths were also constructed. However,

the see returned to Wells and the cathedral
remained unfinished. In the sixteenth
century a rebuilt and remodelled church in
the Perpendicular Gothic style was finally
completed. The beautiful fan vaulting in the
nave was added as late as the nineteenth
century according, it is believed, to the
sixteenth century plans.

Medieval Bath grew wealthy on the wool
trade and the town became an important
centre of the wool industry. Although the
wool trade declined in the sixteenth and
seventeenth centuries, the city was by now
attracting wealthy visitors to take the waters.

In the eighteenth century Bath was
developed into an elegant and fashionable
spa town with neoclassical Palladian
buildings that blended harmoniously with
its Roman baths. The neoclassical style, scale
and grandeur of its public buildings (the
Rooms, the Pump Room, the Circus, and
especially, Royal Crescent) reflect the city's
confidence and ambitions during the reign
of George III.

A group of
exceptional figures
drove Bath's
eighteenth-century
rebirth, turning it into
a city where
architecture and
landscape combined
harmoniously for the
delight of enlightened
cure-takers.

Architect John Wood
planned the city's
Palladian architecture
including the famous
Circus; his son John
oversaw the building
and also planned the
Assembly Rooms and
Royal Crescent.

Quarry-owner Ralph
Allen organized the
supply of the honey-
coloured, dressed
stone Wood needed
for his buildings.

As the city's Master of
Ceremonies, dandy
and gambler Beau
Nash masterminded
Bath's metamorphosis
into the most
fashionable resort in
England.

Pulteney Bridge which
crosses the river Avon
in Bath. Constructed
in the eighteenth
century it is one of the
few bridges in the
world with shops built
into it on both sides.

World Heritage site since

1978 1979 1980 1981 1982 1983 1984 1985 1986 **1987** 1988 1989 1990 1991 1992 1993 1994 1995 1996 1997 1998 1999 2000 2001 2002 2003 2004 2005 2006 2007 2008

Nemrut Dağ
Turkey

Criteria – Human creative genius; Testimony to
cultural tradition; Significance in human history

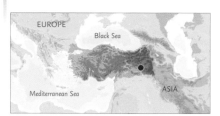

The rulers of Commagene, a kingdom
founded north of Syria and the Euphrates,
left behind several breathtakingly beautiful
funerary sanctuaries. Nemrut Dağ, the most
impressive of all the tomb sites, is that of
Antiochos I of Commagene (69–34 BC). Its
landscaping is one of the most colossal
undertakings of the Hellenistic epoch.
Dominating the summit of Nemrut Dağ is a
conical tumulus of stone chips. This funerary
mound, whose interior layout remains
unknown, is surrounded by artificial terraces.
On the east terrace there is a row of five
colossal seated figures (7 m high)
representing deities, with a lion and an eagle
symmetrically positioned at either end. On
the north side, these stones are decorated
with relief sculptures representing the
Persian ancestors of Antiochos. On the
south side, his Macedonian ancestors
symmetrically face the others.

Head of Antiochus, Nemrut Dağ. ▲

When the empire of Alexander the
Great was breaking up, one of the
kingdoms that formed was
Commagene, which, from 162 BC to
AD 72, existed as a semi-independent
state, its sovereigns defending their
autonomy first against the Seleucids
and then against the Romans.
The monarchs of this dynasty bore
the Greek names of Antiochos or
Mithridates.

World Heritage site since

1978 1979 1980 1981 1982 1983 1984 1985 1986 **1987** 1988 1989 1990 1991 1992 1993 1994 1995 1996 1997 1998 1999 2000 2001 2002 2003 2004 2005 2006 2007 2008

Old Village of Hollókő and its Surroundings
Hungary
Criteria – Traditional human settlement

EUROPE

Adriatic Sea

Black Sea

Hollókő is an exceptional example of a traditional Central European culture that flourished before the agricultural revolution of the twentieth century. This village, which developed mainly during the seventeenth and eighteenth centuries, is a living example of rural life, using traditional farming and forestry techniques. Located about 100 km northeast of Budapest, Hollókő is a small community whose 126 houses and farm buildings, strip-field farming, orchards,

vineyards, meadows and woods cover 1.4 km². As was customary in the region, the first generation of inhabitants settled on either side of the main street. In this one-street village, subsequent generations built their houses at the back of the narrow family plots, thus progressively enlarging the built-up area. The barns were built apart from the village, on the edges of the fields, according to local Palocz custom.

A 1783 decree prohibiting the use of wood for building was ignored, and consequently the village was periodically devastated by fire. The last fire was in 1909 but the houses were rebuilt in the traditional style: half-timbered houses on a stone base with roughcast white-washed walls, enhanced by high wooden pillared galleries with balconies on the street side protected by overhanging porch roofs.

▼

World Heritage site since

1978 1979 1980 1981 1982 1983 1984 1985 1986 **1987** 1988 1989 1990 1991 1992 1993 1994 1995 1996 1997 1998 1999 2000 2001 2002 2003 2004 2005 2006 2007 2008

The Great Wall
China

Criteria – Human creative genius; Interchange of values; Testimony to cultural tradition; Significance in human history; Heritage associated with events of universal significance

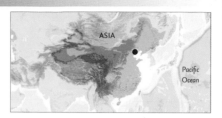

ASIA

Pacific Ocean

This complex, diachronic cultural property is a unique example of a military architectural ensemble serving a single strategic purpose for 2,000 years. Its construction history illustrates advances in defence techniques and adaptation to changing political contexts.

The Wall's testimony to the civilizations of ancient China is illustrated as much by the earlier, tamped-earth sections in Gansu Province as by the famed masonry of the Ming period.

The purpose of the Great Wall was to protect China from military aggression and cultural ingression, and it remains an essential reference in Chinese literature.

Known to the Chinese as the 'Long Wall of Ten Thousand Li', the formidable defensive structures built to ward off invasion is more commonly known as the Great Wall of China. At the end of its nineteen-century-long construction, the Great Wall was the world's largest military structure. Its historic and strategic importance is matched only by its architectural significance.

The building of defensive walls was a common strategy against potential invasion and several were built in China from the eighth century BC onwards. From 220 BC Qin Shi Huang, the first emperor of a unified China, undertook to restore and link the separate sections of the Great Wall which stretched from the region of the Ordos to Manchuria. These were to form the first cohesive defence system against invasions from the north, and by the first century BC, ongoing extensions meant the Wall spanned approximately 6,000 km between Dunhuang in the west and the Bohai Sea in the east.

After the downfall of the Han dynasty in AD 220, construction and maintenance works on the Great Wall were halted: China at that time enjoyed such great military power that the need for a defence policy was no longer felt.

It was the Ming emperors (1368–1644) who, after a long period of conflict that ended with the expulsion of the Mongols, revived the principles of Qin Shi Huang's defence policy and during these centuries 5,650 km of wall were built.

To defend the northern frontier, the Wall was divided into nine zhen, or military districts, and fortresses were built at strategically important points such as passes or fords. Passageways running along the top of the wall made it possible to move troops rapidly and for imperial couriers to travel.

The Great Wall of the Ming is a masterpiece, not only because of the ambition of the undertaking but also the perfection of its construction. The wall constitutes, on the vast scale of a continent, a perfect example of architecture integrated into the landscape.

World Heritage site since

1978 1979 1980 1981 1982 1983 1984 1985 1986 **1987** 1988 1989 1990 1991 1992 1993 1994 1995 1996 1997 1998 1999 2000 2001 2002 2003 2004 2005 2006 2007 2008

Manú National Park
Peru

Criteria – Significant ecological and biological processes; Significant natural habitat for biodiversity

The biological diversity found in the 15,000 km² Manú National Park exceeds that of any other place on Earth. The park is located on the eastern slopes of the Andes and on the Peruvian Amazon, and is situated within the Amazon River basin. It has successive tiers of vegetation rising from 150–4,200 m. The most widespread vegetation types are tropical lowland rainforest, tropical montane rainforest and puna vegetation (grasslands). Despite the high diversity of plant species, the flora of Manú is still poorly known. In the last ten years, 1,147 plant species have been identified within quite a small area, and it is likely many more species will be found. Some 850 species of birds have been identified and rare species, such as the giant otter and the giant armadillo, also find refuge there. Jaguars are often sighted.

The park is inhabited by at least four different native groups: the Machiguenga, the Mascho-Piro, the Yaminahua and the Amahuaca. The forest Indians are nomadic, mostly subsistent on some form of root-crop agriculture on alluvial soils along river banks and lakes, on hunting along water courses and inside the forest, on fishing and on the collection of turtle eggs.

Manú River, Manú National Park. ▲

World Heritage site since

1978 1979 1980 1981 1982 1983 1984 1985 1986 **1987** 1988 1989 1990 1991 1992 1993 1994 1995 1996 1997 1998 1999 2000 2001 2002 2003 2004 2005 2006 2007 2008

Mount Taishan
China

Criteria – Human creative genius; Interchange of values; Testimony to cultural tradition; Significance in human history; Traditional human settlement; Heritage associated with events of universal significance; Natural phenomena or beauty

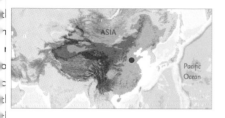

Mount Taishan is one of the birthplaces of Chinese civilization, evidence of human activity dating back 400,000 years to the Palaeolithic Yiyuan Man. Cultural relics include memorial objects, ancient architectural complexes, stone sculptures and archaeological sites of outstanding importance. For over 3,000 years, Chinese emperors of various dynasties have made pilgrimages to Mount Taishan for sacrificial and other ceremonial purposes. Rock inscriptions, stone tablets and temples bear testimony to such visits. Renowned scholars, including Confucius whose home town, Qufu, is only 70 km away, have composed poetry and prose and left their calligraphy on the mountain. Mount Taishan was also an important centre of religious activity for both Buddhism and Taoism. The way in which this rich cultural heritage has been integrated with the natural landscape of the area is considered to be one of China's most precious legacies.

A Temple on Mount Taishan. ▲

The Spring and Autumn Period (770–476 BC) of the Zhou dynasty saw the emergence of two rival states in the area, Qi to the north and Lu to the south. The State of Qi built a 500 km wall as protection from possible invasion by the State of Chu. The ruins of this earliest of great walls in Chinese history are still evident.

World Heritage site since

1978 1979 1980 1981 1982 1983 1984 1985 1986 **1987** 1988 1989 1990 1991 1992 1993 1994 1995 1996 1997 1998 1999 2000 2001 2002 2003 2004 2005 2006 2007 200

Mausoleum of the First Qin Emperor
China

Criteria – Human creative genius; Testimony to cultural tradition; Significance in human history; Heritage associated with events of universal significance

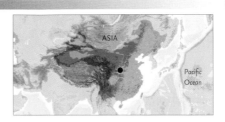

While sinking a well in 1974, three local farmers came upon a pit containing lifesize terracotta statues of warriors. Excavations were begun immediately.

The mausoleum of Qin Shi Huang is the largest preserved in China and contains the famous army of terracotta warriors. It is a unique and stunning architectural ensemble whose layout echoes the urban plan of the capital, Xianyang. The site constitutes one of the most fabulous archaeological reserves in the world.

Qin Shi Huang, the first Emperor of China, arranged for his burial long before his accession. As king of Qin in 247 BC he selected a site at the foot of Mount Li. However, after his accession as emperor in 221 BC, work at his tomb took on extraordinary dimensions.

About 700,000 workers from every Chinese province worked unceasingly to construct a subterranean city within a gigantic mound. The resulting necropolis complex was a scale model of the emperor's palace, the empire and the Earth. Its treasures were safeguarded by automatically triggered weapons designed to kill tomb robbers and its principal craftsmen were walled up alive within the complex to prevent their betraying its secrets.

Qin died in 210 BC and is buried, surrounded by the famous terracotta warriors, at the centre of a complex designed to mirror the urban plan of the capital. The life-size terracotta figures are all different; with their horses, chariots and weapons, they are masterpieces of realism and of great historical interest. According to current estimates, the statue army of the Qin Shi Huang Mausoleum must have represented the exact number of the imperial guards. The emperor's tomb and much of the site remain unexcavated.

The mausoleum's superstructures have disappeared and there remains only a wooded knoll resembling a truncated pyramid on a 350-m-square base. The interior is built within a first square enclosure, with doors in the middle of each of four walls corresponding to the four cardinal points. This in turn is surrounded by a second rectangular enclosure running north to south.

Because of their exceptional technical and artistic qualities, the terracotta warriors and horses and the funerary carts in bronze are major works in the history of Chinese sculpture. The army of statues also bears unique testimony to military organization in China from the fifth to the third century BC, while the direct testimony of their weapons – lances, swords, axes, halberds, bows and arrows – is evident.

Three pits have so far been excavated. Pit 1 contained an army of almost 2,000 warriors, the infantry and cavalry corps standing in battle formation with archers protecting the flanks.

Two other pits to the north were found to contain similar items – 1,500 warriors, carts and horses in Pit 2, and sixty-eight officers and dignitaries and a cart with four horses in Pit 3.

World Heritage site since

1978 1979 1980 1981 1982 1983 1984 1985 1986 **1987** 1988 1989 1990 1991 1992 1993 1994 1995 1996 1997 1998 1999 2000 2001 2002 2003 2004 2005 2006 2007 200

Gros Morne National Park
Canada

Criteria – Natural phenomena or beauty; Major
stages of Earth's history

Typical species found
in Gros Morne are fox
caribou, moose and
arctic hare. Whales
including minke, fin
and pilot, and harbour
seals are some of the
more common
marine mammals
regularly sighted from
the park.

With spectacular scenery, outstanding
geology and diverse ecology, Gros Morne,
located on the western shore of the island of
Newfoundland, provided the evidence that
helped define the theory of plate tectonics.
These exceptional geological formations are
the remnants of an ancient continent and
ocean from hundreds of millions of years
ago and preserved within them is one of the
best and most accessible examples of
exposed ocean crust and mantle material.

A superb fossil assemblage illustrating the
evolution of life from early Cambrian period
through mid-Ordovician times is found in
sedimentary rocks in the park. At Green
Point the rocks and the fossils they contain
have been designated by the International
Union of Geologic Sciences as the world
stratotype representing the boundary
between Cambrian and Ordovician periods
in the geological timescale.

More recent glacial action has exposed the
park's bedrock for scientific study and
resulted in the remarkable scenery of fjords,
glacial valleys, and waterfalls from the alpine
plateau of the Long Range Mountains to the
Gulf of St. Lawrence coastal lowlands with
its estuaries, beaches, sheer cliffs and dunes.
These distinctly different landscapes provide
a range of habitats for a wide variety of flora
and fauna. The park contains more than

thirty separate vegetation communities, is
home to 60 per cent of the plants species
that grow on the island, and is a significant
breeding site for harlequin ducks, blackpoll
warblers, common terns and artic terns.

Glaciated valley of
Western Brook Pond.
▼

World Heritage site since

1978 1979 1980 1981 1982 1983 1984 1985 1986 **1987** 1988 1989 1990 1991 1992 1993 1994 1995 1996 1997 1998 1999 2000 2001 2002 2003 2004 2005 2006 2007 2008

Bahla Fort
Oman

Criteria – Significance in human history

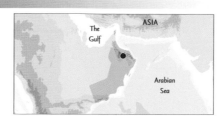

Bahla Fort is an outstanding example of the military architecture of the Sultanate of Oman. The oasis of Bahla owed its prosperity to the Banu Nabhan who, from the mid-twelfth century to the end of the fifteenth century, imposed their rule on the other tribes. Only the ruins of what was a glorious past now remain. Built on a stone base, the adobe walls and towers of the immense fort probably include some structural elements of the pre-Islamic period, but the major part dates from the time of the Banu Nabhan. At the foot of the fort lies the Friday Mosque with its beautiful sculpted mihrab (prayer niche) probably dating back to the fourteenth century. These monuments are inseparable from the small town of Bahla and its souk, palm grove and adobe ramparts surrounding the oasis.

The monuments of Bahla were in a critical state when it was inscribed on the World Heritage List. They had never been restored (thereby conserving a high degree of authenticity), and were not protected by any conservation measures. Major restoration work at the fort and Friday Mosque is now underway.

World Heritage site since

1978 1979 1980 1981 1982 1983 1984 1985 1986 **1987** 1988 1989 1990 1991 1992 1993 1994 1995 1996 1997 1998 1999 2000 2001 2002 2003 2004 2005 2006 2007 2008

Monticello and the University of Virginia in Charlottesville
USA

Criteria – Human creative genius; Significance in human history; Heritage associated with events of universal significance

Thomas Jefferson, better known for his political career, designed Monticello (1769–1809), his plantation home, and his ideal 'academical village' (1817–26), which is still the heart of the University of Virginia. The integration of the building into the landscape, the originality of design, and the refined proportions and decor make Monticello (pictured below) an outstanding example of a neoclassical villa rustica, based on a Roman design, as elaborated by Andrea

Palladio (1508–80). The western façade is dominated by an octagonal dome. The University of Virginia is an unrivalled example of an Enlightenment institution. A half-scale copy of the Pantheon in Rome, which houses the library, dominates the academic village. The ten pavilions housing the professors of the ten schools that made up the university are connected by colonnades that give a feeling of unity to this space.

Thomas Jefferson, author of the American Declaration of Independence and third president of the United States, was also a talented architect. Jefferson's use of an architectural vocabulary based upon classical antiquity symbolizes both the aspirations of the new American republic as the inheritor of European tradition and the cultural experimentation that could be expected as the country matured.

World Heritage site since

1978 1979 1980 1981 1982 1983 1984 1985 1986 **1987** 1988 1989 1990 1991 1992 1993 1994 1995 1996 1997 1998 1999 2000 2001 2002 2003 2004 2005 2006 2007 2008

Ksar of Ait-Ben-Haddou
Morocco

Criteria – Significance in human history;
Traditional human settlement

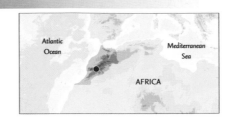

An astonishing loft-fortress overlooks the mountain against which the ksar is located. The lofts (agadir or ighram) are not uncommon in Morocco, but their defensive character is not always as evident with a fortification system linking the loft with the village, conceived as the last bastion of resistance in the event of a siege.

Ait-Ben-Haddou is a striking example of a southern Moroccan ksar, a group of earthen buildings surrounded by high walls, that is a traditional pre-Saharan village community. Inside the defensive walls, which are reinforced by angle towers, each with a zigzag-shaped gate, houses crowd together. Some are modest, others resemble small urban castles with their high angle towers whose upper portion includes decorative motifs in clay brick. There are also public buildings and community areas: collective sheep pens and stables, lofts and silos, a market place, a meeting room for the assembly of family chiefs, a mosque and madrasas. Ait-Ben-Haddou is an extraordinary ensemble of buildings offering a complete panorama of pre-Saharan construction techniques using rammed and moulded earth, clay and brick.

World Heritage site since

1978 1979 1980 1981 1982 1983 1984 1985 1986 **1987** 1988 1989 1990 1991 1992 1993 1994 1995 1996 1997 1998 1999 2000 2001 2002 2003 2004 2005 2006 2007 200

Uluru-Kata Tjuta National Park
Australia

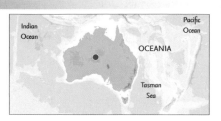

Criteria – Traditional human settlement; Heritage associated with events of universal significance; Natural phenomena or beauty; Major stages of Earth's history

Uluru-Kata Tjuta National Park is included on the World Heritage List for both its natural and cultural values. Formerly called Ayers Rock – Mount Olga National Park, it is located in Australia's Red Centre. It lies in the traditional lands of the Western Desert Aboriginal people, locally known as Anangu. Anangu are part of one of the oldest human societies in the world.

The huge rock formations of Uluru and Kata Tjuta and the surrounding country are part of an important cultural landscape. For Anangu these features are physical evidence of the actions, artefacts and bodies of the ancestral beings (tjukuritja) who travelled the Earth in the creation time. These ancestors, who combined the attributes of humans and animals, journeyed across the landscape creating not only its features, but also Tjukurpa (the law) – the code of behaviour followed by Anangu today.

Uluru and the rock domes of Kata Tjuta dominate the vast red plain, dwarfing the desert oak and spinifex grass of central Australia. Uluru is composed of hard red sandstone, exposed as a result of the folding, faulting and erosion of the surrounding rock. It is 9.4 km in circumference and rises to a relatively flat top that is more than 340 m above the shallow, red sandy dunes around it. Rock art in the caves around its base provides further evidence of the enduring cultural traditions of Anangu.

The thirty-six steep-sided rock domes of Kata Tjuta, lying about 32 kilometres to the west of Uluru, are made up of gently dipping Mount Currie conglomerate. The undulating domes are interspersed with moisture-rich gullies and rocky valleys that are home to rare plants and desert animals. Like Uluru, Kata Tjuta's domes are the visible tips of huge rock slabs that extend far beneath the ground.

The park is home to twenty-one native mammals including the rare hairy-footed dunnart, the sandhill dunnart and the mulgara. More than 170 bird species, 73 species of reptile and at least 7 species of bat have been recorded in the park.

The Uluru monolith.

The park's Aboriginal owners, Anangu, have cared for the landscape for thousands of years using traditional practices governed by Tjukurpa (the law). Aboriginal people learned how to patch burn the country from Tjukurpa, and this knowledge informs an active burning programme that is now a major ecological management tool in the park. Tjukurpa also teaches about the location and care of rock holes and other water sources.

World Heritage site since

1978 1979 1980 1981 1982 1983 1984 1985 1986 **1987** 1988 1989 1990 1991 1992 1993 1994 1995 1996 1997 1998 1999 2000 2001 2002 2003 2004 2005 2006 2007 2008

Group of Monuments at Pattadakal
India

Criteria – Testimony to cultural tradition; Significance in human history

Pattadakal, in Karnataka, represents the high point of an eclectic art which, in the seventh and eighth centuries under the Chalukya dynasty, achieved a harmonious blend of architectural forms from northern and southern India. An impressive series of nine Hindu temples, as well as a Jain sanctuary, can be seen there. One masterpiece from the group stands out – the Temple of Virupaksha, built c. 740 by Queen Lokamahadevi to commemorate her husband's victory over the kings from the South.

The evocative sanctuary ruins in the enclosure may be reached through monumental gates on the west and east sides. In the axis of the courtyard, before the temple, is a pavilion housing a black stone statue of Siva's sacred bull.

◄

Sangameshvara Temple at Pattadakal.

Dja Faunal Reserve
Cameroon

Criteria – Significant ecological and biological processes; Significant natural habitat for biodiversity

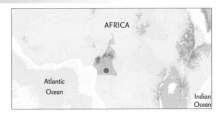

This is one of the largest and best-protected rainforests in Africa, with 90 per cent of its area left undisturbed. Almost completely surrounded by the Dja river, which forms a natural boundary, the reserve is located in a transition zone between the forests of southern Nigeria, southwest Cameroon and the Congo Basin. Except in the southeast, the relief is fairly flat and consists of a succession of round-topped hills. Cliffs in the south are associated with a section of

the river broken up by rapids and waterfalls. Although the area is poorly studied, it is known to have a wide range of primate species. Other mammals include elephant, buffalo, leopard, warthog and pangolin. Reptiles include python, lizard and two species of crocodile, both of which are threatened.

Vegetation in the reserve mainly comprises dense evergreen Congo rainforest, together with swamp vegetation and secondary forest around old villages (which were abandoned in 1946) and recently abandoned cocoa and coffee plantations. A population of pygmies lives in small sporadic encampments in the reserve, maintaining an essentially traditional lifestyle.

World Heritage site since

1978 1979 1980 1981 1982 1983 1984 1985 1986 1987 **1988** 1989 1990 1991 1992 1993 1994 1995 1996 1997 1998 1999 2000 2001 2002 2003 2004 2005 2006 2007 2008

Mount Athos
Greece

Criteria – Human creative genius; Interchange of values; Significance in human history; Traditional human settlement; Heritage associated with events of universal significance; Natural phenomena or beauty

The transformation of a mountain into a sacred place made Mount Athos a unique artistic creation, combining the natural beauty of its site with the expanded forms of architectural creation. Moreover, the monasteries of Mount Athos are a veritable conservatory of masterpieces, including the wall paintings, portable icons, gold objects, embroideries and illuminated manuscripts which each monastery carefully preserves. Mount Athos is the spiritual centre of the Orthodox world, and has exerted a lasting

influence both on it and on the development of religious architecture and monumental painting.

Mount Athos enjoyed an autonomous statute since Byzantine times (from the tenth century onwards). The 'Holy Mountain', which is forbidden to women and children, is also a recognized artistic site. Its monasteries had an influence as far afield as Russia, and its school of painting influenced the history of Orthodox art.

Athos is a self-governing monastic republic within Greece. It includes twenty monasteries, twelve sketes (convents), and about 700 houses, cells or hermitages. Over 1,000 monks live there, either in communities or alone, as well as in the 'desert' of Karoulia where cells cling to the cliff face rising steeply above the sea.

▼

World Heritage site since

1978 1979 1980 1981 1982 1983 1984 1985 1986 1987 **1988** 1989 1990 1991 1992 1993 1994 1995 1996 1997 1998 1999 2000 2001 2002 2003 2004 2005 2006 2007 2008

Old Towns of Djenné
Mali

Criteria – Testimony to cultural tradition;
Significance in human history

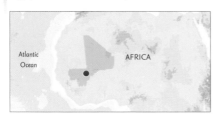

Djenné-Djeno, along with Hambarketolo, Tonomba and Kaniana, bears exceptional witness to the pre-Islamic civilizations on the inland delta of the Niger. Djenné is an outstanding example of an architectural group of buildings illustrating a significant historic period. It has been defined both as 'the most beautiful city of Africa' and 'the typical African city'.

Inhabited since 250 BC, Djenné became a market centre and was an important link in the trans-Saharan gold trade. In the fifteenth and sixteenth centuries it was one of the centres for the propagation of Islam. Its traditional houses, of which nearly 2,000 have survived, are built on hillocks called toguère, as protection from the seasonal Niger floods.

Djenné, which spreads over several toguère, is bisected by a wide avenue. Its Market Place is dominated by the Great Mosque (see picture below). Traditional houses extend out from both sides of this thoroughfare over an ancient land parcel of approximately 2km². The main feature of the domestic architecture, influenced by that of Morocco, is its verticality.

The Great Mosque of Djenné. ▶

World Heritage site since

1978 1979 1980 1981 1982 1983 1984 1985 1986 1987 **1988** 1989 1990 1991 1992 1993 1994 1995 1996 1997 1998 1999 2000 2001 2002 2003 2004 2005 2006 2007 2008

Wet Tropics of Queensland Australia

Criteria – Natural phenomena or beauty; Major stages of Earth's history; Significant ecological and biological processes; Significant natural habitat for biodiversity

The Wet Tropics of Queensland provides an unparalleled living record of the ecological and evolutionary processes that shaped the flora and fauna of Australia over the past 415 million years, first when it was part of the Pangaean landmass, then as the ancient continent Gondwana, and for the past 50 million years an island continent. During this period of evolution the processes of speciation, extinction and adaptation have been determined by history, particularly continental drift and cycles of climatic change.

The site, which stretches along the northeast coast of Australia for some 450 km, from just south of Cooktown to just north of Townsville, is made up largely of tropical rainforests. This area offers a particularly extensive and varied array of plants, marsupials and song birds, along with many rare and endangered animals and plant species.

The rainforests, which make up about 80 per cent of the site, have more plant families with primitive characteristics than any other area on Earth.

The ancestry of Australia's unique marsupials and many of its other animals originated in rainforest ecosystems. The Wet Tropics of Queensland still contains many of their closest surviving members.

World Heritage site since

1978 1979 1980 1981 1982 1983 1984 1985 1986 1987 **1988** 1989 1990 1991 1992 1993 1994 1995 1996 1997 1998 1999 2000 2001 2002 2003 2004 2005 2006 2007 2008

Tower of London
United Kingdom

Criteria – Interchange of values; Significance in human history

The Tower of London is an imposing fortress with many layers of history. Built on the Thames by William the Conqueror to protect his London base and to assert his power over the newly conquered English, it became one of the symbols of royalty in England and is a major reference for the history of medieval military architecture.

The Tower is a complex of fortifications, courtyards and buildings extending over 73,000 m². There are many towers in the ensemble and the impressive White Tower, begun around 1078 and completed around nine years later, is the centrepiece. Although a royal residence for centuries, the White Tower was never intended as the main royal palace but rather as a stronghold, a notion reinforced by the formidable curtain walls, moats and ditches successive kings built around it.

The Crown Jewels are on display in the Tower of London.

The Water Gate entrance was nicknamed Traitors' Gate because prisoners were brought through it to the Tower. Queen Anne Boleyn, Thomas More, Queen Catherine Howard and Princess Elizabeth, later Elizabeth I, entered it as prisoners.

◀

The White Tower stands more than 27 m high and its walls are over 4.5 m thick at their base.

World Heritage site since

1978 1979 1980 1981 1982 1983 1984 1985 1986 1987 **1988** 1989 1990 1991 1992 1993 1994 1995 1996 1997 1998 1999 2000 2001 2002 2003 2004 2005 2006 2007 2008

Manovo-Gounda St Floris National Park
Central African Republic

Criteria – Significant ecological and biological processes; Significant natural habitat for biodiversity

The importance of this park derives from its wealth of flora and fauna. Its vast savannas are home to a wide variety of species: black rhinoceroses, elephants, cheetahs, leopards, wild dogs, red-fronted gazelles and buffalo, while various types of waterfowl are to be found in the northern floodplains.

Some 320 species of bird have been identified in the park, with at least twenty-five species of raptor including bateleur and African fish eagle. There are large seasonal populations of pelican and marabou stork, and many waterbirds and shorebirds.

Paleochristian and Byzantine Monuments of Thessalonika
Greece

Criteria – Human creative genius; Interchange of values; Significance in human history

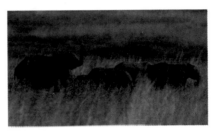

Founded in 315 BC, the provincial capital and sea port of Thessalonika was one of the first bases for the spread of Christianity. Among its Christian monuments are fine churches, some built on the Greek cross plan and others on the three-nave basilica plan. Constructed over a long period, from the fourth to the fifteenth century, they constitute a diachronic typological series, which had considerable influence in the Byzantine world. The mosaics of the rotunda, St Demetrius and St David are among the great masterpieces of early Christian art.

Cosmopolitan and prosperous, Thessalonika grew in commercial and strategic importance during the Roman period and was one of the first bases for the spread of Christianity. St Paul visited twice, in AD 50 and 56, founding a church there.

◄

The Rodonta Temple in Thessalonika.

World Heritage site since

1978 1979 1980 1981 1982 1983 1984 1985 1986 1987 **1988** 1989 1990 1991 1992 1993 1994 1995 1996 1997 1998 1999 2000 2001 2002 2003 2004 2005 2006 2007 2008

Sacred City of Kandy
Sri Lanka

Criteria – Significance in human history;
Heritage associated with events of universal
significance

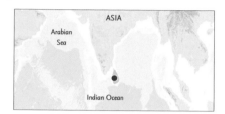

This sacred Buddhist site, popularly known
as the city of Senkadagalapura, was the last
capital of the Sinhala kings, whose
patronage enabled the Dinahala culture to
flourish for more than 2,500 years. It is also
the site of the Temple of the Tooth Relic,
the sacred tooth of the Buddha, which is
a famous pilgrimage site. The Temple of the
Tooth, the palatial complex, and the sacred
city of Kandy are directly associated with the
history of the spread of Buddhism. Kandy,
founded in the fourteenth century, forms the
southern tip of Sri Lanka's 'Cultural Triangle'.
It became the capital of the kingdom in 1592
and remained one of the bastions of
Sinhalese independence until the arrival of
the British in 1815. It is still the religious
capital of Buddhism and a sacred city for
millions of believers.

A Buddhist temple in Kandy. ▲

Enshrined in the Dalada Maligawa
is the relic of the tooth of Buddha.
The ceremonial high point each
year is the splendid ritual on the
feast of Esala Perahera, in which
one of the inner caskets used for
covering the tooth relic is taken in
a grand procession through the
streets of the city.

World Heritage site since

1978 1979 1980 1981 1982 1983 1984 1985 1986 1987 **1988** 1989 1990 1991 1992 1993 1994 1995 1996 1997 1998 1999 2000 2001 2002 2003 2004 2005 2006 2007 2008

Sanctuary of Asklepios at Epidaurus
Greece

Criteria – Human creative genius; Interchange of values; Testimony to cultural tradition; Significance in human history; Heritage associated with events of universal significance

In a small valley in the Peloponnisos, the shrine of Asklepios, the god of medicine, developed in the sixth century BC out of a much earlier cult of Apollo Maleatas. The group of temples and the hospital facilities comprising the Sanctuary of Epidaurus bears exceptional testimony to the healing cults of the Hellenic and Roman worlds. The theatre, the temples of Artemis and Asklepios, the Tholos, the Enkoimeterion (where the sick awaited their cures) and the Propylaea are exceptional examples of Hellenic architecture of the fourth century BC. In particular, the theatre, an architectural masterpiece by Polycletes the Younger of Argos, is famed for its setting and the perfection of its proportions and acoustics. Epidaurus continued to flourish during the Hellenistic period. Despite pillaging by Sulla in 87 BC and by Cilician pirates, the sanctuary prospered during the Roman period.

The temples and the hospital facilities influenced all the sanctuaries in the Hellenic and Roman world. The emergence of modern medicine in a sanctuary originally based on the miraculous healing of supposedly incurable patients is strikingly described in the engraved inscriptions on the remarkable stelae preserved in the museum at Epidaurus.

Theatre at Epidaurus.
▼

World Heritage site since

1978 1979 1980 1981 1982 1983 1984 1985 1986 1987 **1988** 1989 1990 1991 1992 1993 1994 1995 1996 1997 1998 1999 2000 2001 2002 2003 2004 2005 2006 2007 2008

Sinharaja Forest Reserve
Sri Lanka

Criteria – Significant ecological and biological processes; Significant natural habitat for biodiversity

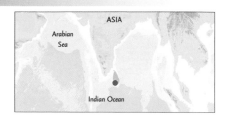

Sinharaja is Sri Lanka's last viable area of primary tropical rainforest. This narrow strip of undulating terrain consists of a series of ridges and valleys. Two main types of forest can be recognized: remnants of Dipterocarpus forest occur in valleys and on their lower slopes; secondary forest and scrub occur where the original forest cover has been removed by shifting cultivation and in other places the forest has been replaced by rubber and tea plantations.

Mesua-Doona forest is the climax vegetation in most of the reserve. More than 60 per cent of the trees are endemic and many of them are considered rare. There is much endemic wildlife, especially birds, and the reserve is also home to over 50 per cent of Sri Lanka's endemic species of mammals and butterflies, as well as many kinds of insects, reptiles and rare amphibians.

The Sinharaja region has long featured in the legends and lore of the people of Sri Lanka. Its name, literally meaning lion (sinha) king (raja), perhaps refers to the original 'king-sized or royal forest of the Sinhalese', a people of the legendary 'lion-race' of Sri Lanka, or to the home of a legendary lion of Sri Lanka.

A spider's web in morning sunlight in Sinharaja Rainforest.
▼

World Heritage site since

1978 1979 1980 1981 1982 1983 1984 1985 1986 1987 **1988** 1989 1990 1991 1992 1993 1994 1995 1996 1997 1998 1999 2000 2001 2002 2003 2004 2005 2006 2007 2008

Medina of Sousse
Tunisia

Criteria – Testimony to cultural tradition;
Significance in human history; Traditional human
settlement

Sousse was an important commercial and
military port during the Aghlabid period in
the ninth century and is an outstanding
example of a town dating from the first
centuries of Islam. Among the earliest
buildings is the ribat, a combined fort and
religious building, that helped defend
Sousse against Byzantine fleets. In 821 its
fortifications were completed with a square
bastion and tower which served both as
a watchtower and as a minaret. Under the

Aghlabids, Sousse rapidly flourished:
significant monuments constructed in this
golden century included the mosque of
Bu Ftata, the first kasbah, and the Great
Mosque. By 859, the town walls neared
completion, and the limits of the medina
were broadly drawn. The surrounding
fortifications made the military function of
the ribat less vital and it reverted fully to its
religious function.

The 'ribat' of Sousse
housed a garrison of
the Defenders of the
Faith, the Mourabitin,
and the austere
architecture reflects its
dual military/
religious function.
A rectangular
enclosure is flanked
by towers and turrets,
with a single gate on
the south and an inner
courtyard rising over
two levels. It has a
mosque on the
southern side.

▼

World Heritage site since

1978 1979 1980 1981 1982 1983 1984 1985 1986 1987 **1988** 1989 1990 1991 1992 1993 1994 1995 1996 1997 1998 1999 2000 2001 2002 2003 2004 2005 2006 2007 2008

Historic Town of Guanajuato and Adjacent Mines
Mexico

Criteria – Human creative genius; Interchange of values; Significance in human history; Heritage associated with events of universal significance

NORTH AMERICA

Pacific Ocean

Caribbean Sea

Tunnels from the silver mines under the town of Guanajuato.

The splendour of the Baroque buildings of Guanajuato is directly linked to the wealth of the mines. The churches of La Compañía (1745–65) and above all La Valenciana (1765–88) are masterpieces of the Mexican Churrigueresque style. The church of La Vanenciana and the mansion of Casa Rul y Valenciana were financed by the most prosperous mines, but even the smaller mines also had churches, palaces or houses close by.

Guanajuato is an outstanding example of an architectural ensemble that incorporates the industrial and economic aspects of a mining operation. Founded by the Spanish in the early sixteenth century, the town became the world's leading silver-extraction centre in the eighteenth century. This past can be seen in its 'subterranean streets' and the 'Boca del Inferno', a mineshaft that plunges a breathtaking 600 m. The town's beautiful Baroque and neoclassical buildings, built with the prosperity from the mines, have influenced buildings throughout central Mexico. The churches of La Compañía and La Valenciana are considered to be among the most beautiful examples of Baroque architecture in Central and South America. Guanajuato was also witness to events which changed the history of the country.

Spanish conquistadores first settled in the region in 1529, and in 1548 they discovered rich outcrops of silver at Guanaxhuata. To protect prospectors, miners and the new settlers, four fortified structures were erected at Marfil, Tepetapa, Santa Ana and Cerro del Cuarto; these settlements formed the nuclei of the later town of Guanajuato. Sprawling through a winding valley at an altitude of 2,084 m, Guanajuato differs from the other colonial towns in New Spain in that it was not laid out on the standard grid plan. Instead, the scattered areas grew together through the spontaneous urbanization of suitable sites on the rough, natural terrain.

Founded when the silver mines were opened, Guanajuato had a symbiotic relationship with them until the nineteenth century. Its growth, the layout of its streets, including the picturesque 'subterranean' streets through which its traffic runs, its plazas, and the construction of hospitals, churches, convents and palaces are all inextricably linked with the industrial history of the region which, with the decline of the Potosí mines in the eighteenth century, became the world's leading silver extraction centre.

The city played a part in the Mexican War of Independence of 1810–21: Miguel Hidalgo, the rebel leader, began his uprising in the state of Guanajuato of which the city is capital.

World Heritage site since

1978 1979 1980 1981 1982 1983 1984 1985 1986 1987 **1988** 1989 1990 1991 1992 1993 1994 1995 1996 1997 1998 1999 2000 2001 2002 2003 2004 2005 2006 2007 2008

Old Town of Galle and its Fortifications
Sri Lanka

Criteria – Significance in human history

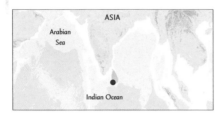

Founded in the sixteenth century by the Portuguese, Galle reached the height of its development in the eighteenth century, before the arrival of the British. It is the best example of a fortified city built by Europeans in south and southeast Asia, showing the interaction between European architectural styles and south Asian traditions. This fortified city, protected by a sea wall finished in 1729, still exists, with few changes. It has an area of 0.52 km² inside the walls defended by fourteen bastions. The city was laid out on a regular grid pattern adapted to the configuration of the terrain: north-south peripheral streets are parallel to the ramparts and not to the central traffic axes. As befitted a fortified city, the Commandant's residence, the arsenal and the powder house were prominent features of the original layout.

The Clock Tower at the Dutch Fort in Galle. ▲

The most salient feature of the architecture at Galle is the use of European models adapted to the geological, climatic, historical, and cultural conditions of Sri Lanka. In the structure of the ramparts, coral is frequently used along with granite.

World Heritage site since

1978 1979 1980 1981 1982 1983 1984 1985 1986 1987 **1988** 1989 1990 1991 1992 1993 1994 1995 1996 1997 1998 1999 2000 2001 2002 2003 2004 2005 2006 2007 2008

Timbuktu
Mali

Criteria – Interchange of values; Significance in human history; Traditional human settlement

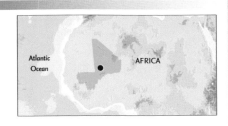

Home of the prestigious Koranic Sankore University and other madrasas, Timbuktu was an important centre for the dissemination of Islam throughout Africa in the fifteenth and sixteenth centuries. Its numerous schools were attended, it is said, by some 25,000 students. Scholars, engineers and architects from various regions in Africa rubbed shoulders with wise men and marabouts in this intellectual and religious centre. Its three great mosques,

Djingareyber, Sankore and Sidi Yahia, recall Timbuktu's golden age. Although continuously restored, these monuments are today under threat from desertification. Apart from the mosques, the World Heritage site comprises sixteen cemeteries and mausolea, essential elements in a religious system as, according to popular belief, they constitute a rampart that shields the city from all misfortune.

Timbuktu is thought to have been founded in the fifth century of the Hegira by a group of Imakcharen Tuaregs who, having wandered 250 km south of their base, established a temporary camp guarded by an old woman, Buktu. Gradually, Tim-Buktu (the place of Buktu) became a small sedentary village at the crossroads of several trade routes.

Henderson Island
United Kingdom

Criteria – Natural phenomena or beauty; Significant natural habitat for biodiversity

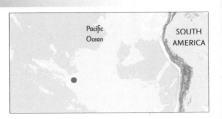

Henderson Island is a raised coral atoll forming part of the Pitcairn Island group in the South Pacific. It is one of the few atolls in the world whose ecology has been practically untouched by human presence. Although it was colonized by Polynesians between the twelfth and fifteenth centuries, this period of settlement had little ecological impact and the island has remained uninhabited in modern times. Its isolated location provides the ideal context

for studying the dynamics of insular evolution and natural selection. The island is arid with only one freshwater spring and has a rugged topography beneath the dense tangled vegetation. It is particularly notable for the ten plant species and four land bird species that are endemic to the island.

With no major landmass within a 5,000 km radius, Henderson Island's remoteness and inhospitable nature have so far effectively ensured its conservation. As a near pristine island ecosystem, it is of immense scientific value.

World Heritage site since

1978 1979 1980 1981 1982 1983 1984 1985 1986 1987 **1988** 1989 1990 1991 1992 1993 1994 1995 1996 1997 1998 1999 2000 2001 2002 2003 2004 2005 2006 2007 200

Archaeological sites of Bat, Al-Khutm and Al-Ayn

Oman

Criteria – Testimony to cultural tradition;
Significance in human history

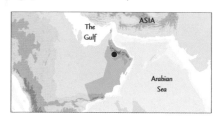

The zone encompassing the settlement and necropolises of Bat is the most complete and best-known site of the third millennium BC, bearing unique witness to the evolution of funeral practices during the first Bronze Age in the Oman peninsula. The settlement and necropolis zones of Bat form a coherent and representative group with two neighbouring contemporary archaeological sites: the tower of Al-Khutm, 2 km west of Bat, and the group of beehive tombs of Qubur Juhhal at Al-Ayn (pictured on the right), 22 km east-southeast of Bat. The twenty-one tombs from the third millennium, aligned on a rocky crest that stands out in the superb mountainous landscape of Jebel Misht to the north, are in a remarkable state of preservation.

Six of twenty-one beehive tombs of Qubur Juhhal at Al-Ayn. ▲

In the settlement zone there are five stone 'towers,' one of which has been entirely excavated. It has been determined that it was built between 2595 and 2465 BC.

From the tower can be distinguished to the east a series of rectangular houses with central courts and, to the north, a vast necropolis.

World Heritage site since

1978 1979 1980 1981 1982 1983 1984 1985 1986 1987 **1988** 1989 1990 1991 1992 1993 1994 1995 1996 1997 1998 1999 2000 2001 2002 2003 2004 2005 2006 2007 2008

Trinidad and the Valley de los Ingenios
Cuba

Criteria – Significance in human history;
Traditional human settlement

Founded by the Spanish in the early sixteenth century in honour of the Holy Trinity, Trinidad is an outstanding example of a colonial city. Towards the end of the eighteenth century, the sugar industry was firmly established in the nearby Valle de Los Ingenios and Trinidad prospered; by 1796 it was the third-largest city in Cuba. The present city owes its charm to its eighteenth and nineteenth century buildings, such as

the Palacio Brunet and the Palacio Cantero. Valle de Los Ingenios is a living museum of the sugar industry, featuring seventy-five ruined sugar mills, summer mansions, barracks, and other facilities related to the field. The famous Manaca-Iznaga Tower, built in 1816, is 45 m high, and the tolling of its bells once marked the beginning and end of working hours on the sugar plantations.

Trinidad was one of the bridgeheads for the conquest of the American continent. It was the departure point for the expeditions led by Francisco Hernández de Córdova in 1517 and by Cortez in 1518.

Santisma Trinidad Church.
▼

World Heritage site since

1978 1979 1980 1981 1982 1983 1984 1985 1986 1987 **1988** 1989 1990 1991 1992 1993 1994 1995 1996 1997 1998 1999 2000 2001 2002 2003 2004 2005 2006 2007 2008

Medieval City of Rhodes
Greece

Criteria – Interchange of values; Significance in human history; Traditional human settlement

The Knights built up the fortifications of Rhodes over two centuries. The ramparts of the medieval city, partially erected on the foundations of the Byzantine enclosure, were constantly maintained and remodelled between the fourteenth and sixteenth centuries. Artillery firing posts were the final features to be added. In the section of the Amboise Gate which was built on the northwestern angle in 1512, the curtain wall was 12 m thick with a 4 m high parapet pierced with gun holes.

Rhodes is an outstanding example of a medieval defensive architectural ensemble. The fortifications of Rhodes, long considered impregnable, exerted an influence throughout the eastern Mediterranean basin in the later Middle Ages. Within its walls, the old town of Rhodes is an important example of traditional human settlement characterized by cultural assimilation.

From 1309–1523 Rhodes was in the possession of the Order of St John of Jerusalem, the Knights Hospitaller. A military hospital order founded to care for sick pilgrims in the Holy Land, the Knights lost their last foothold in Palestine in 1291 with the fall of Acre. In 1309 the Order took Rhodes from the ailing Byzantine Empire, rebuilding the town and transforming it into a stronghold.

The medieval city is located in the Upper Town (Collachium) within a wall 4 km long. Built by the Knights, the town is one of the most beautiful urban ensembles of the Gothic period.

The Order was organized into seven 'Tongues', each having its own seat. The inns of the Tongues of Italy, France, Spain and Provence lined both sides of the principal east-west axis, the famous Street of the Knights, one of the finest testimonies to Gothic urbanism.

The Knights' first hospice was replaced in the late-fifteenth century by the Great Hospital; today the building is used as the archaeological museum. Northwest of the Collachium are the Grand Masters' Palace and St John's Church. At the eastern end of the Street of the Knights, built against the wall, is St Mary's Church which the Knights transformed into a cathedral.

The Lower Town is almost as dense with monuments as the Collachium. In 1522, with a population of 5,000, it was replete with churches, some of Byzantine construction.

Rhodes withstood two invasions in the fifteenth century, firstly by Egyptians and then by Turks. It fell in 1523 to the Ottoman Sultan Suleiman the Magnificent who arrived with a 100,000-strong army, besieging Rhodes for six months.

After 1523 many of the buildings were converted into Islamic mosques but the influence of the Ottoman occupation is seen particularly in combined vernacular architecture, with decorative elements of Ottoman origin. In the Lower Town, Gothic architecture coexists with mosques, public baths and other buildings from the Ottoman period.

World Heritage site since

1978 1979 1980 1981 1982 1983 1984 1985 1986 1987 **1988** 1989 1990 1991 1992 1993 1994 1995 1996 1997 1998 1999 2000 2001 2002 2003 2004 2005 2006 2007 2008

Historic Centre of Lima
Peru

Criteria – Significance in human history

Lima was founded in 1535. The city played a leading role in the history of the New World from 1542, when the Holy Roman Emperor Charles V established the Viceroyalty of Peru to govern most of Spanish South America. However, the creation of the viceroyalties of New Granada and of La Plata in the eighteenth century gradually ended Lima's power.

Although severely damaged by earthquakes, this 'City of the Kings' was, until the mid-eighteenth century, the capital and most important city of the Spanish dominions in South America. The historic centre of Lima bears witness to the architecture and urban development of a Spanish colonial town of great political, economic, and cultural importance.

The historic monuments – religious or public buildings, such as the Torre Tagle Palace – which lie within the perimeter of the World Heritage site date from the seventeenth and eighteenth centuries and are typical examples of Hispano-American Baroque. The architecture of the other buildings is often of the same period and style with the result that, despite the addition of certain nineteenth century constructions such as Art Nouveau Casa Courret, the town's historic nucleus recalls Lima at the time of the Spanish Kingdom of Peru.

Lima Cathedral. ▼

World Heritage site since

1978 1979 1980 1981 1982 1983 1984 1985 1986 1987 **1988** 1989 1990 1991 1992 1993 1994 1995 1996 1997 1998 1999 2000 2001 2002 2003 2004 2005 2006 2007 2008

Canterbury Cathedral, St Augustine's Abbey and St Martin's Church
United Kingdom

Criteria – Human creative genius; Interchange of values; Heritage associated with events of universal significance

Canterbury, in Kent, has been the seat of the spiritual head of the Church of England for nearly five centuries. Three distinct cultural properties are on the World Heritage List: the modest St Martin's Church, the oldest church in England; the ruins of St Augustine's Abbey, which fell into disuse following the dissolution of the community by Henry VIII in 1538; and the superb Christ Church Cathedral, a breathtaking mixture of Romanesque and Perpendicular Gothic, where Archbishop Thomas Becket was murdered in 1170. The cathedral's beauty is enhanced by a set of exceptional stained glass windows which constitute the richest collection in the United Kingdom. These three monuments are milestones in the religious history of the regions of Great Britain before the Reformation.

Christ Church Cathedral (Canterbury Cathedral). ▲

St Martin's Church, located outside the walls of Roman Durovernum, existed in 597 when the monk Augustine was sent from Rome by Gregory the Great to bring Christianity to the Saxon kingdom of Kent. It undoubtedly includes a Roman structure from the fourth century.

World Heritage site since

1978 1979 1980 1981 1982 1983 1984 1985 1986 1987 **1988** 1989 1990 1991 1992 1993 1994 1995 1996 1997 1998 1999 2000 2001 2002 2003 2004 2005 2006 2007 2008

Old City of Salamanca
Spain

Criteria – Human creative genius; Interchange of values; Significance in human history

Salamanca is one of the key centres of a dynasty of architects, decorators and sculptors from Catalonia, the Churriguera. The Churrigueresque style also exerted considerable influence in the eighteenth century in the countries of Latin America. However, the city owes its most essential features to its university. The remarkable group of buildings in the Gothic, Renaissance and Baroque styles which, from the fifteenth to the eighteenth centuries,

grew up around the institution that proclaimed itself 'Mother of Virtues, of Sciences and of the Arts'. The University had already established itself by 1250 as one of the best in Europe. The oldest university building in Salamanca, now the Rectorate, is the former Hospital del Estudio, built in 1413. The main university buildings, Las Escuelas Mayores, are grouped around a central patio and were built between 1415 and 1433.

Beginning with the Roman bridge that spans the Río Tormes southwest of the city, numerous witnesses to the history of Salamanca still stand: the Old Cathedral and San Marcos (twelfth century); the Salina and the Monterrey palaces (sixteenth century); and the **Plaza Mayor** (pictured below), the most sumptuous of the Baroque squares in Spain (eighteenth century).

World Heritage site since

1978 1979 1980 1981 1982 1983 1984 1985 1986 1987 **1988** 1989 1990 1991 1992 1993 1994 1995 1996 1997 1998 1999 2000 2001 2002 2003 2004 2005 2006 2007 2008

Hierapolis-Pamukkale Turkey

Criteria – Testimony to cultural tradition; Significance in human history; Natural phenomena or beauty

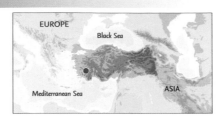

Mineral-laden waters from hot springs at Pamukkale (in Turkish, literally 'cotton castle') have created an unreal landscape, made up of mineral forests, petrified waterfalls and a series of terraced pools. At the end of the second century BC the dynasty of the Attalids, the kings of Pergamon, established the thermal spa of Hierapolis here. The ruins of the baths, temples and other Greek monuments can be seen at the site. The Romans acquired full control in 129 BC and it prospered as a cosmopolitan city where Anatolians, Graeco-Macedonians, Romans and Jews intermingled. The hot springs also served another purpose: the scouring and dyeing of wool. According to Christian tradition, Philip the Apostle converted it and was crucified here by Domitian around the year 87. Christian remains include the cathedral, baptistery, churches and the martyrium of St Philip.

The therapeutic virtues of the waters were exploited with immense hot basins and pools for swimming. Hydrotherapy was accompanied by religious practices, which were developed in relation to local cults. The Temple of Apollo was erected on a fault from which noxious vapours escaped.

Travertine pools and terraces.

▼

World Heritage site since

1978 1979 1980 1981 1982 1983 1984 1985 1986 1987 **1988** 1989 1990 1991 1992 1993 1994 1995 1996 1997 1998 1999 2000 2001 2002 2003 2004 2005 2006 2007 2008

Pre-Hispanic City of Chichen-Itza

Mexico

Criteria – Human creative genius; Interchange of values; Testimony to cultural tradition

Pyramid of Kukulcán, El Castillo.

The buildings of Chichen-Itza illustrate the fusion of the Mayan and Toltec traditions. Specific examples are the Caracol, a circular observatory, and the Pyramid of Kukulcán, El Castillo. Surrounding El Castillo are terraces where the major monumental complexes were built: on the northwest are the Great Ball Court, Skull Wall, Jaguar Temple and the House of Eagles; on the northeast the Temple of the Warriors, Group of the Thousand Columns, market and ball courts; and on the southwest the Tomb of the High Priest.

Chichen-Itza is the most important archaeological vestige of the Maya-Toltec civilization in Yucatán. Its monuments are among the undisputed masterpieces of Mesoamerican architecture because of the beauty of their proportions, the refinement of their construction and the splendour of their sculpted decorations. These monuments exerted an influence throughout the entire Yucatán cultural zone from the tenth to the fifteenth centuries.

Chichen-Itza is the northernmost of the major archaeological sites in Yucatán. Covering more than 3km², it is also one of the largest and richest in monuments. Above all, it is one of the most significant in historical terms because it illustrates two major periods in pre-Hispanic civilizations in the Mesoamerican zone.

The town was established during the Classic period around the early or mid-fifth century AD close to two natural cavities (cenotes or chenes) which facilitated the tapping of underground water. The town that grew up around the sector known as Chichen Viejo already boasted important monuments: the Building of the Nuns, Temple of the Panels and Temple of the Stag, constructed between the sixth and tenth centuries.

The second settlement of Chichen-Itza, and the most important for historians, corresponded to the migration of Toltec warriors from the Mexican plateau towards the south during the tenth century. The Toltec invaders subjugated the local population with a ferocity which even five centuries later the chronicles of the 'sacred books' of the Mayans spoke of. The Toltecs imposed the ritual of human sacrifice which until then was rarely, if at all, practised in the region.

Following the conquest of Yucatán a new style blending the Mayan and Toltec traditions developed, symbolizing the acculturation. Chichen-Itza is a clear illustration of this fusion.

This new architecture, known today as Maya-Yucatec, took from the old local structures the art of stereotomy used on walls and vaults while incorporating certain Toltec elements in the decorations.

After the thirteenth century no major monuments seem to have been constructed at Chichen-Itza and the city rapidly declined in the fifteenth century. In 1556 Bishop Diego de Landa visited the practically abandoned ruins and recorded the legends pertaining to the various monuments. The ruins were not excavated until 1841.

World Heritage site since

1978 1979 1980 1981 1982 1983 1984 1985 1986 1987 **1988** 1989 1990 1991 1992 1993 1994 1995 1996 1997 1998 1999 2000 2001 2002 2003 2004 2005 2006 2007 2008

Meteora
Greece

Criteria – Human creative genius; Interchange of values; Significance in human history; Traditional human settlement; Natural phenomena or beauty

Rising starkly above the Peneas valley in the Thessalian plain, the Meteora, or rock pinnacles, are enormous residual masses of sandstone and conglomerate. Hermits and ascetics probably began settling in this extraordinary area in the eleventh century. In the late-twelfth century a small church called the Panaghia Doupiani or 'Skete' was built at the foot of one of these 'heavenly columns', where monks had already taken up residence. During the fearsome time of political instability in fourteenth-century Thessaly, monasteries were systematically built on top of the inaccessible peaks and, towards the end of the fifteenth century, there were twenty-four of them. They continued to flourish until the seventeenth century. Today, only four monasteries – the Aghios Stephanos, the Aghia Trias, Varlaam and the Meteoron – still house religious communities.

The holy monastery of Saint Nicholas Anapafsa. ▲

The sixteenth-century frescoes in the monasteries built on the Meteora mark a key stage in the development of post-Byzantine painting.

World Heritage site since

978 1979 1980 1981 1982 1983 1984 1985 1986 1987 **1988** 1989 1990 1991 1992 1993 1994 1995 1996 1997 1998 1999 2000 2001 2002 2003 2004 2005 2006 2007 2008

Nanda Devi and Valley of Flowers National Parks
India

Criteria – Natural phenomena or beauty; Significant natural habitat for biodiversity

Nestled high in the west Himalaya, India's Valley of Flowers National Park is renowned for its meadows of endemic alpine flowers and outstanding natural beauty. This richly diverse area is also home to rare and endangered animals, including the Asiatic black bear, snow leopard, brown bear and blue sheep. The gentle landscape of the Valley of Flowers National Park complements the rugged mountain

wilderness of Nanda Devi National Park. Together they encompass a unique transition zone between the mountain ranges of the Zanskar and Great Himalaya, praised by mountaineers and botanists for over a century and in Hindu mythology for much longer.

The area is a vast glacial basin, divided by a series of parallel, north-south oriented ridges which rise up to the encircling mountain rim. Nanda Devi West, India's second-highest mountain, lies on a short ridge projecting into the basin.

Xanthos-Letoon
Turkey

Criteria – Interchange of values; Testimony to cultural tradition

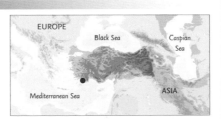

This site, which was the capital of Lycia, illustrates the blending of Lycian traditions and Hellenic influence, especially in its funerary art. The epigraphic inscriptions are crucial for our understanding of the history of the Lycian people and their Indo-European language.

The Lycians were one of the 'Sea Peoples' who invaded the Hittite Empire around 1200 BC. The Greek historian Herodotus, who lived in the fifth century BC, related that they came from Crete to take part in the Trojan War.

◀

Pillar tombs at Xanthos-Letoon, with the 'Harpy Tomb' on the right.

World Heritage site since

1978 1979 1980 1981 1982 1983 1984 1985 1986 1987 **1988** 1989 1990 1991 1992 1993 1994 1995 1996 1997 1998 1999 2000 2001 2002 2003 2004 2005 2006 2007 2008

Strasbourg – Grande Île
France

Criteria – Human creative genius; Interchange of values; Significance in human history

The Grande Île of Strasbourg is an outstanding example of a European medieval city and a unique ensemble of domestic architecture in the Rhine valley of the fifteenth and sixteenth centuries. Surrounded by two arms of the river Ill, the Grande Île (Large Island) is the historic centre of the Alsatian capital. Rising above the high-pitched roofs with multi-storeyed dormer windows, the cathedral and four ancient churches stand out on the skyline. Facing the south transept of the cathedral is the Palais Rohan, built by the Rohan family in 1732–42 as a residence for the cardinals, princes and bishops of the family. The tight network of streets contain public buildings such as the Hôtel de Ville (1585, today the Chamber of Commerce), inns, shops and workshops, as well as elegant town mansions.

The cathedral, with its single spire, is the principal element of the World Heritage site. Goethe considered Notre-Dame de Strasbourg to be the Gothic cathedral par excellence and it had great influence on the development of Gothic sculpture and architecture in Germanic lands.

World Heritage site since

1978 1979 1980 1981 1982 1983 1984 1985 1986 1987**1988**1989 1990 1991 1992 1993 1994 1995 1996 1997 1998 1999 2000 2001 2002 2003 2004 2005 2006 2007 2008

Kairouan
Tunisia

Criteria – Human creative genius; Interchange of values; Testimony to cultural tradition; Traditional human settlement; Heritage associated with events of universal significance

Founded in 670, Kairouan flourished under the Aghlabid dynasty in the ninth century and bears exceptional witness to the civilization of the first centuries of Islam in north Africa. Its rich architectural heritage includes the Great Mosque, with its marble and porphyry columns, the ninth century Mosque of the Three Gates and the Basin of the Aghlabids, filled by water brought by an aqueduct. Kairouan is, moreover, one of the holy cities of Islam. The medina is surrounded by more than 3 km of walls with three gates: its skyline is punctuated by the minarets and the cupolas of its mosques and zawiyas (monasteries), and it has preserved its network of winding streets and courtyard houses. Very few small windows or arched doorways are cut in the exterior walls, but inner walls have larger openings that give on to central courtyards.

The first Islamic place of worship founded in the Maghreb only thirty-eight years after the death of the Prophet Muhammad, is the Zuwiya of Sidi Sahab at Kairouan, where the remains of Abu Djama, one of Muhammad's companions, are kept.

▼

World Heritage site since

1978 1979 1980 1981 1982 1983 1984 1985 1986 1987 1988 **1989** 1990 1991 1992 1993 1994 1995 1996 1997 1998 1999 2000 2001 2002 2003 2004 2005 2006 2007 2008

Archaeological Site of Olympia
Greece

Criteria – Human creative genius; Interchange of values; Testimony to cultural tradition; Significance in human history; Heritage associated with events of universal significance

Olympia bears exceptional testimony to the ancient civilizations of Peloponnisos. The site has been inhabited since prehistoric times. Religious centres of worship succeeded one another during the Hellenic period until, in the tenth century BC, Olympia became a centre for the worship of Zeus. Consecrated to Zeus, the Altis is a major sanctuary that includes the ruins of the two principal temples: the Temple of

Hera (sixth century BC) and the Temple of Zeus (fifth century BC). The sanctuary contained one of the highest concentrations of masterpieces of the ancient Mediterranean world, many of which have been lost. In addition, the site includes the remains of the sports stadia and other structures erected for the Olympic Games, which were held in Olympia every four years beginning in 776 BC.

The significance of the Olympic Games demonstrates the lofty ideals of Hellenic humanism: peaceful competition between free and equal men, whose only ambition is the symbolic reward of an olive wreath. Later, not only athletes but also orators, poets and musicians came to celebrate Zeus at the Games.

Remains of dozens of columns at Olympia.
▼

World Heritage site since

1978 1979 1980 1981 1982 1983 1984 1985 1986 1987 1988 **1989** 1990 1991 1992 1993 1994 1995 1996 1997 1998 1999 2000 2001 2002 2003 2004 2005 2006 2007 2008

Monastery of Alcobaça
Portugal

Criteria – Human creative genius; Significance in human history

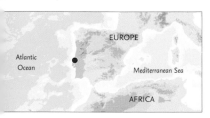

The monastery is an outstanding example of a great Cistercian establishment with the church of Santa Maria d'Alcobaça, a masterpiece of Gothic Cistercian art, and a significant grouping of medieval monastic buildings (cloister and lavabo, chapter room, parlour, dormitory, the monks' room and the refectory). The monastery was given by King Alfonso I to the Cistercians after 1152 on the understanding that they would colonize and work the surrounding lands. In the thirteenth century, while the abbey church and beautiful monastic buildings were under construction, the abbey's intellectual and political influence had already spread throughout the western part of the Iberian Peninsula. The ultimate symbol of its privileged relationship with the Portuguese monarchy is found in the tombs of Dom Pedro and Doña Inés de Castro, dating from 1360, among the most beautiful of Gothic funerary sculptures.

Front entrance of the Monastery of Alcobaça. ▲

The tombs of Inés de Castro and Dom Pedro (Peter I) are the tangible sign of Peter I's rehabilitation of Inés, assassinated at Coimbra on the orders of his father, Alfonso IV. The quality of the sculpture is surpassed by the compelling symbolism of the iconography which evokes human destiny, death and the Christian hope of eternal life.

World Heritage site since

1978 1979 1980 1981 1982 1983 1984 1985 1986 1987 1988 **1989** 1990 1991 1992 1993 1994 1995 1996 1997 1998 1999 2000 2001 2002 2003 2004 2005 2006 2007 200

Mosi-oa-Tunya / Victoria Falls
Zambia and Zimbabwe

Criteria – Natural phenomena or beauty; Major stages of Earth's history

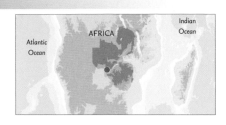

The Victoria Falls are among the most spectacular waterfalls in the world. The Zambezi River, which is more than 2 km wide at this point, plunges noisily down a series of basalt gorges and raises an iridescent mist that can be seen more than 20 km away.

The falls and associated gorges are an outstanding example of river capture and the erosive forces of the water still continue to sculpture the hard basalts. The complex of conservation areas in Zimbabwe covers over 18,500km² excluding forest reserves. The Mosi-oa-Tunya/Victoria Falls National Park abuts Dambwa Forest Reserve in Zambia. The falls are the most significant feature of the park, and when the Zambezi is in full flood (usually February or March) they form the largest curtain of falling water in the world. During these months over 500 million litres of water per minute go over the falls, which are 1,708 m wide, and drop 99 m at Rainbow Falls in Zambia. At low water in November flow can be reduced to around 10 million litres per minute, and the river is divided into a series of braided channels that descend in many separate falls.

The predominant vegetation is mopane forest, with small areas of teak and miombo woodland and a narrow band of riverine forest along the Zambezi. The riverine 'rainforest' within the waterfall splash zone is of particular interest: a fragile ecosystem of discontinuous forest on sandy alluvium, dependent upon maintenance of abundant water and high humidity resulting from the spray plume. There are many tree species within this forest and also some herbaceous species.

Several herds of elephant live in Zambezi National Park, occasionally crossing to the islands and Zambian mainland during the dry season when water levels are low. Small herds of buffalo and wildebeest, as well as zebra, warthog, giraffe, bushpig and hippopotamus are frequently found above the falls. Vervet monkey and chacma baboon are common. Lion and leopard are occasionally seen. Taita falcon breed in the gorges, as do black stork, black eagle, peregrine falcon and augur buzzard.

Victoria Falls forms a geographical barrier between the distinct fish faunas of the upper and middle Zambezi River. Thirty-nine species of fish have been recorded from the waters below the falls.

Below the falls the Zambezi enters a series of gorges which represent locations successively occupied by the falls earlier in their history. Since the uplifting of the Makgadikgadi Pan area two million years ago, the river has been cutting through the basalt, exploiting fissures and forming a series of retreating gorges. Seven previous waterfalls occupied the seven gorges below the present falls, and Devil's Cataract in Zimbabwe is the starting point for a new waterfall that will eventually leave the present lip high above the river in the gorge below.

World Heritage site since

1978 1979 1980 1981 1982 1983 1984 1985 1986 1987 1988 **1989** 1990 1991 1992 1993 1994 1995 1996 1997 1998 1999 2000 2001 2002 2003 2004 2005 2006 2007 2008

Cliff of Bandiagara (Land of the Dogons)
Mali

Criteria – Traditional human settlement; Natural phenomena or beauty

The plateau of Bandiagara is covered in a typically Sudanian savanna vegetation, and a wide range of animal species is found in the region. The cliff and rock habitats support a diversity of species including fox-kestrel, Gabar goshawk, yellow-billed shrike, scarlet-chested sunbird, abundant cliff chats and rock doves.

The Bandiagara site is an outstanding landscape of cliffs and sandy plateaus and one of the main centres for the Dogon culture. The Dogon subsistence farmers did not arrive until the fifteenth and sixteenth centuries, yet the region is rich in unique architecture, from flat-roofed huts to tapering granaries capped with thatch, cliff cemeteries and communal meeting-places. Several age-old social traditions live on in the region, through masks, feasts, rituals, and ceremonies involving ancestor worship. The Dogon people also maintain a strong relationship with the environment with their use of curative and medicinal wild plants and the sacred associations with pale fox, jackal and crocodile. The geological, archaeological and ethnological interest, together with the landscape, make the Bandiagara plateau one of West Africa's most impressive sites.

Banc d'Arguin National Park
Mauritania

Criteria – Significant ecological and biological processes; Significant natural habitat for biodiversity

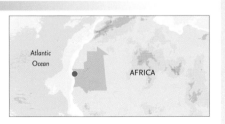

The local people, the Imraguen or Amrig, relate many of their customs to the natural environment. Even their name literally means 'the ones who gather life'. Fishing techniques, unchanged since first recorded by fifteenth-century Portuguese explorers, include the unique symbiotic collaboration with wild dolphins to catch schools of grey mullet.

Banc d'Arguin is located on the Atlantic desert-coast of Mauritania, and the park provides a unique example of a transition zone between the Sahara and the Atlantic. It is a vast area of islands and coastline, with dunes of windblown Saharan sand, shallow coastal pools, mudflats and mangrove swamps. The contrast between the harsh desert environment and the biodiversity of the marine zone has resulted in a land- and seascape of outstanding natural significance. Shallow water vegetation comprises extensive seagrass beds and various seaweeds, offering a favourable habitat for fish. This in turn attracts huge numbers of birds to Banc d'Arguin, which hosts the largest concentration of wintering waders in the world and an extremely diversified community of nesting fish-eating birds.

World Heritage site since

1978 1979 1980 1981 1982 1983 1984 1985 1986 1987 1988 **1989** 1990 1991 1992 1993 1994 1995 1996 1997 1998 1999 2000 2001 2002 2003 2004 2005 2006 2007 2008

Archaeological Site of Mystras
Greece

Criteria – Interchange of values; Testimony to cultural tradition; Significance in human history

EUROPE

Ionian
Sea

Mediterranean Sea

Mystras was built around the fortress erected in 1249 on a hill overlooking Sparta by William of Villehardouin. For almost six centuries, Mystras had a troubled existence and was abandoned in 1832, leaving only the breathtaking medieval ruins, standing in a beautiful landscape. From 1262–1348, Mystras was the seat of the Byzantine military governor. The bishopric of Sparta was transferred to the new city: the Metropolis, dedicated to St Demetrios, was built in 1264, and convents were built and richly decorated. From 1348–1460 Mystras became the capital of the Despotate of Morea. During this period, the zenith of Mystras, the cosmopolitan city was of great importance and its fall to the Turks on 30 May 1460 was seen as being almost equal in importance to the fall of Constantinople in 1453.

The beauty of the churches of Mystras, which, during the fourteenth-century Paleologus Renaissance, had been covered with dramatic frescoes, the renown of its libraries and the glory of its writers (including Georges Gemiste Plethon and Jean Bessarion who brought neo-Platonic humanism to Italy) gave substance to the legend of the 'Wonder of Morea'.

The medieval ruins of Mystras. ▼

World Heritage site since

1978 1979 1980 1981 1982 1983 1984 1985 1986 1987 1988**1989**1990 1991 1992 1993 1994 1995 1996 1997 1998 1999 2000 2001 2002 2003 2004 2005 2006 2007 2008

Buddhist Monuments at Sanchi
India

ASIA

Indian Ocean

South China Sea

Criteria – Human creative genius; Interchange of values; Testimony to cultural tradition; Significance in human history; Heritage associated with events of universal significance

On a hill overlooking a plain, about 40 km from Bhopal, the site of Sanchi comprises a group of Buddhist monuments, monolithic pillars, palaces, temples and monasteries, all in different states of conservation and mainly dating back to the second and first centuries BC. It is the oldest Buddhist sanctuary in existence and was a major centre of Buddhism in India until the twelfth century AD. The principal monument at Sanchi, known as Stupa 1 (pictured on the right), consists of a gigantic mound of sandstone surrounded by

sumptuous porticoes with stone railings. Its hemispherical dome measures 36 m in diameter and is 16 m high. It is particularly famous for the extraordinarily rich decorative work on the four monumental gateways that provide access.

When it was discovered in 1818, Sanchi had lain abandoned for 600 years and was overrun with vegetation. Gradually the hill was cleared, uncovering the ruins of about fifty monuments in one of the most remarkable archaeological complexes in India.

World Heritage site since

1978 1979 1980 1981 1982 1983 1984 1985 1986 1987 1988 1989**1990**1991 1992 1993 1994 1995 1996 1997 1998 1999 2000 2001 2002 2003 2004 2005 2006 2007 2008

Monasteries of Daphni, Hosios Loukas and Nea Moni of Chios
Greece

EUROPE

Ionian Sea

Mediterranean Sea

Criteria – Human creative genius; Significance in human history

Although geographically distant from each other, these three monasteries belong to the same typological series and share the same aesthetic characteristics. The first is in Attica, near Athens, the second in Phocida near Delphi, and the third on an island in the Aegean Sea, near Asia Minor. The churches

are built on a cross-in-square plan with a large dome supported by squinches defining an octagonal space. In the eleventh and twelfth centuries they were decorated with superb marble works as well as mosaics on a gold background, all characteristic of the 'second golden age of Byzantine art'.

The Monastery of Hosios Loukas has an immense central dome, 9 m in diameter, and is one of the most perfect creations of Byzantine architecture. The church is filled with iconographic treasures of a magnitude and coherence rarely equalled.

World Heritage site since

1978 1979 1980 1981 1982 1983 1984 1985 1986 1987 1988 1989 **1990** 1991 1992 1993 1994 1995 1996 1997 1998 1999 2000 2001 2002 2003 2004 2005 2006 2007 2008

Colonial City of Santo Domingo
Dominican Republic

Criteria – Interchange of values; Significance in human history; Heritage associated with events of universal significance

After Christopher Columbus's arrival on the island of Hispaniola in 1492, Santo Domingo became the site of the first cathedral, hospital, customs house and university in the Americas. This Ciudad Colonial (Colonial City), founded in 1498, was laid out on a grid pattern that became the model for almost all town planners in the New World.

Among the Colonial City's most outstanding buildings is the cathedral, constructed between 1514 and 1542. It is the oldest in the Americas and one of the architectural wonders of the original town. The Ozama Fortress and Tower of Homage were built in 1503: this stone group is said to be the oldest formal military outpost still standing in the Americas. The Tower of Homage in the centre of the grounds is an impressive architectural structure that is medieval in style and design.

Bartholomew Columbus, brother of Christopher, founded the city in 1496 on the left bank of the Ozama River. In 1502 a tropical storm destroyed the city and the colony's new governor Nicolás de Ovando decided that it should be completely rebuilt on the site it presently occupies. Its town plan became the model for Spanish towns in the New World.

▼ Museum of the Royal Houses.

World Heritage site since

1978 1979 1980 1981 1982 1983 1984 1985 1986 1987 1988 1989 **1990** 1991 1992 1993 1994 1995 1996 1997 1998 1999 2000 2001 2002 2003 2004 2005 2006 2007 2008

Historic Centre of Saint Petersburg and Related Groups of Monuments
Russian Federation

EUROPE

ASIA

Pacific Ocean

Winter Palace in St Petersburg.

Criteria – Human creative genius; Interchange of values; Significance in human history; Heritage associated with events of universal significance

Known as the 'Venice of the North', with its numerous canals and hundreds of bridges, the city of St Petersburg on Russia's Baltic coast is the result of a vast urban project begun in 1703 under Peter the Great (1672–1725).

St Petersburg's architectural heritage reconciles the very different Baroque and pure neoclassical styles, as can be seen in the Admiralty, the Winter Palace, the Marble Palace and the Hermitage.

The metamorphosis of an inhospitable Baltic coastal area into a superb city where palaces, churches, convents and two-storey stone houses fit in to the urban designs of the Frenchman Alexandre Leblond, was completed in less than twenty years.

The building of Peter the Great's planned capital began in the eighteenth century on the backs of colossal forced labour of Russian soldiers, Swedish and Ottoman prisoners of war, and Finnish and Estonian workers. A network of canals, streets and quays was gradually built up until, by the reign of Catherine the Great (1729–96) at the end of the century, the urban landscape of St Petersburg took on its monumental splendour. An array of foreign architects rivalled one another with audaciousness and splendour in their designs of the

capital's huge palaces and convents and in imperial and princely suburban residences such as Petrodvorets, Lomonosov, Tsarskoie Selo (Pushkin), Pavlovsk and Gatchina.

The impetus of the eighteenth century continued into the nineteenth with astonishing monumental works. In the history of urbanism St Petersburg is probably the only example of a vast project that retained all its logic despite a rapid succession of styles that were reputed to be irreconcilable. Yet from the disparity of styles, an impression of timeless grandeur emerges in this distended historic centre where the greatness of the monuments is on a scale with a landscape free of any background, open to the sea, perpetually swept by sea breezes and criss-crossed by canals running beneath, it is said, more than 400 bridges.

The ensembles designed in St Petersburg and the surrounding area by several international architects exerted great influence in the eighteenth and nineteenth centuries on the development of architecture and monumental arts in Russia and Finland. The site links outstanding examples of Baroque imperial residences with the architectural ensemble of St Petersburg, the Baroque and neoclassical capital par excellence.

Under its previous name of Petrograd, the city was the starting place of the October Revolution of 1917 that brought the Bolsheviks to power in Russia. It was renamed Leningrad as part of the USSR. More than a million of its citizens died of starvation between 1941 and 1944 when the city was besieged for 872 days by the Nazis during the Second World War.

The city voted to readopt the name of St Petersburg (Sankt Peterburg) in 1991 after the fall of Communism.

World Heritage site since

1978 1979 1980 1981 1982 1983 1984 1985 1986 1987 1988 1989 **1990** 1991 1992 1993 1994 1995 1996 1997 1998 1999 2000 2001 2002 2003 2004 2005 2006 2007 2008

Palaces and Parks of Potsdam and Berlin
Germany

Criteria – Human creative genius; Interchange of values; Significance in human history

With 5 km² of parks and 150 buildings constructed between 1730 and 1916, Potsdam's complex of palaces and parks forms an artistic whole, whose eclectic nature reinforces its sense of uniqueness. This series of architectural and landscaping masterpieces was built progressively within a single space, illustrating opposing and reputedly irreconcilable styles, yet without detracting from the harmony of a general composition. The area extends into the

district of Berlin-Zehlendorf, with the palaces and parks lining the banks of the river Havel and Lake Glienicke.

The World Heritage site covers two other ensembles that include parks, chateaux and buildings, in the middle of which stands the Marble Palace, the king's summer residence. In summer 1945 the Chateau of Cecilienhof at the northern end of the park was the site of the Potsdam Conference that decided the fate of postwar Germany.

King Frederick the Great of Prussia (1712–1786) transformed Potsdam from its previous status of garrison town into the 'Prussian Versailles' which was to be his main residence. Sanssouci, a name which reflects the king's desire for intimacy and simplicity, translates the theme of a rustic villa into the marble, mirrors and gold of a Rococo-style palace.

Chinese Tea House in Sanssouci Park, Potsdam. ▼

World Heritage site since

1978 1979 1980 1981 1982 1983 1984 1985 1986 1987 1988 1989 **1990** 1991 1992 1993 1994 1995 1996 1997 1998 1999 2000 2001 2002 2003 2004 2005 2006 2007 2008

Kiev: Saint-Sophia Cathedral and Related Monastic Buildings, Kiev-Pechersk Lavra
Ukraine

Criteria – Human creative genius; Interchange of values; Testimony to cultural tradition; Significance in human history

Designed to rival Hagia Sophia in Constantinople, Kiev's Saint-Sophia Cathedral symbolizes the 'new Constantinople', capital of the Christian principality of Kiev, which was created in the eleventh century. Devastated by the Mongols and the Tatars, Kiev-Pechersk Lavra was almost entirely rebuilt from the seventeenth century onwards. These were times of great prosperity, when pilgrims flocked to the site, and the grounds were filled with numerous Baroque monuments. The Clock Tower and the Refectory Church are two of the main landmarks in a monastic landscape totally transformed by the construction or the renovation of numerous churches. The spiritual and intellectual influence of Kiev-Pechersk Lavra contributed to the spread of Orthodox thought and the Orthodox faith in the Russian world from the seventeenth to the nineteenth centuries.

Today the major elements of the very old historic heritage are Trinity Church, whose twelfth-century structure is hidden by the extremely rich Baroque decor, and especially the catacombs, which include the Near Caves and the Far Caves, whose entrances are respectively at All Saints' Church and at the Church of the Conception of St Anna.

St Sophia Cathedral.
▼

World Heritage site since

1978 1979 1980 1981 1982 1983 1984 1985 1986 1987 1988 1989 **1990** 1991 1992 1993 1994 1995 1996 1997 1998 1999 2000 2001 2002 2003 2004 2005 2006 2007 2008

Kremlin and Red Square, Moscow
Russian Federation

Criteria – Human creative genius; Interchange of values; Significance in human history; Heritage associated with events of universal significance

St Basil's Cathedral.

The citadel of the Kremlin has been inextricably linked to all the most important historical and political events in Russia since it was built between the fourteenth and seventeenth centuries.

Ever since the establishment of the Principality of Moscow in 1263 and the transfer to Moscow of the seat of the Metropolitan see in 1328, the Kremlin of Moscow was the centre of both temporal and spiritual power. Some of its original buildings border Cathedral Square while others, such as the Nativity of the Virgin (1393), were incorporated into the Great Palace when it was rebuilt.

The nucleus expanded northward with the palace of the Patriarchs and the Church of the Twelve Apostles, erected in the seventeenth century, and especially with the Arsenal of Peter the Great which fills the northwest angle of the boundary wall. The triangular palace of the Senate (today the seat of the Council of Ministers) was built by the architect Matvey Kazakov for Empress Catherine II in the northeast sector between the Arsenal and the monasteries of the Miracle and of the Ascension, two splendid structures that were razed in 1932. In the southeast sector Kazakov built for the empress another smaller palace, known as the Nicholas palace, also destroyed in 1932.

The Kremlin has been a major influence in Russian architecture as well as a model for the kremlins, or citadels, in the centre of other Russian cities, such as Pskov, Tula, Kazan and Smolensk.

Red Square, lying beneath the east wall, is closely associated with the Kremlin. At its south end is the famous onion-domed Cathedral of St Basil the Blessed (see photo on the right), one of the most beautiful monuments of Orthodox art. It was originally one of a pair of churches, the other being the Cathedral of Kazan, erected in the vast open area bordering the 'Goum' in 1633 by Prince Pozarsky to commemorate the victory over the Poles. It disappeared in the early 1930s along with several convents in the neighbouring area (Saviour-behind-the-Images, St Nicholas and Epiphany).

The Kremlin contains within its walls a unique series of architectural and artistic masterpieces. These include religious monuments of exceptional beauty such as the Church of the Annunciation, Cathedral of the Dormition, Church of the Archangel and the bell tower of Ivan Veliki; and palaces such as the Great Palace of the Kremlin, the yellow-and-white palace of the tsars which contains the Church of the Nativity of the Virgin, and the Teremnoi Palace.

World Heritage site since

1978 1979 1980 1981 1982 1983 1984 1985 1986 1987 1988 1989 **1990** 1991 1992 1993 1994 1995 1996 1997 1998 1999 2000 2001 2002 2003 2004 2005 2006 2007 2008

Tsingy de Bemaraha Strict Nature Reserve
Madagascar

Criteria – Natural phenomena or beauty;
Significant natural habitat for biodiversity

The fauna of the region has not been studied in any detail. The Tsingy is the only known location for the chameleon and the only protected area where the endemic nesomyine rodent occurs. Other notable threatened species include the goshawk and various species of lemur.

This reserve lies 60–80 km inland from the west coast, north of the Manambolo river. It contains spectacular karstic landscapes and limestone uplands cut into impressive tsingy peaks (sharp-edged and very tall towers of limestone) and a 'forest' of limestone needles. The limestone karst is delimited to the east by abrupt cliffs which rise some 300–400 m above the Manambolo river, creating an impressive canyon. The western slopes rise more gently, and the whole western region of the reserve forms a plateau with rounded hillocks. To the north undulating hills alternate with limestone extrusions, whereas in the south extensive pinnacle formations make access extremely restricted. Vegetation is characteristic of the calcareous karst regions of western Madagascar, with dense, dry, deciduous forest, and extensive savannas and some mangrove swamps.

World Heritage site since

1978 1979 1980 1981 1982 1983 1984 1985 1986 1987 1988 1989 **1990** 1991 1992 1993 1994 1995 1996 1997 1998 1999 2000 2001 2002 2003 2004 2005 2006 2007 2008

Delos
Greece

Criteria – Interchange of values; Testimony to cultural tradition; Significance in human history; Heritage associated with events of universal significance

Delos bears unique witness to the civilizations of the Aegean world from the third millennium BC to the early Christian era. According to Greek mythology, Apollo was born on this tiny island in the Cyclades and it became the major sanctuary dedicated to him, attracting pilgrims from all over Greece. It was a very important cosmopolitan Mediterranean port. Today the island's landscape consists solely of exceptionally extensive and rich ruins

unearthed systematically since 1872. The principal zones are the northeast coastal plain (Sanctuary of Apollo, Agora of the Compitaliasts, Agora of the Delians); the Sacred Lake region (Agora of Theophrastos, Agora of the Italians, the renowned Terrace of the Lions); the Mount Kynthos area (Terrace of the Sanctuaries of the Foreign Gods, Heraion); and the theatre quarter, whose poignant ruins have been overrun by vegetation.

Delos' era of maritime trade ended in 69 BC with the sacking of the island by Athenodoros, and it was then abandoned in the sixth century. It was turned into a quarry site and the columns of its temples were consumed in lime kilns and the walls of its houses left in ruins.

Terrace of the Lions.
▼

World Heritage site since

1978 1979 1980 1981 1982 1983 1984 1985 1986 1987 1988 1989 **1990** 1991 1992 1993 1994 1995 1996 1997 1998 1999 2000 2001 2002 2003 2004 2005 2006 2007 200

Te Wahipounamu – South West New Zealand
New Zealand

Criteria – Natural phenomena or beauty; Major stages of Earth's history; Significant ecological and biological processes; Significant natural habitat for biodiversity

Aoraki (Mount Cook).

Te Wahipounamu, situated in southwest New Zealand, comprises the least-disturbed tenth of New Zealand's landmass, with some 20,000 km² of temperate rainforest, 450 km of alpine communities, and a distinctive fauna. The park provides a habitat for rare and endangered species such as the kea, the world's only alpine parrot, and the takahe, a large, flightless bird. It also contains the best modern representation of the ancient flora and fauna of the original southern supercontinent of Gondwanaland, including Podocarpus species and genera of beech which cover two thirds of the park and can be over 800 years old, flightless kiwis, 'bush' moas and carnivorous Powelliphanta land snails.

The overwhelming mountainous character of the area results from tectonic movement between the Pacific and Indo-Australian plates over the last five million years. High local relief is the result of deep glacial excavation and glaciers are an important feature of the area, especially in the vicinity of Westland and Aoraki/Mount Cook national parks. There have also been substantial post-glacial changes, especially marked in south Westland and the Southern Alps. Erosion is rapid, especially west of the Main Divide. Intense gullying, serrated ridges, and major and minor rock falls are characteristic of this zone. However, glacial landforms are almost entirely intact in Fiordland.

The vegetation is notable for its diversity and essentially pristine condition. A floristically rich alpine vegetation of shrubs, tussocks and herbs extends around the summits of the mountains, from about 1,000 m to the permanent snowline. At warmer lower altitudes, the rainforest is dominated by dense stands of tall podocarps. The wetter, milder west is characterized by luxuriant rainforest and wetlands; the drier, more continental east has more open forest, shrublands and tussock grasslands. The most extensive and least modified natural freshwater wetlands in New Zealand are found in this area. Sizeable open wetlands, including high-fertility swamps and low-fertility peat bogs, are a particular feature of the south Westland coastal plain.

The best-known vegetation chronosequences are those on glacial landforms where the ages of outwash, terrace and higher piedmont surfaces are known. The most impressive landform chronosequence is found in the flights of marine terraces in southern Fiordland.

As the least-modified region on mainland New Zealand, this is the core habitat for many indigenous animals and contains the country's largest population of forest birds. A few mountain valleys in Fiordland harbour the total wild population (about 170 birds) of the takahe, a large, flightless rail believed extinct until 'rediscovered' in 1948. Most of New Zealand's fur seals are found along the southwest coast. Virtually wiped out by sealing in the early 1800s, they currently number about 50,000.

World Heritage site since

1978 1979 1980 1981 1982 1983 1984 1985 1986 1987 1988 1989 **1990** 1991 1992 1993 1994 1995 1996 1997 1998 1999 2000 2001 2002 2003 2004 2005 2006 2007 2008

Kizhi Pogost
Russian Federation

Criteria – Human creative genius; Significance in human history; Traditional human settlement

The pogost of Kizhi (the Kizhi enclosure) is located on one of the many islands in Lake Onega, in Karelia. Two eighteenth-century wooden churches, and an octagonal clock tower, also in wood and built in 1862, can be seen there. These unusual constructions, in which carpenters created a bold visionary architecture, perpetuate an ancient model of parish space and are in harmony with the surrounding landscape.

Kizhi Pogost represents the adaptation of the Orthodox Church parish organization to the challenges posed by the immense distances and far-flung communities of northern Russia. Here, all the buildings needed in the parish's religious life were grouped in one place.

◄

Church of the Transfiguration.

Río Abiseo National Park
Peru

Criteria – Testimony to cultural tradition; Natural phenomena or beauty; Significant ecological and biological processes; Significant natural habitat for biodiversity

Río Abiseo National Park is an outstanding example of ongoing geological processes, biological evolution and man's interaction with the natural environment. The topography is mountainous, often with very steep slopes. The Río Abiseo soils are generally acidic and have never been disturbed by agriculture or timber extraction. The park was created in 1983 to protect the fauna and flora of the rainforests, which includes many rare and important species. Jaguar, spectacled bear, jaguarundi and giant armadillo all thrive here, and the yellow-tailed woolly monkey, previously thought extinct, is found only in this area. Research undertaken since 1985 has already uncovered at altitudes of between 2,500 and 4,000 m, thirty-six previously unknown archaeological sites which give a good picture of pre-Inca society.

Río Abiseo is renowned for its pristine primary cloud forest and highland grasslands (paramo). Archaeological evidence from the park suggests that humans settled in the area from as early as 6000 BC.

World Heritage site since

1978 1979 1980 1981 1982 1983 1984 1985 1986 1987 1988 1989 **1990** 1991 1992 1993 1994 1995 1996 1997 1998 1999 2000 2001 2002 2003 2004 2005 2006 2007 2008

Mount Huangshan
China

Criteria – Interchange of values; Natural phenomena or beauty; Significant natural habitat for biodiversity

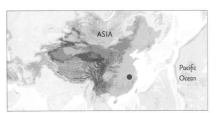

Huangshan, known as 'the loveliest mountain of China', has been acclaimed in art and literature for much of Chinese history. Today it holds the same fascination for those who come on a pilgrimage to the mountain, which is renowned for its magnificent scenery made up of many granite peaks and strangely shaped rocks emerging out of a sea of clouds. It presents a beautiful spectacle, with ridges, gorges, forests, lakes and waterfalls. Many of the lakes having clear blue, turquoise or green waters, while in autumn many of the leaves turn a range of rich colours. The site supports a high diversity of plant and animal species, including some that are threatened. A number of trees are celebrated on account of their age, grotesque shape or precipitously perched position, including 1,000-year-old specimens of Huangshan pine, ginkgo and alpine juniper.

▲
Huangshan is considered to be a prime example of classic Chinese scenery, as typified in Chinese landscape paintings. On 17 June 747, during the Tang dynasty, an imperial order was issued to name it Huangshan (Yellow Mountain). Poets, scholars and artists were among the many visitors, and by the Yuan dynasty (1271–1368) sixty-four temples had been constructed there.

World Heritage site since

1978 1979 1980 1981 1982 1983 1984 1985 1986 1987 1988 1989 1990 **1991** 1992 1993 1994 1995 1996 1997 1998 1999 2000 2001 2002 2003 2004 2005 2006 2007 2008

Paris, Banks of the Seine
France

Criteria – Human creative genius; Interchange
of values; Significance in human history

Masterpieces along
the Seine include
Notre Dame (pictured
right) and Sainte
Chapelle, Louvre,
Palais de l'Institut, Les
Invalides, Place de la
Concorde, École
Militaire,
La Monnaie, Eiffel
Tower (pictured below)
and Palais
de Chaillot.

Some, such as Notre
Dame were definitive
references in the
spread of Gothic
construction, while
the Place de la
Concorde or the vista
at the Invalides
influenced the
development of
European capitals.
The Marais and Île
Saint-Louis have
coherent architectural
ensembles, with
significant examples
of seventeenth and
eighteenth century
Parisian construction:
Hôtel Lauzun and
Hôtel Lambert,
Quai Malaquais and
Quai Voltaire.

The banks of the Seine are studded with
a succession of masterpieces. From the
Louvre to the Eiffel Tower, from the Place de
la Concorde to the Grand and Petit Palais,
the evolution of Paris and its history can be
seen from the river Seine. The Cathedral of
Notre Dame and the Sainte Chapelle are
architectural masterpieces while Baron
Haussmann's wide squares and boulevards
influenced late-nineteenth- and twentieth-
century town planning the world over.

Paris is a river town. Ever since the first
human settlements from prehistoric days
and the village of the Parisii tribes, the Seine
has played both a defensive and an
economic role. The present historic city,
which developed between the sixteenth,
and particularly, the seventeenth centuries
and the twentieth century, shows the
evolution of the relationship between the
river and the people: for defence, for
trading, for promenades, and so on.

The site and the river were gradually
brought under control in a series of
measures: the articulation of the two islets,
Île de la Cité and Île Saint-Louis, with the
bank; the creation of north–south
thoroughfares; installations along the river
course; construction of quays; and the
channelling of the river. Similarly, although
the city's defensive walls have disappeared,
traces of their existence can be seen in the
difference in size and spacing of the
buildings: these are closer together in the
Marais and the Île Saint-Louis, and more
open after the Louvre.

Beyond the Louvre are a number of major
classic constructions laid along three
perpendicular cross-river axes: Palais
Bourbon, Concorde and Madeleine; Les
Invalides and the Grand and Petit Palais;
and the Champ de Mars, École Militaire
and Palais de Chaillot. The ensemble must
be regarded as a geographical and historic
entity. It constitutes a remarkable example
of urban riverside architecture, where the
strata of history are harmoniously
superimposed.

Haussmann's urban plan, which marks
the western part of the city, inspired the
construction of the great cities of the
New World, particularly in Latin America.
The Eiffel Tower and the Palais de Chaillot
are living testimony of the great universal
exhibitions which were of such great
importance in the nineteenth and
twentieth centuries.

World Heritage site since

1978 1979 1980 1981 1982 1983 1984 1985 1986 1987 1988 1989 1990 **1991** 1992 1993 1994 1995 1996 1997 1998 1999 2000 2001 2002 2003 2004 2005 2006 2007 2008

Danube Delta
Romania

Criteria – Natural phenomena or beauty;
Significant natural habitat for biodiversity

EUROPE

Black Sea

The waters of the Danube, which flow into the Black Sea, form the largest and best preserved of Europe's deltas. The reserve is vast in European terms with numerous freshwater lakes interconnected by narrow channels featuring huge expanses of aquatic vegetation. This is the largest continuous marshland in Europe and the second-largest delta (after the Volga), which includes the greatest stretch of reedbeds in the world.

The delta has been classified into twelve habitat types: aquatic, lakes covered with flooded reedbeds; 'plaur', flooded islets; flooded reeds and willows; riverine forest of willows and poplars; cane fields; sandy and muddy beaches; wet meadows; dry meadows; human settlements; sandy and rocky areas; steep banks; and forests on high ground.

Over 300 bird species have been recorded, of which over 176 breed, the most important being cormorant, pygmy cormorant, white pelican and Dalmatian pelican.

The Danube delta is a remarkable alluvial feature constituting critical habitat for migratory birds and other animals. It is the major remaining wetland on the flyway between central and Eastern Europe and the Mediterranean, Middle East and Africa.
▼

World Heritage site since

1978 1979 1980 1981 1982 1983 1984 1985 1986 1987 1988 1989 1990 **1991** 1992 1993 1994 1995 1996 1997 1998 1999 2000 2001 2002 2003 2004 2005 2006 2007 2008

Golden Temple of Dambulla
Sri Lanka

Criteria – Human creative genius; Heritage associated with events of universal significance

A sacred pilgrimage site for twenty-two centuries, this cave monastery, with its five sanctuaries, is the largest and best-preserved cave-temple complex in Sri Lanka. The site was originally occupied by a Buddhist monastic establishment; remains of eighty rock-shelter residences established at that time have been identified. Most probably in the first century BC, the uppermost group of shelters on Dambulla's south face was transformed into shrines.

These transformations continued between the fifth and thirteenth centuries: cave-temples were extended into the sheltering rock, and brick walls constructed to screen the caves. By the end of the twelfth century, with the introduction of sculpture on the upper terrace, the caves assumed their present general form and layout. The Buddhist mural paintings covering an area of 2,100 m² are of particular importance, as are the 157 statues.

One of the site's distinguishing characteristics is the regular renewal of decorated surfaces over time. Conservation measures devoted to stripping back layers of later painting on wall surfaces or sculpture to reveal earlier images would be ignoring the worth of the ongoing tradition which has regularly ensured complete repainting of surfaces.

World Heritage site since

1978 1979 1980 1981 1982 1983 1984 1985 1986 1987 1988 1989 1990 **1991** 1992 1993 1994 1995 1996 1997 1998 1999 2000 2001 2002 2003 2004 2005 2006 2007 200

Thungyai-Huai Kha Khaeng Wildlife Sanctuaries
Thailand

Criteria – Natural phenomena or beauty;
Significant ecological and biological processes;
Significant natural habitat for biodiversity

The site comprises two contiguous wildlife sanctuaries: Thung Yai and Huai Kha Khaeng, which cover more than 6,000 km² along Thailand's western international border with Myanmar (Burma). The sanctuaries contain examples of almost all the forest types of continental southeast Asia and are among the few areas in Asia large enough to support viable populations of large herbivores (300 elephants) and predators (e.g. tigers). They are home to a very diverse array of animals, including 77 per cent of the large mammals, 50 per cent of the large birds and 33 per cent of the land vertebrates to be found in this region. The reason for such exceptional diversity is partly due to its status as one of only two evergreen forest refuges during the driest periods of the Pleistocene glaciations.

At least thirty-four internationally threatened species are found within the confines of the two sanctuaries. It is also home to twenty-two species of woodpecker, more than any other park in the world.

World Heritage site since

978 1979 1980 1981 1982 1983 1984 1985 1986 1987 1988 1989 1990 **1991** 1992 1993 1994 1995 1996 1997 1998 1999 2000 2001 2002 2003 2004 2005 2006 2007 2008

Island of Mozambique
Mozambique

Criteria – Significance in human history;
Heritage associated with events of universal
significance

The island of Mozambique bears important
witness to the establishment and
development of the Portuguese maritime
routes between western Europe and the
Indian subcontinent and thence all of Asia.
The towns and the fortifications on the
island, and on the smaller island of
St Laurent, are an outstanding example of
an architecture in which local traditions,
Portuguese influences, and to a somewhat
lesser extent Indian and Arab influences, are

all interwoven, producing a unified whole.

The incredible architectural unity of the
island derives from the uninterrupted use of
the same building techniques with the same
materials and the same decorative principles.

The island's patrimony also includes its
oldest extant fortress – St Sebastian, built
between 1558 and 1620 – other defensive
buildings and numerous religious buildings,
including many from the sixteenth century.

Portuguese
connections with
Mozambique began
in 1497 when Manuel I
of Portugal asked
Vasco da Gama to
establish a trade route
to India. The explorer
left in July 1497 and in
March 1498 reached
Mozambique where
he was welcomed by
the sultan who thought
the Portuguese were
Muslims. During his
second voyage he
occupied the territories
of Mozambique and
returned to Lisbon in
1503 laden with gold.

Main square of the town of Mozambique. ▼

World Heritage site since

1978 1979 1980 1981 1982 1983 1984 1985 1986 1987 1988 1989 1990 **1991** 1992 1993 1994 1995 1996 1997 1998 1999 2000 2001 2002 2003 2004 2005 2006 2007 2008

Poblet Monastery
Spain

Criteria – Human creative genius; Significance in human history

Lying midway between Tarragona and Lérida, at the foot of the Sierra de Montsant, the Cistercian monastery founded in 1150 by the monks of Fontfroide was transformed into a stronghold in the fourteenth century by Peter IV 'the Ceremonious', King of Aragon, during the War of Castile.

▼

Santa María of Poblet has served as one of the most majestic and austere Cistercian abbeys, as a massive military complex and as a royal residence and pantheon. North of the church is the great cloister with its fountain, chapter room, monks' dormitory, closed cloister, calefactory, refectory and kitchens. The former lay brothers' buildings are on the west and the infirmary to the north. The monastery also has a gatehouse,

guest house, abbot's and prior's lodgings and work buildings. As a fortress, its walls are an excellent example of fourteenth-century military architecture. Poblet is directly associated with the history of the royal houses of Barcelona, Aragon and Castile. Shortly after 1349 Peter IV had a sumptuous dynastic burial place built in the abbey church and part of the monastery was used as a royal residence.

World Heritage site since

1978 1979 1980 1981 1982 1983 1984 1985 1986 1987 1988 1989 1990 **1991** 1992 1993 1994 1995 1996 1997 1998 1999 2000 2001 2002 2003 2004 2005 2006 2007 2008

Historic City of Sucre
Bolivia

Criteria – Significance in human history

Founded in the early sixteenth century, Sucre is an exceptional example of the blending of local architectural traditions with styles imported from Europe. There are many well-preserved sixteenth-century religious buildings, such as San Lázaro, San Francisco and Santo Domingo. The impressive Metropolitan Cathedral was begun in 1559 but not completely finished until 250 years later. In August 1825, the republic of Bolivia declared its independence and the city's name was changed in honour of Mariscal António José de Sucre, who fought against Spanish rule. The House of Freedom is one of Bolivia's most important historical monuments, and many key events that led to independence took place here. It was originally built in 1621 as part of the Convent of the Jesuits.

Sucre was founded as the city of La Plata in 1538. The first capital of Bolivia, its early wealth came from mining activities, but it soon also became a major cultural centre, being home to the Universidad de San Francisco, the Royal Academia Carolina and the seat of the Characas Audiencia, a forerunner of the present Supreme Court.

Aïr and Ténéré Natural Reserves
Niger

Criteria – Natural phenomena or beauty; Significant ecological and biological processes; Significant natural habitat for biodiversity

This is the largest protected area in Africa, covering some 77,000 km², though the area considered a protected sanctuary constitutes only one-sixth of the total area. It includes the volcanic rock mass of the Aïr, a small Sahelian pocket, isolated as regards its climate and flora and fauna, and situated in the Saharan desert of Ténéré. The reserves boast an outstanding variety of landscapes, plant species and wild animals.

The reserve harbours significant populations of the wild relatives of several important crop species: olive, millet and sorghum. Both Sudanese and Mediterranean flora are found above 1,000 m in the sheltered wetter localities in the massifs.

World Heritage site since

1978 1979 1980 1981 1982 1983 1984 1985 1986 1987 1988 1989 1990 **1991** 1992 1993 1994 1995 1996 1997 1998 1999 2000 2001 2002 2003 2004 2005 2006 2007 2008

Shark Bay, Western Australia
Australia

Criteria – Natural phenomena or beauty; Major stages of Earth's history; Significant ecological and biological processes; Significant natural habitat for biodiversity

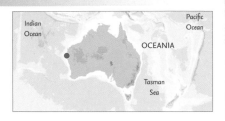

At the most westerly point of the Australian continent, Shark Bay, with its islands and remarkable coastal scenery, is an area of major zoological importance, primarily due to its island habitats being isolated from the disturbances that have occurred elsewhere. The region has three exceptional natural features: its vast seagrass beds, which are the largest and richest in the world; its population of around 10,000 dugongs or 'sea cows'; and its stromatolites, colonies of algae which form hard, dome-shaped deposits and are among the oldest forms of life on Earth. The stromatolites and microbial mats of Hamelin Pool in Shark Bay assist greatly in the understanding of the evolution of the Earth's biosphere. Shark Bay is also home to five species of endangered mammals and has a rich birdlife, with over 230 species, or 35 per cent of Australia's bird species, having been recorded.

The site is renowned for its marine fauna. In addition to its dugong population, humpback whales use the bay as a migratory staging post. Bottlenose dolphin and green turtle also occur in the bay, and endangered loggerhead turtle nest on the beaches. Large numbers of sharks and rays can also be observed.

Old Rauma
Finland

Criteria – Significance in human history; Traditional human settlement

Situated on the Gulf of Bothnia, Rauma is one of the oldest harbours in Finland. Built around a Franciscan monastery, where the mid-fifteenth-century Holy Cross Church still stands, it is an outstanding example of an old Nordic city constructed in wood. Although ravaged by fire in the late seventeenth century, it has preserved its ancient vernacular architectural heritage.

Old Rauma is the largest and one of the most beautiful historical wooden towns in the Nordic countries. The majority of the buildings in the old city have been sensitively restored as part of a comprehensive development plan.

World Heritage site since

1978 1979 1980 1981 1982 1983 1984 1985 1986 1987 1988 1989 1990 **1991** 1992 1993 1994 1995 1996 1997 1998 1999 2000 2001 2002 2003 2004 2005 2006 2007 2008

Historic Centre of Morelia
Mexico

Criteria – Interchange of values; Significance in human history; Heritage associated with events of universal significance

Built in the sixteenth century, Morelia is an outstanding example of urban planning, combining the ideas of the Spanish Renaissance with the Mesoamerican experience. In 1537 a Franciscan monastery was established near the Indian village of Guayangareo. In 1541, it became the new provincial capital, renamed Valladolid. Although fifty noble families settled here, as did many Europeans over the next few centuries, the population remained predominantly of Indian origin. Valladolid

became a bishop's see in 1580. At the same time the College of St Nicholas Obispo (founded in 1540 at Patzcuaro), the oldest institution of higher learning in Mexico, transferred here. Well-adapted to its hilly site, streets still follow the original grid layout. More than 200 historic buildings, including twenty public buildings and twenty-one churches, all in the region's characteristic pink stone, reflect the town's architectural history.

As an intellectual centre Valladolid was among the principal towns in Mexico's fight for independence in the early nineteenth century. Two of the leading figures in the struggle were both priests: Miguel Hidalgo and José Maria Morelos. In honour of the latter, a native of Valladolid, the town's name was changed to Morelia in 1828.

Small courtyard plaza with fountain in the historic centre of Morelia. ▼

World Heritage site since

1978 1979 1980 1981 1982 1983 1984 1985 1986 1987 1988 1989 1990 **1991** 1992 1993 1994 1995 1996 1997 1998 1999 2000 2001 2002 2003 2004 2005 2006 2007 200

Komodo National Park
Indonesia

Criteria – Natural phenomena or beauty;
Significant natural habitat for biodiversity

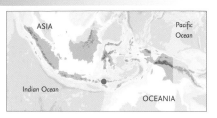

These volcanic islands in southern
Indonesia, of which Komodo is the largest,
are best known for the Komodo monitor,
the world's largest living lizard, whose
appearance and aggressive behaviour have
led to them being called 'Komodo dragons'
(see photo below). They exist nowhere else
in the world and are of great interest to
scientists studying the theory of evolution.
Mammals include primates such as crab-
eating macaque. Introduced species, such as
wild boar, as well as feral domestic animals,
form important prey species for the
Komodo monitor. The seas around the
islands are reported to be among the most
productive in the world due to upwelling
and a high degree of oxygenation resulting
from strong tidal currents. The rugged
hillsides of dry savanna and pockets of
thorny green vegetation contrast starkly
with the brilliant white sandy beaches and
the blue waters surging over coral.

It is thought that the
islands have long
been settled due to
their strategic
importance and the
existence of sheltered
anchorages and
supplies of fresh water
on Komodo and
Rinca. The evidence
of early settlement is
further supported by
the recent discovery
of Neolithic graves,
artefacts and megaliths
on Komodo Island.

World Heritage site since

978 1979 1980 1981 1982 1983 1984 1985 1986 1987 1988 1989 1990 **1991** 1992 1993 1994 1995 1996 1997 1998 1999 2000 2001 2002 2003 2004 2005 2006 2007 2008

Royal Domain of Drottningholm
Sweden

Criteria – Significance in human history

With its castle, its perfectly preserved theatre, its Chinese pavilion and gardens, the ensemble at Drottningholm is the finest example of an eighteenth-century north European royal residence inspired by the Palace of Versailles. The royal domain of Drottningholm (meaning 'Queen's Island') is located on Lake Mälaren, outside Stockholm, and the island's name acknowledges the closely interwoven history of the castle with the different

queens of Sweden. The castle was begun in the seventeenth century but work on it continued throughout the eighteenth century. In 1922, when it once again became a royal residence, it was restored and a large part of the furnishings and decorations of the eighteenth century were returned. The Chinese pavilion, built in 1769 to replace a wooden pavilion from 1753, is considered one of the most important examples of this type of structure conserved in Europe.

In the early twentieth century the theatre, built in 1766, was restored to its original appearance and refurbished with the original fittings, even the stage sets. It is a unique example of a European theatre of the eighteenth century that has conserved its original state. The sophisticated theatrical machinery is still fully intact, permitting quick changes of scene with the curtain up.

Drottningholm Castle. ▼

World Heritage site since

1978 1979 1980 1981 1982 1983 1984 1985 1986 1987 1988 1989 1990 **1991** 1992 1993 1994 1995 1996 1997 1998 1999 2000 2001 2002 2003 2004 2005 2006 2007 2008

Cathedral of Notre-Dame, Former Abbey of Saint-Rémi and Palace of Tau, Reims
France

Criteria – Human creative genius; Interchange of values; Heritage associated with events of universal significance

The outstanding handling of new architectural techniques in the thirteenth century and the harmonious marriage of sculptural decoration with architecture, made Notre-Dame in Reims a masterpiece of Gothic art. The perfection of the architecture and the sculptural ensemble of the cathedral were such that numerous edifices were influenced by it, particularly in regions of Germany. The former abbey still has its beautiful ninth century nave in which lie the remains of Archbishop Saint Rémi

(440–533) who baptized Clovis, the unifier and first King of the Franks. The former archepiscopal palace known as the Tau Palace, which played an important role in religious ceremonies, was almost entirely rebuilt in the seventeenth century.

The cathedral, the archepiscopal palace and the old Abbey of Saint-Rémi are directly linked to the history of the French monarchy, as the place of coronation of the kings of France.

The thirteenth century builders of Reims, perhaps conscious of erecting the cathedral for the coronation of the kings of France, enhanced its structural elements with greater lightness and made more openings in the walls to allow a maximum of light to filter through the stained glass and illuminate the sacred space within.

▼

World Heritage site since

1978 1979 1980 1981 1982 1983 1984 1985 1986 1987 1988 1989 1990 **1991** 1992 1993 1994 1995 1996 1997 1998 1999 2000 2001 2002 2003 2004 2005 2006 2007 2008

Prambanan Temple Compounds
Indonesia

Criteria – Human creative genius; Significance in human history

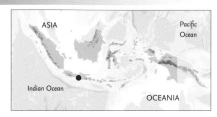

Dedicated to the three great Hindu divinities, Siva, Brahma and Vishnu, Prambanan Temple or Lorojonggrang (Slender Maiden) Temple, as it is known locally, is an outstanding example of Siva art in Indonesia.

Prambanan is a magnificent Hindu temple complex in Java, built in the ninth century and designed as three concentric squares. In all there are 224 temples in the entire complex. The inner square contains sixteen temples, the most significant being the imposing 47-m-high Siva temple flanked to the north by the Brahma temple and to the south by the Vishnu temple. Each temple is decorated with reliefs illustrating the Ramayana. There are also three smaller temples for the animals who serve them (Bull for Siva, Eagle for Brahma and Swan for Vishnu). The compound was deserted soon after it was completed, possibly owing to the eruption of nearby volcano, Mount Merapi. The neighbouring Buddhist complex at Sewu comprises a central temple surrounded by a multitude of minor temples and, surprisingly, shares many design attributes with Prambanan.

Prambanan Temple.
▼

World Heritage site since

1978 1979 1980 1981 1982 1983 1984 1985 1986 1987 1988 1989 1990 **1991** 1992 1993 1994 1995 1996 1997 1998 1999 2000 2001 2002 2003 2004 2005 2006 2007 2008

Serra da Capivara National Park
Brazil
Criteria – Testimony to cultural tradition

The park area has rich archaeological elements but is especially remarkable because of the rock art paintings that decorate its caves and shelters.

Many of the numerous rock shelters in the Serra da Capivara National Park are decorated with cave paintings, some more than 25,000 years old. They are an outstanding testimony to one of the oldest human communities of South America. The paintings' iconography reveals major aspects of the religious beliefs and practices of the area's ancient inhabitants. In total, over 300 archaeological sites have been found within the park. Certain geological formations and palaeofauna that included giant sloths, horses, camelids and early llamas indicate that the Ice Age environment was quite different from the existing semi-arid conditions. Fragments of broken wall found in the Pedra Furada shelter appear to be the oldest traces of rock art in South America; they have been dated to 26,000–22,000 BC.

Ujung Kulon National Park
Indonesia
Criteria – Natural phenomena or beauty;
Significant natural habitat for biodiversity

As a result of human and natural impact, the primary lowland rainforest (**pictured below**) that is the area's natural vegetation now covers only 50 per cent of the park. The single most notable event in this regard was the Krakatoa volcanic eruption of 1883.

This national park, located in the extreme southwest tip of Java on the Sunda shelf, includes the Ujung Kulon peninsula and several offshore islands and encompasses the nature reserve of Krakatoa. In addition to its natural beauty and geological interest – particularly for the study of inland volcanoes – it contains the largest remaining area of lowland rainforests in the Java plain. Several species of endangered plants and animals can be found there, the Javan rhinoceros being the most under threat.

World Heritage site since

1978 1979 1980 1981 1982 1983 1984 1985 1986 1987 1988 1989 1990 **1991** 1992 1993 1994 1995 1996 1997 1998 1999 2000 2001 2002 2003 2004 2005 2006 2007 2008

Historic Town of Sukhothai and Associated Historic Towns
Thailand

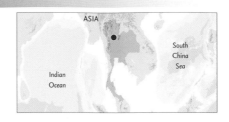

Criteria – Human creative genius; Testimony to cultural tradition

Sukhothai was the capital of the first Kingdom of Siam in the thirteenth and fourteenth centuries. The great civilization of Sukhothai absorbed numerous influences and ancient local traditions to create what is known as the 'Sukhothai style'. The historic town of Sukhothai has a number of fine monuments illustrating the beginnings of Thai architecture, including the monastery (wat) Mahathat, with its royal temple and its cemetery; Sra Si Wat, with its two stupas,

their graceful lines reflected in the water of the town's biggest reservoir; and an impressive prang (reliquary tower) from a somewhat later period. In addition it still has a large part of its fortifications. The associated towns of Si Satchanm, famous for its ceramics, and Kamohena Pet (wall of diamonds), significant for its strategic importance, are also included in the designation of the site.

Ramkhamhaeng (c. 1280–1318) was one of the most important Thai sovereigns, as he brought Sukhothai extensive territory through his military victories. He invented the Siamese alphabet (Khmer script), imposed strict observance of the Buddhist religion and instituted a military and social organization copied from his vanquished neighbours, the Khmers.

Sukhothai Buddha statue in Sukhothai.
▼

World Heritage site since

1978 1979 1980 1981 1982 1983 1984 1985 1986 1987 1988 1989 1990 **1991** 1992 1993 1994 1995 1996 1997 1998 1999 2000 2001 2002 2003 2004 2005 2006 2007 2008

Borobudur Temple Compounds
Indonesia

Criteria – Human creative genius; Interchange of values; Heritage associated with events of universal significance

One of the statues of Buddha and openwork 'stupas' on the circular terraces.

Borobudur is one of the greatest Buddhist monuments in the world. Founded by a king of the Saliendra dynasty, it was built to honour the glory of both the Buddha and its founder.

This harmonious temple complex was built on several levels around a hill which forms a natural centre. The first level above the base comprises five square terraces, graduated in size and forming the base of a pyramid. Above this level are three concentric circular platforms crowned by the main stupa (relic mound) to which stairways provide access. The base and balustrades enclosing the square terraces are decorated in reliefs sculpted in the stone. They illustrate the different phases of the soul's progression towards redemption and episodes from the life of Buddha. The circular terraces are decorated with seventy-two openwork stupas, each containing a statue of Buddha.

Stylistically the art of Borobudur is a tributary of Indian influences (Gupta and post-Gupta styles). The walls of Borobudur are sculptured in bas-reliefs, extending over a total length of 6 km. It has been hailed as the largest and most complete ensemble of Buddhist reliefs in the world, unsurpassed in artistic merit, each scene an individual masterpiece.

This colossal temple was built between AD 750 and 842, more than 300 years before Cambodia's Angkor Wat and 400 years before work had begun on the great European cathedrals. Little is known about its early history except that a huge army of workers toiled in the tropical heat to shift and carve the 60,000 m³ of stone. At the beginning of the eleventh century AD, because of the political situation in Central Java, divine monuments in that area, including the Borobudur Temple, became completely neglected and given over to decay. The sanctuary was exposed to volcanic eruption and other ravages of nature. The temple was not rediscovered until the nineteenth century. A first restoration campaign, supervised by Theodor van Erp, was undertaken shortly after the turn of the century. A second, more recent one (1973–82) was funded by UNESCO.

The name Borobudur is believed to have been derived from the Sanskrit words 'vihara Buddha uhr', meaning 'the Buddhist monastery on the hill'.

Borobudur temple is located in Muntilan, Magelang, about 42 km from Yogyakarta city.

It lay forgotten for centuries under a blanket of volcanic ash and vegetation. Its rediscovery in 1814 was due in large part to Thomas Stamford Raffles, the British Lieutenant-Governor of Java. The temple was fully unearthed in 1835.

World Heritage site since

1978 1979 1980 1981 1982 1983 1984 1985 1986 1987 1988 1989 1990 **1991** 1992 1993 1994 1995 1996 1997 1998 1999 2000 2001 2002 2003 2004 2005 2006 2007 2008

Fortress of Suomenlinna
Finland
Criteria – Significance in human history

Located on islands off Helsinki, Suomenlinna is one of the largest maritime fortresses in the world. Built in the second half of the eighteenth century by Sweden, when Finland was part of the Swedish realm, the purpose was to link and fortify several islands so that entry into the city's harbour could be controlled. The work began in 1748 under the supervision of the Swedish Admiral Augustin Ehrensvärd, who adapted

Vauban's theories to the very special geographical features of Helsinki. By the time of his death in 1772, Ehrensvärd had produced the chain of forts, collectively called Sveaborg (Swedish Fortress), that were to protect the approaches to Helsinki. Following Finland's independence (1918), the name was changed to Suomenlinna (Fortress of Finland), and 6 km of walls and 190 buildings have been preserved.

Sveaborg was built to help Sweden counter the ambitions of Russia, whose principal military and naval base in the Gulf of Finland was Kronstadt, commissioned by Peter the Great to protect the city of St Petersburg. However, the fortress was occupied by the Russians after the war of 1808–9 (despite its reputation as being invulnerable).

World Heritage site since World Heritage site since

1978 1979 1980 1981 1982   1978 1979 1980 1981 1982 1983 1984 1985 1986 1987 1988 1989 1990 **1991** 1992 1993 1994 1995 1996 1997 1998 1999 2000 2001 2002 2003 2004 2005 2006 2007 2008

Bourges Ca
France

Criteria – Human cr
human history

The Cathedral of S
built between the l
thirteenth centurie
masterpieces of G
power of Christian
The tympanum, sc
windows are partic
cathedral was buil
harmonious plan:
chapels surroundi
perspective of the
the interior space a
of the building. Th
surrounded by the
the medieval town
when it was compl
materials, althougl
been replaced over
case with all Gothi

The tympanum of th
central portal of the
west façade bears a
grandiose sculptural
representation of the
Last Judgement, in
which Hell swarms
with demons and

Abbey and Altenmünster of Lorsch
Germany

Criteria – Testimony to cultural tradition;
Significance in human history

The religious complex of the ruined Lorsch Abbey with its 1,200-year-old Torhalle (gatehouse) and associated artworks comprises a rare architectural document of the Carolingian era with impressively preserved sculptures and paintings of that period. It gives architectural evidence of the awakening of the West to the spirit of the early and high Middle Ages under the first king and emperor, Charlemagne, and is a reminder of the past grandeur of an abbey founded around 760–4. The monastery's zenith was probably in 876 when, on the death of Louis II the German, it became the burial place for the Carolingian kings of Germany. The monastery flourished throughout the tenth century, but in 1090 was ravaged by fire.

During the Thirty Years' War, in 1620–1 the Spanish armies pillaged the monastic buildings, which had been in a state of abandon since the Reformation. Only the Torhalle, part of the Romanesque church, insignificant vestiges of the medieval monastery, and classical buildings dating from the period when the Electors of Mainz administered Lorsch, still survive today.

World Heritage site since

1978 1979 1980 1981 1982 1983 1984 1985 1986 1987 1988 1989 1990 1991 **1992** 1993 1994 1995 1996 1997 1998 1999 2000 2001 2002 2003 2004 2005 2006 2007 2008

Kasbah of Algiers
Algeria

Criteria – Interchange of values; Traditional human settlement

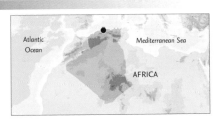

The Kasbah of Algiers is a unique kind of medina, or Islamic city. It preserves very interesting traditional houses in which the ancestral Arab lifestyle and Muslim customs have blended with other architectural traditions. The Kasbah also contains the remains of the citadel, old mosques and Ottoman-style palaces. This tightly integrated urban structure is associated with a deep-rooted sense of community.

It stands in one of the finest coastal sites on the Mediterranean, overlooking the islands where a Phoenician trading-post was established in the fourth century BC. This unique natural site has helped define the city's characteristic winding streets and ancient alleys, while the wealth that its position helped create is reflected in the extreme richness of the interior decoration of the houses.

The first schemes for safeguarding the Algiers Kasbah were approved in the early 1970s. A sensitive redevelopment plan is currently under way, introducing modern comfort without upsetting the traditional urbanism and architecture and restoring the Kasbah's original functions: residential, commercial and cultural quarters.

World Heritage

1978 1979 1980 1

World Heritage site since

1978 1979 1980 1981 1982 1983 1984 1985 1986 1987 1988 1989 1990 1991 **1992** 1993 1994 1995 1996 1997 1998 1999 2000 2001 2002 2003 2004 2005 2006 2007 2008

Pueblo
USA
Criteria – Sig

Taos is a ren
traditional, ¡
ensemble th
shows the tr
constructior
clusters of h
brick. Room
of lower uni
Access to up
holes in the

Jiuzhaig
and His
China
Criteria – Na

Stretching o
part of Sichu
Jiuzhaigou v
than 4,800 n
diverse fores
landscapes a
their series o
and spectacu
species also i
number of er
species, inclu
Sichuan takir

Wulingyuan Scenic and Historic Interest Area
China
Criteria – Natural phenomena or beauty

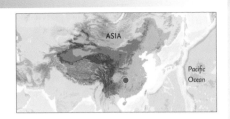

A spectacular area stretching over more than 260 km² in China's Hunan Province, the site is dominated by more than 3,000 narrow sandstone pillars and peaks, many over 200 m high. Between the peaks lie ravines and gorges with streams, pools and waterfalls. The site also contains a number of karst features, notably some forty caves, many of them with spectacular calcite deposits. There are two spectacular natural bridges: Xianrenqiao (Bridge of the Immortals) and Tianqiashengkong (Bridge Across the Sky), which at 357 m above the valley floor may be the highest natural bridge in the world. In addition to the striking beauty of the landscape, the region is also notable for being home to a number of endangered plant and animal species.

This beautiful area is a vital refuge for a number of animal species that are globally threatened with extinction. These include the Chinese giant salamander, Asiatic wild dog, Asiatic black bear, clouded leopard, leopard and Chinese water deer.

◄

Sandstone pillars and peaks in Wulingyuan Scenic and Historic Interest Area.

Old City of Zamość
Poland
Criteria – Significance in human history

Zamość was the personal creation of the Hetman (head of the army) Jan Zamysky, on his own lands. Located on the trade route linking western and northern Europe with the Black Sea, the town was conceived from the beginning as an economic centre based on trade. Modelled on Italian theories of the 'ideal city' and built by the Paduan architect Bernando Morando, Zamość is a perfect example of a late-sixteenth-century Renaissance town. To populate the city, Zamysky brought in merchants of various nationalities and displayed great religious tolerance to encourage people to settle there: they included Ruthenes (Slavs of the Orthodox Church), Turks, Armenians and Jews, among others. Moreover, he endowed the town with its own academy (1595), modelled on Italian cities.

Zamość escaped the destruction suffered by many other Polish towns during the Second World War, and retains its original layout and fortifications. It has a large number of buildings that blend Italian and central European architectural traditions, and is an outstanding example of Polish architecture and urbanism of the sixteenth and seventeenth centuries.

World Heritage site since

1978 1979 1980 1981 1982 1983 1984 1985 1986 1987 1988 1989 1990 1991 **1992** 1993 1994 1995 1996 1997 1998 1999 2000 2001 2002 2003 2004 2005 2006 2007 2008

Historic Monuments of Novgorod and Surroundings
Russian Federation

Criteria – Interchange of values; Significance in human history; Heritage associated with events of universal significance

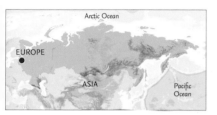

Situated on the ancient trade route between Central Asia and northern Europe, Novgorod was Russia's first capital in the ninth century. As an outstanding cultural and spiritual centre, birthplace of the national style of stone architecture and one of the oldest national schools of painting, the town of Novgorod influenced the development of Russian art as a whole throughout the Middle Ages. The broad range of monuments conserved in Novgorod makes it a veritable 'conservatory' of Russian architecture of the Middle Ages and later periods. Foremost amongst these are the Kremlin with its fifteenth-century fortifications, the church of St Sophia from the mid-eleventh century, the Church of the Transfiguration, decorated with frescoes at the end of the fourteenth century by Theophanes the Greek, and other monuments from the twelfth to nineteenth centuries.

▲ The church of St John the Baptist, Novgorod.

The most ancient Russian Old Church Slavonic manuscripts (eleventh century) were written at Novgorod, including an autonomous historiography (as early as the twelfth century) and, in particular, the first complete translation into Slavonic of the Old and New Testaments (late fifteenth century).

World Heritage site since

1978 1979 1980 1981 1982 1983 1984 1985 1986 1987 1988 1989 1990 1991 **1992** 1993 1994 1995 1996 1997 1998 1999 2000 2001 2002 2003 2004 2005 2006 2007 2008

Historic Centre of Prague
Czech Republic

Criteria – Interchange of values; Significance in human history; Heritage associated with events of universal significance

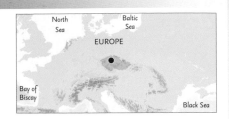

The Charles Bridge and bridge tower on the Stare Město side of the river. Construction started in 1357 and finished in the early fifteenth century. Until 1841 it was the only bridge over the river Vltava. There are thirty seventeenth-century baroque-style statues lining the parapets of the bridge.

Prague is an urban architectural ensemble of outstanding quality, in terms of both its individual monuments and its townscape, and one that is deservedly world famous. The historic centre admirably illustrates the process of continuous urban growth from the Middle Ages to the present day. Its important role in the political, economic, social and cultural evolution of central Europe from the fourteenth century onwards and the richness of its architectural and artistic traditions made it a major model for urban development for much of central and eastern Europe.

The historic city of Prague comprises three separate cities: the Old Town (Stare Město), the Lesser Town (Malá Strana) and the New Town (Nove Město). Prague quickly became the capital of the Bohemian state after its founding on the banks of the Vltava in the ninth century. The twelfth century brought considerable expansion with the building of a monastery at Strahov and a new stone bridge across the river, which led to the growth of the Stare Město.

Growth continued in the mid-fourteenth century when the Nove Město was founded. Under Charles IV, the Holy Roman Emperor (1316–78), the city enjoyed a golden age as the imperial capital and became a major

centre of culture, attracting artists and architects from across Europe and notably Italy.

A disastrous fire in 1541 destroyed much of the settlement on the left bank of the Vltava, and in the rebuilding Renaissance styles predominated. Decline came at the end of the Thirty Years' War in 1648 and it was not until the end of the century that the city recovered, a period that saw the vigorous development of High Baroque. Urban development from 1880 onwards resulted in the demolition of many old buildings, notably in the Jewish Quarter on the right bank. However, the city benefited from the construction of a large number of outstanding buildings in contemporary style.

Prague is rich in monuments from all periods. Of particular importance are Prague Castle, the Cathedral of St Vitus, Hradčany Square, the Gothic Charles Bridge (pictured on the right), the Romanesque Rotunda of the Holy Rood, the Gothic arcaded houses round the Old Town Square, the High Gothic Minorite Church of St James in the Stare Město, and the late-nineteenth-century buildings and town plan of the Nove Město.

The role of Prague in the medieval development of Christianity in central Europe was outstanding, as was its formative influence in the evolution of towns. The city's political status attracted from all over Europe architects and artists who contributed to its wealth of treasures. Since the reign of Charles IV, Prague has also been the intellectual and cultural centre of the region, a status enhanced by the founding of Charles University in the fifteenth century, and the city is closely associated with such names as Wolfgang Amadeus Mozart and Franz Kafka.

World Heritage site since

1978 1979 1980 1981 1982 1983 1984 1985 1986 1987 1988 1989 1990 1991 **1992** 1993 1994 1995 1996 1997 1998 1999 2000 2001 2002 2003 2004 2005 2006 2007 2008

Ban Chiang Archaeological Site
Thailand

Criteria – Testimony to cultural tradition

In addition to farming and metal production, skills were developed in house construction and pottery manufacture. The equipment of burials and the presence of grave goods indicate both social complexity and prosperity.

Ban Chiang was the centre of a remarkable phenomenon of human cultural, social, and technological evolution in the fifth millennium BC, which occurred independently in this area of southeast Asia and spread widely over the whole region. It is without question the most important prehistoric settlement so far discovered in southeast Asia. It presents the earliest evidence for true farming in the region and for the manufacture and use of metals: its long cultural sequence, size and economic status has no parallel in any other

contemporary site in the region. Although occupation ended at Ban Chiang in the third century AD, it was the principal settlement in this area of the Khorat plateau and has given its name to a distinctive archaeological culture.

◄
Artefacts on display at the Ban Chiang National Museum.

Pythagoreion and Heraion of Samos
Greece

Criteria – Interchange of values; Testimony to cultural tradition

The Temple of Hera at Samos is fundamental to an understanding of classical architecture. The stylistic and structural innovations in each of its successive phases strongly influenced the design of temples and public buildings throughout the Greek world.

Many civilizations have inhabited the small Aegean island of Samos since the third millennium BC. The remains of Pythagoreion, an ancient fortified port, can still be seen, as well as the Heraion or Temple of Hera. The fortifications round the ancient town date back to the classical period, with Hellenistic additions. Excavations have revealed much of the street plan of the city, together with its

aqueduct, sewage system, public buildings, sanctuaries and temples, agora, public baths, stadium and town houses. One of the most famous features is the Eupalineio, a tunnel running for 1,040 m through the mountainside to bring water to the city. The complex around the Heraion includes altars, smaller temples and statue bases, along with the remains of a fifth-century Christian basilica.

World Heritage site since

1978 1979 1980 1981 1982 1983 1984 1985 1986 1987 1988 1989 1990 1991 **1992** 1993 1994 1995 1996 1997 1998 1999 2000 2001 2002 2003 2004 2005 2006 2007 2008

Butrint
Albania
Criteria – Testimony to cultural tradition

Inhabited since prehistoric times, Butrint has been the site of a Greek colony, a Roman city and an early Christian bishopric. It became an important trading city and reached its height in the fourth century BC. The fortifications date from the sixth century BC and the hill on which the acropolis stands is encircled by a wall built from huge stone blocks. The amphitheatre, from the third century BC, has stone banks of seating, of which twenty-three rows have been preserved, while the theatre is situated at the foot of the acropolis, close by two temples, one of which is dedicated to Asklepios, the Greek god of medicine. Under Roman rule the city fell slowly into decay. Following a period of prosperity under Byzantine administration, it was abandoned in the late Middle Ages after marshes formed in the area.

Excavations started at the beginning of the twentieth century. The mud and vegetation that covered Butrint had protected it and the entire city was revealed almost intact. Many objects have been excavated – plates, vases, ceramic candlesticks – as well as sculptures, including the remarkable 'Goddess of Butrint' that embodies the Greek ideal of physical beauty.

World Heritage site since

1978 1979 1980 1981 1982 1983 1984 1985 1986 1987 1988 1989 1990 1991 **1992** 1993 1994 1995 1996 1997 1998 1999 2000 2001 2002 2003 2004 2005 2006 2007 2008

Angkor
Cambodia

Criteria – Human creative genius; Interchange of values; Testimony to cultural tradition; Significance in human history

ASIA

Indian Ocean

South China Sea

Angkor Wat Temple. ▶

Khmer art, as developed at Angkor, had a profound influence over much of southeast Asia and played a fundamental role in its distinctive evolution. Khmer architecture evolved largely from that of the Indian subcontinent, from which it soon became clearly distinct as it developed its own special characteristics; some of these evolved independently while others were acquired from neighbouring cultural traditions. The result was a new artistic horizon in oriental art and architecture.

Angkor is one of the most important archaeological sites in southeast Asia. Stretching over some 400 km², Angkor Archaeological Park contains the magnificent remains of the different capitals of the Khmer Empire, from the ninth to the fifteenth century. It includes the famous Temple of Angkor Wat (see photo on the right) and the Bayon Temple with its countless sculptural decorations at Angkor Thom. In total, there are over a hundred temples throughout the site.

In the early ninth century, Jayavarman II united the two states that covered the territory of modern Cambodia, laying the foundations of the Khmer Empire that was to be the major power in southeast Asia for five centuries. Jayavarman's son Yashovarman established Yashodapura (later called Angkor), permanent capital of the Khmer Empire until the fifteenth century.

The first city had the fundamental elements of a Khmer capital: a defensive bank and ditch; a brick- or stone-built state temple at the centre; and a wooden palace. A large reservoir was another essential feature of a Khmer capital and this, now known as the Eastern Baray, was added a decade later with a third temple built in its centre.

In the 960s, Rajendravarman built a second capital at Angkor; the state temple was situated at Pre Rup. He also built the Eastern Mebon temple on an artificial island in the Eastern Baray, and the exquisite temple of Banteay Srei. Rajendravarman's son Jayavarman V abandoned Pre Rup in favour of a new location with its state temple at Ta Kev, which was consecrated around 1000. Shortly afterwards he was overthrown by Suryavarman I, who was responsible for erecting the formidable fortifications around his Royal Palace and state temple, the Phimeanakas, and also for the construction of the great Western Baray. In 1050 his successor created a new and more impressive state temple, the Baphuon.

The accession of Suryavarman II in 1113 brought the next great phase of building. He was responsible for the greatest Khmer monument, Angkor Wat, set within an extensive enclosure and dedicated to Vishnu.

A period of internal instabilily after Suryavarman's death was ended in the 1180s by Jayavarman VII who celebrated his military success by creating yet another capital at Angkor Thom and launching an unprecedented building campaign. His state temple was the towering Bayon, dedicated to Buddha.

World Heritage site since

1978 1979 1980 1981 1982 1983 1984 1985 1986 1987 1988 1989 1990 1991 **1992** 1993 1994 1995 1996 1997 1998 1999 2000 2001 2002 2003 2004 2005 2006 2007 2008

Historic Centre of Telč
Czech Republic

Criteria – Human creative genius; Significance in human history

The houses in Telč, which stands on a hilltop, were originally built of wood. After a fire in the late fourteenth century, the town was rebuilt in stone, surrounded by walls and further strengthened by a network of artificial ponds. The resulting town is an outstanding example of Renaissance town planning and architecture. Baroque elements were introduced by the Jesuits, who built a college (1651–65) and the Church

of the Name of Jesus (1666–7). At the same time Baroque gables were added to the façades of some of the houses in the triangular marketplace; Rococo and classical elements also followed in later remodelling. The result is a public place of great beauty as well as great cultural importance, with a dazzling display of architecture.

The later Middle Ages in central Europe saw the plantation of planned settlement and expansion into areas of virgin forest. Telč is an architectural and artistic ensemble of outstanding quality and the best-preserved surviving example of such settlements.

Cultural and Historic Ensemble of the Solovetsky Islands
Russian Federation

Criteria – Significance in human history

The Solovetsky archipelago comprises six islands in the western part of the White Sea, covering 300 km². They have been inhabited since the fifth century BC and important traces of a human presence from as far back as the fifth millennium BC can be found there. The archipelago has been the site of fervent monastic activity since the fifteenth century, and there are several churches dating from the sixteenth to the nineteenth century.

There are several detached monasteries at Solovetsky: four on Solovetsky Island; the early-seventeenth-century Trinity monastery on Anzer Island; a sixteenth-century complex on Big Zayatsky Island; and St Sergius Monastery, founded in the sixteenth century on Big Muksalma Island.

◄

The Solovetsky Monastery.

World Heritage site since

1978 1979 1980 1981 1982 1983 1984 1985 1986 1987 1988 1989 1990 1991 **1992** 1993 1994 1995 1996 1997 1998 1999 2000 2001 2002 2003 2004 2005 2006 2007 2008

Fraser Island
Australia

*Criteria – Natural phenomena or beauty;
Significant ecological and biological processes;
Major stages of Earth's history*

Indian Ocean

Pacific Ocean

OCEANIA

Tasman Sea

Stretching over 120 km along the eastern coast of Queensland, Fraser Island is the largest sand island in the world. It is a place of exceptional beauty, with long white beaches flanked by strikingly coloured sand cliffs, tall rainforests and numerous freshwater lakes. The massive sand deposits provide a continuous record of climatic and sea level changes over the past 700,000 years. The highest dunes on the island reach up to 260 m above sea level. A surprising variety of vegetation types grow on the island, ranging from coastal heath to subtropical rainforests, with trees up to 50 m high. Birds are the most abundant form of animal life with over 230 species being recorded. It is a particularly important site for migratory wading birds. Few mammal species are present on the island. The most common are bats, particularly flying foxes.

The earliest date for the occupation of Fraser Island is currently 1,500–2,000 years ago. The Badtjala and Kabi Kabi groups of Aboriginal people have cultural and other traditional affiliations with the area. European contact was sporadic and limited to explorers, escaped convicts and shipwreck survivors.

View from Indian Head, Fraser Island.
▼

World Heritage site since

1978 1979 1980 1981 1982 1983 1984 1985 1986 1987 1988 1989 1990 1991 **1992** 1993 1994 1995 1996 1997 1998 1999 2000 2001 2002 2003 2004 2005 2006 2007 2008

El Tajin, Pre-Hispanic City
Mexico

Criteria – Testimony to cultural tradition;
Significance in human history

El Tajin, occupied from 800–1200, became
the most important centre in northeast
Mesoamerica after the fall of the
Teotihuacan Empire. It has survived as an
outstanding example of the grandeur and
importance of the pre-Hispanic cultures of
Mexico. Its cultural influence extended all
along the Gulf of Mexico and penetrated
into the Maya region and the high plateaus
of central Mexico. Its architecture, which is
unique in Mesoamerica, is characterized by
elaborate columns of carved reliefs and key-
pattern friezes. The 'Pyramid of the Niches'
rises in six steps to a temple at the top, with
each storey having rows of square niches.
It is a masterpiece of ancient Mesoamerican
architecture, which reveals the astronomical
and symbolic significance of the building.
The site was abandoned when the region
came under the rule of the powerful
Mexico-Tenochtitlan kingdom.

The artistic,
architectural, and
historical importance
of El Tajin combines
to make this a highly
significant site. It has
been extensively
excavated in recent
years and as a result
it is probably better
understood than many
of the more famous
pre-Hispanic sites in
Mexico.

▼

World Heritage site since

1978 1979 1980 1981 1982 1983 1984 1985 1986 1987 1988 1989 1990 1991 **1992** 1993 1994 1995 1996 1997 1998 1999 2000 2001 2002 2003 2004 2005 2006 2007 2008

Huanglong Scenic and Historic Interest Area
China
Criteria – Natural phenomena or beauty

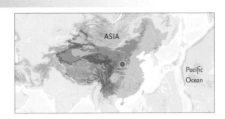

ASIA

Pacific Ocean

Situated in the northwest of Sichaun Province, the Huanglong valley boasts extensive areas of precipitous mountain scenery, snow-covered for much of the year. Xuebaoding, or Snow Mountain Peak, is permanently snow-covered and bears the easternmost glacier in China. Huanglong lies close to the intersection of four floral regions: Eastern Asia, Himalaya, and the subtropical and tropical zones of the Northern Hemisphere. In addition to its

mountain landscape, diverse forest ecosystems can be found, as well as spectacular limestone formations, waterfalls and hot springs. The area also has a large number of endangered mammals including giant panda, golden snub-nosed monkey, brown bear, Asiatic black bear, leopard, Pallas' cat, Asiatic wild dog, Szechwan takin, mainland serow, common goral, argali and three species of deer.

The area has many spectacular steep-sided gorges and forested waterways. Calcite deposition has resulted in areas of travertine pools; algae and bacteria often proliferate, colouring the pools in deep shades from orange and yellow to green and blue.

▼

World Heritage site since

1978 1979 1980 1981 1982 1983 1984 1985 1986 1987 1988 1989 1990 1991 **1992** 1993 1994 1995 1996 1997 1998 1999 2000 2001 2002 2003 2004 2005 2006 2007 2008

White Monuments of Vladimir and Suzdal
Russian Federation

Criteria – Human creative genius; Interchange of values; Significance in human history

The towns of Vladimir and Suzdal hold an important place in Russia's architectural history. Vladimir, founded in 1108, contains an important group of religious and secular monuments. The single-domed Cathedral of the Assumption (1158) contains frescoes by the master painters Andrei Rublev and Daniil Chernii (1408) to replace those destroyed by the Mongols in 1238. The Golden Gate (1164) formed part of the twelfth-century defences.

It is a cubic tower with a church dedicated to the Deposition of the Holy Robe on top. The exterior of the Cathedral of St Demetrius (1194–7) is noteworthy for over 1,000 stone carvings on the general theme of King David. Suzdal is dominated by the Cathedral of the Nativity, built in the thirteenth century and reconstructed in the sixteenth century, with its five-domed top and thirteenth-century Golden Doors.

The other buildings included within the site are the Princely Castle in Bogolyubovo, the Church of the Intercession in Vladimir, the Monastery of Our Saviour and St Euthymius in Suzdal and the Church of Sts Boris and Gleb, near Suzdal, the first church in Russia to be built from white limestone.

Church of the Intercession in Vladimir.
▼

World Heritage site since

1978 1979 1980 1981 1982 1983 1984 1985 1986 1987 1988 1989 1990 1991 1992 **1993** 1994 1995 1996 1997 1998 1999 2000 2001 2002 2003 2004 2005 2006 2007 2008

Route of Santiago de Compostela
Spain

Criteria – Interchange of values; Significance in human history; Heritage associated with events of universal significance

This route from the border between France and Spain was – and still is – taken by pilgrims to Santiago de Compostela in Galicia. Some 1,800 buildings along the route, both religious and secular, are of great historic interest.

Pilgrimages were an essential part of western European spiritual and cultural life in the Middle Ages and the routes they took were equipped with facilities for the spiritual and physical well-being of pilgrims. The Route of St James of Compostela has preserved the most complete material record in its ecclesiastical and secular buildings, settlements and civil-engineering structures.

The route played a fundamental role in encouraging cultural exchanges between the Iberian peninsula and the rest of Europe during the Middle Ages. It remains a testimony to the power of the Christian faith among people of all social classes and from all over Europe.

Two access routes into Spain from France enter at Roncesvalles (Valcarlos Pass) and Canfranc (Somport Pass) and merge west of Pamplona. The route passes through five autonomous communities and 166 towns and villages, and includes over 1,800 historic buildings. In many cases the modern road runs parallel to the ancient route. Thousands of pilgrims follow it on foot or bicycle every year.

World Heritage site since

1978 1979 1980 1981 1982 1983 1984 1985 1986 1987 1988 1989 1990 1991 1992 **1993** 1994 1995 1996 1997 1998 1999 2000 2001 2002 2003 2004 2005 2006 2007 2008

Historic Town of Zabid
Yemen

Criteria – Interchange of values; Significance in human history; Heritage associated with events of universal significance

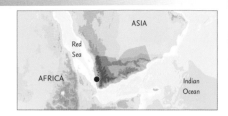

Zabid's domestic and military architecture and its urban plan make it an outstanding archaeological and historical site. Besides being the capital of Yemen from the thirteenth to the fifteenth century, the city played an important role in the Arab and Muslim world for many centuries because of its Islamic university.

Zabid has a remarkable network of streets and alleys, some as little as 2 m wide, which spreads over the town. Occasionally this labyrinth opens out into small squares, but the only large open space is that in front of the citadel.

◄

The Great Mosque of Zabid (also known as the Al-Asha'ir Mosque).

Coro and its Port
Venezuela

Criteria – Significance in human history; Traditional human settlement

With its earthen constructions unique to the Caribbean, Coro is the only surviving example of a rich fusion of local traditions with Spanish Mudéjar and Dutch architectural techniques. One of the first colonial towns founded in 1527, it has some 602 historic buildings.

Unlike other towns on this coast, even Coro's public buildings are of earthen construction, not stone. It has conserved its original layout and early urban landscape, presenting a remarkable record of the earliest years of Spanish colonization.

◄

La Casa de las Ventanas de Hierro (The House of the Iron Windows) which dates from 1765. Its Baroque doorway is 8m tall.

World Heritage site since

1978 1979 1980 1981 1982 1983 1984 1985 1986 1987 1988 1989 1990 1991 1992 **1993** 1994 1995 1996 1997 1998 1999 2000 2001 2002 2003 2004 2005 2006 2007 2008

The Sassi and the park of the Rupestrian Churches of Matera
Italy

Criteria – Testimony to cultural tradition; Significance in human history; Traditional human settlement

The Sassi of Matera and their park are the most outstanding, intact example of a rock-cut settlement in the Mediterranean region, perfectly adapted to its terrain and ecosystem.

Matera's development was due to its geological setting. A belt of soft tufa is located between 350 m and 400 m above the valley bed, and this also contains two natural depressions; it was here that the settlement grew up. The clay plateau above was reserved for agriculture and grazing livestock.

The Matera region has been inhabited since the Palaeolithic period, while later settlements illustrate a number of significant stages in human history. The harsh landscape fostered an independence of spirit which was resistant to successive waves of invaders after the Byzantine period. The area was also very attractive to monastic and utopian communities.

The earliest house form was a simple cave in the tufa with a closing wall formed from the excavated blocks. This developed into a vaulted room built out into the open space, making it available for adaptation and extension. Groups of dwellings around a common courtyard evolved into a vicinato (neighbourhood), with shared facilities such as a cistern.

▼

World Heritage site since

1978 1979 1980 1981 1982 1983 1984 1985 1986 1987 1988 1989 1990 1991 1992 **1993** 1994 1995 1996 1997 1998 1999 2000 2001 2002 2003 2004 2005 2006 2007 2008

Yakushima
Japan

Criteria – Natural phenomena or beauty;
Significant ecological and biological processes

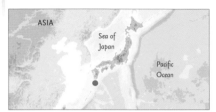

Located in the interior of Yaku Island, at the northern end of the Ryukyu archipelago, Yakushima exhibits a rich flora, with some 1,900 species and subspecies. Of these, ninety-four are endemic, mostly concentrated in the central high mountains. Yakushima is almost 2,000 m high and is the highest mountain in southern Japan. Several other peaks are over 1,800 m with mountain ridges over 1,000 m surrounding these central high peaks. Of great significance to the area is the presence of indigenous Japanese cedar, known colloquially as 'sugi', which can reach more than 1,000 years of age: specimens younger than 1,000 years are known as 'Kosugi'; older specimens, which may reach 3,000 years, are known as 'Yakusugi'. Traditionally, the Island Mountains have been considered to have a spiritual value and the 'Yakusugi' were revered as sacred trees.

The fauna of the island is diverse, with sixteen mammal species. Four mammal subspecies are endemic to the island and a further four are endemic to both Yaku Island and the neighbouring island of Tanegashima. Among the 150 bird species present, four, including Ryukyu robin and Japanese wood pigeon, have been designated as Natural Monuments.

World Heritage site since

1978 1979 1980 1981 1982 1983 1984 1985 1986 1987 1988 1989 1990 1991 1992 **1993** 1994 1995 1996 1997 1998 1999 2000 2001 2002 2003 2004 2005 2006 2007 2008

Archaeological Ensemble of the Bend of the Boyne
Ireland

Criteria – Human creative genius; Testimony to cultural tradition; Significance in human history

The three main prehistoric sites of the Brú na Bóinne Complex – Newgrange, Knowth and Dowth – are situated on the north bank of the river Boyne 50 km north of Dublin. These three great burial mounds are surrounded by about forty satellite passage-graves, creating a great prehistoric funerary landscape. The passage tomb complex in particular represents a spectacular survival of the embodiment of a set of ideas and beliefs of outstanding historical significance unequalled in its European counterparts. The site's ritual significance attracted later monuments, both in protohistory and in the Christian period. Its importance has been further enhanced by the fact that the river Boyne communicates both with the Celtic Sea and the heartland of Ireland, giving it considerable economic and political significance.

Nowhere else in the world is found the continuity of settlement and activity associated with a megalithic cemetery such as that which exists at Brú na Bóinne. The passage tomb complex represents a spectacular survival of the embodiment of a set of ideas and beliefs.

Whale Sanctuary of El Vizcaino
Mexico

Criteria – Significant natural habitat for biodiversity

Located in the central part of the peninsula of Baja California, the sanctuary contains some exceedingly interesting ecosystems. The coastal lagoons of Ojo de Liebre and San Ignacio are an exceptional reproduction and wintering site for grey whales as well as other mammals such as harbour seal, California sea lion, northern elephant seal and blue whale. The lagoons have a series of shallow, sandy bays and saltwater inlets as well as extensive mangroves, with dune communities, bushes and halophytic vegetation surrounding them. These rich ecosystems are important refuges for wintering wildfowl, and birds such as osprey and peregrine falcon also live within the site. The coastal zone is a vital habitat for approximately twenty threatened animal species, including four species of the endangered marine turtle.

As well as an abundance of wildlife, there are a number of prehistoric sites of importance on the peninsula. There are also petroglyphs, wall paintings and ancient ruined structures, together with evidence of the early colonisation from Europe.

World Heritage site since

1978 1979 1980 1981 1982 1983 1984 1985 1986 1987 1988 1989 1990 1991 1992 **1993** 1994 1995 1996 1997 1998 1999 2000 2001 2002 2003 2004 2005 2006 2007 2008

Baroque Churches of the Philippines
Philippines

Criteria – Interchange of values; Significance in human history

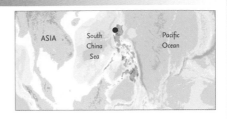

The churches are outstanding examples of the Philippine interpretation of the Baroque style and had an important influence on later church architecture in the region. They represent the fusion of European church design and construction with local materials and decorative motifs to form a new church-building tradition. The four churches, the first of which was built by the Spanish in the late sixteenth century, are located in Manila (San Agustín in Intramuros),

Santa Maria (Nuestra Señora de la Asuncion), Paoay (San Agustín) and Miag-ao (Santo Tomas de Villanueva). The sumptuous façade of the Church of Santo Tomas de Villanueva epitomizes the Filipino transfiguration of western decorative elements, with the figure of St Christopher on the pediment dressed in native clothes, carrying the Christ Child on his back, and holding on to a coconut palm for support.

Unlike other town churches in the Philippines, which conform to the Spanish tradition of siting them on the central plaza, the Church of Nuestra Señora de la Asuncion in Santa Maria with its convento are on a hill surrounded by a defensive wall.

◄
San Agustín in Intramuros Church, Manila.

World Heritage site since

1978 1979 1980 1981 1982 1983 1984 1985 1986 1987 1988 1989 1990 1991 1992 **1993** 1994 1995 1996 1997 1998 1999 2000 2001 2002 2003 2004 2005 2006 2007 2008

Humayun's Tomb, Delhi
India

Criteria – Interchange of values; Significance in human history

This tomb is of particular cultural significance as it is the earliest surviving example of the Mughal scheme of the garden-tomb on the Indian subcontinent. The tomb of Humayun, second Mughal Emperor of India, was built by his widow in 1569–70, fourteen years after his death, at a cost of 1.5 million rupees. It was later used for the burial of various members of the

ruling family and contains some 150 graves. It has aptly been described as the necropolis of the Mughal dynasty. The tomb itself is in the centre of a large garden, laid out in *char baah* (four-fold) style, with pools joined by channels. It inspired several major architectural innovations, culminating in the construction of the Taj Mahal a century later.

Humayun had travelled widely in the Islamic world, notably in Persia and central Asia, and brought back ideas that were applied by the architect of his tomb. The tomb and its surrounding structures are substantially in their original state, and interventions in recent times have been minimal and of high quality.

World Heritage site since

1978 1979 1980 1981 1982 1983 1984 1985 1986 1987 1988 1989 1990 1991 1992 **1993** 1994 1995 1996 1997 1998 1999 2000 2001 2002 2003 2004 2005 2006 2007 2008

Birka and Hovgården
Sweden

Criteria – Testimony to cultural tradition;
Significance in human history

There are no standing remains of the settlement at Birka itself, but its location is vividly indicated by the so-called 'Black Earth,' composed of layers of human occupation and the remains of wooden structures.

Birka-Hovgården is one of the most complete and undisturbed examples of a Viking trading settlement of the eighth–tenth centuries AD. It is also important as the site of the first Christian congregation in Sweden, founded in 831 by St Ansgar. The proto-town of Birka occupies much of the western part of the island of Björkö. The surface evidence is confined mostly to the ramparts of the hill fort, the long ramparts of the town wall, traces of harbours and stone jetties along the shore, and some 3,000 burial mounds and stone settings surrounding the main settlement. Its location on a small island has preserved the entire site from modern development and exploitation. Hovgården is situated on the neighbouring island of Adelsö.

Villages with Fortified Churches in Transylvania
Romania

Criteria – Significance in human history

Lacking the resources of the European nobility and rich merchants, who were able to fortify entire towns, the Transylvanian Saxons chose to create fortresses round their churches, enclosing storehouses within the enceintes to enable them to withstand long sieges.

These Transylvanian villages with their fortified churches provide a vivid picture of the cultural landscape of southern Transylvania. The seven villages inscribed on the World Heritage List, founded by the Transylvanian Saxons, are characterized by a specific land-use system, settlement pattern and organization of the family farmstead that have been preserved since the late Middle Ages. They are dominated by their fortified churches, which illustrate building styles from the thirteenth to the sixteenth century.

◄

The Fortified Church in Viscri, Transylvania.

World Heritage site since

1978 1979 1980 1981 1982 1983 1984 1985 1986 1987 1988 1989 1990 1991 1992 **1993** 1994 1995 1996 1997 1998 1999 2000 2001 2002 2003 2004 2005 2006 2007 2008

Historic Centre of Bukhara
Uzbekistan

Criteria – Interchange of values; Significance in
human history; Heritage associated with events
of universal significance

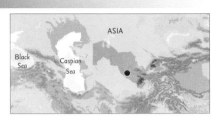

Before the Arab
conquest, Bukhara
was one of the largest
cities of central Asia.
It became a major
cultural centre of the
Caliphate of Baghdad
in 709, and in 892 the
capital of the
independent Samanid
Kingdom. A time of
great economic growth
came to an end with
the sack of the city in
1220 by the Mongol
horde of Genghis
Khan.

▼

Bukhara, which is situated on the Silk Route,
is more than 2,000 years old. It is the most
complete example of a medieval city in
Central Asia, with an urban fabric that has
remained largely intact. Monuments of
particular interest include the famous tomb
of Ismail Samani, a masterpiece of tenth-
century Muslim architecture, the decorated
brick minaret of Poi-Kalyan from the
eleventh century, the Magoki Mosque and
the Chasma Ayub Shrine, along with a large
number of seventeenth-century madrasas.
The historic part of the city, which is in effect
an open-air museum, combines the city's
long history in a single ensemble. It should
be stressed, however, that the real
importance of Bukhara lies not in its
individual buildings but rather in its overall
level of urban planning and architecture.

World Heritage site since

1978 1979 1980 1981 1982 1983 1984 1985 1986 1987 1988 1989 1990 1991 1992 **1993** 1994 1995 1996 1997 1998 1999 2000 2001 2002 2003 2004 2005 2006 2007 2008

Town of Bamberg
Germany

Criteria – Interchange of values; Significance in human history

North Sea

Baltic Sea

● EUROPE

The World Heritage site covers the three centres of settlement that coalesced when the town was founded. These are the Bergstadt, with the cathedral and its precincts; the Inselstadt, defined by the two-arms of the Regnitz River; and the Theuerstadt, a late medieval area of market gardens with scattered houses and large open spaces.

Bamberg is an outstanding and representative example of an early medieval town in central Europe, both in its plan and in its many surviving ecclesiastical and secular buildings. The town was laid out according to medieval planning rules as a cross, with the churches of St Michael, St Stephen, St Gangolf, and St Jacob at the four cardinal points. During its period of greatest prosperity, from the twelfth century onwards, the architecture of Bamberg strongly influenced urban construction in central Europe. This prosperity continued into the later Middle Ages, being helped by the fact that it was the starting point for shipping on the river Main, as well as a renowned cultural centre. In the late eighteenth century it was the centre of the Enlightenment in southern Germany, with eminent philosophers and writers such as Hegel and Hoffmann living there.

▼ Altes Rathaus (Old Town Hall) in Bamberg.

World Heritage site since

1978 1979 1980 1981 1982 1983 1984 1985 1986 1987 1988 1989 1990 1991 1992 **1993** 1994 1995 1996 1997 1998 1999 2000 2001 2002 2003 2004 2005 2006 2007 2008

Jesuit Missions of La Santísima Trinidad de Paraná and Jesús de Tavarangue
Paraguay
Criteria – Significance in human history

The Spanish Crown granted the frontier zone of Paraguay to the Jesuits in 1609, and they created thirty *reducciones* (settlements) in the Rio de la Plata basin, each with its own mission. One of the Jesuits' objectives became the protection of the Indians against the abuses of the colonial *encomienda* system of tribute or labour, which reduced them to a condition of virtual slavery; at the same time they would be brought into the

Christian Church and educated into a sedentary form of life. La Santísima Trinidad was the most ambitious of these missions and the capital of the Guayrá area. Designed by noted Jesuit architect Juan Bautista Primoli, it was constructed in stone in 1706 and has a fine dome and elaborate decoration.

La Santísima Trinidad, the best preserved of the three churches, is of great symbolic importance, because its decoration reflects the spirit of its conception, with its fusion of Christian and native artistic elements.

Maulbronn Monastery Complex
Germany
Criteria – Interchange of values; Significance in human history

Founded in 1147, the Cistercian Maulbronn Monastery is considered the most complete and best-preserved medieval monastic complex north of the Alps. The basic medieval layout and structure of the central complex is virtually complete. Only the monks' refectory and the lay brethren's dormitories have undergone transformations since the Reformation, in order to adapt them for use as a Protestant seminary. Surrounded by fortified walls, the

main buildings were constructed between the twelfth and sixteenth centuries. The monastery's church, mainly in Transitional Gothic style, had a major influence in the spread of Gothic architecture over much of northern and central Europe. Its original wooden beams were replaced by Gothic vaulting in 1424, incorporating Romanesque traditions into the Cistercian requirements of austerity and renunciation.

The Cistercian Order was notable for its innovations in hydraulic engineering and at Maulbronn there exists an elaborate system of reservoirs, irrigation canals and drains to provide water for the community, for fish farming and for irrigating its farmland.

World Heritage site since

1978 1979 1980 1981 1982 1983 1984 1985 1986 1987 1988 1989 1990 1991 1992 **1993** 1994 1995 1996 1997 1998 1999 2000 2001 2002 2003 2004 2005 2006 2007 2008

Engelsberg Ironworks
Sweden
Criteria – Significance in human history

Self-contained estates like Engelsberg comprised not only technical installations but also administrative and residential buildings for management and workers, including those who worked on the associated farm. This site's notable buildings include the master gardener's house, the brewery, stables, a coach-house, smiths' cottages and a monumental slagstone barn.

Engelsberg is an outstanding example of an influential European industrial complex of the seventeenth–nineteenth centuries, with important technological remains and the associated administrative and residential buildings intact. It is the best preserved and most complete example of the type of iron-working estate that made Sweden the economic leader in this field for two centuries. Local peasants had been mining ore and smelting since the thirteenth century, but it was not until the introduction of the waterwheel to power the furnace and hammer bellows in the later Middle Ages that the iron industry began to significantly develop. The first bar-iron forge was operating at Engelsberg in the closing years of the sixteenth century, and by the mid-seventeenth century the scale of operations there was substantial.

Monastery of Horezu
Romania
Criteria – Interchange of values

The church of Bolnica, which is a subgroup of the main monastery, was founded by Princess Maria, wife of Constantin Brancovan. It has an unusual mural, on the subject of the life of the good monk.

Founded in 1690 by the Cantacuzene Prince Constantine Brancovan, the monastery of Horezu in Wallachia is a masterpiece of the Brancovan style. It is laid out according to the precepts of the Athonite Order around the catholicon, which is enclosed by a wall and surrounded by a series of skites (daughter houses of the main monastery). The overall layout is symmetrical on an east-west axis, the skites forming a cruciform plan. The monastery is famous for its architectural purity and balance, the richness of its sculptural detail, the treatment of its religious compositions, its votive portraits and its painted decorative works. The school of mural and icon painting established at the monastery in the eighteenth century was famous throughout the Balkan region.

World Heritage site since

1978 1979 1980 1981 1982 1983 1984 1985 1986 1987 1988 1989 1990 1991 1992 **1993** 1994 1995 1996 1997 1998 1999 2000 2001 2002 2003 2004 2005 2006 2007 2008

Himeji-jo
Japan

Criteria – Human creative genius; Significance in human history

Himeji-jo is the finest surviving example of early seventeenth-century Japanese castle architecture, comprising eighty-three buildings with highly developed systems of defence and ingenious protection devices dating from the beginning of the Shogun period. It is a masterpiece of construction in wood. The centre of the complex is the Tenshu-gun, consisting of a main keep and three subsidiary keeps, with connecting structures. This is surrounded by a system of watchtowers, gates and plastered earthen walls. Set on a low hill, it is visible from every part of the city. The main keep (Dai-Tenshu) has six interior storeys and a basement. The striking appearance of this great wooden structure with its white plastered walls is the source of the name by which it is often known, the Castle of the White Heron (Shirasagi-jo).

Many castles were built in Japan in the early years ▶ of the Shogun period. Most of these have subsequently been demolished and others were destroyed during the Second World War. Of the handful that survives, Himeji-jo is the most complete and unaltered, largely thanks to the efforts of army officers after the Meiji restoration.

World Heritage site since

1978 1979 1980 1981 1982 1983 1984 1985 1986 1987 1988 1989 1990 1991 1992 **1993** 1994 1995 1996 1997 1998 1999 2000 2001 2002 2003 2004 2005 2006 2007 2008

Archaeological Ensemble of Mérida
Spain

Criteria – Testimony to cultural tradition;
Significance in human history

The colony of Augusta Emerita, which became present-day Mérida in Estremadura, was founded in 25 BC at the end of the Spanish Campaign and was the capital of Lusitania. It is an excellent example of a provincial Roman capital during the empire and in the years afterwards. The main monuments in the World Heritage site are the Guadiana bridge (two sections of arches linked by a large pier); the amphitheatre, for 15,000 spectators; the classic Vitruvian theatre, set into a low hill and inaugurated under M. Agrippa; the Temple of Diana, probably from the early years of the first century AD; the alleged 'Arch of Trajan,' which may have been an entrance gate to the original town or, more likely, to the enceinte of the Temple of Diana; and the Circus, one of the largest in the Roman world.

The aqueducts and other elements of Roman water management are especially well preserved and complete Apart from the aqueducts, the site includes three dams and various stretches of underground water channels. The Proserpina and Cornalvo dams, both still functioning, are the most remarkable surviving examples of Roman water management systems.

The Vitruvian Theatre in Mérida. ▼

World Heritage site since

1978 1979 1980 1981 1982 1983 1984 1985 1986 1987 1988 1989 1990 1991 1992 **1993** 1994 1995 1996 1997 1998 1999 2000 2001 2002 2003 2004 2005 2006 2007 2008

Complex of Hué Monuments
Vietnam

Criteria – Testimony to cultural tradition;
Significance in human history

Established as the capital of unified Viet Nam in 1802, Hué was not only the political but also the cultural and religious centre under the Nguycn dynasty until 1945. The integrity of the town layout and building design make it an exceptional specimen of late feudal urban planning. Four citadels or defended enclosures made up the city: Kinh Thanh (Capital City), for official administrative buildings; Hoang Thanh (Imperial City), for royal palaces and shrines; Tu Cam Thanh (Forbidden Purple City), for the royal residences; and Dai Noi (Inner City), defended by brick walls and a moat. A fifth fortress, Tran Hai Thanh, was constructed a little later to protect the capital against assault from the sea. The Perfume River, the main axis, divides the capital in two.

The new capital was planned in accordance with ancient oriental philosophy and Vietnamese tradition. The relationship between the five cardinal points (centre, west, east, north, south), five natural elements (earth, metal, wood, water, fire), and five basic colours (yellow, white, blue, black, red) underlies the conception of the city, and is reflected in the names of some important features.

Wall and gate of Tu Duc Tomb, Hué. ▼

World Heritage site since

1978 1979 1980 1981 1982 1983 1984 1985 1986 1987 1988 1989 1990 1991 1992 **1993** 1994 1995 1996 1997 1998 1999 2000 2001 2002 2003 2004 2005 2006 2007 2008

Churches of Moldavia
Romania

Criteria – Human creative genius; Significance in human history

With their painted exterior walls, decorated with fifteenth- and sixteenth-century frescoes that are considered masterpieces of Byzantine art, seven churches around Suceava in northern Moldavia are unique in Europe. Far from being merely wall decorations, the paintings represent complete cycles of religious murals on all façades. Their outstanding composition, elegant outline and harmonious colours blend perfectly with the surrounding

landscape. At the Church of St George of the former Voronet Monastery, founded by Stephen the Great, the naos and sanctuary were painted between 1488 and 1496 and the narthex in 1552. The walls and the vault of the exonarthex are covered by the 365 scenes of the Calendar of Saints. The exterior murals depict traditional scenes and the famous Last Judgement is on the western wall.

A Christian tradition of decorating the exteriors of churches was adopted and extended in Moldavia. It had its own specific iconography, dominated by certain themes – the Church hierarchy, the Last Judgement and the Tree of Jesse.

Tubbataha Reef Marine Park
Philippines

Criteria – Natural phenomena or beauty; Significant ecological and biological processes; Significant natural habitat for biodiversity

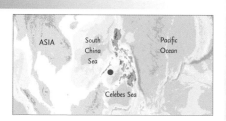

The Tubbataha Reef Marine Park comprises two atolls, North and South Reef, separated by an 8 km channel. The North Reef is a large oblong-shaped platform 2 km wide and completely enclosing a sandy lagoon some 24 m deep. The most prominent feature is the North Islet which serves as a nesting site for birds and marine turtles. Steep and often perpendicular walls extending to a depth of 40–50 m characterize the seaward face of the reef.

The South Reef is a small triangular-shaped reef 1–2 km wide. Like the North Reef, it consists of a shallow platform enclosing a sandy lagoon. South Islet is also used as a nesting site by birds and marine turtles.

Tubbataha has a diverse coral assemblage, with species representing forty-six genera. Forty-six bird species have been recorded and 379 species of fish. Sightings of sharks and rays are common. There are no permanent inhabitants on the reefs, other than during the fishing season, when fishermen establish temporary shelters.

World Heritage site since

1978 1979 1980 1981 1982 1983 1984 1985 1986 1987 1988 1989 1990 1991 1992 **1993** 1994 1995 1996 1997 1998 1999 2000 2001 2002 2003 2004 2005 2006 2007 2008

Qutb Minar and its Monuments, Delhi
India

Criteria – Significance in human history

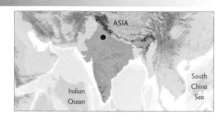

This important collection of Islamic buildings is dominated by the red sandstone tower of Qutb Minar, begun around 1202. In its present form it consists of five storeys, with each storey separated by balconies. The first three storeys are each decorated differently, the lowest being of alternating angular and rounded flutings, the second with rounded flutings alone, and the third with angular flutings alone. It is 72.5 m high, tapering from 14.32 m in diameter at its base

to 2.75 m at its peak. The surrounding area contains funerary buildings, notably the magnificent Alai-Darwaza Gate, the masterpiece of Indo-Muslim art built in 1311, and two mosques, including the Quwwatu'l-Islam (Might of Islam), the oldest in northern India, built of materials reused from some twenty Brahman temples. This mosque consists of a courtyard, cloisters, and a prayer hall.

The Iron Pillar in the mosque compound is 7.02 m tall, 0.93 m of which is below ground. It bears a Sanskrit inscription from the fourth century AD. It is built up of many hundreds of small wrought-iron blooms welded together and is the largest known composite iron object from so early a period.

Ancient carved stone cloisters in Qutb Minar.
▼

World Heritage site since

1978 1979 1980 1981 1982 1983 1984 1985 1986 1987 1988 1989 1990 1991 1992 **1993** 1994 1995 1996 1997 1998 1999 2000 2001 2002 2003 2004 2005 2006 2007 2008

Historic Town of Banská Štiavnica and the Technical Monuments in its Vicinity
Slovakia

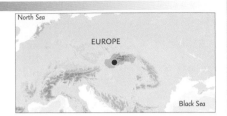

Criteria – Significance in human history; Traditional human settlement

The old medieval mining centre of Banská Štiavnica blends into the surrounding landscape, which contains unique and historic evidence of mining and metallurgical activities. The fifteenth century was a time of immense prosperity: defences were built round the town, the parish church was rebuilt and fortified, and many new houses were built, some of which were, in the sixteenth century, converted into Renaissance 'palaces'. Technological progress continued, and in 1627 Banská

Štiavnica saw the first use of gunpowder in mining, an important breakthrough, and much work on the application of water power in deep mining and on ancillary processes was carried out, particularly in the eighteenth century. During this period, Banská Štiavnica became the most important centre for precious-metal mining in the Habsburg Empire, and many leading engineers and metallurgists from all over Europe were working in the town.

Banská Štiavnica is the oldest mining town in Slovakia; its town seal of 1275 is the earliest known bearing a mining emblem. It lies on the steep slopes of the Glanzenberg and Paradajz mountains. Its ore deposits have been exploited since the late Bronze Age, and a document of 1156 referred to it as the 'land of miners'.

◄ Nový Zámok (The New Castle), which was built in 1564–71 is located on a hill overlooking the town. It was originally a watch tower, but is now used as a museum.

World Heritage site since

1978 1979 1980 1981 1982 1983 1984 1985 1986 1987 1988 1989 1990 1991 1992 **1993** 1994 1995 1996 1997 1998 1999 2000 2001 2002 2003 2004 2005 2006 2007 2008

Buddhist Monuments in the Horyu-ji Area
Japan

Criteria – Human creative genius; Interchange of values; Significance in human history; Heritage associated with events of universal significance

There are around forty-eight Buddhist monuments in the Horyu-ji Area, in Nara Prefecture. Several date from the late seventh or early eighth century, making them some of the oldest surviving wooden buildings in the world. These masterpieces of wooden architecture are important not only for the history of art, since they illustrate the adaptation of Chinese Buddhist architecture and layout to Japanese culture, but also for the history of religion, since their construction coincided

with the introduction of Buddhism to Japan from China by way of the Korean peninsula.

These temples, monasteries and associated buildings are the earliest Buddhist monuments in Japan, dating from shortly after the religion's introduction to the country in the mid-sixth century AD and they had a profound influence on Japanese religious architecture.

◄

A 'tsukubai' (water basin) in one of the Buddhist Temples in the Horyu-ji Area.

Shirakami-Sanchi
Japan

Criteria – Significant ecological and biological processes

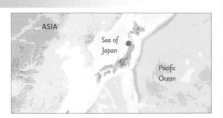

Situated in the mountains of northern Honshu, the area includes the last remaining virgin stand of Siebold's beech forest, which once covered the hills and mountain slopes of northern Japan. The protected area covers about one-third of the Shirakami Mountains which rise to just over 1,200 m and comprise a maze of steep-sided hills and deep valleys. The area is a

refuge for many typical Honshu flora and fauna. The eighty-seven bird species include golden eagle, which has a limited breeding record and is endangered in Japan. Three nesting pairs of black woodpecker, also endangered, are found in the core zone. Hodgson's hawk eagle, has also been recorded in the site as well as Japanese serow. Japanese black bear is common.

The area is a wilderness with no access trails or man-made facilities and its beech forest is virtually undisturbed. Wildlife is fully protected, apart from bears: the region's bear hunters, known as Matagi, use special hunting techniques and faith ceremonies.

World Heritage site since

1978 1979 1980 1981 1982 1983 1984 1985 1986 1987 1988 1989 1990 1991 1992 **1993** 1994 1995 1996 1997 1998 1999 2000 2001 2002 2003 2004 2005 2006 2007 2008

Royal Monastery of Santa María de Guadalupe
Spain

Criteria – Significance in human history;
Heritage associated with events of universal
significance

The monastery is an outstanding repository
of four centuries of Spanish religious
architecture. It symbolizes two significant
events in world history that occurred in 1492:
the reconquest of the Iberian peninsula by
the Catholic Kings, and Christopher
Columbus' arrival in the Americas. The
monastery was, and still remains, a centre of
pilgrimage as well as a cultural centre of the
highest order. Its hospitals and its medical
school were renowned, as was its
scriptorium and its library, containing a rich
collection of documents. Many famous
artists were attracted to Guadalupe,
including Juan de Sevilla, Francisco de
Zurbarán, Vicente Carducho and Luca
Giordano. The harmony between the
buildings and the works of art that it
contains confers outstanding value upon
the ensemble.

The monastery
overlooks a valley
surrounded by high
mountains and
enhanced by
abundant vegetation.
Its famous statue of
the Virgin became a
powerful symbol of
the Christianization
of much of the New
World.

Joya de Cerén Archaeological Site
El Salvador

Criteria – Testimony to cultural tradition;
Significance in human history

Joya de Cerén was a pre-Hispanic farming
community that was buried under an
eruption of the Laguna Caldera volcano
c. AD 600. Because of the exceptional
condition of the remains, they provide an
insight into the daily lives of the Central
American populations who worked the land
at that time. Twelve structures have been
excavated, including living quarters,
storehouses, workshops, kitchens and a
communal sauna. Cerén is thought to have
been home to about 200 people, although
no human remains have been found. The
buildings are grouped into compounds that
include structures for sleeping, storage,
cooking and handicrafts. The specialised
structures include a sweat house, a large
communal building, and two which may
have been used by specialists such as a
shaman or a healer.

A warning earthquake
apparently gave
residents time to flee
but the subsequent
volcanic eruption was
so sudden that
everyday artefacts,
from garden tools and
bean-filled pots to
sleeping mats and
religious items, were
found still in place
around the buildings.

World Heritage site since

978 1979 1980 1981 1982 1983 1984 1985 1986 1987 1988 1989 1990 1991 1992 **1993** 1994 1995 1996 1997 1998 1999 2000 2001 2002 2003 2004 2005 2006 2007 2008

Spišský Hrad and its Associated Cultural Monuments
Slovakia
Criteria – Significance in human history

Spišský Hrad (castle) and its three related sites contain one of the largest number of thirteenth-and fourteenth-century military, political and religious buildings in Eastern Europe. The castle stands on a dramatic hill rising out of the plain of western Slovakia. Construction of the present castle began in the thirteenth century. The town of Spišské Podhradie was founded at the base of the castle mound.

Its first church, destroyed in a Tatar raid, was rebuilt in 1258–73. The street pattern was laid out formally in the fourteenth century and extended in the fifteenth century. Following a fire, most of the houses were rebuilt in Renaissance style. Spišskà Kapitula, a unique fortified ecclesiastical complex of buildings, is based around the Cathedral of St Martin, which was started in 1285. Zehra is one of the earliest Slovak settlements in the region.

▲

The castle, one of the largest in Eastern Europe, is renowned for its Romanesque and Gothic architecture. It consists of the upper keep and its courtyard; two inner baileys with internal fortified access gates; the outer bailey, with the main entrance gate and remains of the garrison's quarters; and a large barbican area, now mostly ruined.

World Heritage site since

1978 1979 1980 1981 1982 1983 1984 1985 1986 1987 1988 1989 1990 1991 1992 **1993** 1994 1995 1996 1997 1998 1999 2000 2001 2002 2003 2004 2005 2006 2007 2008

Vlkolínec
Slovakia

Criteria – Significance in human history;
Traditional human settlement

Vlkolínec, situated in central Slovakia, is a remarkably intact settlement of forty-five vernacular buildings, providing a living example of a traditional central European village. It has preserved its ancient appearance with remarkable fidelity. Although most buildings date from the nineteenth century, Vlkolínec has retained its medieval layout. The characteristic houses of Vlkolínec are situated on the street frontages of narrow holdings, with stables, smaller outbuildings, and barns ranged behind them. The houses are of a traditional timber construction with log walls on stone footings, the walls being coated with clay and whitewashed or painted blue. The parcels of land that surround Vlkolínec retain the elongated strip-shape characteristic of medieval land allotment over most of feudal Europe. Outside these lie areas of common land and forest.

The first recorded settlement at Vlkolínec was in the fourteenth century. A decree of 1630 suggests that its name derives from the important charge laid upon the villagers to maintain the wolf-pits in good order.

A typical Vlkolínec building.
▼

World Heritage site since

1978 1979 1980 1981 1982 1983 1984 1985 1986 1987 1988 1989 1990 1991 1992 **1993** 1994 1995 1996 1997 1998 1999 2000 2001 2002 2003 2004 2005 2006 2007 2008

Historic Centre of Zacatecas
Mexico

Criteria – Interchange of values; Significance in human history

Founded in 1546 after the discovery of a rich silver lode, Zacatecas reached the height of its prosperity in the sixteenth and seventeenth centuries. Built on the steep slopes of a narrow valley, the town has breathtaking views and many important buildings. The cathedral, built between 1730 and 1760, dominates the centre of the town. It is a highly decorated Baroque structure with exceptional façades and other features that reflect the absorption of indigenous ideas and techniques into Roman Catholic iconography. The Jesuit church of Santo Domingo has a quiet beauty which contrasts with the Baroque flamboyance of the college alongside it. Its massive dome and towers provide a counterpoint to the nearby cathedral. It now houses a new Fine Art Museum.

Zacatecas became the economic centre for the region, with a system of forts (presidios), villages and agricultural estates (haciendas) for defence and supply. It was also the base for colonization and the spread of Christianity further to the north; first the Convent of San Francisco and later the College of Guadalupe were responsible for establishing over seventy missions, as far north as Texas and California.

Rock Paintings of the Sierra de San Francisco
Mexico

Criteria – Human creative genius; Testimony to cultural tradition

From c. 100 BC to AD 1300, the Sierra de San Francisco in the El Vizcaino reserve in Baja California was home to a people who have since disappeared but who left one of the most outstanding collections of rock paintings in the world. The paintings are remarkably well-preserved because of the dry climate and the inaccessibility of the site. Showing human figures and many animal species, and illustrating the relationship between humans and their environment, they reveal a highly sophisticated culture. Their composition and size, as well as the precision of the outlines and the variety of colours, but especially the number of sites, make this an impressive testimony to a unique artistic tradition.

The prehistoric rock art of the region was first reported by the Jesuit Father Francisco Javier Clavijero in a publication in Rome in 1789 and some 400 sites have so far been registered.

World Heritage site since

1978 1979 1980 1981 1982 1983 1984 1985 1986 1987 1988 1989 1990 1991 1992 **1993** 1994 1995 1996 1997 1998 1999 2000 2001 2002 2003 2004 2005 2006 2007 2008

Architectural Ensemble of the Trinity Sergius Lavra in Sergiev Posad
Russian Federation
Criteria – Interchange of values; Significance in human history

The Trinity Sergius Lavra is an outstanding and remarkably complete example of a working Orthodox monastery (lavra) of the fourteenth-eighteenth centuries and has exerted a profound influence on architecture in Russia. It was founded in the 1330s by St Sergius and fortified in 1540–60. Among its most important buildings are the Trinity Cathedral, completed in 1425, containing the renowned icon of 'The Trinity' by Rublev; The Church of the Holy Spirit (Dukhovskaya), a four-pillared church with three apses and a single dome built from white limestone; the Cathedral of the Assumption (see photo on the right), a towering structure echoing the Cathedral of the Assumption in the Moscow Kremlin; and the Belfry, which is the tallest building in the complex, begun in 1740 at the instigation of Catherine the Great. The monks' cells, two-storey stone buildings, were built up against the fortress walls in the sixteenth and seventeenth centuries.

Cathedral of the Assumption. ▲

With Peter the Great's consolidation of power, a number of new buildings in the Baroque style were added to the monastery, including the refectory chamber with the Church of St Sergius; a stone palace for the Tsar (Chertogi); the Church over the Gates and the Chapel over the Well.

World Heritage site since

1978 1979 1980 1981 1982 1983 1984 1985 1986 1987 1988 1989 1990 1991 1992 1993 **1994** 1995 1996 1997 1998 1999 2000 2001 2002 2003 2004 2005 2006 2007 2008

Vilnius Historic Centre
Lithuania

Criteria – Interchange of values; Significance in human history

The Lithuanian capital Vilnius has had a profound influence on the cultural and architectural development of much of Eastern Europe. Despite invasions and partial destruction, it has preserved an impressive complex of Gothic, Renaissance, Baroque and classical buildings, as well as its medieval layout, and is an exceptional example of an Eastern European town that evolved organically over several centuries.

The conversion of Lithuania to Christianity in 1387 opened Vilnius to the Western world.

The town grew against the background of a series of major fires from the fifteenth to the eighteenth centuries. The notable churches that exist today date from the seventeenth century, but it was the successive reconstructions that gave the town many of its buildings of special character, including the cathedral, town hall and palaces. Many of the surviving earlier buildings were rebuilt or refurbished in the Baroque style.

The historic centre comprises the areas of the three castles (Upper, Lower and Curved) and the area that was encircled by the medieval wall. The plan is basically circular, radiating out from the original castle site. The street pattern is typically medieval, with small streets dividing it into irregular blocks, but with large squares inserted in later periods.

World Heritage site since

1978 1979 1980 1981 1982 1983 1984 1985 1986 1987 1988 1989 1990 1991 1992 1993 **1994** 1995 1996 1997 1998 1999 2000 2001 2002 2003 2004 2005 2006 2007 2008

Bwindi Impenetrable National Park
Uganda

Criteria – Natural phenomena or beauty;
Significant natural habitat for biodiversity

Bwindi Park covers 320 km² and is known for its exceptional diversity of flora and fauna. Its forest gets the name 'impenetrable' from the dense cover of herbs, vines and shrubs inhabiting its valley floor. The forest is the most diverse in East Africa for tree species (more than 163) and ferns (more than 104). Sixteen species have only a very restricted distribution in southwest Uganda, and one species, Lovoa swynnertonii, is globally threatened.

Located in the Kigezi Highlands of southwest Uganda at the junction of the plain and mountain forests, Bwindi is characterized by steep hills and narrow valleys. The steepness of the slopes means that the soils are very susceptible to erosion in areas where trees are cleared. The park constitutes an important water catchment area, serving the surrounding densely populated agricultural land. Three major tributaries of the Ishasha River drain into Lake Edward to the north; and the Ndego, Kanyamwabo and Shongi rivers flow southwards towards Lake Mutanda.

Bwindi has one of the richest fauna communities in East Africa, including over 214 species of forest bird, 7 species of diurnal primate, 120 species of mammals and 202 species of butterfly. Highly significant is the presence of almost half of the world's population of mountain gorillas and many other endangered species.

The park is also an important locality for the conservation of Afromontane fauna, in particular those endemic to the mountains of the western rift valley. Overall, Bwindi contains nine globally threatened species: mountain gorilla, common chimpanzee, l'Hoest's monkey Cercopithecus l'hoesti, endangered species of African elephant, African green broadbill, Grauer's rush warbler, Chaplin's flycatcher, African giant swallowtail and cream-banded swallowtail. Buffalo were poached to extinction in the late 1960s, as were leopard more recently.

Although the wider Kigezi region may have been occupied from as early as 37,000 years ago, no archaeological sites are known inside the park. The earliest evidence of forest clearance dates back 4,800 years, most likely due to the presence of the Batwa hunter-gatherer people manipulating vegetation with fire. This is the earliest evidence for cultivation anywhere in tropical Africa.

Approximately 10,000 families, all Bantu, cultivate the land immediately surrounding the park. Commercial logging has never taken place in Bwindi due to the rugged terrain.

World Heritage site since

1978 1979 1980 1981 1982 1983 1984 1985 1986 1987 1988 1989 1990 1991 1992 1993 **1994** 1995 1996 1997 1998 1999 2000 2001 2002 2003 2004 2005 2006 2007 2008

City of Luxembourg: its Old Quarters and Fortifications
Luxembourg

Criteria – Significance in human history

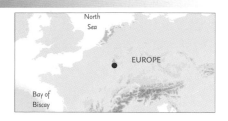

Luxembourg was one of Europe's greatest fortified sites from the sixteenth century until 1867. Its position made it a strategic and military prize and it was repeatedly reinforced by successive powers as it passed from one to another: the Holy Roman Emperors, the House of Burgundy, the Habsburgs, the French and Spanish kings, and the Prussians. Until their partial demolition, the fortifications were a fine example of military architecture spanning several centuries.

The old quarter extends westwards from the Bock promontory with its honeycomb of seventeenth- and eighteenth-century casemates. The Marché-aux-Poissons was the first open space in the town and the Church of Saint Michel, located there, originates from the tenth century. Notre Dame Cathedral is an outstanding example of Netherlands late-Gothic architecture, and the sixteenth-century Grand Ducal Palace stands at the heart of the old town.

Luxembourg's defences were such that it earned the nickname of the 'Gibraltar of the North'. However, the European powers agreed to the Grand Duchy's perpetual neutrality in the 1867 Treaty of London; this led to the demolition of the walls and fortifications and the transformation of its 1.8km^2 fortress into an open city.

The Stiechen Bridge.
▼

World Heritage site since

1978 1979 1980 1981 1982 1983 1984 1985 1986 1987 1988 1989 1990 1991 1992 1993 **1994** 1995 1996 1997 1998 1999 2000 2001 2002 2003 2004 2005 2006 2007 2008

Historic Ensemble of the Potala Palace, Lhasa
China

Criteria – Human creative genius; Significance in human history; Heritage associated with events of universal significance

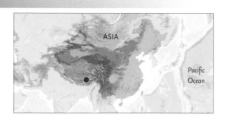

The Potala Palace, winter palace of the Dalai Lama since the seventh century, symbolizes Tibetan Buddhism and its central role in the traditional administration of Tibet. Also founded in the seventh century, the Jokhang Temple Monastery is an exceptional Buddhist religious complex. Norbulingka, the Dalai Lama's former summer palace, is a masterpiece of Tibetan art from the eighteenth century. The architectural beauty and originality of these sites, and their rich ornamentation and harmonious integration in a striking landscape, add to their historic and religious interest.

The Potala Palace complex, comprising the White and Red Palaces with their ancillary buildings, stands on Red Mountain in the Lhasa Valley at an altitude of 3,700 m. The White Palace holds the throne of the Dalai Lama and his personal apartments, while the Red Palace contains chapels and the stupa tombs of previous Dalai Lamas.

The Temple Monastery, in the centre of the old town of Lhasa, comprises an entrance porch, a courtyard and a Buddhist hall, surrounded by accommodation for monks, and storehouses. Norbulingka (treasure garden) is located on the bank of the Lhasa River about 2 km from the Potala Palace. The site consists of a large garden with palaces, halls and pavilions.

The Potala Palace.
▼

World Heritage site since

1978 1979 1980 1981 1982 1983 1984 1985 1986 1987 1988 1989 1990 1991 1992 1993 **1994** 1995 1996 1997 1998 1999 2000 2001 2002 2003 2004 2005 2006 2007 2008

Los Katíos National Park
Colombia

Criteria – Significant ecological and biological processes; Significant natural habitat for biodiversity

The region was previously inhabited by the Kuna, an indigenous group forced to migrate to Panama because of inter-tribal fighting with the Katío-Embera group, from which the park took its name.

Extending over 720 km² in northwestern Colombia, the park comprises two main regions: the mountains of the Serranía del Darién in the west, and in the east the floodplain of the Atrato River, the fastest-flowing river in the world. Lowland swamp forests cover approximately half of the park, while the remainder is lowland through to montane tropical rainforest. The wetlands of the Atrato floodplain are of special interest, and cativo is one of the typical species of tree: it can reach 50 m. An exceptional biological diversity is found in the park, which is home to many threatened animal species and endemic plants. More than 450 species of bird have been recorded, along with some 550 species of vertebrate, excluding fish.

Jelling Mounds, Runic Stones and Church
Denmark

Criteria – Testimony to cultural tradition

The present church was preceded by at least three churches built from wood, all of which were destroyed by fire. Mural paintings dating from around 1100 (and thus the earliest in Denmark) came to light on the walls of the chancel in 1874–5.

The Jelling complex, and especially the pagan burial mounds and the two runic stones, are outstanding examples of pagan Nordic culture. Jelling was a royal manor in the tenth century, during the reign of King Gorm and Queen Thyre. After the death of Thyre, her husband raised a stone in her memory and laid out a joint funerary monument consisting of two large mounds. On his death he was buried in the chamber of the north mound, which may already have contained Thyre's remains. After bringing Denmark and Norway together and introducing Christianity into Denmark, their son Harald Bluetooth set up a stone between the two mounds proclaiming his achievements and built an impressive wooden church, in which the remains of his father were re-interred.

World Heritage site since

1978 1979 1980 1981 1982 1983 1984 1985 1986 1987 1988 1989 1990 1991 1992 1993 **1994** 1995 1996 1997 1998 1999 2000 2001 2002 2003 2004 2005 2006 2007 2008

Church of the Ascension, Kolomenskoye
Russian Federation
Criteria – Interchange of values

The Church of the Ascension was built in 1532 on the imperial estate of Kolomenskoye, near Moscow, to celebrate the birth of the prince who was to become Tsar Ivan IV 'the Terrible'. As the first example of a traditional wooden tent-roofed church in stone and brick, it represents an imaginative and innovative advance in Russian Orthodox church design. Its ground plan is an equal-armed cross and it is unusual in that it has no apse. The interior of the church is small, as the walls are 3–4 m thick, but it is open to the top of the roof, 41 m above. The corners are decorated with pilasters which repeat, with some variations, the decoration of the exterior. With an overall height of 62 m and its very thick walls, the structure retains the elegance of its striking silhouette.

The so-called Italian (Alevisovsky) small brick, ▶ introduced by Italian architects at the end of the fifteenth century, was used for building the church. Carved details are in white limestone from Myachkovo, a suburb of Moscow.

World Heritage site since

1978 1979 1980 1981 1982 1983 1984 1985 1986 1987 1988 1989 1990 1991 1992 1993 **1994** 1995 1996 1997 1998 1999 2000 2001 2002 2003 2004 2005 2006 2007 2008

City of Vicenza and the Palladian Villas of the Veneto
Italy

Criteria – Human creative genius; Interchange of values

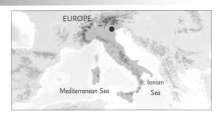

Founded in the second century BC in northern Italy, Vicenza prospered under Venetian rule from the early fifteenth to the end of the eighteenth centuries. But it was the advent of Andrea Palladio (1508–80) that gave Vicenza its enduring form. Palladio was profoundly influenced by his study of the surviving monuments of classical Rome. For Vicenza, he created both public (Basilica, Loggia del Capitaniato, Teatro Olimpico) and private buildings. A total of twenty-six

individual buildings or parts of buildings known to have been designed or reconstructed by Palladio make up the World Heritage site – twenty-three in the city itself and three villas in its immediate environs. Palladio's buildings had a decisive influence on the development of architecture and inspired a distinct architectural style known as Palladian, which spread to other European countries, as well as North America.

The ancient town plan is still recognizable in the modern town, Corso Palladio being the decumanus maximus and Contra Porti the cardo maximus. Among the surviving public buildings erected from the time of Augustus are the remains of the theatre, now incorporated into a more recent structure, and sections of the aqueduct to the north of the city.

The Villa Rotunda.
▼

World Heritage site since

978 1979 1980 1981 1982 1983 1984 1985 1986 1987 1988 1989 1990 1991 1992 1993 **1994** 1995 1996 1997 1998 1999 2000 2001 2002 2003 2004 2005 2006 2007 2008

Ancient Building Complex in the Wudang Mountains
China

Criteria – Human creative genius; Interchange of values; Heritage associated with events of universal significance

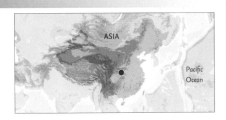

The palaces and temples which form the nucleus of this group of secular and religious buildings exemplify the architectural and artistic achievements of China's Yuan, Ming and Qing dynasties. Situated in the scenic valleys and on the slopes of the Wudang mountains in Hubei Province, the site, which was built as an organized complex during the Ming dynasty fourteenth–seventeenth centuries, contains Taoist buildings from as early as the seventh century. It represents the highest standards of Chinese art and architecture over a period of nearly 1,000 years.

Ming Emperor Zhu Di had, in the twelve years after his enthronement, 20,000 men working on construction in the Wudang Mountains. In that time they added nine palaces, nine temples, seventy-two cliff temples, thirty-six monasteries, and over one hundred stone bridges to the complex.

◄

Temple on Tianzhu Feng (Pillar of Heaven), Wudang Mountains.

Australian Fossil Mammal Sites (Riversleigh / Naracoorte)
Australia

Criteria – Major stages of Earth's history; Significant ecological and biological processes

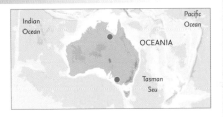

These two sites are representative of the development of Australia's mammal fauna during the Cenozoic era (65 million years ago to the present).

The faunal assemblages of Riversleigh's fossil fields have profoundly altered understanding about Australia's Middle Cenozoic vertebrate diversity. They span a record of mammalian evolution over twenty million years, providing the first records for many distinctive groups of living mammals, as well as many other unique and now extinct Australian mammals such as 'marsupial lions'.

Naracoorte also opens a window into a significant period of the Earth's history on a continent dominated by marsupials. Its assemblage also spans the probable time of arrival of humans in Australia and thus is of additional value in helping to unravel the complex relationships between humans and their environment.

The Pleistocene fossil vertebrate deposits of Victoria Fossil Cave at Naracoorte are considered to be, in terms of both volume and diversity, Australia's largest and best preserved and one of the richest deposits in the world. Tens of thousands of specimens representing ninety-nine vertebrate species have been recovered, ranging in size from very small frogs to buffalo-sized marsupials.

World Heritage site since

1978 1979 1980 1981 1982 1983 1984 1985 1986 1987 1988 1989 1990 1991 1992 1993 **1994** 1995 1996 1997 1998 1999 2000 2001 2002 2003 2004 2005 2006 2007 2008

Doñana National Park
Spain

Criteria – Natural phenomena or beauty;
Significant ecological and biological processes;
Significant natural habitat for biodiversity

In the context of a crowded and long-inhabited continent, Doñana National Park in Andalusia is one of the few national parks in Europe that can match the international significance met by parks in other parts of the world. The park is notable for the great diversity of its biotopes, especially lagoons, marshlands, fixed and mobile dunes, scrub woodland and maquis. It is home to five threatened bird species. It is one of the largest heronries in the Mediterranean region and is the wintering site for more than 500,000 water fowl each year. In general the state of conservation of the park is satisfactory, but it does face numerous threats including agricultural development, tourism, poaching and over-grazing.

Doñana National Park has been a testing ground for conservation in Spain and has become very well known throughout Europe as a result of the controversies faced and the innovative management approaches that have been taken. As the main threats have been averted thus far and as restoration activities are under way, the future for the park looks encouraging.

Rwenzori Mountains National Park
Uganda

Criteria – Natural phenomena or beauty;
Significant natural habitat for biodiversity

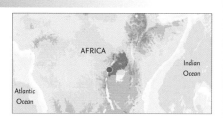

Covering nearly 1,000 km² in western Uganda, the park comprises the main part of the Rwenzori mountain chain, which includes Africa's third highest peak, Mount Margherita (5,109 m). The region's glaciers, waterfalls and lakes make it one of Africa's most beautiful alpine areas. The park has many natural habitats of endangered species and a rich and unusual flora comprising, among other species, the giant heather.

There are several major vegetation zones in the park: broken montane forest below 2,400 m; bamboo forest up to 3,000 m; a tree-heath zone of giant heathers up to 3,800 m; and Afro-alpine moorland up to 4,400 m. The park contains eighty-nine species of bird, four species of diurnal primate and fifteen species of butterfly. No people live within the park, although cultivation is evident in many places up to its border.

The highest reaches of the Rwenzori Mountains are covered by snowfields and glaciers which provide a permanent source of water for the surrounding areas.

In the east, the park is contiguous with Virunga National Park in the Democratic Republic of the Congo.

World Heritage site since

1978 1979 1980 1981 1982 1983 1984 1985 1986 1987 1988 1989 1990 1991 1992 1993 **1994** 1995 1996 1997 1998 1999 2000 2001 2002 2003 2004 2005 2006 2007 2008

Canaima National Park
Venezuela

Criteria – Natural phenomena or beauty; Major stages of Earth's history; Significant ecological and biological processes; Significant natural habitat for biodiversity

Canaima National Park extends over 30,000 km² in southeastern Venezuela along the border between Guyana and Brazil. Roughly 65 per cent of the park is covered by table mountain (tepui) formations, resulting in a unique landscape of great geological interest. The sheer cliffs and waterfalls, including the world's highest waterfall (Angel Falls, 980 m), make this a spectacular place. A main road from Ciudad Bolivar runs along the eastern border of the park, bisecting its southeast corner and providing easy access for tourists. There are no other metalled roads within the park, the western section being accessible only by air. The fauna is diverse: 118 mammal species, 550 birds, 72 reptiles and 55 amphibians have been recorded. Canaima was established as a national park in 1962 and its size was doubled to the present area in 1975.

The forests and savanna of Canaima have been occupied for 10,000 years by various groups of Amerindians of the Carib family, collectively known as the Pemon. Two archaeological sites, containing various hand-fashioned stone tools estimated to be 9,000 years old, have been found in the park.

Angel Falls. ▲

World Heritage site since

1978 1979 1980 1981 1982 1983 1984 1985 1986 1987 1988 1989 1990 1991 1992 1993 **1994** 1995 1996 1997 1998 1999 2000 2001 2002 2003 2004 2005 2006 2007 2008

Earliest 16th-Century Monasteries on the Slopes of Popocatepetl
Mexico

Criteria – Interchange of values; Significance in human history

These fourteen monasteries stand on the slopes of Popocatepetl, an active volcano, to the southeast of Mexico City. They are all built to a similar plan, with an atrium, church, and monastic buildings set around a small courtyard or patio. They are in an excellent state of conservation and are good examples of the architectural style adopted by the first missionaries – Franciscans, Dominicans and Augustinians – who converted the indigenous populations to Christianity in the early sixteenth century. Between 1525 and 1570 more than 100 monasteries were built in this region. By the end of the century over 300 had been established. In the late sixteenth century, many of the monasteries were taken over by the regular clergy and converted into parish churches.

These early monasteries represent an example of a new architectural concept in which open spaces are of renewed importance. They served as an important architectural model for a large number of smaller establishments known as 'missions' rather than monasteries. These were established as far away as the United States.

Rock Carvings in Tanum
Sweden

Criteria – Human creative genius; Testimony to cultural tradition; Significance in human history

The outstanding artistic qualities and vivid scenic compositions of Tanum's rock art make it a unique expression of Bronze Age existence. The elaborate motifs illustrate everyday life, warfare, cult, and religion. Some of the panels were obviously planned in advance. Northern Bohuslän is a land of granite bedrock, parts of which were scraped clean as the ice cap slowly moved northwards, leaving gently curved rock faces exposed. These were the 'canvases' selected by the Bronze Age artists, sited just above the shoreline of the period that began in 1500 BC, i.e. 25–29 m above today's sea level. There are at least 1,500 known rock-carving sites in northern Bohuslän, each with a number of images, but new examples are regularly coming to light as research continues.

The carvings vary from 1 mm deep to as much as 30 or 40 mm. It is suggested that the more deeply engraved figures were of greater symbolic significance and therefore required to be visible to larger gatherings of people.

World Heritage site since

1978 1979 1980 1981 1982 1983 1984 1985 1986 1987 1988 1989 1990 1991 1992 1993 **1994** 1995 1996 1997 1998 1999 2000 2001 2002 2003 2004 2005 2006 2007 2008

Temple and Cemetery of Confucius and the Kong Family Mansion in Qufu

China

Criteria – Human creative genius; Significance in human history; Heritage associated with events of universal significance

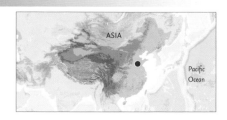

The temple, cemetery and family mansion of Confucius, the great philosopher, politician and educator of the sixth–fifth centuries BC, are located at Qufu in Shandong Province. Two years after his death, Confucius's house in Qufu was consecrated as a temple, within which were preserved his clothing, musical instruments, carriage and books. The temple was rebuilt in AD 153, and repaired and renovated several times in subsequent centuries. Today it comprises more than

100 buildings. The cemetery contains Confucius's tomb and the remains of more than 100,000 of his descendants. The small house of the Kong family developed into a gigantic aristocratic residence, of which 152 buildings with 480 rooms remain. The Qufu complex of monuments has retained its outstanding artistic and historic character due to the devotion of successive Chinese emperors over more than 2,000 years.

Over 1,000 stelae recording imperial donations and sacrifices from the Han dynasty onwards are preserved within the temple, along with outstanding examples of calligraphy and other forms of documentation, all priceless examples of Chinese art. There are many fine carved stones, among the most important being the Han stone reliefs (206 BC–AD 220).

World Heritage site since

1978 1979 1980 1981 1982 1983 1984 1985 1986 1987 1988 1989 1990 1991 1992 1993 **1994** 1995 1996 1997 1998 1999 2000 2001 2002 2003 2004 2005 2006 2007 2008

Lines and Geoglyphs of Nasca and Pampas de Jumana

Peru

Criteria – Human creative genius; Testimony to cultural tradition; Significance in human history

The geoglyphs of Nasca and the pampas of Jumana, which were scratched on the ground between 500 BC and AD 500, are among archaeology's greatest enigmas because of their quantity, nature, size and continuity. The geoglyphs depict living creatures, stylized plants and imaginary beings, as well as geometric figures several kilometres long. Their concentration and juxtaposition and cultural continuity demonstrate that this was an important and long-lasting activity. They are believed to have had ritual astronomical functions.

The Nasca geoglyphs are located in the arid Peruvian coastal plain 400 km south of Lima, and cover 450 km², both in the desert and in the Andean foothills. These are covered with ferruginous sand, and gravel which has acquired a dark patina from weathering. Removal of the gravel reveals the underlying lighter coloured stratum, which contrasts strongly with the darker gravels.

The geoglyphs fall generally into two categories. The first group is representational, depicting animals, birds, insects, plants, fantastic figures and even everyday objects.

The second group comprises the lines, generally straight and criss-crossing in all directions. Some stretch several kilometres and depict geometrical shapes. Others are so-called 'tracks' which appear to have been laid out to accommodate large numbers of people.

World Heritage site since

1978 1979 1980 1981 1982 1983 1984 1985 1986 1987 1988 1989 1990 1991 1992 1993 **1994** 1995 1996 1997 1998 1999 2000 2001 2002 2003 2004 2005 2006 2007 2008

Skogskyrkogården
Sweden

Criteria – Interchange of values; Significance in human history

In 1912 an international architectural competition was held to create a new cemetery on the site of former gravel pits overgrown with pine trees. It was won by two young architects, Asplund and Lewerentz, whose design blended vegetation and architectural elements to create a landscape that is finely adapted to its function. Their sources were not 'high' architecture or landscape design but ancient and medieval Nordic burial archetypes. Nonetheless, skilful use was made of elements from Mediterranean antiquity, such as the Via Sepulchra at Pompeii. The intervention of footpaths, meandering freely through the woodland, is minimal. Graves are laid out without excessive alignment or regimentation among the natural forest. This dignified design has had a profound influence in many countries around the world.

The cemetery design stands out for its intense romantic naturalism, turning the untouched Nordic forest into the dominant experience. The evocation of raw Nordic wilderness constituted a radical departure in landscape architecture as well as cemetery layout in the early twentieth century.

Bagrati Cathedral and Gelati Monastery
Georgia

Criteria – Significance in human history

The construction of Bagrati Cathedral, named after Bagrat III, the first king of united Georgia, started at the end of the tenth century and was completed in the early years of the eleventh century. Although partly destroyed by the Turks in 1691, its ruins still lie in the centre of Kutaisi. Richly ornamented capitals and fragments of piers and vaulting are scattered throughout the interior. The Gelati Monastery, whose main buildings were erected between the twelfth and seventeenth centuries, is a well-preserved complex, with wonderful mosaics and wall paintings. The main church has an interior surmounted by a large dome which beautifully combines space and solemnity, with light streaming in from many windows. The cathedral and monastery epitomise the flowering of medieval architecture in Georgia.

Gelati was not simply a monastery – it was also a centre of science and education, and the Academy established there in the reign of David IV (1073–1125) was one of the most important centres of culture in ancient Georgia.

World Heritage site since

1978 1979 1980 1981 1982 1983 1984 1985 1986 1987 1988 1989 1990 1991 1992 1993 **1994** 1995 1996 1997 1998 1999 2000 2001 2002 2003 2004 2005 2006 2007 2008

Mountain Resort and its Outlying Temples, Chengde
China

Criteria – Interchange of values; Significance in human history

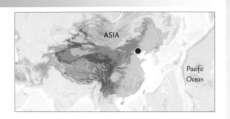

The Mountain Resort, the Qing dynasty's summer palace, in Hebei Province, was built between 1703 and 1792. It is a vast complex of palaces and administrative and ceremonial buildings. Temples of various architectural styles and imperial gardens blend harmoniously into a landscape of lakes, pastureland and forests. In addition to its aesthetic interest, the Mountain Resort is a rare historic vestige of the final development of feudal society in China.

Each year the Emperor would bring his ministers, royal troops, family and concubines, to hunt at Mulan. To accommodate this entourage of several thousand people, twenty-one temporary palaces were built, among them the Mountain Resort and its Outlying Temples.

◄

Xumifosou Zhi Miao (Temple of Happiness and Longevity.

Petäjävesi Old Church
Finland

Criteria – Significance in human history

The Petäjävesi Evangelical Lutheran Old Church is a building of considerable global importance as an example of northern timber church architecture and of the skills of the peasant population. European architectural trends have influenced the external form and the ground plan of the church, but they have been applied masterfully to traditional log construction. The church combines the layout of a Renaissance central church conception and older forms derived from Gothic groined ceilings. The church suffered a period of neglect between 1879 and the 1920s, which was actually a blessing in disguise. When restoration began the historical importance of the building had been recognised and only traditional techniques and materials were used. As a result the level of authenticity is exceptionally high.

This unique log church was built between 1763 and 1765 on a peninsula where Lakes Jamsa and Petäjävesi meet. The location was specifically chosen so that the congregation would be able to reach it by boat or over the ice in the winter.

World Heritage site since

1978 1979 1980 1981 1982 1983 1984 1985 1986 1987 1988 1989 1990 1991 1992 1993 **1994** 1995 1996 1997 1998 1999 2000 2001 2002 2003 2004 2005 2006 2007 2008

Historical Monuments of Mtskheta
Georgia

Criteria – Testimony to cultural tradition;
Significance in human history

The historic churches of Mtskheta, former capital of Georgia, are outstanding examples of medieval religious architecture in the Caucasus. They show the high level of art and culture in the vanished Kingdom of Georgia, which played an outstanding role in the medieval history of its region.

Mtskheta's fortunes had risen and fallen long before 1801, when Georgia became part of Russia. The town's strategic advantages brought settlement as early as the third millennium BC. The kingdom of

Kartli-Iberia evolved from the collapse of Alexander's Greek Empire, and Mtskheta was its capital until the sixth century AD.

The Armaztsikhe (citadel and royal residence) was at the heart of the city and fortified quarters allocated to specialized trades clustered around it, making up 'Great Mtskheta'. Today its ruins stand with the remains of earlier monuments, including a temple and tomb on Bagineti Mountain.

Many of Mtskheta's early Christian monuments survive to the present day. They include the eleventh-century Svetitskhoveli cathedral; the Mtskhetis Jvari (Church of the Holy Rood), the most sacred place in Georgia; and Samtavro (the Place of the Ruler), also built in the eleventh century, which contains the graves of Mirian, the first Georgian Christian king, and his wife.

Church of Alaverdi.
▼

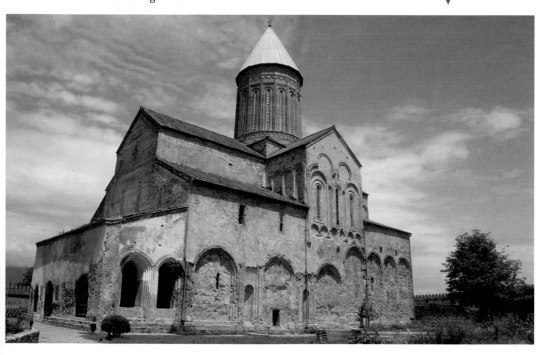

World Heritage site since

1978 1979 1980 1981 1982 1983 1984 1985 1986 1987 1988 1989 1990 1991 1992 1993 **1994** 1995 1996 1997 1998 1999 2000 2001 2002 2003 2004 2005 2006 2007 2008

Historic Monuments of Ancient Kyoto (Kyoto, Uji and Otsu Cities)

Japan

Criteria – Interchange of values; Significance in human history

ASIA

Sea of Japan

Pacific Ocean

Built in AD 794 on the model of the capitals of ancient China, Kyoto was the imperial capital of Japan from its foundation until the mid-nineteenth century. As the centre of Japanese culture for more than 1,000 years, Kyoto illustrates the development of Japanese wooden architecture, particularly religious architecture, and the art of Japanese gardens, which has influenced landscape gardening the world over.

Both Chinese culture and Buddhism were having a profound influence on Japan when the capital moved to Kyoto, then named Heian-kyo, in AD 794. Aristocratic society clustered around the imperial court for the four centuries of the Heian period (794–1192). By the end of this period, however, the military samurai class was growing in power, and civil war started in 1185. It led to the establishment of a samurai military regime at Kamakum, although the imperial court remained at Kyoto. The Sekisui-in at Kozan-ji is the best example of the residential architecture of this period, which ended in 1332 with the establishment of the Muromachi Shogunate. This period saw the building of large temples of the Rinzai Zen sect, such as Temyu-ji, and the creation of Zen gardens, of which that at Saiho-ji is a representative example.

The Muromachi Shogunate reached its height at the end of the fourteenth century; this was reflected in buildings such as the villa of Shogun Ashikaga Yoshimitsu, which later became the Buddhist temple Rokuon-ji. Garden design was refined into pure art, as demonstrated by the garden of the abbot's residence at Ryoan-ji.

Much of Kyoto was destroyed in the Onin War (1467–77), but it was rebuilt by a new urban merchant class who replaced the aristocrats who had fled during the war. The centre of power moved to Edo (present-day Tokyo), and in Kyoto the strong castle of Ngo-jo was built at the heart of the city.

The political stability of the late sixteenth century saw a new spirit of confidence among both merchants and the military, reflected in the opulence and boldness of the architecture; the Sanpo-in residential complex and garden at Daigo-jo are examples of this. The following century saw Heian temples and shrines, such as Kiyomimdera, being restored in traditional style. During this period the supremacy of Kyoto as a centre of pilgrimage became established.

Kinkaku-ji, the 'Golden Pavilion', Kyoto.

Properties on the World Heritage site that date from the foundation of Heian-kyo in the late-eighth century are Karmwakeikauchi-jinja (Shinto shrine), Amomioya-jinja (Shinto shrine), Kyo-o-gokoku-ji To-ji (Buddhist temple), Kiyornim-dera (Buddhist temple), and Enryaku-ji (Buddhist temple); the two large Buddhist temples of Daigo-ji and Ninna-ji are representative of the early Heian period.

The Japanese government's introduction of the first ordinance for the protection of antiquities in 1871, and the Ancient Shrines and Temples Preservation Law of 1897, marked the beginning of the important protection and conservation programmes of modern Japan.

World Heritage site since

1978 1979 1980 1981 1982 1983 1984 1985 1986 1987 1988 1989 1990 1991 1992 1993 **1994** 1995 1996 1997 1998 1999 2000 2001 2002 2003 2004 2005 2006 2007 2008

Ha Long Bay
Vietnam

Criteria – Natural phenomena or beauty; Major
stages of Earth's history

Ha Long Bay in the Gulf of Tonkin includes
some 1,600 islands and islets that form a
spectacular seascape of limestone pillars.
Because of their precipitous nature, most
of the islands are uninhabited. The site's
outstanding scenic beauty is complemented
by its great biological interest.

Ha Long Bay is known as a drowned karst
landscape due to the exceptional grouping
of its limestone karst features which have
been subject to repeated regression and

transgression of the sea over time. The
limestones of Ha Long Bay have been
eroded into a mature landscape of fengcong
(clusters of conical peaks) and fenglin
(isolated towers) karst features, modified by
sea invasion at a later stage.

The area provides a unique and extensive
reservoir of data for the future
understanding of geoclimatic history and
the nature of karst processes in a complex
environment.

The smaller islands
are 'fenglin' towers
50–100 m high. Many
have sheer faces on all
or most sides and
continue to evolve
through rock falls and
large slab failures.

A distinctive feature
of Ha Long Bay is the
abundance of lakes
and limestone caves
within the larger
limestone islands.
▼

World Heritage site since

1978 1979 1980 1981 1982 1983 1984 1985 1986 1987 1988 1989 1990 1991 1992 1993 **1994** 1995 1996 1997 1998 1999 2000 2001 2002 2003 2004 2005 2006 2007 2008

Collegiate Church, Castle, and Old Town of Quedlinburg
Germany
Criteria – Significance in human history

Quedlinburg, in the Land of Sachsen-Anhalt, was a capital of the East Franconian German Empire at the time of the Saxonian-Ottonian ruling dynasty. The importance of Quedlinburg rests on three main elements: the preservation of the medieval street pattern; the wealth of urban vernacular buildings, especially timber-framed houses of the sixteenth and seventeenth centuries; and the important Romanesque collegiate church of

St Servatius. The area comprises the historic town enclosed within the city walls, consisting of the old (tenth century) and new (twelfth century) towns, the Westendorf district with the collegiate church and the buildings of the imperial foundation, St Wipert's Church, and the Münzenberg, the hill on which a Benedictine monastery was founded in 946.

The original urban layout is remarkably well preserved. The town's nucleus is the castle hill, with its administrative and religious buildings, around which settlements of craftsmen and traders quickly grew to service the requirements of the rulers and their households.

Völklingen Ironworks
Germany
Criteria – Interchange of values; Significance in human history

The ironworks, which cover some 6,000 m², dominate the city of Völklingen. They are the only intact example, in the whole of western Europe and North America, of an integrated ironworks that was built and equipped in the nineteenth and twentieth centuries and has remained intact. Völklingen was the first ironworks in the world to use blast-furnace gas on a large scale to drive enormous blowers providing blast to the furnaces. By the end of the

nineteenth century Völklingen was one of Europe's most productive works and Germany's largest producer of steel beams. From the end of the Second World War until pig-iron production ceased in 1986, only minor modernisation and maintenance took place, giving the site the appearance of an ironworks of the 1930s.

The complex contains installations covering every stage in the pig-iron-production process, from raw-materials handling and processing equipment for coal and iron ore through to blast-furnace iron production. Historically the plant was a model for many other similar installations throughout the world.

World Heritage site since

1978 1979 1980 1981 1982 1983 1984 1985 1986 1987 1988 1989 1990 1991 1992 1993 **1994** 1995 1996 1997 1998 1999 2000 2001 2002 2003 2004 2005 2006 2007 2008

City of Safranbolu
Turkey

Criteria – Interchange of values; Significance in
human history; Traditional human settlement

Safranbolu consists of
four distinct districts:
the market place area
of the inner city,
known as Çukur
(The Hole); the area
of Kıranköy; Bağlar
(The Vineyards); and
an area of more
recent settlement
outside the historic
area. The original
Turkish settlement
was located
immediately south
of the citadel and
developed to the
southeast.

Safranbolu is a typical Ottoman city that
has survived to the present day. From the
thirteenth century to the early twentieth
century, Safranbolu was an important
caravan station on the main east-west trade
route. The Old Mosque, Old Bath and
Süleyman Pasha Medrese were built in 1322.
During its apogee in the seventeenth
century, Safranbolu's architecture
influenced urban development throughout
much of the Ottoman Empire. Many
buildings survive from this period, including
the Cinci Inn with its sixty guest rooms
(1640–8), Koprülü Mosque (1661) and Let
Pasha Mosque (1796), as well as many stores,
stables and baths. Following the advent of
the railway in the early twentieth century,
the town underwent a period of economic
deprivation until the building of the
Karabük steelworks, which provided a great
deal of employment in the region.

World Heritage site since

1978 1979 1980 1981 1982 1983 1984 1985 1986 1987 1988 1989 1990 1991 1992 1993 **1994** 1995 1996 1997 1998 1999 2000 2001 2002 2003 2004 2005 2006 2007 2008

Pilgrimage Church of St John of Nepomuk at Zelená Hora
Czech Republic
Criteria – Significance in human history

This pilgrimage church stands at Zelená Hora in Moravia. Constructed on a star-shaped plan, it is the most unusual work by the great architect Jan Blazej Santini, whose highly original Baroque and Neo-Gothic style displays great imagination and inventiveness. Work on the pilgrimage church began in 1719, three years before the formal canonization of John of Nepomuk. The construction of the main structure was completed by 1721.

The number five is dominant in the layout and proportions demonstrated by the five-pointed star shape for the church and the ten-pointed star for the surrounding cloisters that contain five chapels and five entrances.

The main impression given by the interior is its loftiness. The central space opens into five niches; of these, four are partitioned horizontally and the fifth, on the east, is filled by the main altar.

The church retains many of its original furnishings, which include the main altar, designed by Santini and representing the celebration of St John of Nepomuk in heaven; and the four side altars, also designed by Santini and depicting the four Evangelists.

World Heritage site since

1978 1979 1980 1981 1982 1983 1984 1985 1986 1987 1988 1989 1990 1991 1992 1993 1994 **1995** 1996 1997 1998 1999 2000 2001 2002 2003 2004 2005 2006 2007 2008

Historic Centre of Avignon: Papal Palace, Episcopal Ensemble and Avignon Bridge
France

Criteria – Human creative genius; Interchange of values; Significance in human history

Avignon was the seat of the papacy in the fourteenth century and an exceptional group of monuments testify to the city's importance. The austere Gothic palace-fortress of the Palais des Papes dominates the historic centre that includes the Petit Palais, the Romanesque Cathedral of Notre-Dame-des-Doms and the remains of the twelfth-century Avignon Bridge.

In 1309 the Frenchman Bertrand de Got, who was declared Supreme Pontiff as Clement V, refused to go to Rome, choosing instead to install himself temporarily in the Dominican Convent at Avignon. The seat of the papacy did not return to Rome until 1417.

Clement's successor, John XXII (1316-34), moved to the former bishop's palace, which was converted into a papal palace. Benedict XII (1334–42) gradually demolished this building and replaced it with what is now known as the Old Palace, covering the northern part of the present site. It was Benedict's successor, Clement VI (1342–52), who was to complete the ensemble. Clement entrusted the interior decoration to the famous Italian painter Matteo Giovannetti. He also supervised the work of French and Italian painters within the palace.

In 1793 the French Revolutionary Convention decided to demolish this 'Bastille du Midi', but the massive building defied their efforts. Ownership passed to the town in 1810 and it was put at the disposal of the Minister of War who used it as a barracks until 1906, when it was returned to the town.

The Cathedral of Notre-Dame-des-Doms lies to the north of the Palais des Papes. The present building replaced the earlier group of episcopal buildings in the twelfth century.

The Petit Palais on the western side of the Place du Palais was built as a cardinal's residence. It was acquired by John XXII in 1336 to compensate the bishop for the demolition of his palace to allow the construction of the papal palace. The Petit Palais was continuously expanded in the fourteenth and fifteenth centuries.

Saint Bénézet Bridge, the Pont d'Avignon, is one of the most important medieval bridges in Europe. Originally built in the twelfth century, it spanned 900 m across the Rhône but suffered several collapses in the following centuries. It was not rebuilt after a flood in 1668 swept much of it away, and only four of its original twenty-two arches remain.

The Palais des Papes comprises two parts: to the north the Palais Vieux (Old Palace) of Benedict XII and to the south the Palais Neuf (New Palace) of Clement VI. The main courtyard is situated between the two palace buildings.

The palace also houses the pontiffs' private rooms, including robing rooms, bedrooms and studies.

The day room of Clement IV, the Chambre du Cerf, gives access to the Great Chapel of the Palais Neuf; its heavy vault is braced by a massive flying buttress that spans the neighbouring street.

World Heritage site since

1978 1979 1980 1981 1982 1983 1984 1985 1986 1987 1988 1989 1990 1991 1992 1993 1994 **1995** 1996 1997 1998 1999 2000 2001 2002 2003 2004 2005 2006 2007 2008

Carlsbad Caverns National Park
USA

Criteria – Natural phenomena or beauty; Major stages of Earth's history

This karst landscape in New Mexico comprises eighty-one recognized caves that are outstanding not only for their size but also for the profusion, diversity and beauty of their mineral formations.

The park covers a segment of the Capitan Reef. An extensive cave system has developed within the reef as a result of sulphuric acid dissolution. Of the known caves, Carlsbad Cavern is the largest, and Lechuguilla Cave is the most extensive and decorated in the world.

The Capitan Reef complex dates back to the Permian period, some 225–280 million years ago. The exposed sections of this reef lying within the park are among the best preserved in the world accessible for scientific study.

The caves are noted for their migratory bat species. Various species of fungi and bacteria growing inside are of particular scientific and medical interest.

Carlsbad differs from the other existing World Heritage caves in its huge chambers, far larger than others, and for its decorative minerals. Since its initial exploration in 1985, Lechuguilla Cave has been strictly managed, allowing only closely monitored visits by researchers. This cave is noteworthy as an underground laboratory where geological processes can be studied in a virtually undisturbed environment.

Doll's Theatre, Big Room, Carlsbad Caverns National Park. ▼

World Heritage site since

1978 1979 1980 1981 1982 1983 1984 1985 1986 1987 1988 1989 1990 1991 1992 1993 1994 **1995** 1996 1997 1998 1999 2000 2001 2002 2003 2004 2005 2006 2007 2008

Cultural Landscape of Sintra
Portugal

Criteria – Interchange of values; Significance in human history; Traditional human settlement

The structures of Sintra and the Serra harmonize indigenous flora with a refined and cultivated landscape created by man as a result of literary and artistic influences.

The Royal Palace is the dominant architectural feature of Sintra. One of its most important features is the facing with tiles (azulejos), the finest example of this Mudejar technique on the Iberian Peninsula.

Pena Palace, Sintra.

In the nineteenth century Sintra became the first centre of European Romantic architecture. Ferdinand II turned a ruined monastery into a castle, where this new sensitivity was displayed in the use of Gothic, Egyptian, Moorish and Renaissance elements, and in the creation of a park blending local and exotic species of trees. Other fine dwellings, built along the same lines in the surrounding mountains, created a unique combination of parks and gardens, which influenced the development of landscape architecture throughout Europe.

The Serra de Sintra, Ptolemy's 'Mountain of the Moon', encloses various significant parks and gardens. Although almost all the built heritage was destroyed in the 1755 earthquake, there are some outstanding court and military buildings, religious architecture and archaeological sites. These include the Royal Palace and the Palace of Ribafrias, Pena Palace, Quinta de Regaleira, the Town Hall and Trinity Convent.

World Heritage site since

1978 1979 1980 1981 1982 1983 1984 1985 1986 1987 1988 1989 1990 1991 1992 1993 1994 **1995** 1996 1997 1998 1999 2000 2001 2002 2003 2004 2005 2006 2007 2008

Historic Centre of Naples
Italy

Criteria – Interchange of values; Significance in human history

EUROPE

Mediterranean Sea

Ionian Sea

Naples is one of the most ancient cities in Europe, whose contemporary urban fabric preserves the elements of its long and eventful history. Its street pattern, its wealth of historic buildings from many periods, and its setting on the Bay of Naples give it an unparalleled value and one that has had a profound influence in Europe and beyond.

Much of the significance of Naples is due to its urban fabric, which represents twenty-five centuries of growth. The street layout in the earliest parts of the city owes much to its classical origins.

Naples has retained the imprint of the successive cultures that emerged in Europe and the Mediterranean basin. This makes it a unique site, with a wealth of outstanding churches, such as Santa Chiara and San Lorenzo Maggiore, and monuments such as the Castel Nuovo.

Above ground, little survives of the Greek town founded in 470 BC, but important archaeological discoveries have been made since the Second World War. Three sections of the original Greek town walls are visible in the northwest. The surviving Roman remains, notably the large theatre, cemeteries and catacombs, are more substantial.

Castel Nuovo.
▼

World Heritage site since

978 1979 1980 1981 1982 1983 1984 1985 1986 1987 1988 1989 1990 1991 1992 1993 1994 **1995** 1996 1997 1998 1999 2000 2001 2002 2003 2004 2005 2006 2007 2008

Schokland and Surroundings
Netherlands

Criteria – Testimony to cultural tradition; Traditional human settlement

Schokland and its surroundings are outstanding examples of the prehistoric and historic occupation of a typical wetland, especially in relation to the reclamation and occupation of peat areas.

Schokland was a peninsula that by the fifteenth century had become an island. Occupied and then abandoned as the sea encroached, it had to be evacuated in 1859. But following the draining of the Zuider Zee, it has, since the 1940s, formed part of the land reclaimed from the sea. Schokland has vestiges of human habitation going back to prehistoric times. It symbolizes the heroic, age-old struggle of the people of the Netherlands against the encroachment of the sea.

Virgin Komi Forests
Russian Federation

Criteria – Natural phenomena or beauty; Significant ecological and biological processes

The forests form a haven for many threatened mammal species including wolf, otter, beaver, sable, wolverine and lynx. Flying squirrels, brown bears, elks and pine martens are also present.

The Virgin Komi Forests cover 32,800 km² of tundra and mountain landscape, and constitute one of the most extensive areas of virgin boreal forest remaining in Europe. The forest is dominated by lowlands in the west that rise to form the glaciated northern Ural mountains in the east. The vegetation of the lowlands comprises marshes and flood plain islands. Boreal forest extends from the marshes to the foothills of the Urals and is superseded by subalpine scrub woodlands, meadows, tundra and bedrock. This vast area of conifers, aspens, birches, peat bogs, rivers and natural lakes has been monitored and studied for over fifty years. It provides valuable evidence of the natural processes affecting biodiversity in the taiga.

World Heritage site since

1978 1979 1980 1981 1982 1983 1984 1985 1986 1987 1988 1989 1990 1991 1992 1993 1994 **1995** 1996 1997 1998 1999 2000 2001 2002 2003 2004 2005 2006 2007 2008

Waterton Glacier International Peace Park
Canada and USA

Criteria – Natural phenomena or beauty;
Significant ecological and biological processes

Waterton Lake in Canada.

The diverse habitats support a rich variety of mammals, birds, reptiles, amphibians, fish and invertebrates. These include an intact suite of predators including grey wolves, grizzly bears, cougars, lynx, fox, coyotes, fishers and wolverines.

Both parks strive to protect their shared ecosystem through cooperative management, not only between themselves, but also with other neighbours.

In 1932 Waterton Lakes National Park in Alberta, Canada, was joined with Glacier National Park in Montana, United States, to form the world's first International Peace Park. Situated on the border between the two countries and offering outstanding scenery, the combined park is exceptionally rich in plant and mammal species, and in prairie, forest, alpine and glacial features. Both parks feature long, narrow, glacial lakes and colourful, ancient rocks. The Peace Park celebrates the peace and goodwill existing along the world's longest undefended border, as well as a spirit of cooperation which is reflected in wildlife and vegetation management, search and rescue programmes, and joint interpretive programmes, brochures and exhibits.

The Waterton-Glacier stratigraphic record spans more than 1,600 million years of sedimentary and tectonic evolution. Local topography is dominated by the 2,500 m peaks of the Rocky Mountains, and the park is roughly split lengthwise by the Continental Divide. The dominant landforms are typical of mountain glaciation. The mountains were largely shaped by glacial erosion, whereas the rolling grasslands are a result of glacial deposition. The Peace Park's natural processes: fire, wind, flooding and glaciation continue to shape the landscape.

The region's climate also contributes to the richness of life found in the park. The area is influenced by two opposing systems, the Arctic Continental and the Pacific Maritime. The headwaters of three major continental watersheds are also protected within the boundaries of the Peace Park.

Five ecoregions are found within the Waterton-Glacier International Peace Park: alpine, subalpine, montane, foothills parkland, and grasslands. The alpine ecoregion is found above 2,100 m on the west slope and 1,800 m on the east. While it is not well vegetated, it has striking summer displays of wildflowers. Coniferous forest is typical of the cooler subalpine ecoregion, characterized by such species as lodgepole and ponderosa pine, subalpine fir and Englemann spruce. This 'snow forest' is the largest ecoregion in the Peace Park. The montane ecoregion occurs at low to mid elevations. It is a mix of dry grasslands and mixed poplar and coniferous forests. Douglas fir, white spruce and western larch are distinctive trees of this ecoregion. The foothills parkland ecoregion serves as a transition between prairie grasslands and the coniferous forest zone, and is mix of aspens and grasslands. Commonly known as bunchgrass prairie, a typical grass in this ecoregion is rough fescue.

World Heritage site since

1978 1979 1980 1981 1982 1983 1984 1985 1986 1987 1988 1989 1990 1991 1992 1993 1994 **1995** 1996 1997 1998 1999 2000 2001 2002 2003 2004 2005 2006 2007 2008

Roskilde Cathedral
Denmark

Criteria – Interchange of values; Significance in human history

Built in the twelfth and thirteenth centuries, this was Scandinavia's first Gothic cathedral to be constructed of brick and it encouraged the spread of this style throughout northern Europe. It has been the mausoleum of the Danish royal family since the fifteenth century. The original structure was Romanesque; however, when only the eastern half had been built, the plan was changed, the remainder to be under Gothic influence. The transept was located further back and the towers planned for the choir were moved to the west end. Work was virtually complete by around 1275, apart from the north tower, which was finished at the end of the fourteenth century. Porches and side chapels were added up to the end of the nineteenth century. The building thus provides a clear overview of the development of European religious architecture.

Many of the medieval furnishings of the cathedral disappeared at the Reformation, and more were sold at a notorious auction in 1806. The outstanding piece of what remains is the reredos, a masterpiece of Dutch religious art dating from around 1560. It is a triptych, probably from Antwerp, and bears scenes from the life of Christ.

World Heritage site since

1978 1979 1980 1981 1982 1983 1984 1985 1986 1987 1988 1989 1990 1991 1992 1993 1994 **1995** 1996 1997 1998 1999 2000 2001 2002 2003 2004 2005 2006 2007 2008

Ferrara, City of the Renaissance, and its Po Delta
Italy

Criteria – Interchange of values; Testimony to cultural tradition; Significance in human history; Traditional human settlement; Heritage associated with events of universal significance

Ferrara is an outstanding example of a planned Renaissance city which has retained its urban fabric virtually intact. Among the great Italian cities, it is the only one to have an original plan that is not derived from a Roman layout. It did not develop from a central area but rather on a linear axis, along the banks of the Po River, with longitudinal streets and many cross streets, around which the medieval city was organized.

Throughout the sixteenth century the city was planned with the aim of making it a future capital. Its evolution came to an end after the seventeenth century under papal administration, and the city did not undergo any extensions for almost three centuries. The developments in town planning expressed in Ferrara had a profound influence on urban design throughout the following centuries.

In the fifteenth and sixteenth centuries the city became an intellectual and artistic centre that attracted the greatest minds of the Italian Renaissance. Here, Piero della Francesca, Jacopo Bellini and Andrea Mantegna decorated the palaces of the House of Este, and the humanist concept of the 'ideal city' came to life in the neighbourhoods built by Biagio Rossetti in accordance with the new principles of perspective.

Castello Estense. ▼

World Heritage site since

1978 1979 1980 1981 1982 1983 1984 1985 1986 1987 1988 1989 1990 1991 1992 1993 1994 **1995** 1996 1997 1998 1999 2000 2001 2002 2003 2004 2005 2006 2007 2008

Rapa Nui National Park
Chile

Criteria – Human creative genius; Testimony to cultural tradition; Traditional human settlement

Ahu Tongariki, the largest ahu on Rapa Nui.

Several moai are still in an uncompleted condition in the quarries, providing valuable information about their manufacture. Some have large cylindrical pieces of red stone known as pukao as headdresses: these are believed to denote special ritual status. There is a clear stylistic evolution in the form and size of the moai, from the earlier small, round-headed and round-eyed figures to the best-known large, elongated figures with carefully carved fingers, nostrils, long ears and other features.

Rapa Nui, the indigenous name of Easter Island, bears witness to a unique cultural phenomenon. A society that settled there around AD 300 established a powerful and original tradition of monumental sculpture and architecture free from outside influence. From the tenth to the sixteenth centuries this society built shrines and erected enormous stone figures known as moai, creating an unrivalled cultural landscape that continues to fascinate people throughout the world.

The island was settled around AD 300 by Polynesians, probably from the Marquesas, who brought with them a wholly Stone Age society. The high cultural level of this society is best known by its moai and ceremonial shrines (ahu); it is also noteworthy for a form of pictographic writing (rongo rongo), so far undeciphered. All the cultural elements in Rapa Nui before the arrival of Europeans in the eighteenth century indicate that there were no other incoming groups.

Between the tenth and sixteenth centuries the island community expanded steadily, and settlements were set up along the coastline. However, an economic and social crisis in the sixteenth century, attributable to over-population and environmental deterioration, resulted in constant warfare between two separate groups. The warrior class that evolved gave rise to the Birdman cult which superseded the statue-building religion and toppled most of the moai and ahu.

On Easter Sunday 1722 Jacob Roggeveen of the Dutch East India Company chanced upon the island and gave it its European name. It was annexed to Chile in 1888.

Rapa Nui's most famous archaeological features are the moai, believed to represent sacred ancestors who watch over the villages and ceremonial areas. They range from 2 m to 20 m in height and are for the most part carved from the scoria or solidified lava, and lowered down the slopes into previously dug holes.

The ahu vary considerably in size and form. There are certain constant features: notably a raised rectangular platform of large worked stones filled with rubble; a ramp often paved with rounded beach pebbles; and a levelled area in front of the platform. Some have moai on them, and there are tombs in a number of the ahu in which skeletal remains have been discovered. The ahu are generally located on the coast and orientated parallel to it.

World Heritage site since

1978 1979 1980 1981 1982 1983 1984 1985 1986 1987 1988 1989 1990 1991 1992 1993 1994 **1995** 1996 1997 1998 1999 2000 2001 2002 2003 2004 2005 2006 2007 2008

Caves of Aggtelek Karst and Slovak Karst
Hungary and Slovakia
Criteria – Major stages of Earth's history

These caves are remarkable for having the world's highest stalagmite and an ice-filled abyss, which considering the territory's height above sea level, is a unique phenomenon for central Europe.

The variety of formations and the fact that they are concentrated in a restricted area means that the 712 caves currently identified make up a typical temperate-zone karstic system. Because they display an extremely rare combination of tropical and glacial climatic effects, they make it possible to study geological history over tens of millions of years.

◄

Demänovská Cave of Freedom, Slovak Karst Caves.

Messel Pit Fossil Site
Germany
Criteria – Major stages of Earth's history

The Eocene epoch was a remarkable period in the evolution of life on Earth, when mammals became firmly established in all the principal land ecosystems. They also reinvaded the seas, e.g. whales, and took to the air, e.g. bats.

One of the world's four most significant fossil sites, Messel Pit gives the most complete view of the living environment of the Eocene, between 57 million and 36 million years ago. During this period of geological time, North America, Europe and Asia were in continuous land contact and the partial explanation of current distribution patterns is provided by the Eocene fossil record. Messel provides particularly rich information about the early

stages of the evolution of mammals and includes exceptionally well-preserved mammal fossils, ranging from fully articulated skeletons to the contents of stomachs of animals of this period. Mammals were not the only component of the fauna – large quantities of birds, reptiles, fish, insects and plant remains all contribute to an extraordinary fossil assemblage.

World Heritage site since

1978 1979 1980 1981 1982 1983 1984 1985 1986 1987 1988 1989 1990 1991 1992 1993 1994 **1995** 1996 1997 1998 1999 2000 2001 2002 2003 2004 2005 2006 2007 2008

Seokguram Grotto and Bulguksa Temple
Korea, Republic of

Criteria – Human creative genius; Significance in human history

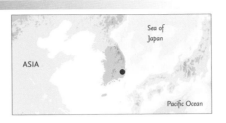

ASIA

Sea of Japan

Pacific Ocean

Established in the eighth century on the slopes of Mount T'oham, the Seokguram Grotto consists of an antechamber, a corridor and a main rotunda. It is built from granite and features thirty-nine Buddhist engravings on the main wall and the principal sculpture of the Buddha in the centre. This monumental statue is posed in the bhumisparsha mudra position, the gesture with which the historical Buddha summoned the Earth as witness to his realization of Enlightenment. With the surrounding portrayals of gods,

Bodhisattvas and disciples, all realistically and delicately sculpted in high and low relief, it is considered a masterpiece of Buddhist art in the Far East. The Temple of Bulguksa, built in 774, and the Seokguram Grotto form a religious architectural complex of exceptional significance.

Multi-coloured lanterns decorate the Dabotap Pagoda, which is located in the Bulguksa Temple. The lanterns are to celebrate Buddha's birthday.
▼

The main Sakyamuni Buddha figure is 3.45 m high, and set on a lotus flower-shaped pedestal. The hair is tightly curled and there is a distinct usnisa, the protuberance on the top of the head symbolizing Supreme Wisdom. Beneath the broad forehead the eyebrows are shaped like crescent moons and the half-closed eyes gaze towards the East Sea.

World Heritage site since

1978 1979 1980 1981 1982 1983 1984 1985 1986 1987 1988 1989 1990 1991 1992 1993 1994 **1995** 1996 1997 1998 1999 2000 2001 2002 2003 2004 2005 2006 2007 2008

San Agustín Archeological Park
Colombia
Criteria – Testimony to cultural tradition

The wealth of megalithic statuary from the archaeological site in San Agustín Archeological Park bears vivid witness to the artistic creativity of the pre-Hispanic culture that flowered in the hostile tropical environment of the northern Andes from the first to the eighth centuries. During this period there was considerable social consolidation. The concentration of power in the hands of the chiefs made possible the production of gigantic works: hundreds of elaborate stone statues were carved, some in complex relief and large in size. Altogether some 300 enormous sculptures of gods, warriors and mythical beasts were created, in styles ranging from abstract to realist, and they make up the largest group of religious monuments in South America. The huge monumental platforms, terraces and mounds, and the temple-like architecture reflect a complex system of religious and magical beliefs.

The principal archaeological monuments are Las Mesitas, containing artificial mounds, terraces, funerary structures and stone statues; the Fuente de Lavapatas, a religious monument carved in the stone bed of a stream; and the Bosque de Las Estatuas, where there are examples of stone statues from the whole region.

World Heritage site since

1978 1979 1980 1981 1982 1983 1984 1985 1986 1987 1988 1989 1990 1991 1992 1993 1994 **1995** 1996 1997 1998 1999 2000 2001 2002 2003 2004 2005 2006 2007 2008

Hanseatic Town of Visby
Sweden

Criteria – Significance in human history;
Traditional human settlement

Atlantic
Ocean

Scandinavia

North
Sea

EUROPE

A former Viking site on the island of
Gotland, Visby was the main centre of the
Hanseatic League in the Baltic from the
twelfth to the fourteenth centuries. During
this period, German merchants settled in
the town, followed by Russian and Danish
traders. Guild houses and churches were
built, and the earlier small wooden buildings
were replaced by large stone houses, built in
parallel rows eastwards from the harbour.

As a result, Visby changed from a simple
Gotland village into an impressive
international town, enclosed by a strong
defensive wall, and increasingly divorced
from its rural hinterland. Its thirteenth-
century ramparts and more than 200
buildings (warehouses and wealthy
merchants' dwellings) from the same period
make it the best-preserved fortified
commercial city in northern Europe.

The best-preserved
medieval warehouse
is the Old Pharmacy
on Strandgatan, with
vaulted rooms on the
ground and top floors,
a latrine cellar, a
medieval well, and
original surrounds on
doors and windows.
Other notable
buildings are von
Lingen's House on
St Hansgatan and
a number of houses
in the narrow streets
running down to the
harbour.

World Heritage site since

1978 1979 1980 1981 1982 1983 1984 1985 1986 1987 1988 1989 1990 1991 1992 1993 1994 **1995** 1996 1997 1998 1999 2000 2001 2002 2003 2004 2005 2006 2007 2008

Historic Centre of Siena
Italy

Criteria – Human creative genius; Interchange of values; Significance in human history

EUROPE

Mediterranean Sea

Ionian Sea

Siena is the embodiment of a medieval city. Its character and quality are preserved to a remarkable degree, and its influence on art, architecture and town planning in the Middle Ages, both in Italy and elsewhere in Europe, was immense. The whole city of Siena, built around the Piazza del Campo, was devised as a work of art that blends into the surrounding landscape.

The historic centre is delimited by a 7-km fortification of ramparts dating from the fourteenth to the sixteenth centuries, the route of which follows the contours of the three hills on which the city stands. The walls, which have been enlarged on several occasions, also include part of the 25-km network of galleries (bottini), which evacuate the spring waters distributed by the public fountains. The main fountains, mostly from the thirteenth century, are veritable buildings in their own right and constructed like Gothic porticoes.

The historic centre developed along the Y-shaped segments defined by the three main arteries that meet at the Croce del Travaglio, adjoining the Piazza del Campo, and onto which the network of minor roads are grafted. Houses and palaces follow one another in rows along the main streets, creating a characteristic urban space with certain notable elements.

The Piazza del Campo, at the junction of three hills, is one of the most remarkable urban open spaces in Italy. Its formation coincides with the growth of the medieval city and the assertion of communal power. Financial and commercial activities were concentrated halfway along the Via Francigena, the entire lengths of the present-day Via dei Banchi Sopra and Via dei Banchi Sotto, and the market place proper was located in the Piazza del Campo, at that time divided in two sectors.

At the end of the twelfth century, the communal government decided to unite the two sectors to create a unique semicircular open space, and promulgated a series of ordinances that regulated not only commercial activities, but also the services and dimensions of the houses, in order to make the façades around the piazza uniform. Under the Medici who ruled Siena from the mid-sixteenth century, the piazza became the ideal setting for spectacular festivals and was opened up to the Palio, the famous horse race between teams from the different quarters of the city.

The Palazzo Pubblico, the focal point of the Piazza del Campo, was in all probability the model for the Gothic palaces of the great noble or mercantile families (Palazzo Tomei, Palazzo Buonsignore); these are characterized by an increase in breadth, the use of brick, large windows and the so-called 'Guelph' crenellation.

The highest point of the town is crowned by the Cathedral of Santa Maria. The lower part of its façade is the work of Giovanni Pisano. The cathedral's remarkable pavement and pulpit, which was carved by Nicolà Pisano, are extremely well-preserved.

World Heritage site since

1978 1979 1980 1981 1982 1983 1984 1985 1986 1987 1988 1989 1990 1991 1992 1993 1994 **1995** 1996 1997 1998 1999 2000 2001 2002 2003 2004 2005 2006 2007 2008

Town of Luang Prabang
Lao People's Democratic Republic

Criteria – Interchange of values; Significance in human history; Traditional human settlement

Luang Prabang is an outstanding example of the fusion of traditional Lao architecture and urban structures with European colonial building in the nineteenth and twentieth centuries. Its unique and remarkably well-preserved townscape illustrates a key stage in the blending of these two distinct cultural traditions. The political and religious centre of the town is situated on a peninsula formed by the Mekong River and its tributaries, and contains royal and noble residences, religious foundations and commercial buildings. The majority are built from wood (parts of the temples are in stone). The colonial element is characterized by one- or two-storey terraced houses built from brick. The monasteries generally consist of the religious buildings (shrine, chapel, library, stupa, stone post), and monastic buildings (communal buildings, cells, refectory).

The traditional Lao wooden houses are divided into two basic spaces: the private rooms and the public terraces. They are usually raised on wooden piles, giving a space beneath for working and for shelter, for both men and animals. Walling may be of planks or plaited bamboo on a wooden frame.

Wat Xieng Thong Temple in Luang Prabang.
▼

World Heritage site since

1978 1979 1980 1981 1982 1983 1984 1985 1986 1987 1988 1989 1990 1991 1992 1993 1994 **1995** 1996 1997 1998 1999 2000 2001 2002 2003 2004 2005 2006 2007 2008

Crespi d'Adda
Italy

Criteria – Significance in human history;
Traditional human settlement

Crespi d'Adda is an outstanding example of a late nineteenth-century 'company town' that survives remarkably intact, and part of which is still in industrial use. In 1878 Cristoforo Benigno Crespi, an enlightened textile manufacturer, built three-storey multi-family houses for his workers around his mill. When his son, Silvio Benigno Crespi, took over the management in 1889, he completed and modified the project. He turned away from the large multiple-

occupancy blocks in favour of the single-family house with its own garden, which he saw as conducive to harmony, and a defence against industrial strife. In addition to small houses, a hydroelectric power station to supply the workers with free electricity; a clinic; public lavatories and wash houses; a consumer cooperative; a school and small theatre; a sports centre; houses for the local priest and doctor; and other common services were built.

The entire complex is laid out in a geometrically regular form and divided into two parts by the main road from Capriate. The factory, a single, compact block with medieval ornamentation, is on one side of the main road. The houses, constructed within a rectangular grid of roads in three lines, are on the opposite side.

World Heritage site since

1978 1979 1980 1981 1982 1983 1984 1985 1986 1987 1988 1989 1990 1991 1992 1993 1994 **1995** 1996 1997 1998 1999 2000 2001 2002 2003 2004 2005 2006 2007 2008

Rice Terraces of the Philippine Cordilleras
Philippines

Criteria – Testimony to cultural tradition;
Significance in human history; Traditional human
settlement

Terraced rice fields are not uncommon in Asia, and the main differences between the Philippines terraces and those elsewhere are their higher altitudes (between 700 m and 1,500 m above sea level) and the steeper slopes. The high-altitude cultivation is based on the use of a special strain of rice which germinates under freezing conditions and grows chest-high, with non-shattering panicles, to facilitate harvesting on slopes that are too steep to permit the use of animals or machinery of any kind.

The rice terraces of the Philippine Cordilleras are living cultural landscapes devoted to the production of one of the world's most important staple crops, rice. They preserve traditional techniques and forms, dating back many centuries, and still viable today. At the same time, they illustrate a remarkable degree of harmony between mankind and a natural environment of great aesthetic appeal, as well as demonstrating sustainable farming systems in mountainous terrain based on careful use of natural resources.

The rice terraces are the only monuments in the Philippines that show no evidence of having been influenced by colonial cultures. Owing to the difficult terrain, the Cordillera tribes are among the few peoples of the Philippines who have successfully resisted foreign domination and preserved their authentic tribal culture. The history of the terraces is intertwined with that of its people, their culture and traditional practices.

The Philippines culture, alone among southeast Asian cultures, is wholly wood-based and the terraces are the only form of stone construction from the pre-colonial period. Terracing began in the Cordilleras some 2,000 years ago, although scholars disagree about its original purpose.

However, it is evidence of a high level of knowledge of structural and hydraulic engineering among the terrace builders. The knowledge and practices involved in maintaining the terraces are transferred orally from generation to generation, without written records but supported by rituals.

Construction of the terraces demands great care and precision. An underground conduit is placed within the fill for drainage purposes. The groups of terraces blanket the mountainsides, following their contours. Above them, rising to the mountain tops, is a ring of private woods (muyong), intensively managed in conformity with traditional practices, which recognizes a total ecosystem and assures an adequate water supply to keep the terraces flooded. Water is equitably shared and no single terrace obstructs the flow on its way down to the next terrace. There is a complex system of communally maintained dams, sluices, channels and bamboo pipes, which drain into a stream at the bottom of the valley.

Villages are associated with groups of terraces, and consist of groups of single-family tribal dwellings which architecturally reproduce the people's spatial interpretation of their mountain environment.

World Heritage site since

1978 1979 1980 1981 1982 1983 1984 1985 1986 1987 1988 1989 1990 1991 1992 1993 1994 **1995** 1996 1997 1998 1999 2000 2001 2002 2003 2004 2005 2006 2007 2008

Jongmyo Shrine
Korea, Republic of
Criteria – Significance in human history

Jongmyo is the oldest and most authentic of the Confucian royal shrines to have been preserved. Dedicated to the forefathers of the Choson dynasty 1392–1910, the shrine has existed in its present form since the sixteenth century and houses tablets bearing the teachings of members of the former royal family. Ritual ceremonies linking music, song and dance still take place there, perpetuating a tradition that goes back to the fourteenth century. Jongmyo is situated in valleys and surrounded by low hills, with artificial additions created to reinforce the

balance of natural elements on the site as laid out in traditional geomancy. The complex is composed of three sets of buildings, each centred around an important shrine or other religious building.

T'aejo, founder of the kingdom, transferred the seat of government to Hanyang (present-day Seoul) in 1394 and ordered the building of Jongmyo. The spirit tablets of four generations of T'aejo's ancestors were later moved there; additional buildings were added later.

Gough and Inaccessible Islands
United Kingdom
Criteria – Natural phenomena or beauty; Significant natural habitat for biodiversity

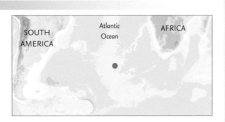

The spectacular cliffs of Gough Island, towering above the remote South Atlantic, make it a strong contender for the title 'most important seabird colony in the world'. At least fifty-four bird species occur here, including 48 per cent of the world's population of the northern rockhopper penguin. Gough is also a major breeding site of the great shearwater with up to three million pairs breeding on the island. The

endangered wandering albatross *Diomedea dabbenena* is virtually restricted to Gough, with up to 2,000 breeding pairs. The last survivors of the southern giant petrel also breed on Gough, with only 100–150 pairs remaining. Inaccessible Island is almost as rich in wildlife, with two bird, eight plant and at least ten invertebrate species endemic to the island.

Gough Island's undisturbed nature makes it invaluable for biological research, which, with weather monitoring, is the only activity permitted on the island. Inaccessible Island is also largely pristine and is one of the few temperate oceanic islands without introduced mammals.

World Heritage site since

1978 1979 1980 1981 1982 1983 1984 1985 1986 1987 1988 1989 1990 1991 1992 1993 1994 **1995** 1996 1997 1998 1999 2000 2001 2002 2003 2004 2005 2006 2007 2008

Kutná Hora: Historical Town Centre with the Church of St Barbara and the Cathedral of Our Lady at Sedlec
Czech Republic

Criteria – Interchange of values; Significance in human history

Kutná Hora, as a result of its silver mines, was one of the most important political and economic centres of Bohemia. Its medieval centre and churches influenced the architecture of central Europe. In the fourteenth century it became a royal city endowed with monuments that symbolized its prosperity: the Church of St Barbara, the Church of St James, the Stone House and the Gothic fountain. The interior of St Barbara's, a jewel of the late Gothic period, is decorated with medieval frescoes depicting the secular life of the medieval mining town of Kutná Hora. The Cistercian Cathedral of Our Lady at Sedlec was restored in line with the Baroque taste of the early eighteenth century. These masterpieces today form part of a well-preserved medieval urban fabric, with some particularly fine private dwellings.

The Church of St Barbara. ▲

The town, built above the steep descent of the Vrchlice Creek, some 60 km east of Prague, prospered from the exploitation of the silver mines, which reached its peak in the fourteenth and fifteenth centuries, when the city became one of the richest in Europe.

World Heritage site since

1978 1979 1980 1981 1982 1983 1984 1985 1986 1987 1988 1989 1990 1991 1992 1993 1994 **1995** 1996 1997 1998 1999 2000 2001 2002 2003 2004 2005 2006 2007 2008

Old and New Towns of Edinburgh
United Kingdom

Criteria – Interchange of values; Significance in human history

Edinburgh Castle in ▶ the Old Town (right) and the eastern end of the New Town (below).

The Scottish capital since 1437, Edinburgh has two distinct areas: the Old Town, dominated by the medieval fortress of Edinburgh Castle; and the neoclassical New Town, whose development from the eighteenth century onwards had a far-reaching influence on European urban planning. The two areas are a remarkable blend of the urban phenomena of organic medieval growth and eighteenth- and nineteenth-century town planning. The successive planned extensions of the New Town and the high quality of the architecture set standards for Scotland and beyond.

Edinburgh Castle was built in the twelfth century on top of a crag-and-tail formation known as Castle Rock, and the main medieval High Street stood on the ridge of the 'tail' that ran downhill from the Castle to the city walls. Other streets, called closes or wynds, led off at right angles on either side. Regular wars with England meant that Edinburgh was well fortified and that building was concentrated largely inside the walls. The pressure on space and the steep fall-off in land on either side of the hill led to the development of multi-storey dwellings from the late medieval period onwards, with some streets effectively being built underground.

By the eighteenth century the town had become so crowded that the city authorities deemed it necessary to expand on the land to the north. A competition to design the new area was won in 1766 by a 26-year-old local architect, James Craig. His design for a formally organized grid of streets, considered to reflect the rational ideas of the Scottish Enlightenment in its order and in its neoclassical styling, was so successful that it was subsequently greatly extended. Part of the extension was Charlotte Square, designed by Robert Adam.

The linking of the Old and New Towns by road and bridge, and the construction of neoclassical buildings in both – such as Adam's design for the Old College of Edinburgh University – harmonized the two contrasting historic areas, giving the city its unique character.

The Chapel of St Margaret was the only part of Edinburgh Castle to survive destruction throughout centuries of wars and sieges. James VI of Scotland and I of England, son of Mary, Queen of Scots, was born in the Castle in 1556.

The Palace of Holyroodhouse, the scene of many events in Scottish history, was originally the guesthouse of Holyrood Abbey. It was transformed into a royal residence by King James IV (1473–1513) and is the official residence of the monarch in Scotland.

World Heritage site since

1978 1979 1980 1981 1982 1983 1984 1985 1986 1987 1988 1989 1990 1991 1992 1993 1994 **1995** 1996 1997 1998 1999 2000 2001 2002 2003 2004 2005 2006 2007 2008

Historic Quarter of the City of Colonia del Sacramento
Uruguay
Criteria – Significance in human history

Founded by the Portuguese in 1680, Colonia del Sacramento was built on the extreme west side of a peninsula by the Río de la Plata. It illustrates the successful fusion of Portuguese, Spanish and Uruguayan styles. In 1704–5, during the War of the Spanish Succession, the town was razed to the ground by the Spanish. The Portuguese began reconstruction immediately and Sacramento became the powerhouse of material, commercial and cultural development in the colony. Its success had a decisive influence on the development of Buenos Aires and its region. The siege of 1777 saw Sacramento incorporated into the Spanish Empire. There are excellent examples of seventeenth to nineteenth-century buildings, ranging from elegant town houses to artisans dwellings, and the town has preserved its wide main streets, large squares, cobbled lanes and intimate open spaces.

Colonia del Sacramento bears remarkable testimony in its layout and its buildings to the nature and objectives of European colonial settlement, particularly during the seminal period at the end of the seventeenth century. It also exercised an unquestioned influence on architectural development on either side of the Río de la Plata.

World Heritage site since

1978 1979 1980 1981 1982 1983 1984 1985 1986 1987 1988 1989 1990 1991 1992 1993 1994 **1995** 1996 1997 1998 1999 2000 2001 2002 2003 2004 2005 2006 2007 2008

Historic Centre of Santa Cruz de Mompox
Colombia

Criteria – Significance in human history; Traditional human settlement

Founded in 1540 on the banks of the river Magdalena, Mompox played a key role in the Spanish colonisation of northern South America. The town was defined by the river, and initially grew only along the banks. Walls were built to protect it during periods of high water and, instead of a central square or plaza, it had three plazas in line, each with its own church, and corresponding with a former Indian settlement. The churches also served as forts in the early years of the city. The historic centre has preserved the harmony and unity of the urban landscape. Most of the buildings are still used for their original purposes, providing an exceptional picture of what a Spanish colonial city was like.

Mompox grew along the banks of the river and was of great logistic and commercial importance: traffic between the port of Cartagena and the interior travelled along the rivers, while overland routes also converged upon the town.

National Archeological Park of Tierradentro
Colombia

Criteria – Testimony to cultural tradition

The hypogea (underground tombs) of Tierradentro are a unique testimony to the everyday life, ritual and burial customs of a stable but now vanished northern Andean pre-Hispanic society. In particular, their carved anthropomorphic representations and polychrome paintings are unique in America. Dating from the sixth to the tenth centuries, these huge burial chambers are up to 12 m wide. There is a symbolic symmetry between the houses of the living above ground and the underground hypogea for the dead, created by a number of elegant elements. This conveys a pleasant aesthetic sensation and evokes a powerful image of the importance of a new stage into which the deceased has entered and the continuity between life and death, between the living and the ancestors.

Tierradentro is also remarkable for its stone statues. They are carved from stone of volcanic origin and represent standing human figures, with their upper limbs placed on their chests. Masculine figures have banded head-dresses, loincloths and various adornments whereas female figures wear turbans, sleeveless blouses and skirts.

World Heritage site since

1978 1979 1980 1981 1982 1983 1984 1985 1986 1987 1988 1989 1990 1991 1992 1993 1994 1995 **1996** 1997 1998 1999 2000 2001 2002 2003 2004 2005 2006 2007 2008

Cologne Cathedral
Germany

Criteria – Human creative genius; Interchange of values; Significance in human history

The cathedral is a High Gothic five-aisled basilica with a projecting transept and two-tower façade. The construction is totally unified. The western section changes in style but this is not perceptible in the overall building. Nineteenth-century work followed medieval forms and techniques faithfully. The original liturgical appointments of the choir include the high altar on a slab of black marble, carved-oak choir stalls (1308–11), painted choir screens (1332–40), fourteen statues on the pillars in the choir (1270–90) and the stained-glass windows, the largest extant cycle of fourteenth-century windows in Europe.

Cologne Cathedral, constructed over six centuries, is a masterpiece of Gothic architecture. Over the years, successive generations of builders were inspired by the same faith and a spirit of absolute fidelity to the original plans. The cathedral is powerful testimony to the strength and persistence of Christian belief in medieval and modern Europe.

The site of the cathedral is thought to have been first used for Christian worship as early as the fourth century AD. Following the Edict of Milan in 313, when the Emperors Constantine and Licinius proclaimed religious freedom in the Roman Empire, a Christian meeting house near the city walls was enlarged into a church. Alongside it were an atrium, a baptistry and a dwelling house, possibly for the bishop. This modest ensemble was greatly extended and enlarged in the following centuries. The result was an immense building, known by the thirteenth century as 'the mother and master of all churches in Germany'.

Despite its generous dimensions, this cathedral was found to be too small to accommodate the throngs of pilgrims who began to visit Cologne from 1164, when the Archbishop of Cologne brought back from Milan the relics of the Three Kings, the Magi who visited Jesus shortly after his birth.

The ambition of the Archbishop Engelbert of Cologne to make his cathedral into one of the most important in the Holy Roman Empire, and a fitting monument to house the sacred relics, led him to urge for the construction of an entirely new building. However, the archbishop was murdered in 1225 and work did not begin until 1248.

By 1560, much of the nave and the four side aisles had been completed, along with the main structure of the lofty south tower of the west end. Despite numerous efforts, the cathedral remained in an uncompleted state for several centuries. Work was delayed when the French seized Cologne in 1794, and only restarted in 1815 after the city passed to Prussia. The neoclassical architect Karl Friedrich Schinkel visited the cathedral in 1816 and sent his talented pupil Ernst Friedrich Zwirner there as cathedral architect. Work did not begin, however, until 1840. Building was finally completed in 1880, 632 years and 2 months after it first began in 1248.

World Heritage site since

1978 1979 1980 1981 1982 1983 1984 1985 1986 1987 1988 1989 1990 1991 1992 1993 1994 1995 **1996** 1997 1998 1999 2000 2001 2002 2003 2004 2005 2006 2007 2008

Lake Baikal
Russian Federation

Criteria – Natural phenomena or beauty; Major stages of Earth's history; Significant ecological and biological processes; Significant natural habitat for biodiversity

More than twenty-five million years old and 1,700 m deep, 31,500 km² Lake Baikal in southeast Siberia is the oldest and deepest lake in the world. It contains 20 per cent of the world's total unfrozen freshwater reserves. Known as the 'Galápagos of Russia', its age and isolation have produced some of the world's richest and most unusual endemic freshwater floras and faunas, which are of exceptional value to evolutionary science, the most notable of which is the Baikal seal, a uniquely freshwater species.

The lake is also surrounded by a system of protected areas that have high scenic and other natural values.

The formation of the basin's geological structures took place during the Palaeozoic, Mesozoic and Cenozoic eras (from 540 million years ago to the present). Various tectonic forces are still ongoing, as evidenced in recent thermal vents in the depths of the lake.

The landscape surrounding the lake basin is exceptionally picturesque, with mountains, forests, tundra, lakes, islands and steppes.

World Heritage site since

1978 1979 1980 1981 1982 1983 1984 1985 1986 1987 1988 1989 1990 1991 1992 1993 1994 1995 **1996** 1997 1998 1999 2000 2001 2002 2003 2004 2005 2006 2007 2008

Hiroshima Peace Memorial (Genbaku Dome)
Japan
Criteria – Heritage associated with events of universal significance

The Hiroshima Peace Memorial is a stark symbol of the most destructive force ever created by humankind; it also expresses the hope for world peace and the ultimate elimination of nuclear weapons. The Hiroshima Prefectural Industrial Promotion Hall was the only structure left standing, albeit in skeletal form, in the area where the first atomic bomb exploded on the morning of 6 August 1945. It was preserved in that state when reconstruction of the city began, and became known as the Genbaku (Atomic Bomb) Dome. In 1966 Hiroshima City Council adopted a resolution that the dome should be preserved in perpetuity. The Peace Memorial Park, in which the dome is the principal landmark, was laid out between 1950 and 1964. Since 1952 the park has been the scene of the annual Hiroshima Peace Memorial Ceremony, held on 6 August.

The Hiroshima Prefectural Industrial Promotion Hall was a three-storey brick building with a five-storey central core topped by an elliptical dome. Located only 150 m from the hypocentre of the explosion, it was almost completely shattered and gutted. However, because the force of the blast came from directly above, the foundations of the core of the building under the dome remained standing.

▼

World Heritage site since

1978 1979 1980 1981 1982 1983 1984 1985 1986 1987 1988 1989 1990 1991 1992 1993 1994 1995 **1996** 1997 1998 1999 2000 2001 2002 2003 2004 2005 2006 2007 2008

Ancient Ksour of Ouadane, Chinguetti, Tichitt and Oualata
Mauritania

Criteria – Testimony to cultural tradition; Significance in human history; Traditional human settlement

These four ancient cities are the only surviving places in Mauritania to have been inhabited since the Middle Ages. They illustrate a traditional way of life centred on the nomadic culture of the people of the western Sahara.

Founded in the eleventh and twelfth centuries to serve the caravans crossing the Sahara, these trading and religious centres became focal points of Islamic culture. Sited on the outskirts of a fertile valley or oasis, their original function was to provide religious instruction, and so they developed around mosques, accompanied by houses for teachers and students. Warehouses were built by traders to safeguard their goods, who needed accommodation for themselves, while inns were provided for those passing through on business. From these elements grew the characteristic form of settlement known as the ksar (plural ksour), with stone architecture and an urban form suited to extreme climatic conditions. Typically, houses with patios crowd along narrow streets around a mosque with a square minaret.

Okapi Wildlife Reserve
Dem. Rep. of the Congo

Criteria – Significant natural habitat for biodiversity
.

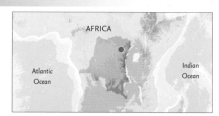

The Pygmy groups that today inhabit the Ituri forest excel in the use and identification of wild plants. Pygmies have a semi-nomadic hunter-gatherer lifestyle and when not hunting with traditional nets or archery, gather insects, fungi, fruits, seeds, plants and honey.

The Okapi Wildlife Reserve occupies about one-fifth of the Ituri forest in the northeast of the Democratic Republic of the Congo. The Congo River basin, of which the reserve and forest are a part, is one of the largest drainage systems in Africa. The reserve contains threatened species of primates and birds and about 5,000 of the estimated 30,000 okapi surviving in the wild. It also has some dramatic scenery, including waterfalls on the Ituri and Epulu rivers. The reserve is inhabited by traditional nomadic pygmy Mbuti and Efe hunters.

◀

Okapi in the Ituri forest.

World Heritage site since

1978 1979 1980 1981 1982 1983 1984 1985 1986 1987 1988 1989 1990 1991 1992 1993 1994 1995 **1996** 1997 1998 1999 2000 2001 2002 2003 2004 2005 2006 2007 2008

Millenary Benedictine Abbey of Pannonhalma and its Natural Environment
Hungary

Criteria – Significance in human history; Heritage associated with events of universal significance

The landscape around the monastic site contains a natural oak forest, and a botanical garden, composed of forest trees and plants, and hedgerow and park species, both native and exotic.

Benedictine monks came in 996 to this sacred mountain in the former Roman province of Pannonia. They established the monastery as the eastern bridgehead of medieval European culture, a role it retained for 1,000 years, with only brief interruptions. The present church, of 1224, is the third on the site; it contains fragments of its predecessors. In 1472 the king took over the monastery and undertook extensive renovations. The present cloister and other buildings with a religious function were built, and the monastery was fortified. However, monastic life became difficult; the monastery was badly damaged by fire and largely abandoned in 1575, to be occupied by the Turks in 1594. The Benedictine community returned in 1638, and the Baroque elements of the monastery, such as the refectory, were added. Today the buildings still house a school and the monastic community.

World Heritage site since

1978 1979 1980 1981 1982 1983 1984 1985 1986 1987 1988 1989 1990 1991 1992 1993 1994 1995 **1996** 1997 1998 1999 2000 2001 2002 2003 2004 2005 2006 2007 2008

Canal du Midi
France

Criteria – Human creative genius; Interchange of values; Significance in human history; Heritage associated with events of universal significance

This 360-km network of navigable waterways linking the Mediterranean and the Atlantic through 328 structures – locks, aqueducts, bridges, tunnels, etc. – is one of the most remarkable feats of civil engineering of modern times. Built between 1667 and 1694, it paved the way for the Industrial Revolution. It was designed by Pierre-Paul Riquet, who was conscious that he was creating a symbol of the power of seventeenth-century France, as well as

a functional communication waterway. He made sure, therefore, that the quality of the architecture on the Canal was worthy of this role. The bridges, locks and associated structures were designed with monumental dignity and simplicity. He was also conscious of the impact of his work on the landscape, and took great pains to ensure that it harmonized with the landscape through which it passed.

One of the canal's most noteworthy features is the Saint-Ferréol dam on the Laudot River in the Montagne-Noire region. This is the largest project on the entire canal and the greatest work of civil engineering of its time.

Canal du Midi, Toulouse.

▼

World Heritage site since

1978 1979 1980 1981 1982 1983 1984 1985 1986 1987 1988 1989 1990 1991 1992 1993 1994 1995 **1996** 1997 1998 1999 2000 2001 2002 2003 2004 2005 2006 2007 2008

Mount Emei Scenic Area, including Leshan Giant Buddha Scenic Area
China

Criteria – Significance in human history; Heritage associated with events of universal significance; Significant natural habitat for biodiversity

The first Buddhist temple in China was built here in Sichuan Province in the first century AD in very beautiful surroundings atop Mount Emei. The addition of other temples turned the site into one of Buddhism's main holy places. Over the centuries, the cultural treasures grew in number. The most remarkable was the Giant Buddha of Leshan (pictured right), carved out of a hillside in the eighth century and looking down on the confluence of three rivers. At 71 m high, it is the largest Buddha in the world.

As one of the four holy lands of Chinese Buddhism, this site is of great historical importance. It's also a vital sanctuary for a many internationally threatened animal species, including lesser (red) panda, Asiatic black bear, and Asiatic golden cat.

Mount Emei is also notable for its very diverse vegetation, ranging from subtropical to sub-alpine pine forests. Some of the trees are more than 1,000 years old.

Upper Svaneti
Georgia

Criteria – Significance in human history; Traditional human settlement

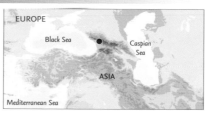

Preserved by its long isolation, the Upper Svaneti region of the Caucasus has a characteristic landscape of small villages dominated by their church towers, set in a natural environment of gorges, alpine valleys and with a backdrop of snow-covered mountains. The unique tower-houses of the area served as dwellings, storehouses and defence posts against the invaders who plagued the region. The village of Chazhashi still has more than 200 of these very unusual buildings. The excellent natural conditions and the unity of architecture and landscape give this region an original quality of its own. The wealth of monumental and minor art (metal work, manuscript illustrations, textiles and embroidery, wood-carving, icon painting, vernacular architecture) is of great importance in the study of Georgia and the Caucasus.

The abundance of towers is the most notable feature of these settlements, especially in the frontier villages. Usually between three and five storeys high with walls that decrease in thickness as they rise, the towers have a slender, tapering profile.

World Heritage site since

1978 1979 1980 1981 1982 1983 1984 1985 1986 1987 1988 1989 1990 1991 1992 1993 1994 1995 **1996** 1997 1998 1999 2000 2001 2002 2003 2004 2005 2006 2007 2008

Historic Centre of Oporto
Portugal

Criteria – Significance in human history

Oporto supported the expeditions of Henry the Navigator, the explorer prince born in the town in the fifteenth century.

The beautiful city of Oporto, built along the hillsides overlooking the mouth of the Douro River, is an exceptional urban landscape with a 2,000-year history.

The historic centre is of high aesthetic value, with evidence of urban development from the Roman, medieval and Almadas periods. The rich and varied civil architecture of the historic centre expresses the cultural values of succeeding periods –

Romanesque, Gothic, Renaissance, Baroque, neoclassical and modern.

The Romans named the town Portus, or port, and its continuous growth, linked to the sea, can be seen in the many and varied monuments, from the cathedral with its Romanesque choir, to the neoclassical Stock Exchange and the typically Portuguese Manueline-style Church of Santa Clara.

In the eighteenth century English entrepreneurs invested in the vineyards of the Douro valley to supply port to the huge English market. Oporto benefited greatly as the shipping port for these wines, as the town's wealth of Baroque buildings demonstrates.

World Heritage site since

1978 1979 1980 1981 1982 1983 1984 1985 1986 1987 1988 1989 1990 1991 1992 1993 1994 1995 **1996** 1997 1998 1999 2000 2001 2002 2003 2004 2005 2006 2007 2008

Belize Barrier Reef Reserve System
Belize

Criteria – Natural phenomena or beauty;
Significant ecological and biological processes;
Significant natural habitat for biodiversity

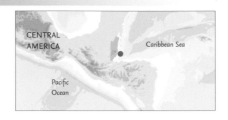

The coastal area of Belize is an outstanding natural system consisting of the largest barrier reef in the northern hemisphere, offshore atolls, several hundred sand cays, mangrove forests, coastal lagoons and estuaries. The system's seven sites illustrate the evolutionary history of reef development and are a significant habitat for threatened species, including marine turtles, manatees and the American marine crocodile.

The approximately 450 sand and mangrove cays confined within the barrier and atolls range in size from small, ephemeral sand spits to larger, permanent islands capable of sustaining human settlements.

A total of 178 terrestrial plants and 247 species of marine flora has been described from the area. There are over 500 species of fish, 65 scleritian corals, 45 hydroids and 350 molluscs in the area, plus a great diversity of sponges, marine worms and crustaceans.

The Belize submarine shelf and its barrier reef represent the world's second-largest reef system and the largest reef complex in the Atlantic-Caribbean area. Outside the barrier are three large atolls: Turneffe Islands, Lighthouse Reef and Glover's Reef. Between the mainland and the barrier reef is an extensive offshore lagoon which increases in width and depth from north to south.

World Heritage site since

1978 1979 1980 1981 1982 1983 1984 1985 1986 1987 1988 1989 1990 1991 1992 1993 1994 1995 **1996** 1997 1998 1999 2000 2001 2002 2003 2004 2005 2006 2007 2008

Bauhaus and its Sites in Weimar and Dessau
Germany

Criteria – Interchange of values; Significance in human history; Heritage associated with events of universal significance

Between 1919 and 1933 the Bauhaus School, based first in Weimar and then in Dessau, revolutionized architectural and aesthetic concepts and practices. The buildings constructed and decorated by the school's professors Walter Gropius, Hannes Meyer, Laszlo Moholy-Nagy and Wassily Kandinsky launched the Modern Movement, which shaped much of the architecture of the twentieth century. The Weimar Bauhaus was obliged to close in 1925 for political reasons. Gropius found support for his cultural and political stance in Dessau, along with the opportunity to create a number of large-scale new buildings. These were situated on the outskirts of the town, and comprise the Bauhaus itself and the Masters' Houses (Meisterhäuser), which served as the residences of the Bauhaus directors and some of its distinguished teachers.

The Haus am Horn in Weimar was built to a design by Georg Muche in 1923 as a model building and exhibit, the first practical statement of the New Building Style. Annexes (a gatehouse, more rooms, a verandah, and a terrace) were added in 1925; however, the original appearance is unchanged. It is the only original Bauhaus building remaining in Weimar.

Lushan National Park
China

Criteria – Interchange of values; Testimony to cultural tradition; Significance in human history; Heritage associated with events of universal significance

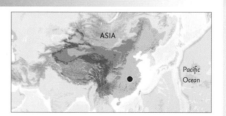

Mount Lushan, in Jiangxi, is one of the spiritual centres of Chinese civilization. Buddhist and Taoist temples, along with landmarks of Confucianism, where the most eminent masters taught, blend effortlessly into a strikingly beautiful landscape. It is an area that has inspired philosophy and art, and into which high-quality cultural properties have been selectively and sensitively integrated up to recent times.

The mountains of Lushan have been the inspiration for some of the finest Chinese classical poetry.

Some 200 historic buildings are scattered over Lushan National Park. The most celebrated is the East Grove Temple complex at the foot of Xianglu Peak. Begun in AD 386, this ensemble was added to progressively over the centuries.

World Heritage site since

1978 1979 1980 1981 1982 1983 1984 1985 1986 1987 1988 1989 1990 1991 1992 1993 1994 1995 **1996** 1997 1998 1999 2000 2001 2002 2003 2004 2005 2006 2007 2008

La Lonja de la Seda de Valencia
Spain
Criteria – Human creative genius; Significance in human history

La Lonja de la Seda de Valencia is an exceptional example of a secular building in late-Gothic style, which dramatically illustrates the power and wealth of one of the great Mediterranean mercantile cities. Built between 1482 and 1533, this group of buildings was originally used for trading in silk – hence its name, the Silk Exchange – and it has always been a centre for commerce. The land occupied by the Lonja is rectangular

in plan. About half of the total area is covered by the main Sala de Contratación; the Tower (including the Chapel), the Consulado, and the large garden complete the ensemble. At the present time, it is still a major trading exchange, now dealing primarily in agricultural products.

The Sala de Contratación is a magnificent hall. The lofty interior is divided into three main aisles by five rows of slender spiral pillars from which spring the elegant vaulting of the roof. It is lit by soaring Gothic windows, the external frames of which, like the doors, are exuberantly ornamented, notably by a series of grotesque gargoyles.

W National Park of Niger
Niger
Criteria – Significant ecological and biological processes; Significant natural habitat for biodiversity

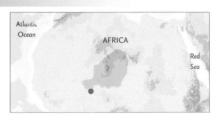

The 'W' National Park, named after the local configuration of the Niger River, is located in a transition zone between savanna and forest landscapes. W hosts important ecosystems that represent the interaction between natural resources and humans since Neolithic times. This interaction has produced characteristic landscapes and plant formations, and a rich biodiversity. The park is known for its large mammals, including aardvarks, baboons, buffalo,

caracal, cheetahs, elephants, hippopotamuses, leopards, lions, serval and warthogs. The wetland area of the park is of international importance for the conservation of birds. A total of 454 plant species has been recorded, including two orchid species found only in Niger. More than seventy mammal and 350 bird species are found in the area.

Many of the park's bird species need large areas for their seasonal migrations. The fact that the park is contiguous to other protected areas in Burkina and Niger makes this area increasingly important to the survival of these species.

World Heritage site since

1978 1979 1980 1981 1982 1983 1984 1985 1986 1987 1988 1989 1990 1991 1992 1993 1994 1995 **1996** 1997 1998 1999 2000 2001 2002 2003 2004 2005 2006 2007 2008

Itsukushima Shinto Shrine
Japan

Criteria – Human creative genius; Interchange of values; Significance in human history; Heritage associated with events of universal significance

The island of Itsukushima, in the Seto inland sea, has been a holy place of Shintoism since the earliest times. The first shrine buildings here were probably erected in the sixth century. The present shrine dates from the twelfth century. The design of the buildings plays on the contrasts in colour and form between mountains and sea, and illustrates the Japanese concept of scenic beauty, which combines nature and human creativity. The architectural style of the north-facing Honsha buildings and the west-facing buildings of the Sessha Marodo-jinja, connected by the *kairo* (roofed corridor), was influenced by the aristocratic dwelling-house style of the Heian period. The frontal view of the buildings, with the mountain as a backdrop, is emphasized, and the entire area resembles a succession of folding screens.

Like many other Shinto shrines that included Buddhist buildings, Itsukushima-jinja lost many of them after the rejection of Buddhism with the Meiji Restoration of 1868. The few that survive in the surrounding hills are considered to be as indispensable to the history of Itsukushima-jinja as its Shinto monuments.

Verla Groundwood and Board Mill
Finland

Criteria – Significance in human history

The Verla Groundwood and Board Mill and its associated residential area is an outstanding, remarkably well-preserved example of the small-scale rural industrial settlements associated with pulp, paper and board production that flourished in northern Europe and North America in the nineteenth and early twentieth centuries. The Industrial Revolution that reached the Kymi river valley in the first half of the 1870s was one of the most dramatic phenomena in the economic history of Finland. Over a very short time dozens of steam sawmills, groundwood mills and board mills were established. The Kymi valley benefited in particular from the construction of timber-floating facilities, a dedicated railway, and the introduction of cooperative floating, enabling logs from the virgin forests of central Finland to be brought to the processing facilities.

Output at Verla gradually diminished throughout the twentieth century, until it was closed down on 18 July 1964, when the last of the old workers retired. The owners decided to preserve the entire complex intact as an industrial heritage museum, just as it had been when the last worker left.

World Heritage site since

1978 1979 1980 1981 1982 1983 1984 1985 1986 1987 1988 1989 1990 1991 1992 1993 1994 1995 **1996** 1997 1998 1999 2000 2001 2002 2003 2004 2005 2006 2007 2008

Early Christian Monuments of Ravenna
Italy

Criteria – Human creative genius; Interchange of values; Testimony to cultural tradition; Significance in human history

Ravenna was the seat of the Roman Empire in the fifth century and then of Byzantine Italy until the eighth century. It has a unique collection of early Christian mosaics and monuments. All eight buildings – the Mausoleum of Galla Placidia, the Neonian Baptistery, the Basilica of Sant'Apollinare Nuovo, the Arian Baptistery, the Archiepiscopal Chapel, the Mausoleum of

Theodoric, the Church of San Vitale and the Basilica of Sant'Apollinare in Classe – were constructed in the fifth and sixth centuries. They show great artistic skill, including a wonderful blend of Graeco-Roman tradition, Christian iconography and oriental and Western styles.

Several of the buildings are unique: for example, the Neonian Baptistery is the finest and most complete example of an early-Christian baptistery; while the Mausoleum of Theodoric is the only surviving tomb of a barbarian king of this period.

◄
The mausoleum of Galla Placidia.

World Heritage site since

1978 1979 1980 1981 1982 1983 1984 1985 1986 1987 1988 1989 1990 1991 1992 1993 1994 1995 **1996** 1997 1998 1999 2000 2001 2002 2003 2004 2005 2006 2007 2008

Defence Line of Amsterdam
Netherlands

Criteria – Interchange of values; Significance in human history; Traditional human settlement

North Sea

EUROPE

Extending 135 km around the city of Amsterdam, this defence line built between 1883 and 1920 is a unique fortification using water. Since the sixteenth century, the people of the Netherlands have used their expert knowledge of hydraulic engineering for defence purposes. Earlier defensive lines were strengthened and co-ordinated into the Stelling, a system based on the intricate polder system of the western part of the

Netherlands. The centre of the country was protected by a network of forty-five armed forts, acting in concert with temporary flooding from polders and an intricate system of canals and locks. The sites of the forts are directly linked with the existing infrastructure of roads, waterways, dykes and settlements and the main defence line runs mainly along pre-existing dykes.

From time immemorial, dykes, sluices and canals have been built in the Netherlands to drain the land. The Stelling van Amsterdam is of outstanding universal value because it elevated and integrated this activity into an extensive defence system which survives intact to this day.

World Heritage site since

1978 1979 1980 1981 1982 1983 1984 1985 1986 1987 1988 1989 1990 1991 1992 1993 1994 1995 **1996** 1997 1998 1999 2000 2001 2002 2003 2004 2005 2006 2007 2008

Pre-Hispanic Town of Uxmal
Mexico

Criteria – Human creative genius; Interchange of values; Testimony to cultural tradition

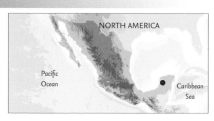

The ruins of the ceremonial structures at Uxmal represent the pinnacle of late-Mayan art and architecture in their design, layout and ornamentation. And the complex of Uxmal and its three related towns of Kabáh, Labná and Sayil admirably demonstrate the social and economic structure of late-Mayan society. Uxmal, in Yucatán, was founded c. AD 700 and had some 25,000 inhabitants. The layout of the buildings, which date from between 700 and 1000, reveals a detailed knowledge of astronomy. The Pirámide del Adivino, as the Spaniards called it – or Pyramid of the Soothsayer – dominates the ceremonial centre, which has well-designed buildings decorated with a profusion of symbolic motifs and sculptures depicting Chaac, the god of rain.

By virtue of its size, the Pirámide del Adivino dominates the ensemble. It is made up of five superimposed elements, two of them reached by monumental stairways on either side of the structure. It is from the Late Classic period and brings together several artistic traditions, including that of the Toltec of Central Mexico.

▼

World Heritage site since

1978 1979 1980 1981 1982 1983 1984 1985 1986 1987 1988 1989 1990 1991 1992 1993 1994 1995 **1996** 1997 1998 1999 2000 2001 2002 2003 2004 2005 2006 2007 2008

Volcanoes of Kamchatka
Russian Federation

Criteria – Natural phenomena or beauty; Major stages of Earth's history; Significant ecological and biological processes; Significant natural habitat for biodiversity

An active crater on Mutnovsky Volcano, which is a complex of four superimposed stratovolcanoes.

The site, in the Russian Far East, comprises six distinct locations. One site is inland: Bystrinsky Nature Park, in the central mountainous spine of the Kamchatka Peninsula. The others, in coastal locations facing east towards the Bering Sea, are Koronotsky Zapovednik; Nalychevo Nature Park; and the contiguous Southern Kamchatka Nature Park and Southern Kamchatka State Nature Reserve. Kluchevskoy Nature Park was added in 2001 as the sixth component of the site.

This is one of the most outstanding volcanic regions in the world, with a high density of active volcanoes, a variety of types and a wide range of related features. The six designated sites group together the majority of volcanic features of the Kamchatka peninsula. The interplay of active volcanoes and glaciers forms a dynamic landscape of great beauty, and the sites contain great species diversity, including the world's largest-known variety of salmonid fish, and exceptional concentrations of sea otter, brown bear and Stellar's sea eagle.

The property represents the most pristine parts of the Kamchatka Peninsula and a remarkable collection of volcanic areas, characteristic of the Pacific Volcanic Ring. This is the surface expression of the subduction of the Pacific Ocean Continental Plate under the Eurasia Plate at a rate of 10 cm annually. There are more than 300 volcanoes in Kamchatka, twenty-nine of which are currently active, including caldera, strata-volcano, somma-volcano and mixed types. The largest in the site is Kronotskaya Sopka (3,528 m). In addition, there is a multitude of thermal and mineral springs, geysers and other phenomena of active volcanism.

Surrounded by sea, the peninsula enjoys a moist and relatively mild climate, which has lead to a lush vegetation cover. The vegetation includes mountain valley taiga forest of birch, larch and spruce; extensive stone-birch forest; riparian forest on alluvial soil of poplars, aspen, alder and willow; peat wetland and extensive coastal wetlands up to 50 km wide; and subalpine shrub and mountain tundra. The region also contains an especially diverse range of Palaearctic flora (including a number of nationally threatened species and at least sixteen endemic species).

The faunal complement is relatively low in diversity, with the Kamchatka Peninsula exhibiting some of the biogeographic qualities of an island. Nevertheless, some species are abundant, including bears, snow ram, northern deer, sable and wolverine, and there is a high level of endemism.

Noteworthy birds include Stellar's sea eagle (the area has 50 per cent of world population), white-tailed eagle, gyrfalcon and peregrine falcon. There are numerous seabird colonies and a concentration of Aleutian tern nests.

Almost all rivers serve as spawning grounds for salmon, a key food-chain species for predatory birds and mammals. All eleven species of salmonid fish coexist in several of Kamchatka's rivers.

World Heritage site since

1978 1979 1980 1981 1982 1983 1984 1985 1986 1987 1988 1989 1990 1991 1992 1993 1994 1995 **1996** 1997 1998 1999 2000 2001 2002 2003 2004 2005 2006 2007 2008

Skellig Michael
Ireland

Criteria – Testimony to cultural tradition;
Significance in human history

This monastic complex, perched since about the seventh century on the steep sides of the rocky island of Skellig Michael, some 12 km off the coast of southwest Ireland, illustrates the extremes of early Christian monasticism. Since the great remoteness of Skellig Michael has, until recently, discouraged visitors, the site is exceptionally well preserved. The principal monastic remains are situated on a sloping shelf on the ridge running north-south on the

northeastern side of the island; the hermitage is on the steeper South Peak. The main monastic enclosure comprises a church, oratories, cells, a souterrain, and many crosses and cross-slabs. It was occupied continuously until the later twelfth century, when a general climatic deterioration led to increased storms in the seas around the island and forced the community to move to the mainland.

There is a tradition that the monastery was founded by St Fionan in the seventh century. It was dedicated to St Michael somewhere between 950 and 1050 and this date fits in well with the architectural style of the oldest part of the existing church, known as St Michael's Church.

World Heritage site since

1978 1979 1980 1981 1982 1983 1984 1985 1986 1987 1988 1989 1990 1991 1992 1993 1994 1995 **1996** 1997 1998 1999 2000 2001 2002 2003 2004 2005 2006 2007 2008

Lednice-Valtice Cultural Landscape
Czech Republic

Criteria – Human creative genius; Interchange of values; Significance in human history

Between the seventeenth and twentieth centuries, the ruling dukes of Liechtenstein transformed their domains in southern Moravia into a striking landscape that married Baroque architecture with countryside fashioned according to English romantic principles of landscape architecture. The realization of this grandiose design began in the seventeenth century with the creation of avenues connecting Valtice with other parts of the estate, and continued throughout the eighteenth century, imposing order on

nature. The early years of the nineteenth century saw the application of the English concept of the designed park. The Chateau of Valtice has medieval foundations, but it underwent successive reconstructions in Renaissance, Mannerist and, most significantly, Baroque style. By contrast, the Lednice Chateau (see photo below) is not widely visible, having begun as a Renaissance villa of around 1570, and then progressively changed and reconstructed to take account of Baroque, Classical and Neo-Gothic fashions.

An important element in the appearance of the area is the very wide range of native and exotic tree species and the planting strategy adopted. The greatest variety is to be found in the parklands which cluster around the two main residences and along the banks of the fishponds between Lednice and Valtice.

World Heritage site since

1978 1979 1980 1981 1982 1983 1984 1985 1986 1987 1988 1989 1990 1991 1992 1993 1994 1995 **1996** 1997 1998 1999 2000 2001 2002 2003 2004 2005 2006 2007 2008

Historic City of Meknes
Morocco

Criteria – Significance in human history

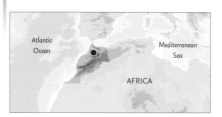

The Historic City of Meknes provides an exceptionally well-preserved example of the urban fabric and monumental buildings of a seventeenth-century Maghreb capital city. Meknes was founded in the eleventh century by the Almoravids as a military settlement. Later, the founder of the Alawite dynasty, Moulay Ismail (1672–1727), made Meknes his capital city and carried out many reconstructions and additions, such as mosques, mausolea and gardens. But his main contribution was the creation of a new imperial city, built in the Hispano-Moorish style. It is impressive in both extent and construction, enclosed by high walls pierced by monumental gates, where the harmonious blending of the Islamic and European styles of the seventeenth-century Maghreb are still evident today. Within the walls are: the palace, with its enormous stables; a military academy; vast granaries; and water storage cisterns.

Some of the fondouks (inns) that cluster around the gates were devoted to specific crafts or trades: for example, the Fondouk Hanna dealt solely in henna, while the Jewish craftsmen worked at the Fondouk Lihoudi. Certain quarters were reserved for specific trades and activities.

A traditional Moroccan palace in Meknes. ▶

World Heritage site since

1978 1979 1980 1981 1982 1983 1984 1985 1986 1987 1988 1989 1990 1991 1992 1993 1994 1995 **1996** 1997 1998 1999 2000 2001 2002 2003 2004 2005 2006 2007 2008

Historic Centre of the City of Pienza
Italy

Criteria – *Human creative genius; Interchange of values; Significance in human history*

It was in this Tuscan town that Renaissance town-planning concepts were first put into practice after Pope Pius II decided, in 1459, to transform the look of his birthplace. He chose the architect Bernardo Rossellino, who applied the principles of his mentor, Leon Battista Alberti. This new vision of urban space was realized in the superb square known as Piazza Pio II and the buildings around it: the Piccolomini Palace,

the Borgia Palace and the cathedral, with its pure Renaissance exterior and an interior in the late-Gothic style of south German churches. Pius II's project also required the building of large houses for the cardinals in his retinue, and work on these began in 1463. Two structures with a social function, the hospital and the inn in front of the church of St Francis, were built on his orders.

The leading humanist Enea Silvio Piccolomini, elected Pope in 1458, was born in Corsignano, situated on a hill southeast of Siena. When he returned there after becoming Pope, he was struck by the extreme misery of its inhabitants, which inspired him to endow his birthplace with new buildings and a new name, and to make it his summer court.

World Heritage site since

1978 1979 1980 1981 1982 1983 1984 1985 1986 1987 1988 1989 1990 1991 1992 1993 1994 1995 **1996** 1997 1998 1999 2000 2001 2002 2003 2004 2005 2006 2007 2008

Historic Monuments Zone of Querétaro
Mexico
Criteria – Interchange of values; Significance in human history

The old colonial town of Querétaro is unusual in having retained the geometric street plan of the Spanish conquerors side by side with the twisting alleys of the Indian quarters. The Otomi, the Tarasco, the Chichimeca and the Spanish lived together peacefully in the town, which is notable for the many ornate civil and religious Baroque monuments from its golden age in the seventeenth and eighteenth centuries. Its first chapel

(La Cruz) was built on a small hillock at the eastern end of the valley. The Plaza de Armas, the seat of government, was arcaded on two sides, and surrounded by government buildings and residences of leading citizens. Many monastic orders established themselves and left behind outstanding Baroque buildings. The many non-religious buildings, again mostly Baroque, are enhanced by the pink stone of Querétaro.

Querétaro was the site of historic events: the peace treaty with the United States was concluded here in 1848, and in 1867 Emperor Maximilian was imprisoned and later executed after the defeat of his army. The National Constitution was signed on 5 February 1917 by all the revolutionary groups after two months of debate in the Teatro de la República.

World Heritage site since

1978 1979 1980 1981 1982 1983 1984 1985 1986 1987 1988 1989 1990 1991 1992 1993 1994 1995 **1996** 1997 1998 1999 2000 2001 2002 2003 2004 2005 2006 2007 2008

Laponian Area
Sweden

Criteria – Testimony to cultural tradition; Traditional human settlement; Natural phenomena or beauty; Major stages of Earth's history; Significant ecological and biological processes

This area, which lies close to the Arctic Circle in northern Sweden, has been occupied continuously by the Saami people since prehistoric times. It is the largest area in the world with an ancestral way of life based on the seasonal movement of livestock (transhumance), the Saami spending the summer in the mountains and the winters in the coniferous forests to the east. There are no permanent settlements occupied throughout the year anywhere in this area. There are two landscape types: an eastern taiga area and a western mountainous landscape, with steep valleys and powerful rivers. Birch, low heath and alpine meadows are found below boulder fields, permanent snowfields and glaciers. The nomadic life based on herding of tame reindeer did not develop until the seventeenth and eighteenth centuries.

Researchers, working on large mammal predators and the white-tailed eagle, indicated that all populations seem to be healthy, with the exception of the wolverine. Also, there are more than 150 bird species and 100 bears, including wanderers, in the area.

A traditional Saami teepee.
▼

World Heritage site since

1978 1979 1980 1981 1982 1983 1984 1985 1986 1987 1988 1989 1990 1991 1992 1993 1994 1995 **1996** 1997 1998 1999 2000 2001 2002 2003 2004 2005 2006 2007 2008

Historic Centre of the City of Salzburg
Austria

Criteria – Interchange of values; Significance in human history; Heritage associated with events of universal significance

Salzburg is of outstanding value as an important example of a European ecclesiastical city-state. The city preserves to a remarkable degree its dramatic townscape, its historically significant urban fabric, and a large number of outstanding ecclesiastical and secular buildings developed from the Middle Ages through to the nineteenth century, when it was a city-state ruled by prince-archbishops. Its flamboyant Gothic art attracted many craftsmen and artists before the city became even better known through the work of the Italian architects Vincenzo Scamozzi and Santini Solari, to whom the historic centre owes much of its Baroque appearance. This meeting point of northern and southern Europe perhaps sparked the genius of Salzburg's most famous son, Wolfgang Amadeus Mozart, whose name has been associated with the city ever since.

Salzburg is rich in buildings from the Gothic period onwards, which combine to create a townscape of great individuality and beauty. The cathedral (St Rupert and St Virgil) is the pre-eminent ecclesiastical building and the spiritual city centre. The present structure dates from 1628 and is the work of Santini Solari, the court master builder. It replaced the former

cathedral, designed by Palladio's pupil Vincenzo Scamozzi, which burnt down in 1598. Solari's cathedral preserves many of Scamozzi's features. Mozart was christened in the cathedral in 1756 at one day old.

The Benedictine Abbey of St Peter, founded in the closing years of the seventh century, contains in its church the only High Romanesque structures in Salzburg, mostly dating from the early twelfth century. The main body of the church has undergone many modifications since then.

The Nonnberg Benedictine Nunnery is the oldest convent north of the Alps, founded around the same time as the Abbey of St Peter. The present massive complex, on the eastern peak of the Mönchberg, is a striking feature of the townscape, with its dominating church roof and Baroque dome.

The Hohensalzburg Fortress, on the steep rock fan overlooking the city, has been continually rebuilt and enlarged since its first foundation as a Roman fort through to the late seventeenth century, when it reached its present extent.

The Archbishop's Residence, begun in the early twelfth century, lies in the heart of the old town. The present layout dates to the major rebuilding carried out in the early seventeenth century.

There is a clear separati visible on the ground ar on the map, between th lands of the prince-archbishops and those of the burghers – former is characterized monumental buildings open spaces, and the lat by small plots fronting c narrow streets, with the only open spaces provid by the three historic markets.

The city skyline is characterized by its profusion of spires and domes, and is dominated by the fortress of Hohensalzburg.

Salzburg cathedral.

World Heritage site since

1978 1979 1980 1981 1982 1983 1984 1985 1986 1987 1988 1989 1990 1991 1992 1993 1994 1995 **1996** 1997 1998 1999 2000 2001 2002 2003 2004 2005 2006 2007 2008

Monasteries of Haghpat and Sanahin
Armenia

Criteria – Interchange of values; Significance in human history

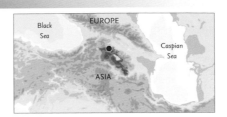

These two Byzantine monasteries represent the highest flowering of Armenian religious architecture, which blends elements of Byzantine ecclesiastical architecture with vernacular architecture of the Caucasian region. The two monastic complexes were important centres of learning. Sanahin was renowned for its school of illuminators and calligraphers. It consists of a large group of buildings on the plateau above the Debet gorge, laid out on two rectangular axes, with their façades facing west. The main church, built in the tenth century, is the Cathedral of the Redeemer. Construction of the main church of the large fortified monastic complex of Haghpat, dedicated to the Holy Cross, began in 966–7 and was completed in 991. The building is complete, apart from some eleventh- and twelfth-century restorations. The monastery also has a gavit, through which access is gained to the church, chapter house and library.

Armenia became independent at the end of the ninth century and Armenian art was revived when the kingdom was consolidated. The two monasteries of Haghpat and Sanahin date from this period, during the prosperity of the Kiurikian dynasty and the Zakarian Princes. They were important centres of learning, housing some 500 monks.

Bell tower of Haghpat Monastery. ▼

World Heritage site since

1978 1979 1980 1981 1982 1983 1984 1985 1986 1987 1988 1989 1990 1991 1992 1993 1994 1995 **1996** 1997 1998 1999 2000 2001 2002 2003 2004 2005 2006 2007 2008

The Trulli of Alberobello
Italy

Criteria – Testimony to cultural tradition;
Significance in human history; Traditional human
settlement

EUROPE

Mediterranean Sea

Ionian
Sea

Alberobello, the city of drystone dwellings
known as trulli, is an exceptional example of
vernacular architecture. Trulli, mostly dating
from before the end of the eighteenth
century, were constructed using roughly
worked limestone boulders and were built
directly on the underlying natural rock.
The walls that form the rectangular rooms
are pierced by small windows. Fireplaces,
ovens and alcoves are recessed into the
thickness of the walls. The stone roofs, which
are circular or oval, are not painted and
develop a patina of mosses and lichens; they
sometimes bear mythological or religious
symbols in white ash. By contrast, the walls
of the trulli must be whitewashed at regular
intervals. The Monti quarter contains
1,030 trulli. Its streets run downhill and
converge at the base of the hill. The Aja
Piccola quarter, with 590 trulli, is less
homogeneous than Monti.

Tradition has it that drystone
walling was imposed upon villagers
so their houses could be quickly
dismantled, either to dispossess
recalcitrant householders or to
avoid house tax (with the houses
being quickly reconstructed once
the tax collector departed).

World Heritage site since

1978 1979 1980 1981 1982 1983 1984 1985 1986 1987 1988 1989 1990 1991 1992 1993 1994 1995 **1996** 1997 1998 1999 2000 2001 2002 2003 2004 2005 2006 2007 2008

Palace and Gardens of Schönbrunn
Austria

Criteria – Human creative genius; Significance in human history

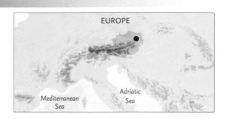

Schönbrunn is a remarkable example of a Baroque royal palace that vividly illustrates the tastes, interests and aspirations of successive Habsburg monarchs. Apart from some minor nineteenth-century additions, the palace and its gardens were built in the eighteenth century. Schönbrunn was designed by the architects Johann Bernhard Fischer von Erlach and Nicolaus Pacassi and is full of outstanding examples of decorative art.

The vast Baroque gardens and their buildings testify to the imperial dimensions and functions of the palace; the courtyard provides access to the Palace Chapel and the Palace Theatre. The orangery on the east side of the main palace building was used by the Empress Maria Theresa to cultivate exotic plants, and the Schönbrunn zoo, founded in 1752 by her husband, is the oldest in the world.

The impressive Great Gallery is elaborately decorated with stucco ornamentation and ceiling frescoes symbolizing the Habsburg Empire, while the Ceremonial Hall is notable for its series of monumental paintings depicting events in the long reign of the Empress Maria Theresa.

World Heritage site since

1978 1979 1980 1981 1982 1983 1984 1985 1986 1987 1988 1989 1990 1991 1992 1993 1994 1995 **1996** 1997 1998 1999 2000 2001 2002 2003 2004 2005 2006 2007 2008

Luther Memorials in Eisleben and Wittenberg
Germany

Criteria – Significance in human history;
Heritage associated with events of universal
significance

These places in Saxony-Anhalt are all
associated with the lives of Martin Luther
and his fellow-reformer Melanchthon.
They include Melanchthon's house in
Wittenberg, the houses in Eisleben where
Luther was born in 1483 and died in 1546,
his room in Wittenberg, the local church
and the castle church where, on 31 October
1517, Luther posted his famous '95 Theses',
thereby launching the Reformation and a
new era in the religious and political history
of the Western world. Because of its
symbolic importance, the famous bronze
door on the north side of the church, where
the Latin text of the 95 Theses is displayed,
is only used on special occasions. The church
houses the tombs of Luther and Melanchthon.

In 1525 Luther broke
with his monastic
vows and married the
former nun, Katharina
von Bora. His
household in
Wittenberg became
the centre for
reformists from all
over Europe, and the
family room that he
created on the first
floor was the setting
for his 'table talks',
which were later
published.

Sangiran Early Man Site
Indonesia

Criteria – Testimony to cultural tradition;
Heritage associated with events of universal
significance

Sangiran is a key site for the understanding
of human evolution. Excavations from 1936
to 1941 led to the discovery of the first hominid
fossil at this site. Later, fifty early human
fossils (Pithecanthropus erectus/Homo erectus)
were found – half of all the world's known
hominid fossils – together with numerous
animal and floral fossils such as rhinoceros,
elephant ivory, buffalo horn, deer horn and
many others. Palaeolithic stone tools found
at Ngebung include flakes, choppers and
cleavers in chalcedony and jasper and, more
recently, bone tools. The site has also produced
Neolithic axes. This evidence indicates that
hominids have inhabited the area for at least
1.5 million years. Nowadays, the region is
entirely devoted to peasant agriculture.

Ever since von
Koenigswald found
flake tools in the
Ngebung village in
1934, the site has
made an immense
contribution to the
study of evolution
over the past million
years by illustrating
the evolution of Homo
erectus. Homo erectus is
important to the study
of the early history of
mankind before the
emergence of the
modern Homo sapiens.

◄
Display of fossil
elephant ivory at
Sangiran.

World Heritage site since

1978 1979 1980 1981 1982 1983 1984 1985 1986 1987 1988 1989 1990 1991 1992 1993 1994 1995 **1996** 1997 1998 1999 2000 2001 2002 2003 2004 2005 2006 2007 2008

Archaeological Site of Aigai (modern name Vergina)
Greece

Criteria – *Human creative genius; Testimony to cultural tradition*

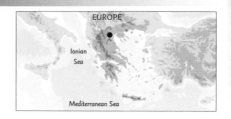

As capital and site of the royal court, Aigai was the most important urban centre in the region from 800 to 400 BC. Philip II was assassinated in the theatre here in 336 BC and Alexander the Great proclaimed king.

The city of Aigai, the ancient first capital of the Kingdom of Macedonia, was discovered in the nineteenth century near Vergina, in northern Greece. The most important building so far discovered is the monumental palace, located on a plateau directly below the acropolis. Sumptuously decorated, its large gallery commanded a view of the whole Macedonian plain. The site also has a necropolis which extends for over 3 km and contains over 300 grave-mounds, some as early as the eleventh century BC. One of the royal tombs in the Great Tumulus contained a solid gold casket with remains identified as those of Philip II. He conquered all the Greek cities, paving the way for his son Alexander and the expansion of the Hellenistic world.

Church Village of Gammelstad, Luleå
Sweden

Criteria – *Interchange of values; Significance in human history; Traditional human settlement*

Luleå Gammelstad illustrates the adaptation of conventional urban design to the special geographical and climatic conditions of a hostile natural environment. The settlement has been shaped by people's religious and social needs rather than economic and geographical forces.

Gammelstad, ('Old Town') at the head of the Gulf of Bothnia, is the best-preserved example of a 'church village', a unique kind of village formerly found throughout northern Scandinavia. The 424 wooden houses, huddled round the early fifteenth-century stone church, were used only on Sundays and at religious festivals to house worshippers from the surrounding countryside who could not return home the same day because of the distance and difficult travelling conditions. The doors, which face the street, are very varied in design, as are the window shutters. Most of the doors bear a pyramid device, a motif from pagan antiquity reinterpreted as a Christian symbol depicting an altar with a sacrificial fire. Gammelstad's church is the largest of its type in northern Scandinavia.

World Heritage site since

1978 1979 1980 1981 1982 1983 1984 1985 1986 1987 1988 1989 1990 1991 1992 1993 1994 1995 **1996** 1997 1998 1999 2000 2001 2002 2003 2004 2005 2006 2007 2008

Historic Walled Town of Cuenca
Spain

Criteria – Interchange of values; Traditional human settlement

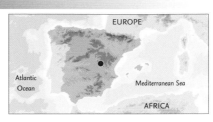

Built by the Moors in a defensive position at the heart of the Caliphate of Cordoba, Cuenca is an unusually well-preserved medieval fortified town. Conquered by the Castilians in the twelfth century, it became a royal town and bishopric. In the upper town some remains of the Moorish fortress still survive among the large aristocratic houses, monasteries, and churches from the medieval, Renaissance and Baroque periods. The twelfth-century cathedral, built

on the site of the former Great Mosque, was the first Gothic cathedral in Spain. Most of its churches and monastic buildings were founded early in the town's history and underwent many additions over the following centuries. The importance of the upper town lies, however, not so much in its individual buildings, as in the townscape that they create, on the fortified site dominating the river valleys.

The private houses near the Episcopal Palace were built in the later medieval period on the spectacular steep bluffs overlooking the bend of the Huécar River. These famous 'casas colgadas' (hanging houses) were rebuilt in the sixteenth century in their present narrow, high form, with two or three rooms on each of three or more floors.

▼

World Heritage site since

1978 1979 1980 1981 1982 1983 1984 1985 1986 1987 1988 1989 1990 1991 1992 1993 1994 1995 **1996** 1997 1998 1999 2000 2001 2002 2003 2004 2005 2006 2007 2008

Castel del Monte
Italy

Criteria – Human creative genius; Interchange of values; Testimony to cultural tradition

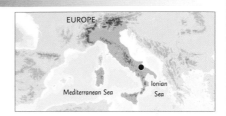

EUROPE

Mediterranean Sea

Ionian Sea

The site offers a fittingly impressive monument to its originator, Frederick II. A polyglot, mathematician, astronomer and scientist, he also founded the University of Naples and brought social and economic stability to his people.

When the Emperor Frederick II built this castle near Bari in the thirteenth century, he imbued it with symbolic significance, as reflected in its location and the mathematical and astronomical precision of its design. A unique masterpiece of medieval military architecture, Castel del Monte is a successful blend of elements from classical antiquity, the Islamic Orient and north European Cistercian Gothic. It is sited on a rocky peak that dominates the surrounding countryside and consists of a regular octagon surrounding a courtyard, with a tower, also octagonal, at each angle. The castle is of special interest because of the absence of features that are common to the overwhelming majority of military monuments of this period such as outer bailey, moat, stables, kitchen, storerooms and chapel.

World Heritage site since

1978 1979 1980 1981 1982 1983 1984 1985 1986 1987 1988 1989 1990 1991 1992 1993 1994 1995 1996 **1997** 1998 1999 2000 2001 2002 2003 2004 2005 2006 2007 2008

Macquarie Island
Australia

Criteria – Natural phenomena or beauty; Major stages of Earth's history

OCEANIA

Indian Ocean

Pacific Ocean

Tasman Sea

The breeding population of royal penguins is estimated at over 850,000 pairs – one of the greatest concentrations of sea birds in the world. The penguins share the island with elephant seals and four species of albatross. The surrounding Nature Reserve and Macquarie Island Marine Park contain one of the world's largest marine highly-protected zones, covering more than 160,000 km².

Macquarie Island is an oceanic island in the Southern Ocean, lying 1,500 km southeast of Tasmania and approximately halfway between Australia and the Antarctic continent. The island, approximately 34-km long and 5.5-km wide, is the exposed crest of the undersea Macquarie Ridge, raised to its present position where the Indo-Australian tectonic plate meets the Pacific plate. It is a site of major geoconservation significance, being the only island in the world composed entirely of oceanic crust and rocks from the Earth's mantle 6 km below the ocean floor that are being actively exposed above sea-level. These unique exposures include excellent examples of pillow basalts and other extrusive rocks. Macquarie Island's beauty lies in its remote and windswept landscape of steep escarpments, lakes, dramatic changes in vegetation, and the vast congregations of wildlife around its dark, dramatic shores.

World Heritage site since

1978 1979 1980 1981 1982 1983 1984 1985 1986 1987 1988 1989 1990 1991 1992 1993 1994 1995 1996 **1997** 1998 1999 2000 2001 2002 2003 2004 2005 2006 2007 2008

Maritime Greenwich
United Kingdom

Criteria – Human creative genius; Interchange of values; Significance in human history; Heritage associated with events of universal significance

The ensemble of buildings at Greenwich, an outlying district of London, and the park in which they are set, symbolize English artistic and scientific endeavour in the seventeenth and eighteenth centuries. The focus of the ensemble is the Queen's House, the work of Inigo Jones, and the first true Renaissance building in Britain, and a striking departure from the architectural forms that preceded it. It was inspired by Italian style, and it was in its turn the direct inspiration for classical houses all over Britain. The Queen's House and its associated buildings have housed the National Maritime Museum since 1937. The complex that was until recently the Royal Naval College was designed by Christopher Wren and is the most outstanding group of Baroque buildings in Britain. Greenwich Royal Park contains the Old Royal Observatory, the work of Wren and the scientist Robert Hooke.

The Old Royal Observatory is situated on the brow of Greenwich Hill and dominates the landscape. It houses an octagonal room which was used by the Royal Society for meetings and dinners, and this is surmounted by the famous time-ball, which indicates Greenwich Mean Time daily at 13.00.

▼

World Heritage site since

1978 1979 1980 1981 1982 1983 1984 1985 1986 1987 1988 1989 1990 1991 1992 1993 1994 1995 1996 **1997** 1998 1999 2000 2001 2002 2003 2004 2005 2006 2007 2008

Mount Kenya National Park/ Natural Forest
Kenya

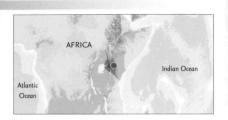

Criteria – Natural phenomena or beauty; Significant ecological and biological processes

At 5,199 m, Mount Kenya is the second-highest peak in Africa. It is an ancient extinct volcano, and during its period of activity, around three million years ago, it is thought to have risen to 6,500 m. There are twelve remnant glaciers on the mountain, all receding rapidly, and four secondary peaks that sit at the head of the U-shaped glacial valleys. With its rugged glacier-clad summits and forested middle slopes, Mount Kenya is one of the most impressive landscapes in East Africa. The evolution of its afro-alpine flora also provide an outstanding example of ecological processes. Mount Kenya is regarded as a holy mountain by all the communities (Kikuyu and Meru) living adjacent to it. They believe that their traditional God (Ngai) and his wife (Mumbi) live on the peak of the mountain and use it for their traditional rituals.

In the lower forest and bamboo zone, mammals include giant forest hog, white-tailed mongoose, elephant, black rhinoceros, and leopard (which have also been seen in the alpine zone). There have also been reported sightings of the golden cat.

Pyrénées – Mont Perdu
France and Spain

Criteria – Testimony to cultural tradition; Significance in human history; Traditional human settlement; Natural phenomena or beauty; Major stages of Earth's history

This outstanding mountain landscape, which spans the contemporary national borders of France and Spain, is centred around the peak of Mount Perdu, a calcareous massif that rises to 3,352 m. The site, with a total area of 306 km², includes two of Europe's largest and deepest canyons on the Spanish side and three major cirque walls on the more abrupt northern slopes within France, classic presentations of these geological landforms. The site is also a pastoral landscape reflecting an agricultural way of life that was once widespread in the upland regions of Europe but now survives only in this part of the Pyrenees. Thus it provides exceptional insights into past European society through its landscape of villages, farms, fields, upland pastures and mountain roads.

The location of the Pyrenees between two seas, their geological structure and climatic asymmetries result in a rich mosaic of vegetation types. The site supports many valuable wildlife species including the marmot and the Spanish ibex, of which there are only three female individuals.

World Heritage site since

1978 1979 1980 1981 1982 1983 1984 1985 1986 1987 1988 1989 1990 1991 1992 1993 1994 1995 1996 **1997** 1998 1999 2000 2001 2002 2003 2004 2005 2006 2007 2008

Historic Centre (Old Town) of Tallinn
Estonia

Criteria – Interchange of values; Significance in human history

Tallinn is an exceptionally well-preserved example of a northern European medieval trading city. The origins of Tallinn date back to the thirteenth century, when a castle was built there by the crusading knights of the Teutonic Order. It developed into a major centre of the Hanseatic League, and its wealth is demonstrated by the opulence of the public buildings and the domestic architecture of the merchants' houses.

The lower town preserves to a remarkable extent the medieval urban fabric of narrow winding streets and fine public and burgher buildings. There are several medieval churches within the city walls. The restored Church of St Nicholas (Niguliste) and the Church of St Olaf (Oleviste) are both in typical basilical form, with lofty vaulting and a precise geometry of form, in what is recognized as the distinctive Tallinn School.

The most prominent feature of the town is the Toompea limestone hill. The western part is occupied by the castle, of which the tower known as Long Hermann, two bastions and the imposing walls survive on the western, northern and eastern sides.

The tower of St Olaf's church overlooks the Old Town.
▼

World Heritage site since

1978 1979 1980 1981 1982 1983 1984 1985 1986 1987 1988 1989 1990 1991 1992 1993 1994 1995 1996 **1997** 1998 1999 2000 2001 2002 2003 2004 2005 2006 2007 2008

Old Town of Lijiang
China

Criteria – Interchange of values; Significance in
human history; Traditional human settlement

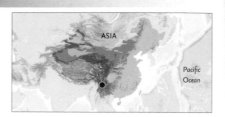

The Lijiang Junmin
prefecture was
established in 1382.
Of the original
286 m-long complex
that was built to house
the administration,
only the Yizi Pavilion,
the Guagbi Tower and
a stone archway
remain.

The group known as
the Yuquan
architectural
structures in the
Heilongtan Park (Jade
Spring Park) dates
from the Ming and
Qing dynasties that
ruled from the
fourteenth century
onwards. Most
notable is the Wufeng
Tower (1601), moved
from the Fugue
Temple of which it
formed part; it is now
one of the major
historical sites in
Yunnan Province.

The Old Town of Lijiang has retained a
historic townscape of high quality and
authenticity. Its architecture is noteworthy
for the blending of elements from several
cultures over many centuries. Lijiang, which
is perfectly adapted to its uneven
topography, also possesses an ancient
water-supply system of great complexity
and ingenuity that still functions effectively
today.

The old town is built on a mountain slope
running from northwest to southeast, facing
a deep river. The northern part was a
commercial district and the main streets
there radiate from the broad thoroughfare
known as Sifangjie, traditionally the
commercial and trading centre of northwest
Yunnan Province. On the west side of the
Sifangjie is the imposing three-storeyed
Kegongfang (Imperial Examination
Archway), flanked by the Western and
Central rivers.

A sluice on the Western River uses the
different levels of the two waterways to
wash the streets, which are paved with slabs
of a fine-grained red breccia. In this unique
form of municipal sanitation, water flows on
to the Shuangshi Bridge where it branches
into three tributaries which subdivide into a
network of channels and culverts to supply
every house in the town. Many springs and

wells in the town supplement the supply.

Such a complex system of watercourses
requires a large number of bridges and the
city has 354 altogether, of several types. It is
from these that Lijiang derives its name, the
'City of Bridges'.

The feature that most represents the
culture of the Naxi people who settled in the
area is the wealth of domestic dwellings in
the city. The basic timber-framed structure
developed into a unique architectural style
with the absorption of elements of Han and
Zang architecture. Most houses are two-
storeyed. The chuandoushi wooden frames
are walled with adobe on the ground floor
and planks on the upper floors; the walls
have stone foundation courses. Exteriors are
plastered and lime-washed, and there are
often brick panels at the corners. Roofs are
tiled and the houses have verandas.

Decoration of the houses is important,
with arches over gateways, screen walls,
external corridors, doors and windows,
courtyards and roof beams. Wooden
elements are elaborately carved with
domestic and cultural imagery – pottery,
musical instruments, flowers, birds, etc. –
and gate arches take several elegant forms.

World Heritage site since

1978 1979 1980 1981 1982 1983 1984 1985 1986 1987 1988 1989 1990 1991 1992 1993 1994 1995 1996 **1997** 1998 1999 2000 2001 2002 2003 2004 2005 2006 2007 2008

Lumbini, the Birthplace of the Lord Buddha
Nepal

Criteria – Testimony to cultural tradition; Heritage associated with events of universal significance

Siddhartha Gautama, the Lord Buddha, was born in 623 BC in the famous gardens of Lumbini, which soon became a place of pilgrimage. Among the pilgrims was the Indian emperor Ashoka, who erected one of his commemorative pillars there. The site is now being developed as a Buddhist pilgrimage centre, where the archaeological remains associated with the birth of the Lord Buddha form a central feature.

Lumbini is situated at the foothills of the Himalaya in modern Nepal. In the Buddha's time, Lumbini was a beautiful garden full of green and shady sal trees, and the site still retains its legendary charm and serenity.

The Sundarbans
Bangladesh

Criteria – Significant ecological and biological processes; Significant natural habitat for biodiversity

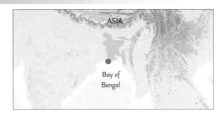

The Sundarbans mangrove forest, one of the largest such forests in the world at 1,400 km², lies on the delta of the Ganges, Brahmaputra and Meghna rivers on the Bay of Bengal. It is adjacent to the border of India's Sundarbans World Heritage site inscribed in 1987. The site is intersected by a complex network of tidal waterways, mudflats and small islands of salt-tolerant mangrove forests, and presents an excellent example of ongoing ecological processes. The area is known for its wide range of fauna, including 260 bird species, the

Bengal tiger and other threatened species such as the estuarine crocodile and the Indian python.

The Sundarbans is a uniquely dynamic landscape, shaped in turn by monsoon rains, flooding, delta formation, and tidal influence. The area also supports one of the largest populations of Royal Bengal Tiger with 350 individuals.

World Heritage site since

1978 1979 1980 1981 1982 1983 1984 1985 1986 1987 1988 1989 1990 1991 1992 1993 1994 1995 1996 **1997** 1998 1999 2000 2001 2002 2003 2004 2005 2006 2007 2008

Portovenere, Cinque Terre, and the Islands (Palmaria, Tino and Tinetto)
Italy

Criteria – Interchange of values; Significance in human history; Traditional human settlement

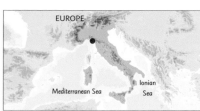

The eastern Ligurian coast between Cinque Terre and Portovenere is a site of outstanding scenic and cultural value. The area covers some 15 km of jagged, steep coastline, which the work of man over the centuries has transformed into an intensively terraced landscape, so as to be able to wrest from nature a few hectares of land suitable for agriculture, such as growing vines and olive trees. Most of these cultivation terraces were built in the twelfth century. The human communities have adapted themselves to this inhospitable terrain by building compact settlements directly on the rock, with winding streets. The use of natural stone for footing gives these settlements their characteristic appearance. They are generally grouped round religious buildings or medieval castles.

The five villages of Cinque Terre date back to the later Middle Ages. Starting from the north, the first is the fortified centre of Monterosso al Mare; then come Vernazza, Corniglia, Manarola and **Riomaggiore** (pictured below). Off the coast at Portovenere are the three islands of Palmaria, Tino and Tinetto, noteworthy for the many remains of early monastic establishments that they contain.

World Heritage site since

1978 1979 1980 1981 1982 1983 1984 1985 1986 1987 1988 1989 1990 1991 1992 1993 1994 1995 1996 **1997** 1998 1999 2000 2001 2002 2003 2004 2005 2006 2007 2008

Historic Fortified City of Carcassonne
France

Criteria – Interchange of values; Significance in human history

The historic city of Carcassonne is an outstanding example of a medieval fortified town, with its massive defences encircling the castle and its fine Gothic cathedral. A fortified settlement has existed on the hill where Carcassonne now stands since pre-Roman times. During the turbulent years of the late third and early fourth centuries, the town was protected by the construction of a defensive wall some 1,200 m long. The twelfth-century count's castle was built over the western part of the Roman wall, and by the end of the thirteenth century the town had assumed its definitive appearance as a medieval fortress. Carcassonne is also of special importance because of the lengthy restoration campaign undertaken during the second half of the nineteenth century by Viollet-le-Duc, one of the founders of the modern science of conservation.

The exterior of the cathedral, like that of most southern French Gothic churches, has no flying buttresses, stability being assured by means of the interior vaulting. It contains some important sculpture, notably the thirteenth-century tomb of Bishop Radulph. The stained glass in the windows of the apse and the transept is of exceptionally high quality.

World Heritage site since

1978 1979 1980 1981 1982 1983 1984 1985 1986 1987 1988 1989 1990 1991 1992 1993 1994 1995 1996 **1997** 1998 1999 2000 2001 2002 2003 2004 2005 2006 2007 2008

Morne Trois Pitons National Park
Dominica

Criteria – Major stages of Earth's history;
Significant natural habitat for biodiversity

Luxuriant tropical forest blends with scenic volcanic features of great scientific interest in this national park, centred on the 1,342 m-high volcano known as Morne Trois Pitons. The landscape is characterized by volcanic piles with precipitous slopes and deeply incised valleys. The so-called Valley of Desolation (Grand Soufriere) contains fumaroles, hot springs, mud pots, sulphur vents and the Boiling Lake, which is the world's second largest of its kind. The valley is a large amphitheatre surrounded by mountains and consisting of at least three separate craters, where steam vents, small ponds and hot springs bubble up through the ground. Boiling Lake is surrounded by cliffs and is almost always covered by clouds of steam. The Morne Trois Pitons National Park covers nearly 70 km² and has the richest biodiversity of the Lesser Antilles.

Other outstanding features in the area include the Emerald Pool, fed by the Middleham Falls Stinking Hole, a lava tube in the middle of the forest; and the Boeri and Freshwater lakes. Freshwater Lake is the largest of Dominica's four freshwater lakes. Boeri Lake is located in the crater of an extinct volcano.

Middleham Falls which feeds the Emerald Pool.

▼

World Heritage site since

1978 1979 1980 1981 1982 1983 1984 1985 1986 1987 1988 1989 1990 1991 1992 1993 1994 1995 1996 **1997** 1998 1999 2000 2001 2002 2003 2004 2005 2006 2007 2008

Cocos Island National Park
Costa Rica

Criteria – Significant ecological and biological processes; Significant natural habitat for biodiversity

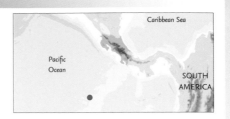

Cocos Island National Park, located 550 km off the Pacific coast of Costa Rica, is the only island in the tropical eastern Pacific with a tropical rainforest. Its position as the first point of contact with the northern equatorial counter-current, and the myriad interactions between the island and the surrounding marine ecosystem, make the area an ideal laboratory for the study of biological processes. The underwater world of the National Park has become famous due to the attraction it holds for divers, who rate it as one of the best places in the world to view large pelagic species such as sharks, rays, tuna and dolphins. Like other oceanic islands, Isla del Coco presents an impoverished flora compared to that of the continent, but has a high number of endemic species (at least seventy species of vascular plant). The vegetation is exuberant and owes its lushness to the heavy rainfalls and rugged relief, which favours condensation.

The island has been known to mariners and cartographers since the first half of the sixteenth century. However, its position was vaguely indicated and therefore could only be located by experienced sailors. Fishermen, pirates, commercial sailors and scientific expeditions arrived at the island searching for fresh water and shelter.

Episcopal Complex of the Euphrasian Basilica in the Historic Centre of Poreč
Croatia

Criteria – Interchange of values; testimony to cultural tradition; significance in human history

The group of religious monuments in Poreč, where Christianity was established as early as the fourth century, constitutes the most complete surviving complex of its type. The basilica, atrium, baptistery and episcopal palace are outstanding examples of religious architecture, while the basilica itself combines classical and Byzantine elements. Erected by Bishop Euphrasia, all these buildings were richly ornamented with mosaics, alabaster, marble, mother-of-pearl and stucco, in the lavish tradition of the Byzantine 'Golden Age' during the reign of Justinian. Later additions to the complex were the Kanonika (Canon's House) of 1257, the sixteenth-century bell tower, and some minor buildings such as the sacristy (fifteenth century) and two chapels (seventeenth and nineteenth centuries respectively).

The Episcopal Complex is an integral part of the historic centre of Poreč, which has preserved to a considerable extent its Roman street pattern, dating from the time when the town was part of the province of Histria.

World Heritage site since

1978 1979 1980 1981 1982 1983 1984 1985 1986 1987 1988 1989 1990 1991 1992 1993 1994 1995 1996 **1997** 1998 1999 2000 2001 2002 2003 2004 2005 2006 2007 2008

Lake Turkana National Parks
Kenya

Criteria – Major stages of Earth's history;
Significant natural habitat for biodiversity

The area around Lake Turkana is mostly semi-desert, with open plains flanked by volcanic formations including Mount Sibiloi, the site of the remains of a petrified forest possibly seven million years old. The most saline of Africa's large lakes, Turkana is an outstanding laboratory for the study of plant and animal communities. The three National Parks are major breeding grounds for the Nile crocodile, hippopotamus and a variety of venomous snakes. Mammals include Burchell's and Grevy's zebras, Grant's gazelle, Beisa oryx, hartebeest, topi, lesser kudu, lion and cheetah. More than 350 species of aquatic and terrestrial bird have been recorded in Lake Turkana, and it serves as a stopover for migrant birds such as warblers, wagtails and little stints.

Extensive palaeontological finds in the parks have contributed more to the understanding of paleo-environments than any other site on the continent. These include evidence of the existence of a relatively intelligent hominid two million years ago. Human fossils include the remains of *Australopithecus robustus, Homo habilis, Homo erectus* and *Homo sapiens.*

World Heritage site since

1978 1979 1980 1981 1982 1983 1984 1985 1986 1987 1988 1989 1990 1991 1992 1993 1994 1995 1996 **1997** 1998 1999 2000 2001 2002 2003 2004 2005 2006 2007 2008

Classical Gardens of Suzhou
China

Criteria – Human creative genius; Interchange of values; Testimony to cultural tradition; Significance in human history; Traditional human settlement

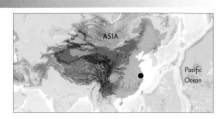

ASIA

Pacific Ocean

Chinese gardens have certain elements in common governing their positioning, layout, scenery, planting, contents and philosophy. The end result is one in which art, nature, and ideas are integrated perfectly to create ensembles of great beauty and harmony.

Located on the lower Yangtze River, Suzhou is spanned by numerous waterways and its streets retain their traditional beauty. Marco Polo called the city 'Venice of the Orient' when he visited in the thirteenth century. The city's many gardens were created by retired bureaucrats and politicians of the Ming and Qing Dynasties.

Classical Chinese garden design, which seeks to recreate natural landscapes in miniature, is nowhere better illustrated than in the nine gardens in the historic city of Suzhou. They are generally acknowledged to be masterpieces of the genre. Dating from the eleventh to the nineteenth centuries, the gardens reflect in their meticulous design the profound metaphysical importance of natural beauty in Chinese culture.

The Canglang Pavilion was built in the early eleventh century. It is reached across a zigzag stone bridge from which the mountains, covered with old trees and bamboo, suddenly become visible.

The Lion Forest Garden was created in 1342 as the Budhi Orthodox Monastery but was detached from the temple in the seventeenth century. It features a series of man-made mountains and an artificial waterfall on steep cliffs.

The sixteenth-century Garden of Cultivation is typical, both in layout and in the design of its buildings, of the classical Ming dynasty garden. A pond takes up a quarter of its area.

The Couple's Garden Retreat dates from the eighteenth century. The East Garden is dominated by a mountain rising from a pool flanked by buildings. The more subdued

West Garden has limestone hills pierced by interlinking caves.

In the Retreat and Reflection Garden, the central feature is the pool, surrounded by elegant buildings and the double-tiered Celestial Bridge.

The Mountain Villa with Embracing Beauty dates from the sixteenth century. It is intensively detailed, with peaks rising to 7 m, dells, paths, caves, stone houses, ravines, precipices and cliff.

The Humble Administrator's Garden has been the site of the residence of Suzhou notables since the second century AD and is one of China's most famous gardens. Its central section is a re-creation of the scenery of the Lower Yangtze.

The Lingering Garden dates from the sixteenth century. It features mountain and lake scenery encircled by buildings and visited by means of a narrow, winding path which gives unexpected views of great beauty.

The Garden of the Master of the Nets is entered from the south through a gateway flanked by enormous carved blocks of stone, designating the owner's court rank. The layout of buildings and gardens is extremely subtle, so that a small area gives the impression of great size and variety.

World Heritage site since

1978 1979 1980 1981 1982 1983 1984 1985 1986 1987 1988 1989 1990 1991 1992 1993 1994 1995 1996 **1997** 1998 1999 2000 2001 2002 2003 2004 2005 2006 2007 2008

Cathedral, Torre Civica and Piazza Grande, Modena
Italy

Criteria – Human creative genius; Interchange of values; Testimony to cultural tradition; Significance in human history

The magnificent twelfth-century cathedral at Modena, the work of two great artists Lanfranco and Wiligelmo, is a supreme example of early Romanesque art. With its piazza and soaring bell tower, the Modena complex exemplifies an architectural complex where religious and civic values are combined. The cathedral was also a large sculpture workshop, brilliantly illustrated by Wiligelmo, particularly in the façade, which is a veritable corpus of the sculptor's work. The Maestri Campionesi – architects and sculptors commissioned to maintain the building from the second half of the twelfth century onwards – made various alterations and improvements to the building. Only minor changes have been made to the Piazza Grande: its quadrangular shape has been preserved and it has been lined on its northern side by the flank of the cathedral.

▲

The Torre Civica, whose tall silhouette is a landmark to travellers approaching the town, is closely linked to the cathedral by two arches. This monumental tower, built from the same materials as the cathedral, consists of six floors emphasized by small blind arcades lit by simple openings, and then by two- and three-light windows on the upper floors.

World Heritage site since

1978 1979 1980 1981 1982 1983 1984 1985 1986 1987 1988 1989 1990 1991 1992 1993 1994 1995 1996 **1997** 1998 1999 2000 2001 2002 2003 2004 2005 2006 2007 2008

Changdeokgung Palace Complex
Korea, Republic of

Criteria – Interchange of values; Testimony to cultural tradition; Significance in human history

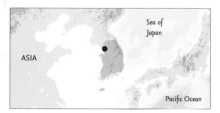

In the early fifteenth century, the Emperor T'aejong ordered the construction of a new palace, which he named Changdeokgung (Palace of Illustrious Virtue). A Bureau of Palace Construction was set up to create the complex in accordance with traditional design principles. These included the palace in front, the market behind, three gates and three courts (administrative court, royal residence court and official audience court). The compound was divided into two parts: the main palace buildings and the Piwon (royal secret garden). The result is an exceptional example of Far Eastern palace architecture and design, in which the buildings are integrated into and harmonized with the natural setting. Changdeokgung Palace had a great influence on the development of Korean architecture, and garden and landscape planning for many centuries.

The garden was landscaped with a series of terraces planted with lawns, flowering trees, flowers, a lotus pool, and pavilions set against a wooded background. There are over 26,000 specimens of 100 indigenous trees in the garden, along with 23,000 planted specimens of 15 imported species.

World Heritage site since

1978 1979 1980 1981 1982 1983 1984 1985 1986 1987 1988 1989 1990 1991 1992 1993 1994 1995 1996 **1997** 1998 1999 2000 2001 2002 2003 2004 2005 2006 2007 2008

Medina of Tétouan (formerly known as Titawin)
Morocco

Criteria – Interchange of values; Significance in human history; Traditional human settlement

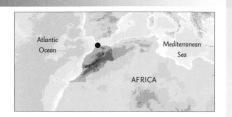

Tétouan was of particular importance in the Islamic period, from the eighth century onwards, since it served as the main point of contact between Morocco and Andalusia. After the Reconquest, the town was rebuilt by Andalusian refugees who had been expelled by the Spanish. This is well illustrated by its art and architecture, which reveal clear Andalusian influence. Although one of the smallest of the Moroccan medinas, Tétouan is unquestionably the most complete and it has been largely untouched by subsequent outside influences.

Tétouan is mentioned by a number of Arab writers of the tenth–twelfth centuries. However it wasn't until the fortifications were rebuilt in the mid-eighteenth century that the medina assumed its current distinctive 'figure-of-eight' configuration.

◀ Main Square, Tétouan.

Heard and McDonald Islands
Australia

Criteria – Major stages of Earth's history; Significant ecological and biological processes

Heard and McDonald Islands are located in the Southern Ocean, approximately 1,700 km from the Antarctic continent and over 4,100 km southwest of Perth. As the only volcanically active subantarctic islands they 'open a window into the Earth', thus providing the opportunity to observe ongoing geomorphic processes. Permanent snow and ice cover about 70 per cent of

Heard Island, and its relatively fast-flowing glaciers respond quickly to changes in climate. The distinctive conservation value of Heard and McDonald Islands – one of the world's rare pristine island ecosystems – lies in the absence of introduced plants and animals, making them especially valuable for scientific reference purposes.

Lying in a remote and stormy part of the globe, Heard and McDonald Islands were unknown to humanity until the nineteenth century. Driving westerly winds create unique weather patterns including spectacular cloud formations and rapid changes in precipitation.

World Heritage site since

1978 1979 1980 1981 1982 1983 1984 1985 1986 1987 1988 1989 1990 1991 1992 1993 1994 1995 1996 **1997** 1998 1999 2000 2001 2002 2003 2004 2005 2006 2007 2008

Mill Network at Kinderdijk-Elshout
Netherlands

Criteria – Human creative genius; Interchange of values; Significance in human history

North
Sea

EUROPE

The Kinderdijk-Elshout mill network bears powerful testimony to the outstanding contribution made by the people of the Netherlands to the technology of handling water. The site illustrates all the typical features associated with this technology – dykes, reservoirs, pumping stations, administrative buildings and a series of beautifully preserved windmills. At one time, there were more than 150 such mills in the Alblasserwaard and Vijfheerenlanden area; this had dropped to seventy-eight in the 1870s, but today the total is only twenty-eight. The World Heritage Site contains nineteen of these mills, mostly dating from the mid-eighteenth century. Although they went out of use in the late 1940s, all nineteen are still maintained in operating condition, functioning as fall-back mills in case of failure of the modern equipment.

Most of the mills are so-called bonnet mills, in which only the top section revolves with the wind. Built from brick or wood, they have large sails that come within 30 cm of the ground, hence their name, 'ground sailers'.

▼

World Heritage site since

1978 1979 1980 1981 1982 1983 1984 1985 1986 1987 1988 1989 1990 1991 1992 1993 1994 1995 1996 **1997** 1998 1999 2000 2001 2002 2003 2004 2005 2006 2007 2008

Archaeological Areas of Pompei, Herculaneum and Torre Annunziata
Italy

Criteria – Testimony to cultural tradition; Significance in human history; Traditional human settlement

EUROPE

Mediterranean Sea

Ionian Sea

The remains of the towns of Pompei and Herculaneum, buried by the eruption of Vesuvius on 24 August AD 79, provide a complete and vivid picture of society and daily life at a specific moment in the past that is without parallel anywhere in the world.

When Vesuvius erupted it engulfed the two flourishing Roman towns and many wealthy villas in the area. Since the mid eighteenth century these have been progressively excavated and made accessible to the public. The vast expanse of the commercial town of Pompei contrasts with the smaller but better-preserved remains of the holiday resort of Herculaneum, while the superb wall paintings of the Villa Oplontis at Torre Annunziata give a vivid impression of the opulent lifestyle enjoyed by the wealthier citizens of the Early Roman Empire.

The main forum of Pompei is flanked by the foundations of several imposing public buildings including the Capitolium (temple), Basilica (courthouse) and public baths. Pompei is renowned for its domestic buildings, ranged along well-paved streets. The earliest type is the atrium house of which the House of the Surgeon, entirely inward looking with a courtyard at its centre, is a good example. The exceptional Villa dei

Misteri (House of the Mysteries) takes its name from the wall paintings in the triclinium (dining room) that depict the initiation rites ('mysteries') of the cult of Dionysus.

A special characteristic of Pompei is its wealth of graffiti. An election was imminent at the time of the eruption and many slogans were scrawled on walls along with other graffiti of a more personal, often scurrilous, nature.

Much less of Herculaneum, built on a promontory overlooking the Bay of Naples, has been uncovered, partly because of the depth to which it was buried. However, the nature of its covering is such that the buildings are better preserved than those of Pompei.

There are several impressive public buildings, and the houses are also remarkable for their extent and decoration. Those fronting on the sea, such as the House of the Deer, have large courtyards and rich decoration. The town is noteworthy for the completeness of its shops, still containing fittings such as enormous wine jars. Of great importance in both towns are the artistic styles represented by sculptures, mosaics and, above all, wall paintings.

Part of Pompei.

Pompei was founded by the southern Italian Osci (Opicians) in the sixth century BC, while Herculaneum was said to have been founded by Hercules. They fell to the Romans in 89 BC.

Both towns came to an abrupt and catastrophic end on 24 August AD 79. The area had recently been shaken by an earthquake and reconstruction work was still in progress when Vesuvius erupted with tremendous violence. Pompei was buried under a thick layer of volcanic ash and stone and Herculaneum disappeared under a pyroclastic flow of volcanic mud.

World Heritage site since

1978 1979 1980 1981 1982 1983 1984 1985 1986 1987 1988 1989 1990 1991 1992 1993 1994 1995 1996 **1997** 1998 1999 2000 2001 2002 2003 2004 2005 2006 2007 2008

Medieval Town of Toruń
Poland

Criteria – Interchange of values; Significance
in human history

Toruń is a small historic trading city that
preserves to a remarkable extent its original
street pattern and outstanding early
buildings. The town owes its origins to the
Teutonic Order, which built a castle there
in the mid-thirteenth century as a base for
the conquest and colonization of Prussia.
It soon developed a commercial role as part
of the Hanseatic League. In the Old and
New Towns, the many imposing public and

private buildings from the fourteenth and
fifteenth centuries, among them the house
of Copernicus, are striking evidence of
Toruń's importance. The Old Town, which
forms the western part of the complex, is
laid out around its central Market Place. The
New Town developed from 1264, to the
north of the castle and the east of the Old
Town, into a centre for crafts and industry.

The Old Town was
fortified progressively
between 1250 and
1300 with a double
wall strengthened by
bastions. These
fortifications were
reconstructed in
1420–49 and partly
dismantled in the
nineteenth century,
but most of the
southern sector, with
gates and towers
facing the river,
survives intact.

Toruń Town Hall.
▼

World Heritage site since

1978 1979 1980 1981 1982 1983 1984 1985 1986 1987 1988 1989 1990 1991 1992 1993 1994 1995 1996 **1997** 1998 1999 2000 2001 2002 2003 2004 2005 2006 2007 2008

Eighteenth-Century Royal Palace at Caserta with the Park, the Aqueduct of Vanvitelli, and the San Leucio Complex
Italy

Criteria – Human creative genius; Interchange of values; Testimony to cultural tradition; Significance in human history

The monumental complex at Caserta, while cast in the same mould as other eighteenth-century royal palaces, is exceptional for the broad sweep of its design. The King of Naples decided in 1750 to build a new royal palace to rival the Palace of Versailles. He employed Luigi Vanvitelli, then engaged in the restoration of St Peter's in Rome. The Royal Palace is rectangular in plan, with four large interior courtyards and contains 1,200 rooms and thirty-four staircases. The main axis of the park is punctuated by a series of Baroque fountains and stretches of water. This magnificent perspective terminates in the Great Fountain, where water cascades down from a height of 150 m. The works at San Leucio, designed to produce silk, are also of outstanding interest because of the idealistic principles underlying its original conception and management.

A waterfall, which is a copy of one at Versailles, in the grounds of the Palace of Caserta. ▲

In 1778 the king decided to begin the production of silk at San Leucio, and the industrial complex included a school, accommodation for teachers, silkworm rooms, and facilities for spinning and dyeing the silk. His regulations of 1789 laid down piecework rates of pay, abolished dowries and prescribed similar clothing for all the workers.

World Heritage site since

1978 1979 1980 1981 1982 1983 1984 1985 1986 1987 1988 1989 1990 1991 1992 1993 1994 1995 1996 **1997** 1998 1999 2000 2001 2002 2003 2004 2005 2006 2007 2008

Palau de la Música Catalana and Hospital de Sant Pau, Barcelona
Spain

Criteria – Human creative genius; Interchange of values; Significance in human history

These two monuments are among the finest contributions to Barcelona's architecture by the Catalan art nouveau architect Lluis Doménech i Montaner. The Palau de la Música Catalana is an exuberant steel-framed structure, full of light and space and decorated by many of the leading designers of the day. Construction began in 1905 and was completed three years later. The Hospital de Sant Pau is equally bold in its design and decoration, while at the same time perfectly adapted to the needs of the sick. Work began in 1901 and was not finally completed until 1930. The hospital is of immense architectural importance because it is the largest hospital complex in Modernist style.

The Hospital de Sant Pau is an outstanding vindication of its creator's maxim that beauty has therapeutic value. For Doménech i Montaner it was essential to give sick people a feeling of well-being and beauty, which would contribute to an early convalescence.

Historic Area of Willemstad, Inner City and Harbour, Netherlands Antilles
Netherlands

Criteria – Interchange of values; Significance in human history; Traditional human settlement

The people of the Netherlands established a trading settlement at a fine natural harbour on the Caribbean island of Curaçao in 1634. The town developed continuously over the following centuries. The modern town consists of several distinct historic districts whose architecture reflects not only European urban-planning concepts but also styles from the Netherlands and from the Spanish and Portuguese colonial towns with which Willemstad engaged in trade.

Willemstad stands out for the diversity of its four historic districts, separated by a natural harbour. Each has its own unique urban morphology, while sharing a distinctive 'tropicalized' Dutch architecture.

World Heritage site since

1978 1979 1980 1981 1982 1983 1984 1985 1986 1987 1988 1989 1990 1991 1992 1993 1994 1995 1996 **1997** 1998 1999 2000 2001 2002 2003 2004 2005 2006 2007 2008

Archaeological Site of Panamá Viejo and Historic District of Panamá

Panama

Criteria – Interchange of values; Significance in human history; Heritage associated with events of universal significance

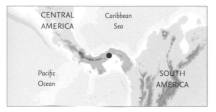

The archaeological site of Panamá Viejo is the site of the oldest European town on the American mainland, founded in 1519 by the conquistador Pedrarías Dávila. It soon became a commercial and administrative centre, as well as an important port and the seat of a Royal Tribunal. Only the climate, being considered unhealthy, prevented the development of the town to the size and importance of Guatemala or Bogotá. The old town was destroyed by fire in 1672, and a new town (the 'Historic District' of Panamá), 8 km to the southwest, replaced it a year later, and the site was abandoned and never rebuilt; it is now a public park where the impressive ruins of the cathedral, churches, water installations, town hall and private houses are preserved.

Cathedral at Panamá Viejo. ▲

The 'Historic District' of Panamá has preserved its original street plan, its architecture and an unusual mixture of Spanish, French and early American styles.

The Salón Bolívar was the venue for the unsuccessful attempt made by Simón Bolívar, El Libertador, in 1826 to establish a Pan-American congress.

World Heritage site since

1978 1979 1980 1981 1982 1983 1984 1985 1986 1987 1988 1989 1990 1991 1992 1993 1994 1995 1996 **1997** 1998 1999 2000 2001 2002 2003 2004 2005 2006 2007 2008

Dougga / Thugga
Tunisia

Criteria – Interchange of values; Testimony to
cultural tradition

Before the Roman annexation of Numidia
in 46 BC, the town of Thugga, built on an
elevated site overlooking a fertile plain, was
the capital of an important Libyco-Punic
state. It flourished under Roman and
Byzantine rule, but declined in the Islamic
period. Its impressive ruins include temples
and sanctuaries, a forum, public baths, a
theatre, an amphitheatre, a circus, a market,
fountains, private houses, shops and
mausoleums. The small rectangular forum,
which is surrounded by a marble colonnade,
is crossed by part of the later Byzantine
fortifications. On one side of it is the
capitolium, dedicated to Jupiter, Juno and
Minerva, and one of the finest buildings of
its type in North Africa. One of the most
significant monuments in Thugga is the
Libyco-Punic mausoleum in the southern
part of the town. This is the only major
monument of Punic architecture still
surviving in Tunisia.

Entrance to the theatre. ▲

The important collection of over
2,000 Libyan, Punic, Greek and
Roman inscriptions from Thugga
has made a decisive contribution
to the deciphering of the Libyan
language; also to knowledge of the
social and municipal life of the
Numidians, and Roman colonial
policy and municipal organization
in its provinces.

World Heritage site since

1978 1979 1980 1981 1982 1983 1984 1985 1986 1987 1988 1989 1990 1991 1992 1993 1994 1995 1996 **1997** 1998 1999 2000 2001 2002 2003 2004 2005 2006 2007 2008

Hallstatt-Dachstein / Salzkammergut Cultural Landscape
Austria

Criteria – Testimony to cultural tradition; Significance in human history

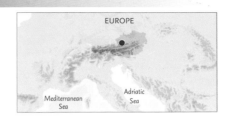

EUROPE

Mediterranean Sea

Adriatic Sea

The typical Hallstatt house is tall and narrow, making maximum use of the restricted space and the steep topography. The lower storeys are constructed in stone, with barrel vaulting supporting timber-framed upper storeys, as is customary in Alpine regions. Only a few preserve the original flat saddleback roofs covered with wooden planks or shingles.

Human activity in the magnificent natural landscape of the Salzkammergut began in prehistoric times, with the salt deposits being exploited as early as the second millennium BC. This resource formed the basis of the area's prosperity up to the middle of the twentieth century, a prosperity that is reflected in the fine architecture of the town of Hallstatt. The name of the medieval town, first recorded in a deed of 1305, is derived from the West German hal (salt) and the Old High German stat (settlement). The site also includes the Dachstein Mountains, rising to some 3,000 m, which form the highest of the karst massifs in the northern limestone Alps. They are notable for the large number of caves they contain, the longest being the Hillatzhöhle (81 km).

World Heritage site since

1978 1979 1980 1981 1982 1983 1984 1985 1986 1987 1988 1989 1990 1991 1992 1993 1994 1995 1996 **1997** 1998 1999 2000 2001 2002 2003 2004 2005 2006 2007 2008

Botanical Garden (Orto Botanico), Padua
Italy

Criteria – Interchange of values; Testimony to cultural tradition

The Botanical Garden of Padua is the original of all botanical gardens throughout the world. It was created in 1545 and still preserves its original layout: a circular central plot, symbolizing the world, surrounded by a ring of water, representing the ocean. Various additions have been made in the intervening centuries – a pumping installation to supply ten fountains in the seventeenth century, four monumental entrances in 1704, and new masonry greenhouses in the late eighteenth and early nineteenth centuries. An arboretum, an English garden with winding paths, and a small hillock (belvedere) were also added around this time. The garden has traditionally collected and grown particularly rare plants, which have then been introduced into the rest of Europe. Currently there are over 6,000 species grown here.

The Botanical Garden also houses two important collections. The library contains more than 50,000 volumes and manuscripts of immense historical and bibliographic importance. The herbarium is the second most extensive in Italy.

Rohtas Fort
Pakistan

Criteria – Interchange of values; Significance in human history

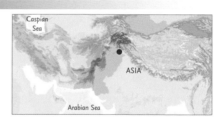

Following his defeat of the Mughal emperor Humayun in 1541, Sher Shah Suri built a strong fortified complex at Rohtas, a strategic site in the north of what is now Pakistan. It was never taken by force and has survived intact to the present day. The main fortifications consist of the massive walls, which extend for more than 4 km; they are lined with bastions and pierced by monumental gateways. Rohtas Fort, also called Qila Rohtas, is an exceptional example of early Muslim military architecture in Central and South Asia.

Rohtas is a complex of defensive works surrounding a small hill alongside the Kahan River. Its stone walls vary according to the terrain, with heights ranging from 10.05 m to 18.28 m and a thickness of up to 12.5 m.

◄
Section of the fortifications of Rohtas Fort.

World Heritage site since

1978 1979 1980 1981 1982 1983 1984 1985 1986 1987 1988 1989 1990 1991 1992 1993 1994 1995 1996 **1997** 1998 1999 2000 2001 2002 2003 2004 2005 2006 2007 2008

Historic City of Trogir
Croatia
Criteria – Interchange of values; Significance in
human history

The plan of
contemporary Trogir
reflects the Hellenistic
layout in the location,
dimensions and
shapes of its
residential blocks.
The two ancient main
streets, the *cardo* and
the *decumanus*, are still
in use, and paving
from the forum has
been located by
excavation at their
intersection.

Trogir is a remarkable example of urban
continuity. The street plan of this island
settlement dates back to the Hellenistic
period and it was embellished by successive
rulers with many fine public and domestic
buildings and fortifications. Its beautiful
Romanesque churches are complemented
by the outstanding Renaissance and
Baroque buildings from the period of
Venetian rule after 1420. The ancient town of
Tragurion (island of goats) was founded as a
trading settlement by Greek colonists in the
third century BC. The town flourished in the
Roman period. Between the thirteenth and
fifteenth centuries much new building took
place, including the cathedral and the
Camerlengo fortress and the reconstruction
of the fortifications. Throughout the town,
and in particular round the ramparts, are the
palaces of leading families. Many of these
rise directly from the foundations of late
Classical or Romanesque buildings.

World Heritage site since

1978 1979 1980 1981 1982 1983 1984 1985 1986 1987 1988 1989 1990 1991 1992 1993 1994 1995 1996 **1997** 1998 1999 2000 2001 2002 2003 2004 2005 2006 2007 2008

Hospicio Cabañas, Guadalajara
Mexico

Criteria – Human creative genius; Interchange
of values; Testimony to cultural tradition;
Significance in human history

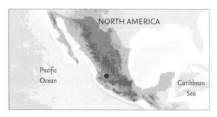

The Hospicio Cabañas was built at the
beginning of the nineteenth century by the
Bishop of Guadalajara, Juan Ruiz de
Cabañas, to provide care and shelter for the
disadvantaged – orphans, old people, the
handicapped and chronic invalids. This
remarkable complex, which incorporates
several unusual features designed
specifically to meet the needs of its
occupants, was created by Manuel Tolsá. The
entire complex is laid out on a rectangular
plan: all the buildings, which are single
storey, except the chapel and the kitchen,
are ranged round twenty-three courtyards.
The great majority of these are arcaded on at
least two sides. In the 1930s, the chapel was
decorated with a superb series of murals,
now considered some of the masterpieces
of Mexican art. They are the work of José
Clemente Orozco, one of the greatest
Mexican muralists of the period.

The growth of the Mexican muralist
movement was a demonstration of
national cohesion and identity. In the
1930s, Orozco was commissioned by
the government to paint the murals
in the chapel of the Hospicio
Cabañas. They represented the multi-
ethnic character of Mexican society
and the allegory of the Man of Fire,
which are among his finest works.

World Heritage site since

1978 1979 1980 1981 1982 1983 1984 1985 1986 1987 1988 1989 1990 1991 1992 1993 1994 1995 1996 **1997** 1998 1999 2000 2001 2002 2003 2004 2005 2006 2007 2008

Las Médulas
Spain

Criteria – Human creative genius; Interchange of values; Testimony to cultural tradition; Significance in human history

Las Médulas gold-mining area is an outstanding example of innovative Roman technology, based on hydraulic power, in which all the elements of the ancient landscape, both industrial and domestic, have survived. The Archaeological Zone of Las Médulas (ZAM) comprises the mines themselves and also large areas where the tailings resulting from the process were deposited. Within the area there are dams used to collect the vast amounts of water needed for the mining process and the intricate canals by means of which the water was conveyed to the mines. Human settlement is represented by villages, of both the indigenous inhabitants and the imperial administrative and support personnel (including army units).

Unlike the situation in other imperial gold-mining areas such as Wales, the workers at Las Médulas were free men, not slaves. Their settlements stand alongside, yet are clearly distinguishable from those which housed the imperial officials and their staff.

San Millán Yuso and Suso Monasteries
Spain

Criteria – Interchange of values; Significance in human history; Heritage associated with events of universal significance

The monastic community founded by St Millán in the mid-sixth century became a place of pilgrimage. A fine Romanesque church built in honour of the holy man still stands at the site of Suso. The monastery consists of a series of hermits' caves, a church, and an entrance porch or narthex. The caves, originally used by the monks, are cut into the southern slope of the mountain. It was here that the first literature was produced in Castilian, and from which one of the most widely spoken languages in the world today is derived. In the early sixteenth century the community was housed in the fine new monastery of Yuso, below the older complex; it is still a thriving community today.

The Codex Aemilianensis 60 was written in the Suso scriptorium in the ninth and tenth centuries by a monk who added Castilian and Basque marginal notes, with a prayer in Castilian: this is the first known example of written Spanish.

World Heritage site since

1978 1979 1980 1981 1982 1983 1984 1985 1986 1987 1988 1989 1990 1991 1992 1993 1994 1995 1996 **1997** 1998 1999 2000 2001 2002 2003 2004 2005 2006 2007 2008

Costiera Amalfitana
Italy

Criteria – Interchange of values; Significance in human history; Traditional human settlement

The layout of the settlements on the Costiera Amalfitana shows an eastern influence, with closely-spaced houses climbing steep hillsides and connected by a maze of alleys and stairs. A distinctive Arab-Sicilian architecture originated and developed in Amalfi.

Intensively settled in the sixth century, Amalfi quickly became a maritime trading power, enjoying a near-monopoly in the Tyrrhenian Sea. Wood, iron, weapons, wine and fruit were traded in eastern markets for spices, perfumes, pearls, jewels, textiles and carpets to sell in the West. However, Amalfi was eclipsed by the power of Pisa in the twelfth century.

Costiera Amalfitana (Amalfi Coast) is an outstanding example of a Mediterranean landscape and an area of great physical beauty and natural diversity. It has been intensively settled by communities since the early Middle Ages and has a number of towns, such as Amalfi and Ravello, with significant architectural and artistic works. Rural areas show the versatility of the inhabitants in adapting their use of the land to the diverse terrain, which ranges from terraced vineyards and orchards on lower slopes to wide upland pastures.

The site covers 112 km² in the Province of Salerno. Its exceptional cultural and natural scenic values are a result of its dramatic topography and historical evolution. The area's natural boundary is the southern slope of the peninsula formed by the Lattari hills. It consists of four main stretches of coast (Amalfi, Atrani, Reginna Maior, Reginna Minor), some minor ones (Positano, Praiano, Certaria, Hercle), with mountain villages and hamlets behind and above them.

The towns and villages of Costiera Amalfitana are characterized by their architectural monuments, such as the Torre Saracena at Cetara; the Romanesque Cathedral of Amalfi and its 'Cloister of Paradise', with their strong oriental influences; the Church of San Salvatore de' Bireto at Atrani; and Ravello with its fine cathedral and the superb Villa Rufolo.

There is an immense diversity of landscapes, ranging from coastal settlements through intensively cultivated lower slopes and large areas of open pastoral land to dramatic high mountains. In addition, there are micro-landscapes of great scientific interest resulting from topographical and climatic variations, and striking natural formations in the limestone karst at both sea level and above.

Inland, the steep slopes rising from the coast are covered with terraces banked with drystone walling and used for the cultivation of citrus and other fruits, olives, vines and vegetables. Further inland, the hillsides are given over to dairy farming, which has ancient roots in the area, based on sheep, goats, cattle and buffalo. In some parts of the Costiera the natural landscape survives intact, with little, if any, human intervention. It supports the traditional Mediterranean flora of myrtle, lentisk, broom and euphorbia. Elsewhere there are stands of trees such as holm oak, alder, beech and chestnut. Other zones shelter pantropical ferns, butterwort, dwarf palms and endemic carnivorous species.

World Heritage site since

1978 1979 1980 1981 1982 1983 1984 1985 1986 1987 1988 1989 1990 1991 1992 1993 1994 1995 1996 **1997** 1998 1999 2000 2001 2002 2003 2004 2005 2006 2007 2008

Ancient City of Ping Yao
China

Criteria – Interchange of values; Testimony to
cultural tradition; Significance in human history

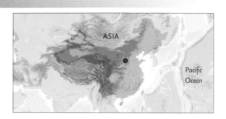

Prosperity derived
from trade and
banking meant that,
in addition to public
buildings and
temples, shops and
monuments, Ping Yao
was endowed with
high-quality, well-built
private houses; these
have largely survived.

The Ancient City of Ping Yao is an
outstanding example of a Han Chinese city
of the Ming and Qing dynasties
(fourteenth–twentieth centuries) that has
retained all its features to an exceptional
degree. It offers a remarkably complete
picture of cultural, social, economic and
religious development during one of the
most seminal periods of Chinese history. In
1370, during the reign of the Ming Emperor
Hong Wu, the city was fortified with a
massive new defensive wall and the internal
layout was greatly altered, reflecting the
strict rules of Han planning. The circuit of
walls measures 6 km in length, the precise
dimension for a city of this grade according
to Han prescriptions. There are six fortified
gates and seventy-two massive bastions
along its length.

World Heritage site since

1978 1979 1980 1981 1982 1983 1984 1985 1986 1987 1988 1989 1990 1991 1992 1993 1994 1995 1996 **1997** 1998 1999 2000 2001 2002 2003 2004 2005 2006 2007 2008

Residences of the Royal House of Savoy
Italy

Criteria – Human creative genius; Interchange of values; Significance in human history; Traditional human settlement

EUROPE

Mediterranean Sea

Ionian Sea

When Emmanuel-Philibert, Duke of Savoy, moved his capital to the small fortified medieval town of Turin in 1562, he began a vast series of building projects, continued in Baroque style by his successors, to demonstrate the power of the ruling house. This outstanding complex of buildings was designed and embellished by the leading architects and artists of the time. It radiates into the surrounding countryside from the Royal Palace (Palazzo Reale) in the centre of

Turin to include many country residences and hunting lodges. The brick Palazzo Reale, late seventeenth century in its present form, was built around a courtyard, with porticoes opening behind onto landscaped gardens. It formed the focus of a 'Command Centre', linked to many buildings, including the Palazzo Chiablese on the Piazza Reale, and buildings on the Piazza Castello (former State Secretariats and the Palazzo Madama).

Guarino Guarini gave his personal touch to Turin's Baroque architecture with the Palazzo Carignano (1679–85), one of the most attractive and impressive seventeenth century-Italian palaces. He placed an oval rotunda in the middle of the main façade, giving it an undulating appearance.

The recently restored ballroom of the Royal Palace (Palazzo Reale).
▼

World Heritage site since

1978 1979 1980 1981 1982 1983 1984 1985 1986 1987 1988 1989 1990 1991 1992 1993 1994 1995 1996 **1997** 1998 1999 2000 2001 2002 2003 2004 2005 2006 2007 2008

Archaeological Area of Agrigento
Italy

Criteria – Human creative genius; Interchange of values; Testimony to cultural tradition; Significance in human history

Agrigento was one of the greatest cities of the ancient Mediterranean world and it has been exceptionally well preserved. Its great row of Doric temples is one of the most outstanding monuments of Greek art. Founded as a Greek colony in the sixth century BC, it reached its height during the rule of the tyrant Thero (488–473). The most substantial remains are from this time. The Temple of Olympian Zeus, of which only the foundations and main altar survive, was one of the largest of all Greek temples. The so-called Temple of Concord is the most impressive surviving Doric temple in the Greek world after the Parthenon in Athens. In addition, there are large excavated areas of the residential Hellenistic and Roman Agrigento, and extensive ancient cemeteries with tombs and monuments from the pagan and Christian periods.

The Valley of the Temples covers most of the built-up part of the ancient city and its public monuments. It is closed by the ridge running parallel to the sea that was assigned, in antiquity, the role of a sacred area.

The area between the acropolis and the temples was laid out in the early fifth century BC.

Temple of Hera (Juno) in Agrigento.
▼

World Heritage site since

1978 1979 1980 1981 1982 1983 1984 1985 1986 1987 1988 1989 1990 1991 1992 1993 1994 1995 1996 **1997** 1998 1999 2000 2001 2002 2003 2004 2005 2006 2007 2008

Archaeological Site of Volubilis
Morocco

Criteria – Interchange of values; Testimony to cultural tradition; Significance in human history; Heritage associated with events of universal significance

The Mauritanian capital, founded in the third century BC, became an important outpost of the Roman Empire after AD 40 and is an exceptionally well-preserved example of a colonial town on the fringes of the Roman Empire. During the Roman period, a town wall, with eight monumental gates, and a new monumental centre including a capitol, basilica and baths, were constructed. The triumphal arch of Caracalla, which spans the decumanus

maximus, marks the boundary between the Punic-Hellenistic town and the extension in the Roman period to the northeast. At the beginning of the reign of Diocletian, in 285, the Romans abruptly abandoned the region, for reasons that remain obscure, and Volubilis entered its 'dark age'. Volubilis was later briefly to become the capital of Idris I, founder of the Idrisid dynasty, who is buried at nearby Moulay Idris.

The buildings of Volubilis are for the most part constructed using the grey-blue limestone quarried nearby on the Zerhoun massif. They are notable for the large number of mosaic floors still in situ. Although they do not attain the artistic level of other North African mosaics, they are lively and varied in form and subject matter.

The triumphal arch of Caracalla.
▼

World Heritage site since

1978 1979 1980 1981 1982 1983 1984 1985 1986 1987 1988 1989 1990 1991 1992 1993 1994 1995 1996 **1997** 1998 1999 2000 2001 2002 2003 2004 2005 2006 2007 2008

Historic Centre of Riga
Latvia
Criteria – Human creative genius; Interchange of values

Riga was independent from 1221 and in 1282 it formed an alliance with Lübeck and Visby to become a member of the Hanseatic League. By the fifteenth century Riga was a typical Hanseatic town, with winding streets and densely-packed houses, a large central market and strong fortifications.

In the sixteenth century the city became embroiled in struggles between Russia, Poland and Sweden. Finally in 1710, it fell to the Russian army, remaining part of the Tsarist Russian Empire until the creation of the first Republic of Latvia in 1918.

The Historic Centre of Riga, while retaining its medieval and later urban fabric relatively intact, is of outstanding value by virtue of the quality and the quantity of its Jugendstil, or German Art Nouveau architecture, unparalleled anywhere in the world, and its nineteenth-century architecture in wood. It has exerted a considerable influence within the Baltic cultural area on subsequent developments in architecture.

Riga was a major centre of the powerful Hanseatic League of north European and Baltic traders, deriving its prosperity in the thirteenth to the fifteenth centuries from the trade with central and eastern Europe. The urban fabric of its medieval centre reflects this prosperity, though most of the earliest buildings were destroyed by fire or war. Riga became an important economic centre in the nineteenth century, when the suburbs surrounding the medieval town were laid out, first with imposing wooden buildings in neoclassical style and then in Jugendstil. These three districts comprise the historic centre of the city.

Few medieval houses are still intact; of these one of the most interesting is the House of the Three Brothers, an impeccably restored group from the fifteenth century. The late-seventeenth-century Reutem's House and Dammnstem's House are more monumental buildings, notable for their interior decorations and fittings and for their impressive façades.

The boulevards are lined with many important public buildings from the nineteenth and early twentieth centuries; they include the National Theatre and the Museum of Latvian Art. The creation of the boulevards coincided with the reign of eclecticism in Europe, and this movement is abundantly represented. Eclecticism allowed architects to produce many flights of fancy, well-illustrated by the House of the Cat on Meistaru Street. The suburbs that developed so rapidly at this time are notable for both their wooden buildings in the classical Russian style and the extraordinary quality of the new buildings.

However, it was Jugendstil, which reached Riga via Finland at the end of the nineteenth century, that provided the suburban area with its most noteworthy feature. There are countless examples, perhaps the most outstanding of which are the works of Mikhail Eisenstein in Alberta Street and Elizabeth Street. National Romanticism evolved into Jugendstil in Latvia, again on the Finnish model. There are some striking examples of this movement in Alberta Street and Brivibas Street.

House of the Blackheads, Riga. ▶

World Heritage site since

1978 1979 1980 1981 1982 1983 1984 1985 1986 1987 1988 1989 1990 1991 1992 1993 1994 1995 1996 1997 **1998** 1999 2000 2001 2002 2003 2004 2005 2006 2007 2008

Summer Palace and Imperial Garden in Beijing
China

Criteria – Human creative genius; Interchange of values; Testimony to cultural tradition

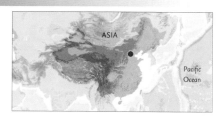

ASIA

Pacific
Ocean

The Imperial Garden and Summer Palace in Beijing – first built in 1750, largely destroyed in the Second Opium War of 1860, and restored in 1886 – is a masterpiece of Chinese garden design. The natural landscape of hills and open water is combined with artificial features such as pavilions, halls, palaces, temples and bridges to form a harmonious ensemble of outstanding aesthetic value. The site covers an area of almost 3km², three-quarters of

which is covered by water. The main framework is supplied by the Hill of Longevity and Kunming Lake, complemented by man-made features. It is designed on a grandiose scale, commensurate with its role as an imperial garden. It is divided into three areas, each with its particular function: political and administrative activities, residence, and recreation and sightseeing. It became a public park in 1924.

The political area is reached by means of the monumental East Palace Gate. The central feature is the Hall of Benevolence and Longevity, an imposing structure with its own courtyard garden. This area connects directly with the residential area, which is made up of three complexes of buildings.

The Seventeen-Arch Bridge.
▼

World Heritage site since

1978 1979 1980 1981 1982 1983 1984 1985 1986 1987 1988 1989 1990 1991 1992 1993 1994 1995 1996 1997 **1998** 1999 2000 2001 2002 2003 2004 2005 2006 2007 2008

Naval Port of Karlskrona
Sweden

Criteria – Interchange of values; Significance in human history

Karlskrona is an outstanding example of a late-seventeenth-century European planned naval city. It was founded in 1680 when Sweden was a major power whose territory included modern Finland, Estonia, Latvia, and parts of north Germany. The plan of Karlskrona integrates strategic imperatives with the classical ideal. The Baroque layout with wide main streets radiating out from a central square lined with majestic public buildings is clearly discernible in the present-day town. The centre of the town is Stortorget (Great Square), at the highest point of the island of Trossö where the two main churches of the town, both dating from the first half of the eighteenth century, Rådhuset (the City Hall) from the same period, and later public buildings such as the Concert Hall, the City Library, and the Post Office, are located.

The naval harbour is to the south of the town, from which it was originally separated by an impressive enclosure wall, only small sections of which survive.

To the south of the Parade Ground is Gamle Varvet (the Old Shipyard). This is made up of a number of fine buildings dating mainly from the late-eighteenth century.

World Heritage site since

1978 1979 1980 1981 1982 1983 1984 1985 1986 1987 1988 1989 1990 1991 1992 1993 1994 1995 1996 1997 **1998** 1999 2000 2001 2002 2003 2004 2005 2006 2007 2008

The Four Lifts on the Canal du Centre and their Environs, La Louvière and Le Roeulx (Hainault)
Belgium

Criteria – Testimony to cultural tradition; Significance in human history

These ingenious lifts consist essentially of two mobile compartments, each supported by a single hydraulic press. When one compartment is at the level of the upper bay, the other is at the lower level. As the first descends, the other rises.

The four hydraulic boat-lifts on this short stretch of the Canal du Centre represent the apogee of the application of engineering technology to the construction of canals. Of the eight hydraulic boat-lifts built around the end of the nineteenth century, the only ones in the world which still exist in their original working condition are these four. Together with the canal itself and its associated structures, they constitute a remarkably well-preserved and complete example of a late nineteenth-century industrial landscape. Lift No. 1 at Houdeng-Gœgnies was completed in 1888. The others were built thirteen years later and incorporate a number of modifications to the basic design, though the operating principle remains the same.

Choirokoitia
Cyprus

Criteria – Interchange of values; Testimony to cultural tradition; Significance in human history

Among the site's most noteworthy finds are the anthropomorphic figurines in stone (and one in clay), which point to the existence at this early period of elaborate spiritual beliefs.

The Neolithic settlement of Choirokoitia, occupied from the seventh to the fourth millennia BC, is one of the most important prehistoric sites in the eastern Mediterranean. The earliest occupation, consisting of circular houses built from mud-brick and stone with flat roofs, was on the eastern side of the hill. It was protected by natural slopes on three sides and a massive wall barring access from the west. A second defensive wall was erected to protect a later extension of the village. The finds from excavations at the site have thrown much light on the evolution of human society in this key region, and since only part of the site has been excavated, it forms an exceptional archaeological reserve for future study.

World Heritage site since

1978 1979 1980 1981 1982 1983 1984 1985 1986 1987 1988 1989 1990 1991 1992 1993 1994 1995 1996 1997 **1998** 1999 2000 2001 2002 2003 2004 2005 2006 2007 2008

La Grand-Place, Brussels
Belgium

Criteria – Interchange of values; Significance in human history

La Grand-Place in Brussels is a remarkably homogeneous body of public and private buildings, dating mainly from the late-seventeenth century. The architecture provides a vivid illustration of the level of social and cultural life of the period in this important political and commercial centre. The earliest written reference to the Nedermarckt (Lower Market), as it was originally known, dates from 1174; the present name came into use in the last quarter of the eighteenth century.

The rectangular outline of today's Grand-Place has developed over the centuries as a result of successive enlargements and other modifications, and did not take on its definitive form until after 1695, when it was restored to its original layout and appearance following bombardment by the French. It has, however, always had seven streets running into it.

The Hôtel de Ville (City Hall), which covers most of the south side of the Grand-Place, consists of a group of buildings around a rectangular internal courtyard. The part facing on to the square is from the fifteenth century, consisting of two L-shaped buildings. The entire façade is decorated with statues dating from the nineteenth century.
▼

World Heritage site since

1978 1979 1980 1981 1982 1983 1984 1985 1986 1987 1988 1989 1990 1991 1992 1993 1994 1995 1996 1997 **1998** 1999 2000 2001 2002 2003 2004 2005 2006 2007 2008

Cilento and Vallo di Diano National Park with the Archeological sites of Paestum and Velia, and the Certosa di Padula
Italy

Criteria – Testimony to cultural tradition; Significance in human history

The Cilento is an outstanding cultural landscape. The dramatic groups of sanctuaries and settlements along its three east-west mountain ridges vividly represent the area's historical evolution: it was a major route not only for trade, but also for cultural and political interaction during the prehistoric and medieval periods. The Cilento was also the boundary between the Greek colonies of Magna Graecia and the indigenous Etruscan and Lucanian peoples. The remains of two major cities from classical times, Paestum and Velia, are found there. Of the monastic properties, the most impressive is the Certosa di San Lorenzo at Padula in the Vallo di Diano. Construction began in 1306, but in its present form it is essentially Baroque, built in the seventeenth and eighteenth centuries.

At Paestum, the Greek city of Poseidonia, a number of exceptional public buildings have been revealed, the most outstanding of which are the three great temples of Hera, Ceres and Poseidon. Less survives of Velia Elea, although its notable Porta Rosa is the oldest and most complete example of a Greek arched town gate.

East Rennell
Solomon Islands

Criteria – Significant ecological and biological processes

East Rennell makes up the southern third of Rennell Island, the southernmost island in the Solomon Island group in the western Pacific. Rennell, 86 km long by 15 km wide, is the largest raised coral atoll in the world. The site includes approximately 370 km² and a marine area extending 5.6 km to sea. A major feature of the island is Lake Tegano, which was the former lagoon on the atoll. The lake, the largest in the insular Pacific (155 km²), is brackish and contains many rugged limestone islands and endemic species. Rennell is mostly covered with dense forest, with a canopy averaging 20 m in height. Combined with the strong climatic effects of frequent cyclones, the site is a true natural laboratory for scientific study. The site is under customary land ownership and management.

Rennell was formed by the uplift of corals which formed on an undersea ridge and then were subject to faulting. The landform is a typical jagged and eroded limestone karst rising to 200 m.

World Heritage site since

1978 1979 1980 1981 1982 1983 1984 1985 1986 1987 1988 1989 1990 1991 1992 1993 1994 1995 1996 1997 **1998** 1999 2000 2001 2002 2003 2004 2005 2006 2007 2008

Temple of Heaven: an Imperial Sacrificial Altar in Beijing
China

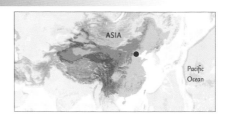

Criteria – Human creative genius; Interchange of values; Testimony to cultural tradition

The Temple of Heaven is a masterpiece of architecture and landscape design which symbolizes the relationship between Earth and Heaven that stands at the heart of Chinese cosmogony, and also the special role played by the emperors within that relationship. The Altar of Heaven and Earth was completed in 1420. The central building was a large rectangular sacrificial hall, where sacrifices were offered to Heaven and Earth, with the Fasting Palace to the southwest. In the ninth year of the reign of Emperor Jiajing (1530) the decision was taken to offer separate sacrifices to Heaven and to Earth, and so the Circular Mound Altar was built to the south of the main hall, for sacrifices to Heaven. The Altar of Heaven and Earth was renamed the Temple of Heaven.

The area occupied by the Temple of Heaven is almost square, the two southern corners being right-angled and those on the north rounded. This symbolizes the ancient Chinese belief that Heaven is round and the Earth square. The main Temple of Heaven, the Circular Mound, repeats the symbolism of the walls, as the central round feature (Heaven) is inside a square enclosure (Earth).

World Heritage site since

1978 1979 1980 1981 1982 1983 1984 1985 1986 1987 1988 1989 1990 1991 1992 1993 1994 1995 1996 1997 **1998** 1999 2000 2001 2002 2003 2004 2005 2006 2007 2008

Ir. D.F. Woudagemaal (D.F. Wouda Steam Pumping Station)
Netherlands

Criteria - Human creative genius; Interchange of values; Significance in human history

North
Sea

EUROPE

The Dutch landscape has been created by battling against water and much of the country would be flooded if it had not been protected over the centuries by dykes and kept dry by a sophisticated water-control system (waterstraat).

For centuries, windmills were used to discharge excess water in the Netherlands. The first steam pump was built in 1825 and the construction of steam-driven pumping stations reached its peak between 1870 and 1885; very few new ones were built after 1900. There were about 700 in operation between 1900 and 1910. Extreme flooding in 1894 led to a decision to reclaim the Lauwerszee and drain the southwestern part of Friesland province. The Wouda Pumping Station at Lemmer, opened in 1920, was the key to this operation. It is the largest steam-pumping station ever built and is still in operation. It represents the high point of the contribution made by Netherlands engineers and architects in protecting their people and land against the natural forces of water.

World Heritage site since

1978 1979 1980 1981 1982 1983 1984 1985 1986 1987 1988 1989 1990 1991 1992 1993 1994 1995 1996 1997 **1998** 1999 2000 2001 2002 2003 2004 2005 2006 2007 2008

Historic Centre of Urbino
Italy

Criteria – Interchange of values; Significance in human history

The great artist Raphael was born in Urbino, in a small fourteenth-century building with a charming interior courtyard. What was probably the artist's first important work, a Madonna and Child, is in the first-floor room where he was born in 1483.

During its short cultural pre-eminence in the fifteenth century, the hill-town of Urbino attracted some of the most outstanding humanist scholars and artists of the Renaissance, who created an exceptional urban setting whose influence spread far into Europe. In the mid-fifteenth century Federico II da Montefeltro, who ruled the city and duchy of Urbino, undertook a radical rebuilding campaign in the city. The walls were rebuilt according to the designs of Leonardo da Vinci. The new Ducal Palace, by Luciano Laurana and Francesco di Giorgio Martini, was inserted into the urban fabric with the minimum of disturbance, incorporating existing medieval structures. Along with the adjacent cathedral, the palace became the model for new buildings in the Renaissance style. Owing to Urbino's economic and cultural stagnation from the sixteenth century onwards, it has preserved much of its Renaissance appearance.

World Heritage site since

1978 1979 1980 1981 1982 1983 1984 1985 1986 1987 1988 1989 1990 1991 1992 1993 1994 1995 1996 1997 **1998** 1999 2000 2001 2002 2003 2004 2005 2006 2007 2008

University and Historic Precinct of Alcalá de Henares
Spain

Criteria – Interchange of values; Significance in human history; Heritage associated with events of universal significance

Founded by Cardinal Jiménez de Cisneros in the early sixteenth century, Alcalá de Henares was the world's first planned university city. Cisneros took over a partly abandoned medieval town and converted it into a city whose function was solely that of a university. This included the creation of houses to lodge professors and students. The little Chapel of St Justus was rebuilt as a church and given the title 'Magistral'. More centres of learning were added

progressively: there were eventually to be twenty-five Colegios Menores, while eight large monasteries were also colleges of the university. Its primary objective was to provide administrators for the Church and for the Spanish Empire, training over 12,000 students in the sixteenth century. From the mid-seventeenth century, the number of students declined, and in 1836 the university was transferred to Madrid.

The Complutense Polyglot Bible (1514–7) illustrates the type of work that began in Alcalá: a masterpiece of typography, it established the bases of modern linguistic analysis as well as the accepted structure for dictionaries. This work was supported by that of Antonio de Nebrija, author of the first European grammar of a Romance language, published in 1492.

World Heritage site since

1978 1979 1980 1981 1982 1983 1984 1985 1986 1987 1988 1989 1990 1991 1992 1993 1994 1995 1996 1997 **1998** 1999 2000 2001 2002 2003 2004 2005 2006 2007 2008

Ouadi Qadisha (the Holy Valley) and the Forest of the Cedars of God (Horsh Arz el-Rab)
Lebanon

Criteria – Testimony to cultural tradition;
Significance in human history

The Qadisha valley is one of the most important early Christian monastic settlements in the world. Its monasteries, many of which are of a great age, stand in dramatic positions in a rugged landscape. Nearby are the remains of the great forest of cedars of Lebanon, highly prized in antiquity for the construction of great religious buildings.

The deep Qadisha Valley is located at the foot of Mount al-Makmal. Its slopes form natural ramparts and their steep cliffs contain rock-cut chapels and hermitages, often surrounded by terraces made by the hermits for growing grain, grapes and olives.

◄
Forest of cedars, Bsharre.

World Heritage site since

1978 1979 1980 1981 1982 1983 1984 1985 1986 1987 1988 1989 1990 1991 1992 1993 1994 1995 1996 1997 1998 **1999** 2000 2001 2002 2003 2004 2005 2006 2007 2008

Robben Island
South Africa

Criteria – Testimony to cultural tradition;
Heritage associated with events of universal significance

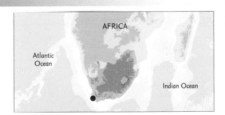

Robben Island, 7km off the coast from Cape Town, was used at various times between the seventeenth and twentieth centuries as a prison, a hospital for socially unacceptable groups and a military base.
Its buildings, particularly those of the late twentieth century such as the maximum security prison for political prisoners, witnessed the triumph of democracy and freedom over oppression and racism.

The most celebrated of Robben Island's prisoners was Nelson Mandela, who was incarcerated there for some twenty years. The last political prisoners left the island in 1991 and the prison closed down finally in 1996; since that time it has been developed as a museum.

◄
Prison buildings, Robben Island.

World Heritage site since

1978 1979 1980 1981 1982 1983 1984 1985 1986 1987 1988 1989 1990 1991 1992 1993 1994 1995 1996 1997 1998 **1999** 2000 2001 2002 2003 2004 2005 2006 2007 2008

Heart of Neolithic Orkney
United Kingdom

Criteria – Human creative genius; Interchange of values; Testimony to cultural tradition; Significance in human history

The Neolithic monuments of Orkney bear unique or exceptional testimony to an important indigenous cultural tradition which flourished for between 500 and 1,000 years but disappeared by about 2000 BC. They are an outstanding example of a type of architectural ensemble and archaeological landscape which illustrates a significant stage of human history, during which the first large ceremonial monuments were built. They are testimony to the cultural achievements of the Neolithic peoples of northern Europe during the period 3000–2000 BC.

The monuments consist of Maes Howe, a large chambered tomb; the Stones of Stenness and the Ring of Brodgar, two ceremonial stone circles; and Skara Brae, a settlement. There are also a number of unexcavated burial, ceremonial and settlement sites. The group constitutes a major prehistoric cultural landscape which gives a graphic depiction of life in this remote archipelago in the far north of Scotland some 5,000 years ago.

Maes Howe is a Neolithic masterpiece, an exceptionally early architectural accomplishment. With its almost classical strength and simplicity it is a unique survivor from 5,000 years ago. It is an expression of genius within a group of people whose other tombs were claustrophobic chambers in smaller mounds.

Passage graves such as Maes Howe were large structures, made from stones erected to form a passage leading from the outer edge of the mound to the chamber containing the remains of the dead. The general orientation of these structures demonstrates their builders' knowledge in respect to seasonal movements.

Stenness is a unique and early expression of the ritual customs of the people who buried their dead in tombs like Maes Howe and lived in settlements like Skara Brae. The Ring of Brodgar is the finest known truly circular late-Neolithic or early Bronze Age stone ring.

Skara Brae has particularly rich surviving remains. It displays remarkable preservation of stone-built furniture and a fine range of ritual and domestic artefacts, which together demonstrate with exceptional completeness the domestic, ritual and burial practices of a now-vanished culture.

When it was built 5,000 years ago, the settlement of Skara Brae was further from the sea than it is now. The settlement was abandoned some 600 years after its construction and most of the houses were emptied of their contents.

In the mid-nineteenth century the remains of Skara Brae were revealed when the overlying sand dune was swept away by a violent storm. Some clearance work took place in 1913, and in 1924 a protective breakwater was built.

Part of the Ring of Brodgar in Orkney at sunset.

World Heritage site since

1978 1979 1980 1981 1982 1983 1984 1985 1986 1987 1988 1989 1990 1991 1992 1993 1994 1995 1996 1997 1998 **1999** 2000 2001 2002 2003 2004 2005 2006 2007 2008

Ibiza, Biodiversity and Culture
Spain

Criteria – Interchange of values; Testimony to cultural tradition; Significance in human history; Significant ecological and biological processes; Significant natural habitat for biodiversity

Ibiza provides an excellent example of the interaction between marine and coastal ecosystems. The dense prairies of oceanic Posidonia (seagrass), an important endemic species found only in the Mediterranean basin, contain and support a diversity of marine life. The island also preserves considerable evidence of its long history. The archaeological sites at Sa Caleta settlement and Puig des Molins necropolis testify to the important role played by the island in the Mediterranean economy, particularly during the Phoenician-Carthaginian period. In 1235, Ibiza town was dominated by Christians, who built the Catalan castle, visible from the inside of the present building, the medieval fortifications, and the Gothic cathedral. The fortified Upper Town, Alta Vila, is an outstanding example of Renaissance military architecture; it had a profound influence on the development of fortifications in the Spanish settlements of the New World.

The Phoenician-Punic cemetery of Puig des Molins is situated in the southwest of the Upper Town. At the beginning of the sixth century BC, the ashes of the dead were placed in a natural grotto after cremation. Later, shafts and funerary chambers were dug, and sarcophagi were lowered through shafts into family sepulchres.

The cathedral and Old Town of Ibiza.
▼

World Heritage site since

1978 1979 1980 1981 1982 1983 1984 1985 1986 1987 1988 1989 1990 1991 1992 1993 1994 1995 1996 1997 1998 **1999** 2000 2001 2002 2003 2004 2005 2006 2007 2008

State Historical and Cultural Park 'Ancient Merv'
Turkmenistan

Criteria – Interchange of values; Testimony to cultural tradition

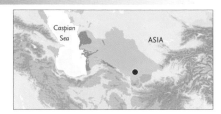

The cities of the vast Merv oasis exerted considerable influence over the cultures of Central Asia and Iran for four millennia. The oasis formed part of the Great Seljuk Empire, which had its capital here. This was one of the principal cities of its time, and its famous libraries attracted scholars from all over the Islamic world. It also had a pronounced influence in the development of architecture, architectural decoration, science and culture. Merv is the oldest and best-preserved of the oasis-cities along the Silk Route in Central Asia. A number of monuments are still visible, particularly from the last two millennia. These include the central Beni Makhan mosque and its cistern, the Buddhist stupa and monastery, and the 'Oval Building' in the northwest quarter.

The fifteenth century walls and moat of the city and citadel are of exceptional interest in that they represent the remarkable continuous record of the evolution of military architecture from the fifth century BC to the fifteenth–sixteenth centuries AD.

Great Kyz Kala, Merv.
▼

World Heritage site since

1978 1979 1980 1981 1982 1983 1984 1985 1986 1987 1988 1989 1990 1991 1992 1993 1994 1995 1996 1997 1998 **1999** 2000 2001 2002 2003 2004 2005 2006 2007 2008

Archaeological Sites of Mycenae and Tiryns
Greece

EUROPE

Ionian
Sea

Mediterranean Sea

Criteria – *Human creative genius; Interchange of values; Testimony to cultural tradition; Significance in human history; Heritage associated with events of universal significance*

The archaeological sites of Mycenae and Tiryns are the imposing ruins of the two greatest cities of the Mycenaean civilization, which dominated the eastern Mediterranean world from the fifteenth to the twelfth centuries BC and played a vital role in the development of classical Greek culture. These two cities are indissolubly linked to the Homeric epics, the Iliad and the Odyssey, which have influenced European

art and literature for more than three millennia. The Palace at Mycenae was constructed on the summit of the hill and surrounded by massive cyclopean walls in three stages (c. 1350, 1250 and 1225 BC respectively). A series of tholos (beehive-shaped tombs) were also built on the slopes of the hill: the Tomb of Aegisthos (c. 1500 BC), the Lion Tholos Tomb (c. 1350 BC), the Tomb of Clytemnestra (c. 1220 BC).

As at Mycenae, the earliest human occupation known at Tiryns is from the Neolithic period. The oldest architectural remains, on the Upper Citadel, are from the early Bronze Age (c. 3000 BC). A new fortified palace complex was constructed in the fourteenth century BC, the defences were extended in the early thirteenth century BC, and the Lower Citadel was also fortified.
Lion Gate at Mycenae.
▼

World Heritage site since

1978 1979 1980 1981 1982 1983 1984 1985 1986 1987 1988 1989 1990 1991 1992 1993 1994 1995 1996 1997 1998 **1999** 2000 2001 2002 2003 2004 2005 2006 2007 2008

Belfries of Belgium and France
Belgium and France
Criteria – Interchange of values; Significance in human history

Belfries are outstanding representatives of civic and public architecture in Europe. Twenty-three belfries in the north of France and the belfry of Gembloux in Belgium were inscribed on the World Heritage list as a group, an extension to the thirty-two Belgian belfries inscribed in 1999 as 'Belfries of Flanders and Wallonia'. Built between the eleventh and seventeenth centuries, invariably in an urban setting, the belfries are potent symbols of the transition from feudalism to mercantile urban society. While Italian, German and English towns mainly chose to build town halls, in part of northwestern Europe greater emphasis was placed on building belfries. Compared with the keep symbol of the seigneurs and the bell-tower symbol of the Church, the belfry symbolizes the power of the aldermen. Over the centuries, they came to represent the influence and wealth of the towns.

Most of the belfries cover the periods of the fourteenth–fifteenth and sixteenth–seventeenth centuries, thereby illustrating the transition in style from Norman Gothic to later Gothic, which then mingles with Renaissance and Baroque forms. In the fourteenth and fifteenth centuries, the belfries abandoned the model of the keep in favour of finer, taller towers, such as those of Dendermonde, Lier and Aalst.

iSimangaliso Wetland Park
South Africa
Criteria – Natural phenomena or beauty; Significant ecological and biological processes; Significant natural habitat for biodiversity

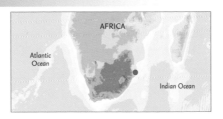

The ongoing fluvial, marine and aeolian processes in the site have produced a variety of landforms, including coral reefs, long sandy beaches, coastal dunes, lake systems, swamps, and extensive reed and papyrus wetlands. The interplay of the park's environmental heterogeneity with major floods and coastal storms, and a transitional geographic location between subtropical and tropical Africa, has resulted in exceptional species diversity and ongoing speciation. The mosaic of landforms and habitat types creates superlative scenic vistas. The site contains critical habitat for a range of species from Africa's marine, wetland and savanna environments.

iSimangaliso Wetland Park (previously known as Greater St Lucia Wetland Park) is the largest and most diverse estuarine system in Africa. It is the only area with coral reefs and has a high number of threatened species, including rhino, hippo and leopard.

World Heritage site since

1978 1979 1980 1981 1982 1983 1984 1985 1986 1987 1988 1989 1990 1991 1992 1993 1994 1995 1996 1997 1998 **1999** 2000 2001 2002 2003 2004 2005 2006 2007 2008

Mountain Railways of India
India

Criteria – Interchange of values; Significance in human history

This site includes three railways, all still fully operational. The Darjeeling Himalayan Railway, opened in 1881, was the first example of a hill passenger railway, its design applying bold and ingenious engineering solutions to the problem of establishing an effective rail link across mountainous terrain. The construction of the Nilgiri Mountain Railway, a 46-km-long metre-gauge single-track railway in Tamil Nadu State was first proposed in 1854, but due to the difficulty of the mountainous location the work only started in 1891 and was completed in 1908. This railway, scaling

an elevation of 326 m to 2,203 m, represented the latest technology of the time. The Kalka Shimla Railway, a 96-km-long, single track working rail link was built in the mid-19th century to provide a service to the highland town of Shimla.

The three Mountain Railways of India are outstanding examples of the interchange of values on developments in technology. The Darjeeling Himalayan Railway helped the area become one of the main tea-growing areas in India and is still the most outstanding example of a hill passenger railway.

◄

The 'Himalayan Bird' or 'Toy Train' on the Darjeeling Himalayan Railway at Dali Monastery.

Historic Centre of Santa Ana de los Ríos de Cuenca
Ecuador

Criteria – Interchange of values; Significance in human history; Traditional human settlement

Santa Ana de los Ríos de Cuenca is set in a valley surrounded by the Andean mountains in the south of Ecuador. This inland colonial town (entroterra), now the country's third city, was founded in 1557 on the rigorous planning guidelines issued thirty years earlier by the Spanish king Charles V. Cuenca still observes the formal orthogonal town plan that it has respected for 400 years.

One of the region's agricultural and administrative centres, it has been a melting pot for local and immigrant populations. Cuenca's architecture, much of which dates from the eighteenth century, was 'modernized' in the economic prosperity of the nineteenth century as the city became a major exporter of quinine, straw hats and other products.

Despite the growth that came with its prosperity, the Andean mountain chains have allowed Cuenca to maintain close contact with its natural environment. The Urban Development Plan of 1982 also helped safeguard the image of the town.

World Heritage site since

1978 1979 1980 1981 1982 1983 1984 1985 1986 1987 1988 1989 1990 1991 1992 1993 1994 1995 1996 1997 1998 **1999** 2000 2001 2002 2003 2004 2005 2006 2007 2008

Historic Fortified Town of Campeche
Mexico

Criteria – Interchange of values; Significance in human history

Campeche is a typical example of a harbour town from the Spanish colonial period in the New World. The historic centre has kept its outer walls and system of fortifications, designed to defend this Caribbean port against attacks from the sea. A chequerboard plan was chosen for the town, with a Plaza Mayor facing the sea and surrounded by government and religious edifices. The area of historic monuments is spread over 1.8 km², including 0.5 km² surrounded by walls, with the town stretching out on each side, following the configuration of the coast. The protected group of buildings consists of two subgroups: one with a high density of buildings of great heritage value, and another, which is not so dense or valuable but which forms a transitional and protective zone.

Among almost 1,000 buildings of historic value are the Cathedral of the Immaculate Conception, several churches, the Toro theatre and the municipal archives. The system of fortifications includes the redoubts of San José and San Miguel, and the batteries of San Lucas, San Matías and San Luís.

A colourful street in Campeche. ▼

World Heritage site since

1978 1979 1980 1981 1982 1983 1984 1985 1986 1987 1988 1989 1990 1991 1992 1993 1994 1995 1996 1997 1998 **1999** 2000 2001 2002 2003 2004 2005 2006 2007 2008

Museumsinsel (Museum Island), Berlin
Germany
Criteria – Interchange of values; Significance in human history

The museum as a social phenomenon owes its origins to the Age of Enlightenment in the eighteenth century. The five museums, built between 1824 and 1930 on the Museumsinsel, a small island in the river Spree, are the realization of a visionary project and show the evolution of approaches to museum design. The project began when the Altes Museum was built to the designs of Karl Friedrich Schinkel in 1824–8. A plan to develop the part of the island behind this museum was drawn up by the court architect, Friedrich August Stüler in 1841, followed by the building of the Neues Museum (1843–7). In 1866 the Nationalgalerie was built and in 1897–1904 the Kaiser-Friedrich-Museum (now the Bodemuseum). Stüler's plan was completed in 1909–30 with the construction of Alfred Messel's Pergamonmuseum.

The 'Amazon' statue by August Kiss, which is one of the two statues that flank the entrance to the Atlas Museum.
▼

The three-winged Pergamonmuseum was built to exhibit the greatly expanded collections of antiquities resulting from German excavations at Pergamon and other Greek sites in Asia Minor as well as those from Mesopotamia formerly housed in the Vorderasiatisches Museum. It rises directly from the river Spree, like the Bodemuseum, with which it is harmonized in scale and proportions.

World Heritage site since

1978 1979 1980 1981 1982 1983 1984 1985 1986 1987 1988 1989 1990 1991 1992 1993 1994 1995 1996 1997 1998 **1999** 2000 2001 2002 2003 2004 2005 2006 2007 2008

My Son Sanctuary
Vietnam

Criteria – Interchange of values; Testimony to cultural tradition

The ruins of My Son vividly illustrate the importance of the Champa Kingdom to the political and cultural history of Southeast Asia. A unique expression of cultural interchange, it marks the introduction of the Hindu architecture of the Indian subcontinent into the area.

Between the fourth and thirteenth centuries a unique culture which owed its spiritual origins to Indian Hinduism developed on the coast of contemporary Vietnam. This is graphically illustrated by the remains of a series of impressive tower-temples located in a dramatic site that was the religious and political capital of the Champa Kingdom for most of its existence.

◄ My Son sanctuary at Tra Kieu.

Discovery Coast Atlantic Forest Reserves
Brazil

Criteria – Significant ecological and biological processes; Significant natural habitat for biodiversity

The rainforests of southern Bahia and northern Espirito Santo have the world's highest density of tree species and perhaps the largest number of trees of Pau Brasil (brazil wood) left on Earth.

The Discovery Coast Atlantic Forest Reserves, in the states of Bahia and Espírito Santo, consist of eight separate protected areas containing 1,120 km² of Atlantic forest and associated shrub (restingas). The rainforests of Brazil's Atlantic coast are the world's richest in terms of biodiversity. The site contains a distinct range of species with a high level of endemism and reveals a pattern of evolution that is not only of great scientific interest but is also of importance for conservation.

World Heritage site since

1978 1979 1980 1981 1982 1983 1984 1985 1986 1987 1988 1989 1990 1991 1992 1993 1994 1995 1996 1997 1998 **1999** 2000 2001 2002 2003 2004 2005 2006 2007 2008

Droogmakerij de Beemster (Beemster Polder)
Netherlands

Criteria – Human creative genius; Interchange
of values; Significance in human history

North
Sea

EUROPE

The Netherlands owes its existence to
reclaimed land like the Beemster Polder.
If no dykes had been constructed and if
there were no drainage of excess water,
65 per cent of the modern country would
be under water. The seventeenth-century
Beemster Polder is the oldest area of
reclaimed land in the country and one of the
most remarkable. It has preserved intact its
well-ordered landscape of fields, roads,
canals, dykes and settlements, laid out in
accordance with classical planning principles.
Draining large areas like Beemster was made
possible by the dramatic improvement in
pumping technology using windmills
driving waterwheels. Pumps were later
converted to steam power, then diesel in
the twentieth century. Now drainage is
carried out by a fully automated electric
pumping station.

The innovative
landscape of the
Beemster Polder had
a profound impact on
reclamation projects
in Europe and
beyond. Its creation
advanced the
interrelationship
between humankind
and water at a crucial
period of social and
economic expansion.

De Rijp, Beemster
Polder.
▼

World Heritage site since

1978 1979 1980 1981 1982 1983 1984 1985 1986 1987 1988 1989 1990 1991 1992 1993 1994 1995 1996 1997 1998 **1999** 2000 2001 2002 2003 2004 2005 2006 2007 2008

Wooden Churches of Maramureş
Romania

Criteria – Significance in human history

The Maramureş churches are outstanding examples of vernacular religious wooden architecture resulting from the interchange of Orthodox religious traditions with Gothic influences. The eight churches of Maramureş are based on traditional timber architecture and stand on bases of stone blocks and pebble fillings. These narrow, high, timber constructions, characterized by their tall, slim clock towers at the western end of the building, show a high level of artistic maturity and craft skills. They are: the Church of the Presentation of the Virgin at the Temple (Bârsana); the Church of Saint Nicholas (Budeşti); the Church of the Holy Paraskeva; the Church of the Nativity of the Virgin; the Church of the Holy Archangels (Plopiş); the Church of the Holy Parasceve (Poienile Izei); the Church of the Holy Archangels (Rogoz); and the Church of the Holy Archangels (Şurdeşti).

The Church of the Holy Parasceve is one of the oldest of the wooden churches of Maramureş (1604), and reveals two phases in the development of such buildings. The first can be seen in the lower part of the walls, with a sanctuary based on a square plan. In the eighteenth century the walls were raised and the interior was decorated with paintings.

Typical wooden church of Maramureş. ▶

World Heritage site since

1978 1979 1980 1981 1982 1983 1984 1985 1986 1987 1988 1989 1990 1991 1992 1993 1994 1995 1996 1997 1998 **1999** 2000 2001 2002 2003 2004 2005 2006 2007 2008

Historic Town of Vigan
Philippines

Criteria – Interchange of values; Significance in human history

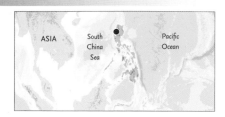

Established in the sixteenth century, Vigan is the best-preserved example of a planned Spanish colonial town in Asia. Its architecture reflects the coming together of cultural elements from elsewhere in the Philippines, from China and from Europe, resulting in a culture and townscape that have no parallel anywhere in east and southeast Asia.

The Mestizo River was central to the development of the town in the sixteenth–nineteenth centuries: large sea-going vessels could berth in the delta and small craft communicated with the interior. It is no longer navigable owing to silting, and so the town is no longer on an island. As the major commercial centre for the region, Vigan traded directly with China. It also supplied goods for shipment to Mexico, and thence onwards to Europe. This trade resulted in constant exchanges of peoples and cultures between the Ilocanos, Filipinos, Chinese, Spanish, and (in the twentieth century) North Americans.

Vigan's layout conforms closely to the traditional Spanish chequerboard plan. What makes it unique amongst colonial towns, however, is the blending of the Latin tradition with strong Chinese, Ilocano and Filipino influences.

The town is located in the delta of the Abra River, off the coastal plain of the China Sea, close to the north-east tip of the island of Luzon. The present-day municipality is divided into nine urban districts and thirty rural villages. Almost half the total area is still in use for agriculture. The Historic Core Zone is defined on two sides by the Govantes and Mestizo rivers.

◄

St Paul's Metropolitan Cathedral, Vigan.

World Heritage site since

1978 1979 1980 1981 1982 1983 1984 1985 1986 1987 1988 1989 1990 1991 1992 1993 1994 1995 1996 1997 1998 **1999** 2000 2001 2002 2003 2004 2005 2006 2007 2008

Desembarco del Granma National Park
Cuba

Criteria – Natural phenomena or beauty; Major stages of Earth's history

The park is situated in and around Cabo Cruz and includes the world's largest and best-preserved systems of marine terraces (both above and below sea level) on calcareous rock, as well as some of the most pristine and impressive coastal cliffs bordering the western Atlantic. It contains examples of most ecosystems present in the region, including the coral reef of Cabo Cruz, seagrass beds and mangroves and old submarine terraces up to 30 m deep. The area is one of the most important centres of plant diversity and endemism in Cuba. It has a remarkable archaeological value as it was the original settlement of groups that belonged to the Taina Culture, to which the local population has strong genetic and spiritual links. Many events related to the Cuban Revolution also took place in the area of Cabo Cruz.

A total of 512 species of plants exist in the site, of which around 60 per cent are endemic. There are also 13 species of mammal, 110 of birds, 44 of reptiles and 7 of amphibians. Several species are of conservation concern including the Caribbean manatee and blue-headed quail-dove.

Hortobágy National Park – the Puszta
Hungary

Criteria – Significance in human history; Traditional human settlement

The cultural landscape of the Hortobágy Puszta consists of a vast area of plains and wetlands in eastern Hungary. Traditional forms of land use, such as the grazing of domestic animals, have been present in this pastoral society for more than two millennia. The oldest surviving structures on the plains are the thirteenth-century stone bridges (including Nine Arch Bridge at Hortobágy, the longest stone bridge in Hungary), and the csárdas which were provincial inns built in the eighteenth and nineteenth centuries to provide food and lodging for travellers. The typical csárda consists of two buildings facing one another, both single-storeyed and thatched or, occasionally, roofed with shingles or tiles. These contained a tavern, guest rooms and provision for horses and carriages.

The Hortobágy National Park, the first national park in Hungary, preserves intact and visible the evidence of its traditional use over more than 2,000 years, representing the harmonious interaction between human beings and nature.

World Heritage site since

1978 1979 1980 1981 1982 1983 1984 1985 1986 1987 1988 1989 1990 1991 1992 1993 1994 1995 1996 1997 1998 **1999** 2000 2001 2002 2003 2004 2005 2006 2007 2008

Dazu Rock Carvings
China

Criteria – Human creative genius; Interchange
of values; Testimony to cultural tradition

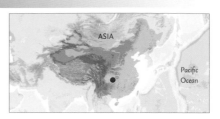

The Dazu carvings represent the pinnacle
of Chinese rock art for their high aesthetic
quality and their diversity of style and subject
matter. Carvings in the steep hillsides of Dazu
County date back to AD 650, in the early years
of the Tang dynasty, but the main period began
in the late ninth century continuing for over
400 years. There are seventy-five protected
sites containing some 50,000 statues. Tantric
Buddhism from India and the Chinese Taoist
and Confucian beliefs came together here
to create a highly original and influential
manifestation of spiritual harmony. They are
remarkable for their subject matter, both
secular and religious, and the light that they
shed on everyday life in China during this
period.

In one area, at Beishan,
there are 264 niches
with statues, one
intaglio painting, and
eight inscribed pillars.
In all there are over
10,000 carvings, more
than half of which
represent Tantric
Buddhism.

World Heritage site since

1978 1979 1980 1981 1982 1983 1984 1985 1986 1987 1988 1989 1990 1991 1992 1993 1994 1995 1996 1997 1998 **1999** 2000 2001 2002 2003 2004 2005 2006 2007 2008

City of Graz – Historic Centre
Austria

Criteria – Interchange of values; Significance in human history

EUROPE

Mediterranean Sea

Adriatic Sea

The historic centre of the city of Graz reflects artistic and architectural movements originating from the Germanic region, the Balkans, and the Mediterranean, for which it served as a crossroads for centuries. Among the hundreds of buildings of great historic and architectural interest are the Mausoleum of Emperor Ferdinand II, started in 1614, whose façade reflects the transition from the Renaissance to the Baroque style, and the Seminary (former Jesuit College), started in 1572, which illustrates the severe Renaissance architecture adopted by the order when it was first established in the German province. Of the original castle where Emperor Frederick III resided, all that remains is a Gothic hall, a late-Gothic chapel, and a double spiral staircase dating back to 1499.

Frederick III built the present cathedral in late-Gothic style (1438–64) alongside a Romanesque church dedicated to St Aegidius. Following the transfer of the bishopric from Seckau to Graz, the church of St Aegidius, used for 200 years as a centre for the Counter-Reformation, became the cathedral of the new diocese in 1786.

World Heritage site since

1978 1979 1980 1981 1982 1983 1984 1985 1986 1987 1988 1989 1990 1991 1992 1993 1994 1995 1996 1997 1998 **1999** 2000 2001 2002 2003 2004 2005 2006 2007 2008

Hoi An Ancient Town
Vietnam

Criteria – Interchange of values; Traditional
human settlement

Hoi An Ancient Town is an exceptionally
well-preserved example of a southeast Asian
trading port dating from the fifteenth to the
nineteenth centuries. Its buildings and
street plan reflect the influences, both
indigenous and foreign, that combined to
produce this unique site. Most of the
buildings are in the traditional architectural
style of the nineteenth and twentieth
centuries. They are aligned along narrow

lanes and include many religious buildings,
such as pagodas, temples, meeting houses,
etc. The rise of other ports on the coast of
Vietnam, in particular Da Nang, and the
silting of Hoi An's harbour, led to its final
eclipse. As a result of this economic
stagnation, it has preserved its early
appearance in a remarkably intact state, the
only town in the country to have done so.

The pagodas are almost
all from the nineteenth
century, although
inscriptions show them
to have been founded
in the seventeenth and
eighteenth centuries.
They conform to a
square layout and
decoration is largely
confined to the
elaborate roofs. In the
case of the larger
examples, they
constituted nuclei of
associated buildings
with religious and
secular functions.

The harbour of Hoi An. ▼

World Heritage site since

1978 1979 1980 1981 1982 1983 1984 1985 1986 1987 1988 1989 1990 1991 1992 1993 1994 1995 1996 1997 1998 **1999** 2000 2001 2002 2003 2004 2005 2006 2007 2008

Sukur Cultural Landscape
Nigeria

Criteria – Testimony to cultural tradition;
Traditional human settlement; Heritage
associated with events of universal significance

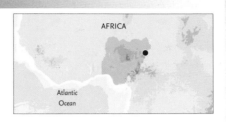

The Sukur Cultural Landscape, with the
Palace of the Hidi Chief on a hill dominating
the villages below, is a remarkably intact
expression of a society and its spiritual and
material culture. Situated on a plateau in
northeastern Nigeria, the area has been
occupied for centuries. There are a number
of shrines and altars, particularly in and
around the Hidi Palace. Complex social
relationships can also be observed in the
disposition of the cemeteries located in the
surrounding hills, while the remains of
many disused iron-smelting furnaces
illustrate an elaborate economic pattern of
production and distribution. Of considerable
social and economic importance are the
subterranean wells, surmounted by conical
stone structures and surrounded by an
enclosure wall.

The landscape of the
Sukur plateau is
characterized by
extensive terracing, of
a type known elsewhere
in Nigeria. While
primarily intended to
provide level areas for
agriculture, the terraces
are also invested with
a spiritual significance,
as shown by the
sacred trees,
entrances and ritual
sites within them.

San Cristóbal de La Laguna
Spain

Criteria – Interchange of values; Significance in
human history

San Cristóbal de La Laguna, in the Canary
Islands, has two nuclei: the original,
unplanned Upper Town; and the Lower Town,
the first ideal unfortified 'city-territory'.
In 1502, a regular town plan based on
Leonardo da Vinci's model for Imola was
drawn up by the Captain General for the
area. Wide major streets linked the public
open spaces and formed the grid on which
smaller streets were superimposed. The
resulting Lower Town expanded rapidly,
attracting the island's ruling classes, and
monastic communities began building.
A piped water supply was installed at the
expense of the Town Council in 1521, and the
first public buildings were constructed.
A number of fine churches and public and
private buildings dating from the sixteenth
to the eighteenth centuries still exist.

San Cristóbal de La
Laguna was founded
in 1497 and was the
first non-fortified
Spanish colonial
town; as such, its
layout was the model
for many colonial
towns in the Americas.

World Heritage site since

1978 1979 1980 1981 1982 1983 1984 1985 1986 1987 1988 1989 1990 1991 1992 1993 1994 1995 1996 1997 1998 **1999** 2000 2001 2002 2003 2004 2005 2006 2007 2008

Litomyšl Castle
Czech Republic
Criteria – Interchange of values; Significance in human history

One of the castle's most striking interior features is its late-eighteenth-century neoclassical theatre. Constructed entirely of wood, it seats 150 in nine loggias and a lower floor. Its original painted decoration, stage decorations and stage machinery have survived intact.

Litomyšl Castle is an outstanding and immaculately preserved example of the arcade castle, a type of building first developed in Italy and modified in the Czech lands to create an evolved form of special architectural quality. It illustrates in an exceptional way the aristocratic residences of central Europe in the Renaissance and their subsequent development under the influence of new artistic movements. Its design and decoration are particularly fine, including the later High Baroque features added in the eighteenth century. It preserves intact the range of ancillary buildings associated with an aristocratic residence of this type, of which the most interesting is the Brewery. Originally constructed as a counterpart to the castle, it stylishly blends elements of High Baroque and neoclassicism.

Historic Centre of the Town of Diamantina
Brazil
Criteria – Interchange of values; Significance in human history

The town has several architectural curiosities, including the Old Market Hall constructed in 1835 and recently restored; the Passadiço, a covered footbridge in blue and white wood; and the chafariz of the Rua Direita, near the cathedral, a sculpted fountain which guarantees that whoever drinks from it will return to Diamantina.

Diamantina, a colonial village set like a jewel in a necklace of inhospitable rocky mountains, recalls the exploits of diamond prospectors in the eighteenth century and testifies to the triumph of human cultural and artistic endeavour over the environment. Inspired by the model of a Portuguese medieval town, the colonisers transposed some of the architectural features of their home country to their adopted land while still respecting the continuity of the first settlement. The architecture is of Baroque inspiration, like most other mining villages in Brazil. Its streets are paved in a uniquely picturesque style and the casario, a regular alignment of eighteenth- and nineteenth-century semi-detached houses, has brightly coloured façades on a white ground, displaying some affiliations with the Portuguese Mannerist architecture.

World Heritage site since

.978 1979 1980 1981 1982 1983 1984 1985 1986 1987 1988 1989 1990 1991 1992 1993 1994 1995 1996 1997 1998 **1999** 2000 2001 2002 2003 2004 2005 2006 2007 2008

Wartburg Castle
Germany

Criteria – Testimony to cultural tradition; Heritage associated with events of universal significance

Wartburg Castle is an outstanding monument of the feudal period in central Europe. It is rich in cultural associations, most notably its role as the place of exile of Martin Luther, who composed his German translation of the New Testament here. In the midst of the forest, the castle occupies a rocky spur that looks down over the city of Eisenach. The castle is reached from the northern end of the spur, by a

tower with a drawbridge, followed by a number of outbuildings which form an outer courtyard. Next follows the lower courtyard, the main features of which are the keep and the palace, onto which the Knights' Baths back. The South Tower marks the farther end of the spur. The centre of the lower courtyard is occupied by a cistern. The outworks are now partially buried or in ruins.

After the Napoleonic wars, a national sentiment emerged which revelled in the image of ancient Germany as symbolized by Wartburg Castle. On the initiative of the Grand Duke of Saxony, the site was completely renovated: the palace was rebuilt from its ruins, the curtain wall restored and the remainder of the buildings reconstructed.

World Heritage site since

1978 1979 1980 1981 1982 1983 1984 1985 1986 1987 1988 1989 1990 1991 1992 1993 1994 1995 1996 1997 1998 **1999** 2000 2001 2002 2003 2004 2005 2006 2007 2008

Península Valdés
Argentina
Criteria – Significant natural habitat for biodiversity

The local population of orcas at Península Valdés has used the beaches of Valdés to develop a unique and spectacular form of shallow-water hunting. The orcas swim quickly towards the shore, chasing sea lions or young elephant seals into the surf where they then grab the prey in their jaws. The whales often beach themselves in the process. Adult orcas have been seen teaching their young how to hunt in this way, sometimes repeatedly pulling their prey off the beach for youngsters to catch.

Península Valdés, in the Argentinean province of Chubut, is a 4,000 km² promontory, protruding 100 km eastwards into the South Atlantic. The 400-km shoreline includes a series of gulfs, rocky cliffs, shallow bays and lagoons with extensive mudflats, sandy and pebble beaches, coastal sand dunes and small islands. The Ameghino Isthmus, which links the peninsula to the rest of South America, has an average width of only 11 km; with the Golfo San José to the north and the Golfo Nuevo to the south, the area has an island quality.

It is a site of global significance for the conservation of marine mammals and the shores and waters around the peninsula are an important habitat for them. An important breeding population of southern right whales uses the protected waters for mating and calving and every year over 1,500 whales visit the península. The stable population of orcas in this area have developed a unique hunting strategy to adapt to local coastal conditions. The southern elephant seal has its most northerly colony here. It reaches peak numbers of over 1,000 individuals and this is the only colony in the world reported to be on the increase. The southern sea lion also breeds here in large numbers. In addition, thirty-three other species of

marine mammals are found in the area.

Terrestrial mammals are abundant, with thirty-three species being reported. Large herds of guanaco can be seen throughout the peninsula. Other species include the mara, an Argentinean endemic, and the red fox, both of which are endangered in other parts of the country.

Península Valdés has a high diversity of birds. There are 181 species present, of which sixty-six are migratory, including the Antarctic pigeon which is considered vulnerable. The peninsula's intertidal mudflats and coastal lagoons are important staging sites for migratory shorebirds. Most numerous is the Magellanic penguin with almost 40,000 active nests distributed among five different colonies.

The tail (fluke) of a submerging right whale just off the coast of Península Valdés.

World Heritage site since

1978 1979 1980 1981 1982 1983 1984 1985 1986 1987 1988 1989 1990 1991 1992 1993 1994 1995 1996 1997 1998 **1999** 2000 2001 2002 2003 2004 2005 2006 2007 2008

Historic Centre of Sighişoara
Romania

Criteria – Testimony to cultural tradition;
Traditional human settlement

Founded in the thirteenth century by German craftsmen and merchants, Sighişoara is a fine example of a small, fortified medieval town which played an important strategic and commercial role on the fringes of central Europe. Its historic centre is composed of the fortified citadel on a steeply sloping plateau and the lower town with its woody slopes below. Apart from nineteenth-century settlements, it has kept its original medieval urban character

and its network of narrow streets. Many of the houses still have a barrel-vaulted basement, workshops on the ground floor, and the living rooms on the upper floors. The citadel plateau is enclosed by a wall, and nine of the original fourteen towers still stand. The imposing Clock Tower dominates the three squares of the historic centre and protects the stairway connecting the upper town and the lower town.

Notable among the monuments in the historic centre of Sighişoara is the Church of St Nicholas, typical of the Gothic architecture of Transylvania. Perched on the hill, it can be reached by a staircase of 175 steps. The decorative sculpture on the façade reflects Central European influences.

World Heritage site since

1978 1979 1980 1981 1982 1983 1984 1985 1986 1987 1988 1989 1990 1991 1992 1993 1994 1995 1996 1997 1998 **1999** 2000 2001 2002

Fossil Hominid Sites of Sterkfontein, Swartkrans, Kromdraai, and Environs
South Africa

Criteria – Testimony to cultural tradition; Heritage associated with events of universal significance

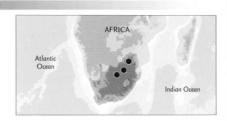

The Sterkfontein area contains a large and scientifically significant group of sites that throw light on the earliest ancestors of humanity. They constitute a vast reserve of scientific data linked to the history of the most ancient periods of humanity.

The Taung Skull Fossil Site, part of the extension to the site inscribed in 1999, is the place where, in 1924, the celebrated Taung Skull – a specimen of the species Australopithecus africanus – was found. Makapan Valley, also in the site, features in its many archaeological caves traces of human occupation and evolution dating back some 3.3 million years. The area contains essential elements that define the origin and evolution of humanity. Fossils found there have enabled the identification of several specimens of early hominids, more particularly of Paranthropus, dating back between 4.5 million and 2.5 million years, as well as evidence of the domestication of fire 1.8 million to 1 million years ago.

Viñales Valley
Cuba

Criteria – Significance in human history

Viñales Valley supplies tobacco for the manufacture of the country's famous cigars. Traditional production methods, including animal traction, are still used after recent experiments showed that mechanical methods lower the quality of the tobacco.

The Viñales Valley is encircled by mountains and its landscape is interspersed with dramatic rocky outcrops. Traditional techniques are still in use for agricultural production, particularly of tobacco. The quality of this cultural landscape is enhanced by the vernacular architecture of its farms and villages, where a rich multi-ethnic society survives, illustrating the cultural development of the islands of the Caribbean, and of Cuba.

World Heritage site since

1978 1979 1980 1981 1982 1983 1984 1985 1986 1987 1988 1989 1990 1991 1992 1993 1994 1995 1996 1997 1998 **1999** 2000 2001 2002 2003 2004 2005 2006 2007 2008

Kalwaria Zebrzydowska: the Mannerist Architectural and Park Landscape Complex and Pilgrimage Park
Poland

Criteria – Interchange of values; Significance in human history

The layout of Kalwaria Zebrzydowska was based on the landscape of Jerusalem at the time of Christ. This used a system of measurement that allowed the urban features of Jerusalem to be reproduced symbolically on the natural landscape.

Kalwaria Zebrzydowska is a breathtaking cultural landscape of great spiritual significance. The Counter-Reformation in the late-sixteenth century led to a flowering in the creation of calvaries in Europe. Mikolaj Zebrzydowski, the Voivod of Cracow, had already completed a private hermitage on this site when he was persuaded by Bernardine (Cistercian) monks to enlarge his original design. Building work on the new calvary started in 1600, eventually covering an extensive landscape complex with many chapels, linked in form and theme to those in Jerusalem. It was conceived as being for the use not only of the local inhabitants but also of believers from elsewhere in Poland and neighbouring countries. Remarkably, its layout has remained virtually unchanged since its conception and it remains to this day a sacred place of pilgrimage.

The Monastery of Kalwaria Zebrzydowska.

▼

World Heritage site since

1978 1979 1980 1981 1982 1983 1984 1985 1986 1987 1988 1989 1990 1991 1992 1993 1994 1995 1996 1997 1998 **1999** 2000 2001 2002 2003 2004 2005 2006 2007 2008

Villa Adriana (Tivoli)
Italy

Criteria – Human creative genius; Interchange of values; Testimony to cultural tradition

The Villa Adriana at Tivoli, near Rome, is a masterpiece that uniquely brings together the highest expressions of the material cultures of the ancient Mediterranean world. Its complex of classical buildings was created in the second century AD by the Roman emperor Hadrian. It was a symbol of a power that was gradually becoming absolute and which distanced itself from the capital. After Hadrian's death in 138,

his successors preferred Rome as their permanent residence, but the villa continued to be enlarged and further embellished. Study of the villa's monuments played a crucial role in the rediscovery of the elements of classical architecture by the architects of the Renaissance and the Baroque period. It also profoundly influenced many nineteenth- and twentieth-century architects and designers.

The Golden Square is one of the most impressive buildings in the complex: the vast peristyle is surrounded by a two-aisled portico with alternate columns in cipollino marble and Egyptian granite.

The Maritime Theatre at Villa Adriana.
▼

World Heritage site since

1978 1979 1980 1981 1982 1983 1984 1985 1986 1987 1988 1989 1990 1991 1992 1993 1994 1995 1996 1997 1998 **1999** 2000 2001 2002 2003 2004 2005 2006 2007 2008

Brimstone Hill Fortress National Park
Saint Kitts and Nevis

Criteria – Testimony to cultural tradition;
Significance in human history

Brimstone Hill Fortress National Park is an outstandingly well-preserved example of seventeenth- and eighteenth-century military architecture in a Caribbean context. Designed by the British and built by African slave labour, the fortress is testimony to European colonial expansion, the African slave trade and the emergence of new societies in the Caribbean. The heart of the fortress is Fort George, the massive masonry structure on one of the twin peaks that dominate the complex and still in an excellent state of repair. It is the earliest surviving British example of the type of fortification known as the 'polygonal system', and one of the finest examples known anywhere in the world. The fortress was abandoned as a result of British defence cuts in 1853. The wooden buildings were auctioned and dismantled and masonry buildings were plundered for their cut stone.

On entering the fortress, the first structure is the Barrier Redoubt. Next comes the North-West Work, which incorporates the stout Magazine Bastion with its associated water catchments and cistern. This is linked to the South-East Work, the main feature of which is the Orillon Bastion. A prominent feature here is the bombproof Ordnance Storehouse.

Area de Conservación Guanacaste
Costa Rica

Criteria – Significant ecological and biological processes; Significant natural habitat for biodiversity

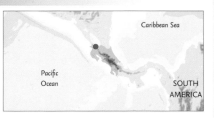

Guanacaste is located in northwestern Costa Rica. It stretches 105 km from the Pacific, across the Pacific coastal lowlands, over three high volcanoes and down into the Atlantic coastal lowlands. It includes the Guanacaste Cordillera and surrounding flatlands and coastal areas. It contains important natural habitats for the conservation of biological diversity, including the best dry-forest habitats from Central America to northern Mexico and key habitats for endangered or rare plant and animal species. The marine area includes various near-shore islands and islets (mostly uninhabited), open ocean marine zones, beaches, rocky coasts, and approximately 20 km of sea turtle nesting beaches and a high diversity of wetland ecosystems (thirty-seven wetlands). The wetland forests are considered to be among the most pristine in Central America and worldwide.

Guanacaste's beaches are of global importance for the protection of Olive ridley sea turtles and leatherback sea turtles, both endangered. The Naranjo and Nancite beaches alone host over 250,000 turtles during the breeding and mating season.

World Heritage site since

1978 1979 1980 1981 1982 1983 1984 1985 1986 1987 1988 1989 1990 1991 1992 1993 1994 1995 1996 1997 1998 **1999** 2000 2001 2002 2003 2004 2005 2006 2007 2008

Dacian Fortresses of the Orastie Mountains
Romania

Criteria – Interchange of values; Testimony to cultural tradition; Significance in human history

Built in the first centuries BC and AD under Dacian rule, these fortresses show an unusual fusion of military and religious architectural techniques and concepts from the classical world and the late-European Iron Age. The Dacians inhabited the central and western part of the region between the Carpathians and the Danube. It was a typical Iron Age culture, practising agriculture, stock-raising, fishing and metal-working, as well as trade with the Graeco-Roman world. These six defensive works, which formed the nucleus of the Dacian Kingdom, were conquered by the Romans at the beginning of the second century AD. Their extensive and well-preserved remains stand in spectacular natural surroundings and give a dramatic picture of a vigorous and innovative civilization.

The system developed by the Dacians to defend their capital, Sarmizegetusa Regia, was composed of three distinct fortified elements: sites on dominant physical features, fortresses, and linear defences.

The circular sanctuary at Sarmizegetusa Regia. ▼

World Heritage site since

1978 1979 1980 1981 1982 1983 1984 1985 1986 1987 1988 1989 1990 1991 1992 1993 1994 1995 1996 1997 1998 **1999** 2000 2001 2002 2003 2004 2005 2006 2007 2008

Mount Wuyi
China

Criteria – Testimony to cultural tradition;
Heritage associated with events of universal
significance; Natural phenomena or beauty;
Significant natural habitat for biodiversity

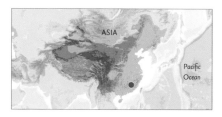

Mount Wuyi is the most outstanding area for
biodiversity conservation in southeast China.
It contains what is probably the largest and
best preserved areas of humid subtropical
forest in the world, and is a refuge for a large
number of ancient, relict species, many of
them endemic to China. The riverine landscape
of the Nine Bend River (lower gorge) is also
of exceptional scenic quality in its
juxtaposition of smooth rock cliffs with clear,
deep water. Its serene beauty is complemented
by numerous temples and monasteries,
many now in ruins. These provided the
setting for the development and spread of
neo-Confucianism, a doctrine that played
a dominant role in the countries of eastern
and southeastern Asia for many centuries
and influenced philosophy and government
over much of the world.

Mount Wuyi has
received international
recognition for its
high diversity and
large numbers of rare
and unusual fauna.

Endangered species
include Chinese tiger,
clouded leopard, black
muntjac and Chinese
giant salamander.

World Heritage site since

1978 1979 1980 1981 1982 1983 1984 1985 1986 1987 1988 1989 1990 1991 1992 1993 1994 1995 1996 1997 1998 1999 2000 2001 2002 2003 2004 2005 2006 2007 2008

Bronze Age Burial Site of Sammallahdenmäki
Finland

Criteria – Testimony to cultural tradition;
Significance in human history

EUROPE

The Sammallahdenmäki cemetery includes thirty-three burial cairns and is the largest and best cairn site in all Finland. It provides a unique insight into the funerary practices and social and religious structures of northern Europe more than three millennia ago. Of the cairns, twenty-eight can be securely dated to the early Bronze Age. They lie along the crest and upper slopes of a 700-m-long ridge, and are disposed in several distinct clusters. The structures were built using granite boulders quarried from the cliff face or collected from the site itself. They can be classified into several different groups according to their shapes and sizes: small low round cairns, large mound-like cairns, and round walled cairns. They enclose cists made from stone slabs.

The site is associated with sun worship rituals, a cult that spread from Scandinavia over the entire region. At the time the hill of Sammallahdenmäki was completely bare of trees and was probably chosen for its unimpeded view of the sea and its openness to the sun in all directions.

Miguasha National Park
Canada

Criteria – Major stages of Earth's history

The palaeontological site of Miguasha Park, in southeastern Quebec on the southern coast of the Gaspé peninsula, is considered to be the world's most outstanding illustration of the Devonian Period known as the 'Age of Fishes'. Dating from 380 million years ago, the Upper Devonian Escuminac Formation represented here contains five of the six fossil fish groups associated with this period. Its paramount importance is due to it having the greatest number and best-preserved fossil specimens of the lobe-finned fishes that gave rise to the first four-legged, air-breathing terrestrial vertebrates – the tetrapods.

The site is characterized not only by the number of its fossil specimens but also by their exceptional condition, which allows further study: for example, of soft body parts represented in gill imprints, digestive traces and cartilaginous elements of skeleton.

World Heritage site since

1978 1979 1980 1981 1982 1983 1984 1985 1986 1987 1988 1989 1990 1991 1992 1993 1994 1995 1996 1997 1998 **1999** 2000 2001 2002 2003 2004 2005 2006 2007 2008

Jurisdiction of Saint-Emilion
France

Criteria – Testimony to cultural tradition;
Significance in human history

The Jurisdiction of Saint-Emilion is an outstanding example of a historic vineyard landscape that has survived intact and remains active to the present day. Viticulture was introduced to this fertile region by the Roman emperor Augustus. He created the province of Aquitania in 27 BC and established the first vineyards by grafting new varieties of grape on the Vitis biturica that grew naturally in the region. This industry intensified in the Middle Ages when the Saint-Emilion area benefited from its location on the pilgrimage route to Santiago de Compostela. Many churches, monasteries and hospices were built there from the eleventh century onwards. It was granted the special status of a 'jurisdiction' during the period of English rule in the twelfth century.

In the eighteenth century the quality of the wines from the region was recognized as exceptional. Saint-Emilion has also been noteworthy for its innovations, such as the establishment of the first wine syndicate in 1884 and the first cooperative cellars in the Gironde in 1932.

World Heritage site since

1978 1979 1980 1981 1982 1983 1984 1985 1986 1987 1988 1989 1990 1991 1992 1993 1994 1995 1996 1997 1998 **1999** 2000 2001 2002 2003 2004 2005 2006 2007 2008

Western Caucasus
Russian Federation

Criteria – Significant ecological and biological processes; Significant natural habitat for biodiversity

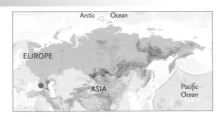

The Western Caucasus, extending over 2,750 km² of the extreme western end of the Caucasus mountains and located 50 km northeast of the Black Sea, is one of the few large mountain areas of Europe that has not experienced significant human impact. Its subalpine and alpine pastures have only been grazed by wild animals, and its extensive tracts of undisturbed mountain forests, extending from the lowlands to the subalpine zone, are unique in Europe. The site has a great diversity of ecosystems, with important endemic plants and wildlife, and is the place of origin and reintroduction of the mountain subspecies of the European bison.

The lack of human disturbance in this area has allowed ecological processes to continue naturally over the millennia. This offers exceptional opportunities for studying both competitive interactions between grazing animals and predator/prey interactions.

Puerto-Princesa Subterranean River National Park
Philippines

Criteria – Natural phenomena or beauty; Significant natural habitat for biodiversity

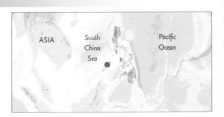

This park features a spectacular limestone karst landscape with an underground river. One of the river's distinguishing features is that it emerges directly into the sea, and its lower portion is subject to tidal influences. The area also represents a significant habitat for biodiversity conservation. The site contains a full 'mountain-to-sea' ecosystem and has some of the most important forests in Asia.

The spectacular karst landscape has both surface karst features and an extensive 8.2-km-long underground river system – one of the most unique of its type in the world, with chambers up to 120 m wide and 60 m high.

World Heritage site since

1978 1979 1980 1981 1982 1983 1984 1985 1986 1987 1988 1989 1990 1991 1992 1993 1994 1995 1996 1997 1998 **1999** 2000 2001 2002 2003 2004 2005 2006 2007 2008

Laurisilva of Madeira
Portugal

Criteria – Significant ecological and biological processes; Significant natural habitat for biodiversity

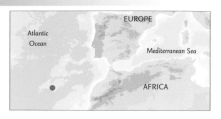

The laurel forest's ancient trees in the valley bottoms, waterfalls and cliffs provide spectacular scenery and have great ecological value. By collecting and retaining moisture they help protect the micro-climate and maintain water supplies.

The Laurisilva of Madeira is an outstanding relict of a virtually extinct flora of great interest. Fossil evidence shows that laurisilva flora once covered much of southern Europe in the Tertiary era, 15–40 million years ago, but this vegetation is now confined to the Azores, Madeira and the Canary Islands. The laurisilva on Madeira is the largest area surviving, with approximately 150 km² within the 270 km² Madeira Nature Reserve, and is in very good condition, with around 90 per cent believed to be primary forest. The laurisilva of Madeira is notable for its biological diversity with at least sixty-six vascular plant species endemic to the island and many endemic animal species such as the Madeiran long-toed pigeon.

World Heritage site since

1978 1979 1980 1981 1982 1983 1984 1985 1986 1987 1988 1989 1990 1991 1992 1993 1994 1995 1996 1997 1998 **1999** 2000 2001 2002 2003 2004 2005 2006 2007 2008

Historic Centre (Chorá) with the Monastery of Saint John 'the Theologian' and the Cave of the Apocalypse on the Island of Pátmos

Greece

Criteria – Testimony to cultural tradition; Significance in human history; Heritage associated with events of universal significance

The small island of Pátmos in the Dodecanese is reputed to be where St John the Theologian wrote both his Gospel and the Apocalypse, two of Christianity's most sacred works.

The Monastery of Hagios Ioannis Theologos (Saint John the Theologian) and the Cave of the Apocalypse was founded in the late-tenth century on the island of Pátmos, the northernmost island of the Dodecanese group. Together with the associated medieval settlement of Chorá, the monastery constitutes an exceptional example of a traditional Greek Orthodox pilgrimage centre of outstanding architectural interest. The town of Chorá is one of the few settlements in Greece that have evolved uninterruptedly since the twelfth century. There are few other places in the world where religious ceremonies that date back to the early Christian times are still being practised unchanged.

The Monastery of Saint John 'the Theologian'.
▼

World Heritage site since

1978 1979 1980 1981 1982 1983 1984 1985 1986 1987 1988 1989 1990 1991 1992 1993 1994 1995 1996 1997 1998 **1999** 2000 2001 2002 2003 2004 2005 2006 2007 2008

Lorentz National Park
Indonesia

Criteria – Major stages of Earth's history;
Significant ecological and biological processes;
Significant natural habitat for biodiversity

Lorentz National Park, 25,000 km²,is the largest protected area in southeast Asia. It is the only protected area in the world to incorporate a continuous, intact transect from snowcap to tropical marine environment, including extensive lowland wetlands. Located at the meeting-point of two colliding continental plates, the area has a complex geology with ongoing mountain formation as well as major sculpting by glaciation. The area also contains fossil sites which provide evidence of the evolution of life on New Guinea, a high level of endemism and the highest level of biodiversity in the region.

The park has two very distinct zones: the swampy lowlands and the high mountain area of the central cordillera. This zone is one of only three equatorial regions of sufficiently high altitude to retain permanent ice.

Shrines and Temples of Nikko
Japan

Criteria – Human creative genius; Significance in human history; Heritage associated with events of universal significance

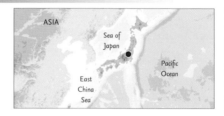

The shrines and temples of Nikko, together with their natural surroundings, have for centuries been a sacred site known for its architectural and decorative masterpieces. They are closely associated with the history of the Tokugawa Shoguns.

The shrines and temples and their environment are associated with the Shinto perception of the relationship of man with nature, in which mountains and forests have a sacred meaning. This religious practice is still very much alive today.

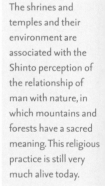

◄
The Tashogu Temple at Nikko.

World Heritage site since

1978 1979 1980 1981 1982 1983 1984 1985 1986 1987 1988 1989 1990 1991 1992 1993 1994 1995 1996 1997 1998 **1999** 2000 2001 2002 2003 2004 2005 2006 2007 2008

Archaeological Monuments Zone of Xochicalco
Mexico

Criteria – Testimony to cultural tradition;
Significance in human history

Xochicalco is an exceptionally well-preserved and complete example of a fortified political, religious and commercial centre from the Epiclassic period of Mesoamerica. Its architecture and art represent the fusion of cultural elements from different parts of Mesoamerica, from the troubled period of 650–900 that followed the break-up of the great Mesoamerican states such as Teotihuacan,

Monte Albán, Palenque and Tikal. The intensive cultural regrouping that followed the breakdown of these earlier political structures fuelled the city's growth. The city was built on a series of natural hills. The highest of these was the core of the settlement, with many public buildings, but evidence of occupation has been found on six of the lower hills surrounding it.

The scale of the city's engineering work is substantial. Terracing and massive retaining walls have created a series of open spaces defined by platforms and pyramidal structures. They are linked by a complex system of staircases, terraces and ramps to create a main north-south communication axis through the city.

World Heritage site since

1978 1979 1980 1981 1982 1983 1984 1985 1986 1987 1988 1989 1990 1991 1992 1993 1994 1995 1996 1997 1998 **1999** 2000 2001 2002 2003 2004 2005 2006 2007 2008

Atlantic Forest South-East Reserves
Brazil

Criteria – Natural phenomena or beauty;
Significant ecological and biological processes;
Significant natural habitat for biodiversity

The twenty-five reserves that make up this site contain some of the best and most extensive examples of the remaining Atlantic forests in Brazil and display the biological wealth and evolutionary history of the one of the world's richest and most endangered habitats. From mountains covered by dense forests down to wetlands, and to coastal islands with isolated mountains and dunes, the area comprises a rich natural environment of great scenic beauty. Partially isolated since the Ice Age, the Atlantic forests have evolved into a complex ecosystem with exceptionally high endemism (70 per cent of the tree species, 85 per cent of the primates and 39 per cent of the mammals). More than 450 tree species per hectare can be found in some areas, indicating that the diversity of woody plants in the region is larger than in the Amazon rainforest.

The reserves contain a highly diverse fauna with several species of conservation concern. There are some 120 species of mammals, including jaguar (pictured below), ocelot, bush dog, La Plata otter, twenty species of bat and various species of endangered primate, notably muriqui and brown howler monkey. The newly discovered black-faced lion tamarin is endemic to the area.

World Heritage site since

1978 1979 1980 1981 1982 1983 1984 1985 1986 1987 1988 1989 1990 1991 1992 1993 1994 1995 1996 1997 1998 **1999** 2000 2001 2002 2003 2004 2005 2006 2007 2008

Cueva de las Manos, Río Pinturas
Argentina
Criteria – Testimony to cultural tradition

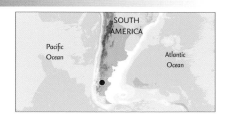

The hunting scenes in the caves depict animals and human figures interacting in a dynamic and naturalistic manner. Different hunting strategies are depicted, with animals being surrounded, ambushed or attacked by hunters using bolas, weighted throwing weapons.

The Cueva de las Manos, Río Pinturas, contains an exceptional assemblage of cave art, executed between 13,000 and 9,500 years ago. It takes its name 'Cave of the Hands' from the stencilled outlines of human hands in the cave, but there are also many depictions of animals, such as guanacos, which are still common in the region, as well as hunting scenes. The people who were responsible for the paintings may have been the ancestors of the historic hunter-gatherer communities of Patagonia found by European settlers in the nineteenth century.

World Heritage site since

1978 1979 1980 1981 1982 1983 1984 1985 1986 1987 1988 1989 1990 1991 1992 1993 1994 1995 1996 1997 1998 1999 **2000** 2001 2002 2003 2004 2005 2006 2007 2008

Walled City of Baku with the Shirvanshah's Palace and Maiden Tower
Azerbaijan
Criteria – Significance in human history

The Maiden Tower is a unique monument of Azerbaijan architecture rising to eight storeys. This astonishing cylindrical structure was built in two periods, with the bottom three storeys built as early as the seventh or sixth centuries BC as an astronomical observatory or fire temple.

Built on a site inhabited since the Palaeolithic period, the Walled City of Baku reveals evidence of Zoroastrian, Sasanian, Arabic, Persian, Shirvani, Ottoman, and Russian presence in cultural continuity. The Inner City (Icheri Sheher) has preserved much of its twelfth-century defensive walls. The twelfth-century Maiden Tower (Giz Galasy) is built over earlier structures dating from the seventh to sixth centuries BC, and the fifteenth-century Shirvanshahs' Palace is one of the pearls of Azerbaijan's architecture.

◄ Maiden Tower.

World Heritage site since

1978 1979 1980 1981 1982 1983 1984 1985 1986 1987 1988 1989 1990 1991 1992 1993 1994 1995 1996 1997 1998 1999 **2000** 2001 2002 2003 2004 2005 2006 2007 2008

Kronborg Castle
Denmark

Criteria – Significance in human history

North
Sea

Baltic
Sea

EUROPE

Located on a strategically important site commanding the Sund – the stretch of water between Denmark and Sweden – the royal castle of Kronborg at Helsingør is of immense symbolic value to the Danish people. Work began on the construction of this outstanding Renaissance castle in 1574. In 1629 Kronborg was devastated by fire, only the walls being left standing. Christian IV immediately commissioned the restoration

of the castle, largely returning it to its original appearance. Under Frederik III and Christian V large fortifications were built, the outer defensive works were considerably enlarged under Frederik IV, and the castle itself underwent substantial restoration and alteration. In 1785 it passed to the military and has remained intact to the present day. It is world-renowned as Elsinore, the setting of Shakespeare's Hamlet.

In its original form the Banqueting Hall had a magnificently carved and gilded ceiling and its walls were hung with tapestries. Only fourteen of the tapestries, prepared for the north wall and depicting Danish kings, have survived; of these, seven are on display at Kronborg, the remainder being in the National Museum in Copenhagen.

World Heritage site since

1978 1979 1980 1981 1982 1983 1984 1985 1986 1987 1988 1989 1990 1991 1992 1993 1994 1995 1996 1997 1998 1999 **2000** 2001 2002 2003 2004 2005 2006 2007 2008

Greater Blue Mountains Area
Australia

Criteria – Significant ecological and biological processes; Significant natural habitat for biodiversity

The Greater Blue Mountains consists of a mostly forested landscape on a sandstone plateau that commences about 60 kms west of central Sydney. The property is made up of eight protected areas: the Blue Mountains, Wollemi, Yengo, Nattai, Kanangra-Boyd, Gardens of Stone and Thirlmere Lakes National Parks, and the Jenolan Caves Karst Conservation Reserve. The plateau has enabled the survival of a rich diversity of plant and animal life by providing a refuge from climatic changes during recent geological history. It is particularly noted for its wide representation of eucalypt communities, ranging from wet and dry sclerophyll forests to mallee heathlands, and for the diversity of species of eucalypts. There are 101 species of eucalypt (over fourteen per cent of the global total) in the Greater Blue Mountains, twelve of which are believed to occur only in the Sydney sandstone region.

The Greater Blue Mountains also contains ancient, relict species of global significance. The most famous of these is the recently discovered Wollemi pine, a 'living fossil' dating back to the age of the dinosaurs. Thought to have been extinct for millions of years, the few surviving trees of this ancient species are known only from three small populations located in remote gorges within the Area.

Wollondilly Lookout, Nattai National Park. ▼

World Heritage site since

1978 1979 1980 1981 1982 1983 1984 1985 1986 1987 1988 1989 1990 1991 1992 1993 1994 1995 1996 1997 1998 1999 **2000** 2001 2002 2003 2004 2005 2006 2007 200

Garden Kingdom of Dessau-Wörlitz
Germany
Criteria – Interchange of values; Significance in human history

The Garden Kingdom of Dessau-Wörlitz is an outstanding example of the application of the philosophical principles of the Age of the Enlightenment to the design of a landscape that integrates art, education and economy in a harmonious whole. The first essays in grand landscape design began in 1683 with the plan for Oranienbaum, which unified town, palace, and park. The resulting complete Baroque ensemble, with obvious Dutch connections deriving from its designer, Cornelis Ryckwaert, has survived to this day. Further developments on these lines took place around 1700 with the reclamation of marshy areas along the river Elbe and the creation of planned villages and farmsteads. The final major stage of design was the work done by Prince Leopold III Friedrich Franz over the entire principality.

Prince Leopold III Friedrich Franz (1740–1817) developed existing landscape designs into something far more remarkable. By the time he died, virtually the entire principality had become a unified garden whose characteristic features have been preserved.

Three Castles, Defensive Wall and Ramparts of the Market-Town of Bellinzone
Switzerland
Criteria – Significance in human history

The Bellinzone site consists of a group of fortifications grouped around the castle of Castelgrande, which stands on a rocky peak looking out over the entire Ticino valley. Running from the castle, a series of fortified walls protect the ancient town and block the passage through the valley. A second castle, Montebello, forms an integral part of the fortifications; a third but separate castle, Sasso Corbaro, was built on an isolated rocky promontory southeast of the other fortifications.

The Bellinzone ensemble is the sole remaining example in the entire Alpine region of medieval military architecture.

◄
The towers of the castle of Castelgrande.

World Heritage site since

1978 1979 1980 1981 1982 1983 1984 1985 1986 1987 1988 1989 1990 1991 1992 1993 1994 1995 1996 1997 1998 1999 **2000** 2001 2002 2003 2004 2005 2006 2007 2008

uKhahlamba / Drakensberg Park
South Africa

Criteria – Human creative genius; Testimony to cultural tradition; Natural phenomena or beauty; Significant natural habitat for biodiversity

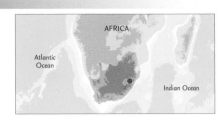

Drakensberg Park, or uKhahlamba Park, is the largest protected area on The Great Escarpment of the southern African subcontinent. It comprises a northern and a significantly larger southern section. There is considerable variation in topography, including vast basalt and sandstone cliffs, deep valleys, intervening spurs and extensive plateau areas. Among a total of 2,153 species of plant, there are a large number of internationally and nationally threatened species. The fauna includes 48 mammal, 296 bird, 48 reptile, 26 amphibian and 8 fish species. This spectacular natural site also contains many caves and rock-shelters with the largest and most concentrated group of paintings in Africa south of the Sahara, made by the San people over a period of 4,000 years.

The vegetation in the park is influenced by topography and the effects of climate, soil, geology, slope, drainage and fire. It is altitudinally zoned, forming three belts coinciding with the main topographical features: the river valley system, the spurs and the summit plateau.

The Giant's Castle Game Reserve in the central region of the Drakensberg Park.
▼

World Heritage site since

1978 1979 1980 1981 1982 1983 1984 1985 1986 1987 1988 1989 1990 1991 1992 1993 1994 1995 1996 1997 1998 1999 **2000** 2001 2002 2003 2004 2005 2006 2007 2008

Stone Town of Zanzibar
Tanzania

Criteria – Interchange of values; Testimony to cultural tradition; Heritage associated with events of universal significance

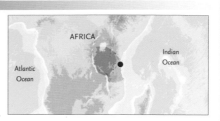

The restored old dispensary, now the Stone Town Cultural Centre.

The historical evolution of the Stone Town is illustrated by its street pattern, one of narrow, winding streets resulting from the unplanned building of houses and shops. There are few public open spaces as many of the houses have their own enclosed spaces. The principal construction material is coralline-rag stone set in a thick, lime mortar and then plastered and lime-washed.

The vernacular architecture of the Stone Town is preponderantly of two-storey buildings with long, narrow rooms around an open courtyard, reached through a narrow corridor.

The Stone Town of Zanzibar is a fine example of the Swahili coastal trading towns of east Africa. It retains its urban fabric and townscape virtually intact and contains many fine buildings that reflect its culture, which has brought together and homogenized disparate elements of the cultures of Africa, Arabia, India and Europe over more than a millennium.

For many centuries there was intense trading activity between Asia and Africa, and this is illustrated admirably by the architecture and urban structure of the Stone Town. Zanzibar also has great symbolic importance in the history of slavery because it was one of the main slave-trading ports in east Africa and a key location in the Arab slave trade.

Zanzibar was one of a loose confederation of small coastal city states known as the Zenj bar (Black Empire) which operated in the eighth–tenth centuries; the name derives from the Perso-Arabic word meaning 'the coast of the blacks'. Arab settlers were present in Zanzibar by the tenth century.

The Portuguese established a presence in the region at the end of the fifteenth century, using it as a base for exploration, but they were expelled by Omani Arabs in the seventeenth century. The Arabs formed Zanzibar's ruling elite until the twentieth century.

The Omanis traded in spices, ivory and slaves, and the city became the largest slave-trade port in east Africa: in the nineteenth century, an estimated 50,000 slaves a year were passing through Zanzibar. As a consequence the city's rulers and merchants grew very rich and embellished the Stone Town, the oldest part of the city, with palaces and fine mansions. Its characteristic Swahili architecture was overwhelmed by the Omanis' style of massively built multi-storey blocks in mortared coral and with flat roofs. The Minaret Mosque dates from this period. Another architectural component came from India with the addition of wide verandas.

Modern urban development began during the reign of Sultan Barghash (1870–88). His most notable contribution to the Stone Town architecture was the House of Wonders, but his greatest legacy was the provision of piped water to the town. The final phase of architectural development came in 1890 when Zanzibar became a British protectorate. The British imported colonial architecture while also introducing features derived from the Islamic architecture of Istanbul and Morocco.

World Heritage site since

1978 1979 1980 1981 1982 1983 1984 1985 1986 1987 1988 1989 1990 1991 1992 1993 1994 1995 1996 1997 1998 1999 **2000** 2001 2002 2003 2004 2005 2006 2007 2008

Land of Frankincense
Oman

Criteria – Testimony to cultural tradition;
Significance in human history

This group of archaeological sites in Oman represents the production and distribution of frankincense, one of the most important luxury items of trade in antiquity, from the Mediterranean and Red Sea regions to Mesopotamia, India and China. They constitute outstanding testimony to the civilization that flourished in southern Arabia from the Neolithic to the late Islamic periods. The Oasis of Shishr and the entrepôts of Khor Rori and Al-Baleed are excellent examples of medieval fortified settlements in the Persian Gulf region and were important sites on the frankincense trade routes. The sources of ancient frankincense can be identified with the three areas in the Dhofar region in which the frankincense tree is still to be found, and are represented by the Frankincense Park of Wadi Dawkah.

The port of Sumhuram/Khor Rori was founded at the end of the first century to control the trade in Dhofar incense. Indian seamen who had brought cotton cloth, corn and oil in exchange for incense overwintered there, waiting for the favourable monsoon winds to take them home.

Monumental gate, Khor Rori.
▼

World Heritage site since

1978 1979 1980 1981 1982 1983 1984 1985 1986 1987 1988 1989 1990 1991 1992 1993 1994 1995 1996 1997 1998 1999 **2000** 2001 2002 2003 2004 2005 2006 2007 2008

Curonian Spit
Lithuania and Russian Federation

Criteria – Traditional human settlement

Human habitation of this elongated sand dune peninsula, 98 km long and 0.4–4 km wide, dates back to prehistoric times. Throughout this period it has been threatened by the natural forces of wind and waves. Its survival to the present day has been made possible only as a result of ceaseless human efforts to combat the erosion of the Spit, dramatically illustrated by continuing stabilization and reforestation projects. The most significant element of the Spit's cultural heritage is represented by the old fishing settlements. The earliest of these were buried in sand when the woodland cover was removed. Those that have survived are all along the coast of the lagoon. At the end of the nineteenth century more elaborate buildings – lighthouses, churches, schools and villas – began to be erected alongside the simpler vernacular houses.

The surviving buildings of cultural significance are the houses of fishermen built during the nineteenth century. In their original form they were built from wood and thatched with reeds. A homestead consisted of two or three buildings: a dwelling house, a cattle shed, and a smokehouse for curing fish.

Sand dunes on the Curonian Spit.
▼

World Heritage site since

1978 1979 1980 1981 1982 1983 1984 1985 1986 1987 1988 1989 1990 1991 1992 1993 1994 1995 1996 1997 1998 1999 **2000** 2001 2002 2003 2004 2005 2006 2007 2008

Imperial Tombs of the Ming and Qing Dynasties
China

Criteria – Human creative genius; Interchange of values; Testimony to cultural tradition; Significance in human history; Heritage associated with events of universal significance

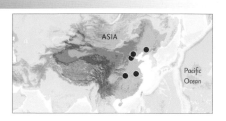

The Imperial Tombs of the Ming and Qing Dynasties are four groups of tombs in five provinces of eastern China. The tombs, designed in keeping with the Chinese principles of geomancy, provide outstanding evidence of Chinese beliefs and traditions from the fourteenth century onwards. The sites are characterized by the attempt to achieve harmony between the natural environment and the buildings.

A plain or broad valley was chosen, with the perspective of a mountain range to the north, against which the tombs would be built, with a lower elevation to the south. The site had to be framed on the east and west by chains of hills, and feature at least one waterway. A main access road several kilometres in length, known as the Way of the Spirits, led to the site.

The western Qing tomb contains fourteen imperial tombs and two building complexes: the Yongfu Tibetan Buddhist temple and the temporary palace where the imperial family resided when it came to honour its ancestors. The site of eastern Qing tombs contains fifteen mausolea in which 161 bodies were buried – emperors, empresses, concubines and princesses.

Western Qing tomb. ▼

World Heritage site since

1978 1979 1980 1981 1982 1983 1984 1985 1986 1987 1988 1989 1990 1991 1992 1993 1994 1995 1996 1997 1998 1999 **2000** 2001 2002 2003 2004 2005 2006 2007 2008

Historical Centre of the City of Arequipa

Peru

Criteria – Human creative genius; Significance in human history

The Historical Centre of Arequipa, built in a form of soft volcanic rock called sillar, represents a masterly integration of European and native building styles. This combination of influences is illustrated by the city's robust walls, archways and vaults, courtyards and open spaces, and the intricate Baroque decoration of its façades. The World Heritage site consists of forty-nine original blocks of the Spanish layout and twenty-four blocks from the colonial period and the nineteenth century. The core of the historic town is the Plaza de Armas (Plaza Mayor) with its mid-nineteenth-century cathedral, the most important neoclassical religious building in the country. At one corner of the Plaza are the church and cloisters of La Compañia, the most representative ensemble of the Baroque mestizo period at the end of the eighteenth century.

Façade of Iglesias de los Jesuitas, Arequipa. ▲

The merit of Arequipa architecture is also evident in the profusion of its *casonas*, characteristic well-proportioned vernacular houses. The Historical Centre contains some 500 *casonas*, of which over 250 are listed for protection. The heavy structures have been enhanced with ornamental designs in large thick rounded frames or deep protrusions and sculptures on flat surfaces.

World Heritage site since

1978 1979 1980 1981 1982 1983 1984 1985 1986 1987 1988 1989 1990 1991 1992 1993 1994 1995 1996 1997 1998 1999 **2000** 2001 2002 2003 2004 2005 2006 2007 2008

City of Verona
Italy

Criteria – Interchange of values; Significance in human history

The historic city of Verona was founded in the first century BC. It prospered under the rule of the Scaliger family in the thirteenth and fourteenth centuries and as part of the Republic of Venice from the fifteenth to eighteenth centuries. Verona has preserved a remarkable number of monuments from antiquity, and from the medieval and Renaissance periods. The core of the city consists of the Roman town in the loop of the Adige River, where the remains include city gates, the theatre and the Amphitheatre Arena, the second-largest after the Colosseum in Rome. The heart of Verona is the ensemble consisting of the Piazza delle Erbe and the Piazza dei Signori with their historic buildings, including the Palazzo del Comune, Palazzo del Governo, Loggia del Consiglio, Arche Scaligere and Domus Nova.

The cathedral (Duomo) was first built in the sixth century but rebuilt in the twelfth century after an earthquake. The façade, completed in the fourteenth century, is in Verona marble and has bas-reliefs representing sacred and profane episodes of different types. There is a fine twelfth-century cloister with arcades on double colonnades.

The Duomo and Adige River, Verona.
▼

World Heritage site since

1978 1979 1980 1981 1982 1983 1984 1985 1986 1987 1988 1989 1990 1991 1992 1993 1994 1995 1996 1997 1998 1999 **2000** 2001 2002 2003 2004 2005 2006 2007 2008

Ruins of León Viejo
Nicaragua

Criteria – Testimony to cultural tradition;
Significance in human history

León Viejo is one of the oldest Spanish colonial settlements in the Americas. The town was founded in 1524 by Francisco Hernández de Córdoba, who was sent from Panamá to conquer the Pacific zone northwards to Tezoatega (now the village of El Viejo). It developed, like many colonial towns in Latin America, round a central plaza, on the extreme northeast shore of what was to be called the Lake of León.

The town reached its peak of development around 1545, although it was still relatively small, with its Spanish population not exceeding some two hundred. The town was laid out on a grid pattern and excavations have uncovered remains of the Cathedral, the Convent of La Merced and the Royal Foundry.

The murder of Bishop Antonio de Valdivieso in 1550 seemed to mark a turning point in its fortunes: it was widely believed to have put a curse on the town, which suffered from both natural and economic disasters in the years that followed.

An eruption of the nearby volcano, Momotombo, in 1578, combined with raging inflation drove the richer inhabitants away. By 1603 there were only ten houses remaining. The final blow came on 11 January 1610, when a severe earthquake destroyed what was still standing, and the decision was taken to move the city to a site near the village of Subtiava.

Early Christian Necropolis of Pécs (Sopianae)
Hungary

Criteria – Testimony to cultural tradition;
Significance in human history

In the fourth century, a remarkable series of decorated tombs were constructed in the cemetery of the Roman provincial town of Sopianae, modern Pécs. The World Heritage site consists of sixteen funerary monuments which are important both structurally and architecturally, since they were built as underground burial chambers with memorial chapels above the ground. The tombs are important also in artistic terms, since they are richly decorated with Christian-themed murals of outstanding quality. One of the most remarkable is Burial chamber I (Peter-Paul) which is cut into the slope of the Mecsek hills. Discovered in 1782, this late fourth-century chamber consists of an above-ground chapel, the subterranean burial chamber proper with fine religious wall paintings, and a small vestibule leading to the burial chamber.

The burial chambers illustrate the unique early Christian sepulchral art and architecture of the northern and western Roman provinces. One of the most exceptional chambers has a niche carved above the sarcophagus with a painting of a wine pitcher and glass, symbolising the thirst of the soul journeying to the netherworld.

World Heritage site since

1978 1979 1980 1981 1982 1983 1984 1985 1986 1987 1988 1989 1990 1991 1992 1993 1994 1995 1996 1997 1998 1999 **2000** 2001 2002 2003 2004 2005 2006 2007 2008

Churches of Chiloé
Chile

Criteria – Interchange of values; Testimony to cultural tradition

The most typical feature of these buildings is the tower façade, on the side facing the esplanade, made up of an entrance portico, the gable wall or pediment, and the tower itself. Most are of two or three storeys, with hexagonal or octagonal drums to reduce wind resistance.

The Churches of Chiloé are outstanding examples of the successful fusion of European and indigenous cultural traditions to produce a unique form of wooden architecture. The mestizo culture, resulting from Jesuit missionary activities in the seventeenth and eighteenth centuries, has survived intact in the Chiloé archipelago, and achieves its highest expression in these wooden churches. By the end of the nineteenth century over 100 churches had been built; between fifty and sixty survive to the present day, and fourteen of these constitute the World Heritage site: Achao (Quinchao); Quinchao; Castro; Rilán (Castro); Nercón (Castro); Aldachildo (Puqueldón); Ichuac (Puqueldón); Detif (Puqueldón); Vilipulli (Chonchi); Chonchi; Tenaún (Quemchi); Colo (Quemchi); San Juan (Dalcahue); and Dalcahue. The traditional Chiloé churches are located near the shore, facing an esplanade.

Blaenavon Industrial Landscape
United Kingdom

Criteria – Testimony to cultural tradition; Significance in human history

Big Pit, the last substantial working colliery at Blaenavon, closed in 1980. Big Pit is now a museum of coal mining of international significance, and one of only two mining museums in the United Kingdom where visitors can be taken underground.

The area around Blaenavon is evidence of the pre-eminence of south Wales as the world's major producer of iron and coal in the nineteenth century. All the necessary elements can still be seen – coal and ore mines, quarries, a primitive railway system, furnaces, workers' homes, and the social infrastructure of the community. During the 1840s and 1850s, as population numbers increased with the influx of migrant workers, the scattered housing of the workers and the associated school, church and chapels were complemented by the evolution of a town with a variety of urban functions. There were three principal clusters of buildings in the area, one around the ironworks, one along the east-west axis, now King Street, and the other around St Peter's Church.

World Heritage site since

1978 1979 1980 1981 1982 1983 1984 1985 1986 1987 1988 1989 1990 1991 1992 1993 1994 1995 1996 1997 1998 1999 **2000** 2001 2002 2003 2004 2005 2006 2007 2008

Historic Town of St George and Related Fortifications, Bermuda
United Kingdom
Criteria – Significance in human history

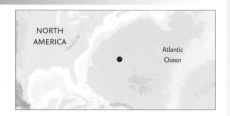

The historic town of St George is an outstanding example of a fortified colonial town dating from the early seventeenth century. It is the oldest English town in the New World. St George was a garrison town from its earliest days, and military installations developed on its eastern side. The first of many barracks were built on Barrack Hill in 1780, followed by other buildings such as hospitals and a chapel.

These were constructed in standard British military style but using local materials. The town's fortifications graphically illustrate the development of English military engineering from the seventeenth to the twentieth centuries, being adapted to take account of the development of artillery over this period. The fortifications continued to serve until the coastal defence came to an end in 1956.

The architecture of Bermuda has changed little since the end of the seventeenth century. The simple, well-proportioned houses, of one or two storeys, have roofs of stone slabs painted white. Some of the houses, such as Bridge House, the Hunter Building, or Whitehall, are impressive mansions, dating from the nineteenth century and embellished with imposing balconies and verandas.

Archaeological Ensemble of Tárraco
Spain
Criteria – Interchange of values; Testimony to cultural tradition

Tárraco, modern-day Tarragona, was a major administrative and mercantile city in Roman Spain and the centre of the Imperial cult for all the Iberian provinces. It was endowed with many fine buildings, and parts of these have been revealed in a series of exceptional excavations. Although most of the remains are fragmentary, many preserved beneath more recent buildings, they present a vivid picture of the grandeur of this Roman provincial capital.

The Roman town was sited on a hill with the seat of the provincial government at its crest and on two terraces created below. The town is rich in important buried remains, including some buildings that are completely preserved.

◄ The Roman Circus.

World Heritage site since

1978 1979 1980 1981 1982 1983 1984 1985 1986 1987 1988 1989 1990 1991 1992 1993 1994 1995 1996 1997 1998 1999 **2000** 2001 2002 2003 2004 2005 2006 2007 2008

The Loire Valley between Sully-sur-Loire and Chalonnes
France

Criteria – Human creative genius; Interchange of values; Significance in human history

Château de Chambord.

The Loire basin occupies a huge area in central and western France stretching from the southern part of the Massif Central to an Atlantic coast estuary. Along the Loire, between Orléans and Angers, the valley is characterized by low cliffs of tufa and limestone. The valley has a long history of periodic catastrophic flooding, carefully recorded as stone-cut water levels at numerous places along its route, and even today its inhabitants live under perennial threat of severe inundation. Much contemporary river management is orientated to minimizing that risk.

The Loire Valley is an outstanding cultural landscape of great beauty, containing historic towns and villages, the great architectural monuments of the chateaux and cultivated lands formed by two millennia of interaction between their population and the physical environment, primarily the river Loire itself. It is noteworthy for the quality of its architectural heritage, in its historic towns such as Blois, Chinon, Orléans, Saumur and Tours, but in particular in its world-famous castles, such as the Château de Chambord (see photo on the right).

For most of its length within the World Heritage site, the Loire is confined by dykes. Its banks are punctuated at intervals of only a few kilometres by a series of villages, small towns and cities. Land use is varied, from urban density through intense horticulture to vineyards (some reliant on flooding) and hunting forest.

Roman impact on the landscape was massive, and it still strongly influences settlement location, urban form and road communications. The Loire was one of the most important arteries for communications and trade in Gaul. In the late Roman period around 372, St Martin, Bishop of Tours, founded an abbey at Marmoutier and this served as the model for other monastic settlements in the Loire Valley in the centuries that followed.

The sanctuary at Tours was an early pilgrimage centre and the area's many monasteries served as focal points for settlement in the Middle Ages, as did local fortified settlements.

The Loire Valley was a frontier zone during the Hundred Years' War (1337–1453) and the scene of many confrontations between the French and English. Castles were rebuilt and extended to become massive fortresses, the forerunners of the chateaux of today. The ever-present danger to Paris from the English resulted in the royal court spending long periods at Tours. With the end of the war the valley proved an ideal place for humanism and the Renaissance to take root in France: this involved the dismantling of fortresses and their reconstruction as palaces for pleasure and recreation.

The seventeenth–eighteenth centuries saw the development of a secular commercial economy based on industry and trade. The romantic representation of the valley in the nineteenth century by writers and painters led to the Loire becoming a magnet for tourists, first from France, then Europe and the rest of the world.

World Heritage site since

1978 1979 1980 1981 1982 1983 1984 1985 1986 1987 1988 1989 1990 1991 1992 1993 1994 1995 1996 1997 1998 1999 **2000** 2001 2002 2003 2004 2005 2006 2007 2008

Central Amazon Conservation Complex
Brazil

Criteria – Significant ecological and biological processes; Significant natural habitat for biodiversity

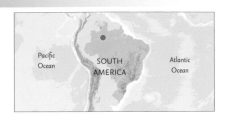

The Central Amazon Conservation Complex makes up the largest protected area in the Amazon Basin (over 60,000 km²) and is one of the planet's richest regions in terms of biodiversity. The site is made up of Jaú National Park, Demonstration area of Mamairauá Sustainable Development Reserve, Amanā Sustainable Development Reserve and the Anavilhanas Ecological Station. It includes an important sample of annually flooded (várzea) ecosystems, igapó forests, lakes and channels, which take the form of a constantly evolving aquatic mosaic that is home to the largest array of electric fish in the world. Numerous species of conservation concern live within the park, including jaguar, giant otter, Amazonian manatee, South American river turtle and black cayman. The Complex contains approximately 60 per cent of the species of fish reported to exist within the Negro River watershed, and 60 per cent of the birds recorded from the central Amazon.

The site contains the nine-tier waterfall of the Carabinani River and also includes a significant proportion of the black-water drainage system, the headwaters of which are located primarily in the Guiane Shield. Its dark colour results from organic acids released into the water through the decomposition of organic matter and the lack of terrestrial sediments.

The Amazon Basin from space.
▼

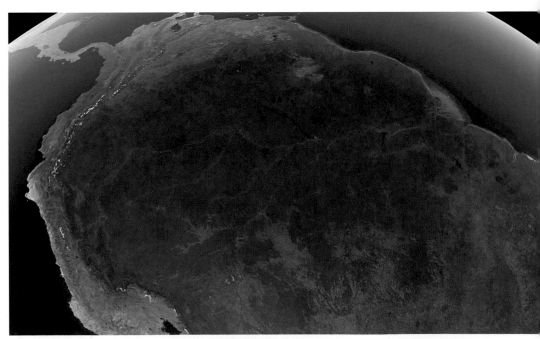

World Heritage site since

1978 1979 1980 1981 1982 1983 1984 1985 1986 1987 1988 1989 1990 1991 1992 1993 1994 1995 1996 1997 1998 1999 **2000** 2001 2002 2003 2004 2005 2006 2007 2008

Isole Eolie (Aeolian Islands)
Italy
Criteria – Major stages of Earth's history

EUROPE

Mediterranean Sea

Ionian Sea

The Isole Eolie (Aeolian Islands) are located off the northern coast of Sicily and provide an outstanding record of volcanic island-building and volcanic phenomena. The group consists of seven islands (Lipari, Vulcano, Salina, Stromboli, Filicudi, Alicudi and Panarea) and five small islets (Basiluzzo, Dattilo, Lisca Nera, Bottaro and Lisca Bianca) in the vicinity of Panarea. Studied since at least the eighteenth century, the islands have provided the science of

vulcanology with examples of two types of eruption (Vulcanian and Strombolian). A total of 900 plant species have been recorded in the Aeolian Islands, including four endemic species. About forty bird species have been recorded, including ten under the Sicilian Red List of threatened bird species. The islands are also important for migrant bird species, and are an 'Important Bird Area' for congregatory species identified by Birdlife International.

Panarea is the smallest of the islands. It has a remarkable variety of differing environments in comparison with the other islands, especially in terms of flora, and is a fascinating site for naturalists. Stromboli is the only island in the archipelago that has permanent volcanic activity.

Stromboli.
▼

World Heritage site since

1978 1979 1980 1981 1982 1983 1984 1985 1986 1987 1988 1989 1990 1991 1992 1993 1994 1995 1996 1997 1998 1999 **2000** 2001 2002 2003 2004 2005 2006 2007 2008

Island of Saint-Louis
Senegal
Criteria – Interchange of values; Significance in human history

Founded as a French colonial settlement in the seventeenth century, Saint-Louis was urbanized in the mid-nineteenth century. It was the capital of Senegal from 1872 to 1957 and played an important cultural and economic role in the whole of West Africa. The island consists of a North quarter and a South quarter, with the Place Faidherbe in the centre. It is situated in a magnificent lagoon formed by the two arms of the Senegal River. The regular town plan, the system of quays, and the characteristic colonial architecture give Saint-Louis its distinctive appearance. The main historic buildings include the ancient fort, now the Governor's Palace, which marks the centre of the island. Situated next to it is the cathedral, completed in 1828.

In 1957, Saint-Louis ceased being capital of Senegal. This meant the departure of the French garrison and their families and the closure of offices and shops, with a drastic reduction in the French population. At present the city has revived its economy, based on fishing, agriculture and tourism.

Gochang, Hwasun and Ganghwa Dolmen Sites
Korea, Republic of
Criteria – Testimony to cultural tradition

The Gochang, Hwasun and Ganghwa sites contain the highest density and greatest variety of dolmens in the Republic of Korea, and indeed of any country. Dolmens are found in western China and the coastal areas of the Yellow Sea basin. They arrived in the Korean Peninsula with the Bronze Age and usually consist of two or more undressed stone slabs supporting a huge capstone. They were simple burial chambers, erected over the bodies or bones of Neolithic and Bronze Age worthies. The cemeteries at Gochang, Hwasun and Ganghwa contain many hundreds of excellent examples, preserving important evidence of how the stones were quarried, transported and raised and of how dolmen types changed during the second and first millennia BC in northeast Asia.

The dolmens would originally have been covered by earth mounds (barrows), but these gradually disappeared as a result of weathering and animal action. They may also have been platforms on which corpses were exposed to permit excarnation to take place, leaving bones for burial in collective or family tombs.

Ganghwa dolmen.

World Heritage site since

1978 1979 1980 1981 1982 1983 1984 1985 1986 1987 1988 1989 1990 1991 1992 1993 1994 1995 1996 1997 1998 1999 **2000** 2001 2002 2003 2004 2005 2006 2007 2008

The Cathedral of St James in Šibenik
Croatia

Criteria – Human creative genius; Interchange of values; Significance in human history

EUROPE

Adriatic Sea

Tyrrhenian Sea

The Cathedral of St James in Šibenik (built 1431–1535) bears witness to the considerable exchanges in the field of monumental arts between northern Italy, Dalmatia and Tuscany in the fifteenth and sixteenth centuries. The three architects who succeeded one another in the construction of the Cathedral – Francesco di Giacomo, Georgius Mathei Dalmaticus and Niccolò di Giovanni Fiorentino – developed a structure built entirely from stone, using unique construction methods for the vaulting and the dome of the cathedral. There is a close correspondence between the interior and exterior forms of the building. The decorative elements of the cathedral, such as a remarkable frieze containing seventy-one sculptured faces of men, women, and children, also illustrate the successful fusion of Gothic and Renaissance art.

The roofing of the aisles, as well as that of the apses and the dome, is made from stone 'tiles'. These roofing tiles are laid side by side with their horizontal edges overlapping, and the joints are made by the perfect fit. On the dome the tiles are held in place by stone wedges fitted with great precision.

World Heritage site since

1978 1979 1980 1981 1982 1983 1984 1985 1986 1987 1988 1989 1990 1991 1992 1993 1994 1995 1996 1997 1998 1999 **2000** 2001 2002 2003 2004 2005 2006 2007 200

Ischigualasto / Talampaya Natural Parks
Argentina
Criteria – Major stages of Earth's history

Six geologic formations make up the Triassic basin, the earliest of which are the Talampaya and Tarjados formations, red sandstone that forms the impressive cliffs of the Talampaya National Park. The remaining formations are composed of lake beds, swamps, river channels and flood plain deposits. These formations contain the abundant vertebrate and flora fossils.

Talampaya National Park and the contiguous Ischigualasto Provincial Park straddle the border between the provinces of San Juan and La Rioja in northwestern Argentina. The site constitutes almost the entire sedimentary basin known as the Ischigualasto-Villa Union Triassic basin. It was formed by layers of continental sediments deposited by rivers, lakes and swamps over the entire Triassic period (245–208 million years ago). The sediments contain fossils of a wide range of plants and animals including the ancestors of mammals and dinosaurs. They constitute the world's most complete continental fossil record known from the Triassic, revealing the evolution of vertebrates as well as the environments they lived in during this period. Some fifty-six genera of fossil vertebrates have been recorded from the area, including fish, amphibians, and a great variety of reptiles.

World Heritage site since

1978 1979 1980 1981 1982 1983 1984 1985 1986 1987 1988 1989 1990 1991 1992 1993 1994 1995 1996 1997 1998 1999 **2000** 2001 2002 2003 2004 2005 2006 2007 2008

Historic Centre of Shakhrisyabz
Uzbekistan

Criteria – Testimony to cultural tradition;
Significance in human history

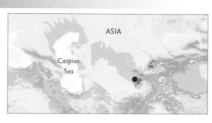

ASIA

Caspian
Sea

The historic centre of Shakhrisyabz contains
a collection of exceptional monuments and
ancient quarters which bear witness to the
city's secular development, and particularly
to the period of its apogee during the
Timurid rule of the fifteenth–sixteenth
centuries. The site consists of a number of
monuments, including the grandiose Ak-
Sarai Palace, begun in 1380; the Dorus

Saodat complex, housing the royal
mausoleum and religious areas; and the
eighteenth-century Chor-su bazaar and
baths. The dimensions of the magnificent
Ak-Sarai Palace , known as the 'White
Palace,' can be deduced from the size of the
gate-towers, traces of which still survive:
two towers each 50 m in height, and an arch
with a span of 22 m.

The Dorus Saodat
complex was designed
as a place of burial for
the ruling family and
contained, in addition
to the tombs
themselves, a prayer
hall, a mosque, and
accommodation for the
religious community
and pilgrims. The main
façade was faced with
white marble. The tomb
of Timur, also of white
marble, is a masterpiece
of the architecture of
this period.

The two gate towers of Ak-Sarai Palace. ▼

World Heritage site since

1978 1979 1980 1981 1982 1983 1984 1985 1986 1987 1988 1989 1990 1991 1992 1993 1994 1995 1996 1997 1998 1999 **2000** 2001 2002 2003 2004 2005 2006 2007 200

High Coast / Kvarken Archipelago
Finland and Sweden
Criteria – Major stages of Earth's history

Displacement of settlements by land uplift has created a relict cultural landscape: Stone Age remains from 5000 BC are 150 m above sea level, while those of the Bronze and Iron Ages are at 30 m and 15 m respectively.

The High Coast, Sweden, and the Kvarken Archipelago, Finland, are situated in the Gulf of Bothnia, a northern extension of the Baltic Sea. The 5,600 islands of the Kvarken Archipelago feature unusual ridged washboard moraines, 'De Greer moraines', formed by the melting of the continental ice sheet, 10,000 to 24,000 years ago. The Archipelago is continuously rising from the sea in a process of rapid glacio-isostatic uplift, whereby the land, previously pressed down under the weight of a glacier, lifts at rates that are among the highest in the world. As a consequence islands appear and unite, peninsulas expand, and lakes evolve from bays and develop into marshes and peat fens. The High Coast has also been largely shaped by the combined processes of glaciation, glacial retreat and the emergence of new land from the sea. Since the last retreat of the ice from the High Coast 9,600 years ago, the uplift has been in the order of 285 m which is the highest known 'rebound'. The site affords outstanding opportunities for the understanding of the important processes that formed the glaciated and land uplift areas of the Earth's surface.

Monastery of Geghard and the Upper Azat Valley
Armenia
Criteria – Interchange of values

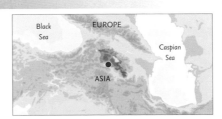

The monastery, founded in the fourth century and later rebuilt, was famous for its relics, most notably the spear that wounded Jesus on the Cross, and for relics of the Apostles Andrew and John.

The monastery of Geghard contains a number of churches and tombs, most of them cut into the rock, which illustrate the very peak of Armenian medieval architecture. The complex of medieval buildings is set into a landscape of great natural beauty, surrounded by towering cliffs at the entrance to the Azat Valley.

World Heritage site since

1978 1979 1980 1981 1982 1983 1984 1985 1986 1987 1988 1989 1990 1991 1992 1993 1994 1995 1996 1997 1998 1999 **2000** 2001 2002 2003 2004 2005 2006 2007 2008

Kinabalu Park
Malaysia

Criteria – Significant ecological and biological processes; Significant natural habitat for biodiversity

Kinabalu Park is dominated by Mount Kinabalu (4,095 m), the highest mountain between the Himalaya and New Guinea. It has a wide range of habitats, from rich tropical lowland and hill rainforest to tropical mountain forest, sub-alpine forest and scrub on the higher elevations. It has been designated as a 'Centre of Plant Diversity for Southeast Asia' and is exceptionally rich in species, with examples of flora from the Himalaya, China, Australia and Malaysia. Wildlife is also diverse with ninety species of lowland mammal and twenty-two others found in the montane zone. Four species of primate occur and 326 bird species have been recorded. Half of all Borneo's bird, mammal and amphibian species – and two-thirds of its reptiles – are represented in the park, including many rare and endangered species.

The park has between 5,000–6,000 vascular plant species, 1,000 of which are orchids. Rafflesia, a rare parasitic plant, is also found. The mountain flora has diverse 'living fossils' such as the celery pine and the trig-oak, the evolutionary link between oaks and beeches.

Summit of Mount Kinabalu.
▼

World Heritage site since

1978 1979 1980 1981 1982 1983 1984 1985 1986 1987 1988 1989 1990 1991 1992 1993 1994 1995 1996 1997 1998 1999 **2000** 2001 2002 2003 2004 2005 2006 2007 2008

Assisi, the Basilica of San Francesco and Other Franciscan Sites
Italy

Criteria – Human creative genius; Interchange of values; Testimony to cultural tradition; Significance in human history; Heritage associated with events of universal significance

EUROPE

Mediterranean Sea · Ionian Sea

Basilica of San Francesco. ▶

The Eremo delle Carceri on Mount Subasio were originally a series of caves where Francis and his companions came to pray. A small convent was later built on the site.

San Damiano is a monastic complex, essential for the understanding of the religious awakening of Francis; it was also Clare's convent and the place of her death.

Santa Maria degli Angeli is a sixteenth-century Renaissance church, built to protect the original chapel of Porziuncola, from where Francis sent his order to their mission and the place where he died.

The medieval hill town of Assisi is the birthplace of Saint Francis. It is a unique example of continuity of a city-sanctuary from its Umbrian-Roman and medieval origins to the present, represented in the cultural landscape, the religious ensembles and traditional land use. The town's medieval art masterpieces have made it a fundamental reference point for Italian and European art and architecture. It is also closely associated with the work of the Franciscan Order whose message has contributed significantly to developments in spirituality, art and architecture in the world.

Assisi stands on the hill of Asio at the foot of Mount Subasio. The most important event in its history was the life and work of Francis of Assisi (1182–1226), who initiated the Franciscan Order. His companion Clare, also later canonized, founded the sister order to the Franciscans, the Order of St Clare. After the canonization of Francis in 1228 it was decided to build a monumental church in his honour. This construction was followed by the Basilica of Santa Chiara to honour St Clare.

The Basilica of San Francesco is an outstanding example of an architectural ensemble that has significantly influenced the development of art and architecture.

Its construction began in 1228. The lower basilica is entered through an exquisite Gothic portal, its interior completely covered with frescoes. The upper basilica, entered through a loggia, has a magnificent east front in white limestone, with a large rose window in the centre. The interior walls are decorated with a series of paintings by Giotto and Cimabue relating to the faith and life of the saint.

The Cathedral of San Rufino probably dates from the eighth century; it was rebuilt around 1036 as a cathedral. The west front is a masterpiece of Umbrian Romanesque architecture, connected with the cathedral and the church of San Pietro of Spoleto.

The construction of the Basilica of Santa Chiara started in 1257. The structure is characterized by three flying buttresses and the interior is painted with a cycle of frescoes by several artists, illustrating the legend of St Clare.

Originally built outside the city walls, the Benedictine abbey of San Pietro is recorded from 1029; in the mid-twelfth century it adopted the Cluniac reform and it passed later to the Cistercians.

World Heritage site since

1978 1979 1980 1981 1982 1983 1984 1985 1986 1987 1988 1989 1990 1991 1992 1993 1994 1995 1996 1997 1998 1999 **2000** 2001 2002 2003 2004 2005 2006 2007 2008

Tiwanaku: Spiritual and Political Centre of the Tiwanaku Culture
Bolivia

Criteria – Testimony to cultural tradition;
Significance in human history

The ruins of Tiwanaku bear striking witness to the power of the empire that played a unique role in the development of the Andean pre-Hispanic civilization. The buildings are exceptional examples of the ceremonial and public architecture and art of one of the most important manifestations of the civilizations of the region. Tiwanaku began as a small settlement around 1200 BC. It was self-sufficient, with a non-irrigated form of farming based on frost-resistant crops, essential at this high altitude, producing tubers such as potatoes, oca and cereals, notably quinoa. In more sheltered locations near Lake Titicaca, maize and peaches were also cultivated. The inhabitants lived in rectangular adobe houses that were linked by paved streets. Following the introduction of copper metallurgy and improved irrigation, Tiwanaku's influence grew, reaching its apogee between 500 and 900 when it dominated a large area of the southern Andes and beyond.

The most imposing monument at Tiwanaku is the temple of Akapana. Originally a pyramid with seven superimposed platforms, its retaining walls rose to a height of over 18 m and were clad in distinctive blue stone.

World Heritage site since

1978 1979 1980 1981 1982 1983 1984 1985 1986 1987 1988 1989 1990 1991 1992 1993 1994 1995 1996 1997 1998 1999 **2000** 2001 2002 2003 2004 2005 2006 2007 2008

Mir Castle Complex
Belarus

Criteria – Interchange of values; Significance in human history

North Sea

EUROPE

Black Sea

The Mir Castle complex vividly symbolizes the history of Belarus and, as such, it is one of the major national symbols of the country. It lies in a fertile region in the geographical centre of Europe, at the crossroads of the most important trade routes, and was, at the same time, at the epicentre of crucial European and global military conflicts between neighbouring powers with different religious and cultural traditions. The construction of this castle, in the Gothic style, began at the end of the fifteenth century. It was subsequently extended and reconstructed, first in the Renaissance and then in the Baroque style. After being abandoned for nearly a century and suffering severe damage during the Napoleonic period, the castle was restored at the end of the nineteenth century, with the addition of a number of other elements and the landscaping of the surrounding area as a park. Its present form is graphic testimony to its often turbulent history.

The Mir Castle complex is situated on the bank of a small lake at the confluence of the Miryanka River and a small tributary. Its impressive fortified walls feature four exterior corner towers rising to five storeys and a six-storey external gate tower.

▼

World Heritage site since

1978 1979 1980 1981 1982 1983 1984 1985 1986 1987 1988 1989 1990 1991 1992 1993 1994 1995 1996 1997 1998 1999 **2000** 2001 2002 2003 2004 2005 2006 2007 2008

Rietveld Schröderhuis (Rietveld Schröder House)
Netherlands

Criteria – Human creative genius; Interchange of values

The Rietveld Schröder House in Utrecht was commissioned by Ms Truus Schröder-Schräder, designed by the architect Gerrit Thomas Rietveld, and built in 1924. This small family house, with its interior, the flexible spatial arrangement, and the visual and formal qualities, was a manifesto of the ideals of the De Stijl group of artists and architects in the Netherlands in the 1920s, and has since been considered one of the icons of the Modern Movement in architecture.

The design and building of the house took place simultaneously. The owner and architect commissioned a full photographic documentation of the architecture to ensure that the new approach of the De Stijl group was presented to reflect their intended ideas.

Noel Kempff Mercado National Park
Bolivia

Criteria – Significant ecological and biological processes; Significant natural habitat for biodiversity

The National Park is one of the largest (15,230 km²) and most intact parks in the Amazon Basin. With an altitudinal range from 200 m to nearly 1,000 m, it is the site of a rich mosaic of habitat types from Cerrado savanna and forest to upland evergreen Amazonian forest. Located on the border with Brazil, the site includes a large section of the Huanchaca Plateau and surrounding lowlands. Several rivers have their sources on the plateau and form spectacular waterfalls. The park boasts an evolutionary history dating back over a billion years to the Precambrian period. An estimated 4,000 species of flora, as well as over 600 bird species and viable populations of many globally endangered or threatened vertebrate species, live in the park.

The park has five distinct ecosystems and its outstanding habitat diversity favours the existence of a wide range of wildlife. More than 130 species of animals live in the park including many of the rare mammals of Bolivia such as river otters and jaguars.

World Heritage site since

1978 1979 1980 1981 1982 1983 1984 1985 1986 1987 1988 1989 1990 1991 1992 1993 1994 1995 1996 1997 1998 1999 **2000** 2001 2002 2003 2004 2005 2006 2007 2008

Holy Trinity Column in Olomouc
Czech Republic

Criteria – Human creative genius; Significance in human history

The Olomouc Holy Trinity Column is one of the most exceptional examples of the apogee of central European Baroque artistic expression. In the reconstruction (1648–1650) following the Thirty Years' War the city of Olomouc took on a new appearance. Many impressive public and private buildings were constructed in a local variant of the prevailing style, which became known as 'Olomouc Baroque'. The most characteristic expression of this style was a group of monuments (columns and fountains), of which the Holy Trinity Column is the crowning glory. Erected in the early years of the eighteenth century and rising to a height of 35 m, it is decorated with many fine religious sculptures, including the work of the distinguished Moravian artist Ondrej Zahner.

The Column forms a triumphal stone statement of the creeds of both Christianity and of citizenship. Faith and religious tradition are harmoniously combined with the idea of the city – its traditions, protection and civil administration. ▶

World Heritage site since

1978 1979 1980 1981 1982 1983 1984 1985 1986 1987 1988 1989 1990 1991 1992 1993 1994 1995 1996 1997 1998 1999 **2000** 2001 2002 2003 2004 2005 2006 2007 2008

Gunung Mulu National Park
Malaysia

Criteria – Natural phenomena or beauty; Major stages of Earth's history; Significant ecological and biological processes; Significant natural habitat for biodiversity

Gunung Mulu National Park on the island of Borneo protects an exceptional range of natural phenomena. The 529 km² park contains 17 vegetation zones, exhibiting some 3,500 species of vascular plants, 80 species of mammal and 270 species of bird (including 24 Borneon endemics). The park is considered to be one of the richest sites in the world for palms with 109 species of 20 genera identified. The park is dominated by Gunung Mulu, a 2,377-m-high sandstone pinnacle. Another outstanding karst feature is the 'pinnacles', 50-m-high sharp blades of rock that project through the rainforest canopy. The site also has a high concentration of large cave passages which provide a major wildlife spectacle of cave swiftlets and bats. Three million wrinkled-lipped freetail bats inhabit one cave alone.

The park is important for its karst features. There are at least 295 km of explored caves, estimated to be at least 2–3 million years old. Sarawak Chamber, which is 600 m by 415 m and 80 m high, is the largest known cave chamber in the world.

King's Chamber in Wind Cave in Gunung Mulu National Park.
▼

World Heritage site since

1978 1979 1980 1981 1982 1983 1984 1985 1986 1987 1988 1989 1990 1991 1992 1993 1994 1995 1996 1997 1998 1999 **2000** 2001 2002 2003 2004 2005 2006 2007 2008

Palmeral of Elche
Spain
Criteria – Interchange of values; Traditional human settlement

The Palmeral (a landscape of date palm groves) of Elche was laid out towards the end of the tenth century AD, when the Muslim city of Elche was also erected. Much of the Iberian peninsula was Arab at this time and the Palmeral represents a unique transference of Arab agricultural practices to the European continent. Cultivation of date palms in Elche is known at least since Iberian times, dating from around the fifth century BC, but the Arabs introduced more systematic practices, including an elaborate irrigation system which is still functioning. Arab geographers and European travellers throughout history have testified to this exceptional example of progressive landscaping.

Palms form an essential component of the culture of Elche, manifesting their influence on society in many ways, including the processions on Palm Sunday, the Night of the Kings and on the town's coat of arms. ▶

World Heritage site since

1978 1979 1980 1981 1982 1983 1984 1985 1986 1987 1988 1989 1990 1991 1992 1993 1994 1995 1996 1997 1998 1999 2000 2001 2002 2003 2004 2005 2006 2007 2008

Agricultural Landscape of Southern Öland
Sweden

Criteria – Significance in human history;
Traditional human settlement

The interaction between man and the natural environment in the south of Öland, an island in the Baltic sea, is of unique universal value. Human beings have lived here for some five thousand years and adapted their way of life to the physical constraints of the island. As a consequence, the landscape is unique, with abundant evidence of continuous human settlement from prehistoric times to the present day.

Land use has not changed significantly since then, with arable farming and animal husbandry still remaining the principal economic activity. The southern part of the island is dominated by Stora Alvaret, one of the largest limestone pavements in Europe. Its medieval pattern of villages and field systems is still clearly visible, a very rare occurrence in northern Europe.

The prosperity of the island, due in no small measure to its situation on the main trading route through the Kalmar Sound, is reflected in the imposing stone churches built in the twelfth century, such as those at Hulterstad and Resmö. They were fortified as defence against attacks from marauders.

Monastic Island of Reichenau
Germany

Criteria – Testimony to cultural tradition;
Significance in human history; Heritage
associated with events of universal significance

The island of Reichenau on Lake Constance preserves the traces of the Benedictine monastery, founded in 724, which exercised remarkable spiritual, intellectual and artistic influence. The churches of St Mary and St Marcus, St Peter and St Paul, and St George, mainly built between the ninth and eleventh centuries, provide a panorama of early medieval monastic architecture in central Europe. Their wall paintings bear witness to impressive artistic activity.

The Monastery of Reichenau was a highly significant artistic centre in tenth- and eleventh-century Europe, as illustrated by its monumental wall paintings and illuminations, and it became a famous centre for teaching and creativity in literature, science, and the arts.

◀

Church of St George in Reichenau.

World Heritage site since

1978 1979 1980 1981 1982 1983 1984 1985 1986 1987 1988 1989 1990 1991 1992 1993 1994 1995 1996 1997 1998 1999 **2000** 2001 2002 2003 2004 2005 2006 2007 2008

Wachau Cultural Landscape
Austria

Criteria – Interchange of values; Significance in human history

EUROPE

Mediterranean Sea

Adriatic Sea

Several impressive castles dominate the towns and the Danube valley, and there are many architecturally and artistically significant ecclesiastical buildings spread throughout both townscape and landscape.

The Wachau, a stretch of the Danube valley between Melk and Krems, is an outstanding example of a riverine landscape bordered by mountains in which material evidence of its long historical evolution has survived to a remarkable degree. The architecture, the human settlements, and the agricultural use of the land in the Wachau vividly illustrate a basically medieval landscape which has evolved organically and harmoniously over time. Clearance of the natural forest cover by man began in the Neolithic period, although radical changes in the landscape did not take place until around 800, when the Bavarian and Salzburg monasteries began to cultivate the slopes of the Wachau, creating the present-day landscape pattern of vine terraces.

Vineyards near Spitz in Wachau.
▼

World Heritage site since

1978 1979 1980 1981 1982 1983 1984 1985 1986 1987 1988 1989 1990 1991 1992 1993 1994 1995 1996 1997 1998 1999 **2000** 2001 2002 2003 2004 2005 2006 2007 2008

Gyeongju Historic Areas
Korea, Republic of

Criteria – Interchange of values; Testimony to
cultural tradition

ASIA

Yellow
Sea

Sea of
Japan

Pacific Ocean

The Korean peninsula
was ruled for almost
1,000 years by the
Shilla dynasty, and the
sites and monuments
in and around their
capital, Kyongju, bear
outstanding
testimony to their
cultural
achievements.

The Gyeongju Historic Areas contain a
remarkable concentration of outstanding
examples of Korean Buddhist art, in the
form of sculptures, reliefs, pagodas, and the
remains of temples and palaces from the
flowering, in particular between the seventh
and tenth centuries, of this unique form of
artistic expression. There are three major
areas ('belts') at Gyeongju. Mount Namsan
Belt has a large number of prehistoric and
historic remains. Wolsong Belt is home to
the ruined palace site of Wolsong, the
Kyerim woodland (which legend identifies
as the birthplace of the founder of the
Kyongju Kim clan), and the Ch'omsongdae
Observatory. Tumuli Park Belt consists of
three groups of royal tombs. Excavations
here have produced rich grave-goods of
gold, glass and fine ceramics.

Buddha relief
carved in rock on
Mount Namsan.
▼

World Heritage site since

1978 1979 1980 1981 1982 1983 1984 1985 1986 1987 1988 1989 1990 1991 1992 1993 1994 1995 1996 1997 1998 1999 **2000** 2001 2002 2003 2004 2005 2006 2007 2008

Gusuku Sites and Related Properties of the Kingdom of Ryukyu
Japan

Criteria – Interchange of values; Testimony to cultural tradition; Heritage associated with events of universal significance

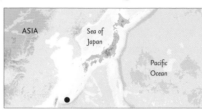

For several centuries the Ryukyu Islands served as a centre of economic and cultural interchange between southeast Asia, China, Korea and Japan, and this is vividly demonstrated by the surviving monuments. In the tenth–twelfth centuries, Ryukyuan farming communities began to enclose their villages with simple stone walls for protection. From the twelfth century onwards powerful groups known as aji began to emerge. They enlarged the defences of their settlements, converting them into fortresses and adopting the term gusuku to describe these formidable castles. The ruins of the castles, on imposing elevated sites, are evidence for the social structure over much of that period, while the sacred sites provide testimony to the survival of an ancient form of religion.

The Stone Gate of the Sonohyan Shrine was erected in 1519 by Shô Shin. It was the guardian shrine of the Ryukyu Kingdom, where prayers were offered for peace and security at annual ritual ceremonies, and represents the unique style of Ryukyu stone architecture.

Ensemble of the Ferrapontov Monastery
Russian Federation

Criteria – Human creative genius; Significance in human history

The Ferrapontov Monastery, in the Vologda region of northern Russia, is an exceptionally well-preserved and complete example of a Russian Orthodox monastic complex of the fifteenth–seventeenth centuries, a period of great significance in the development of the unified Russian state and its culture. There are six major elements in the complex. The Cathedral of the Nativity of the Virgin (1490) is the earliest and the nucleus; its interior is graced by the magnificent wall paintings of Dionisy, the greatest Russian artist of the end of the fifteenth century. This was followed by the Church of the Annunciation, the Treasury Chamber and ancillary buildings. In the seventeenth century the Gate Church, the Church of St Martinian and the bell tower were added.

In the nineteenth century a diminished area of the complex was enclosed by a brick wall. It reopened as a convent for nuns in 1904, but this was closed in 1924. It currently serves as the museum of the frescoes of Dionisy, opened in the first half of the twentieth century, but greatly enlarged and improved since 1975.

World Heritage site since

1978 1979 1980 1981 1982 1983 1984 1985 1986 1987 1988 1989 1990 1991 1992 1993 1994 1995 1996 1997 1998 1999 **2000** 2001 2002 2003 2004 2005 2006 2007 2008

Ciudad Universitaria de Caracas
Venezuela

Criteria – Human creative genius; Significance in human history

The architecture of the university involves the use of spatial elements that have been extracted from Venezuelan colonial architecture, such as bright colours, latticed windows for ventilation and internal gardens of copious tropical vegetation.

The City University of Caracas is an outstanding example, and one of the best in the world, of the modern urban, architectural and artistic concepts of the early twentieth century. Built to the design of the architect Carlos Raúl Villanueva between 1940 and 1960, the university campus integrates the large number of buildings and functions into a clearly articulated ensemble which is at the same time protected from light and heat. It includes masterpieces of modern architecture and visual arts, such as the Aula Magna with the 'Clouds' of Alexander Calder, the Olympic Stadium, and the Covered Plaza. Villanueva's project is characterized by the application of modern technology, the audacity of the forms, and the use of bare concrete structures, conceived as sculptures.

Catalan Romanesque Churches of the Vall de Boí
Spain

Criteria – Testimony to cultural tradition; Significance in human history

The Arab invasion of the Iberian Peninsula never penetrated the valleys of the high Pyrenees, but they were exposed to outside cultural influences, particularly in the eleventh century from Lombardy, a noted centre of Romanesque architecture.

The narrow Vall de Boí is screened by the high peaks of the Beciberri/Punta Alta massif, in the high Pyrenees. Its picturesque small villages are set amid woodland and meadows. Each village in the valley contains a Romanesque church, and is surrounded by a pattern of enclosed fields. There are extensive seasonally-used grazing lands on the higher slopes. The churches of the Vall de Boí are an especially pure and consistent example of Romanesque art in a virtually untouched rural setting. In the Middle Ages there was an influx of culture and money from outside the area. Most of the churches in the valley were built then, at the instigation of a single family, at the same time as historical Catalonia was being created.

World Heritage site since

1978 1979 1980 1981 1982 1983 1984 1985 1986 1987 1988 1989 1990 1991 1992 1993 1994 1995 1996 1997 1998 1999 **2000** 2001 2002 2003 2004 2005 2006 2007 2008

Bardejov Town Conservation Reserve
Slovakia

Criteria – Testimony to cultural tradition;
Significance in human history

The fortified town of Bardejov provides exceptionally well-preserved evidence of the economic and social structure of trading towns in medieval central Europe. The plan, buildings and fortifications of the town illustrate the typical urban complex that developed in central Europe in the Middle Ages at major points along the great trade routes of the period. Bardejov is situated on a floodplain terrace of the Topl'a River, in northeastern Slovakia in the hills of the Beskyd Mountains. From the first quarter of the eighteenth century, Slovaks and Hassidic Jews came into Bardejov in large numbers, and among other remarkable features is the town's small Jewish quarter. Centred around the Great Synagogue, a fine building constructed in 1725–1747, the complex also contains ritual baths, a kosher slaughterhouse and a meeting building, now a school.

Historic wooden church in Bardejov Town Conservation Reserve. ▲

The town centre is dominated by the rectangular main square, closed on three sides by forty-six burgher houses with typical narrow frontages. On the fourth side is the parish church of St Egidius, together with the town school.

World Heritage site since

1978 1979 1980 1981 1982 1983 1984 1985 1986 1987 1988 1989 1990 1991 1992 1993 1994 1995 1996 1997 1998 1999 **2000** 2001 2002 2003 2004 2005 2006 2007 2008

Historic and Architectural Complex of the Kazan Kremlin
Russian Federation

Criteria – Interchange of values; Testimony to cultural tradition; Significance in human history

Built on an ancient site, the Kazan Kremlin consists of an outstanding group of historic buildings dating from the sixteenth to nineteenth centuries. It is exceptional testimony of the Kazan Khanate and is the only surviving Tatar fortress with traces of the original town-planning conception. It is, furthermore, an excellent example of a synthesis of Tatar and Russian influences in architecture, integrating different cultures (Bulgar, Golden Horde, Tatar, Italian and Russian), as well as showing the impact of Islam and Christianity. It was conquered by Ivan the Terrible in 1552 and became the Christian See of the Volga Land, and is still an important place of pilgrimage.

The oldest building within the Kazan Kremlin complex is the sixteenth-century Annunciation Cathedral, constructed of local light sandstone.

The tower of the ▶ Kazan Kremlin.

World Heritage site since

1978 1979 1980 1981 1982 1983 1984 1985 1986 1987 1988 1989 1990 1991 1992 1993 1994 1995 1996 1997 1998 1999 **2000** 2001 2002 2003 2004 2005 2006 2007 2008

Archaeological Site of Atapuerca
Spain

Criteria – Testimony to cultural tradition; Traditional human settlement

The caves of the Sierra de Atapuerca provide unique testimony of the origin and evolution both of the existing human civilisation and of other cultures that have disappeared. The rich fossil record ranges from the earliest human beings in Europe, living nearly one million years ago, right up to remains of modern man. They represent an exceptional reserve of data, the scientific study of which provides priceless information about the appearance and the way of life of these remote human ancestors. The Galería del Silex site contains more than fifty painted and engraved panels with geometrical motifs, hunting scenes, and anthropomorphic and zoomorphic figures. Excavation has revealed human remains (largely young adults and children) and ceramic fragments, identified as being related to sacrificial activities.

The evolutionary line or lines from the African ancestors of modern man are documented in the Sierra de Atapuerca. The earliest human fossil remains in Europe, dating from around 800,000 years ago, were found in the Gran Dolina site here.

Jesuit Block and Estancias of Córdoba
Argentina

Criteria – Interchange of values; Significance in human history

The Jesuit Block in Córdoba, heart of the former Jesuit Province of Paraguay, contains the core buildings of the Jesuit system: the university, the church and residence of the Society of Jesus, and the college. Along with the five estancias, or farming estates, they are exceptional examples of the fusion of European and indigenous cultures during a seminal period in South America. The university is arranged round a central open space (originally a botanical garden), and constructed in stone and brick, with spacious colonnades around the courtyard. The Society of Jesus Church is a massive domed structure with two squat towers at the west end. The interior is richly decorated, the retablo of the main altar and the pulpit being outstanding examples of Baroque.

Córdoba itself, established by Jerónimo Luis de Cabrera in 1573, was laid out on the standard Spanish colonial chequerboard pattern. In common with other religious orders, the Jesuits were allocated one of the seventy blocks of the original city.

World Heritage site since

1978 1979 1980 1981 1982 1983 1984 1985 1986 1987 1988 1989 1990 1991 1992 1993 1994 1995 1996 1997 1998 1999 **2000** 2001 2002 2003 2004 2005 2006 2007 2008

Mount Qingcheng and the Dujiangyan Irrigation System
China

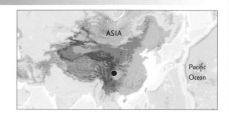

Criteria – Interchange of values; Significance in human history; Heritage associated with events of universal significance

Construction of the Dujiangyan irrigation system began in the third century BC. This system still controls the waters of the Minjiang River and distributes it to the fertile farmland of the Chengdu plains. Mount Qingcheng was the birthplace of Taoism, which is celebrated in a series of ancient temples.

In AD 142 the philosopher Zhang Daoling founded the doctrine of Taoism on Mount Qingcheng. During the Jin dynasty (265–420) several Taoist temples were built on the mountain and the teachings of Taoism were disseminated widely from here throughout China.

◄

The Dujiangyan irrigation system crossing the Minjiang River.

Major Town Houses of the Architect Victor Horta (Brussels)
Belgium

Criteria – Human creative genius; Interchange of values; Significance in human history

The four major town houses – Hôtel Tassel, Hôtel Solvay, Hôtel van Eetvelde, and Maison and Atelier Horta – located in Brussels and designed by the architect Victor Horta, one of the earliest initiators of Art Nouveau, are some of the most remarkable pioneering works of architecture of the end of the nineteenth century. The stylistic revolution represented by these works is characterised by their open plan, the diffusion of light, and the brilliant joining of the curved lines of decoration with the structure of the building.

The Horta buildings brilliantly illustrate the transition from the nineteenth to the twentieth centuries in art, thought and society. Of the four houses, the Hôtel Solvay is the best preserved, with its interior intact and its utilities in functional order.

◄

Hôtel van Eetvelde.

World Heritage site since

978 1979 1980 1981 1982 1983 1984 1985 1986 1987 1988 1989 1990 1991 1992 1993 1994 1995 1996 1997 1998 1999 **2000** 2001 2002 2003 2004 2005 2006 2007 2008

Roman Walls of Lugo
Spain

Criteria – Significance in human history

The walls of Lugo were built in the later part of the third century to defend the Roman town of Lucus. The entire circuit survives intact and is the finest example of late Roman fortifications in western Europe. Despite the strength of its fortifications, Lugo was unable to resist the Suevi when they swept into the Iberian peninsula in the early fifth century and destroyed the town by fire. They were to be dislodged in turn by the Visigoths, who captured the town in 457 and settled it once again. The irresistible Moorish invasion of Spain saw Lugo overwhelmed and sacked in 714, but it was recaptured for Christendom by Alfonso I of Asturias in 755 and restored by Bishop Odarius. The town was to be ravaged once again in 968 by the Normans, on their way to the Mediterranean, and it was not restored until the following century.

Of the original interval towers, forty-six have survived intact, and there are a further thirty-nine that are wholly or partly dismantled. There are ten gates: five ancient and five recent. One of the best preserved is the Miñá Gate, which still has its original vaulted arch set between two towers, in characteristic Roman form.

Archaeological Landscape of the First Coffee Plantations in the South-East of Cuba
Cuba

Criteria – Testimony to cultural tradition; Significance in human history

The production of coffee in eastern Cuba during the nineteenth and early twentieth centuries resulted in the creation of a unique cultural landscape, illustrating a significant stage in the development of this form of agriculture. The site consists of the remains of 171 coffee plantations on the rugged slopes of the Sierra Maestra. The traditional plantation consists of a number of basic elements: at its centre is the residence of the owner, surrounded by much more modest accommodation for the slaves, both domestic and agricultural. The main industrial element was the terraced drying floor (secadero), on which the coffee beans were steeped in water in preparation for processing. The secaderos are recognizable as large sunken areas surrounded by low walls and linked with cisterns or water channels.

In the late nineteenth century, coffee production began in other parts of Latin America, such as Brazil, Colombia and Costa Rica. New techniques were introduced, based on developed agricultural systems. The early plantations in eastern Cuba found themselves unable to compete in the growing world markets and they gradually closed down.

World Heritage site since

1978 1979 1980 1981 1982 1983 1984 1985 1986 1987 1988 1989 1990 1991 1992 1993 1994 1995 1996 1997 1998 1999 **2000** 2001 2002 2003 2004 2005 2006 2007 200

Historic Centre of Brugge
Belgium

Criteria – Interchange of values; Significance in human history; Heritage associated with events of universal significance

Brugge Belfry.

The most important of Brugge's squares are the Burg and the Grand Place. For 1,000 years the Burg square has symbolized the alliance of religious and civic authorities. The Grand Place is the site of the halls, the belfry and the Waterhalle, symbolizing municipal autonomy.

From the Middle Ages the architecture of Brugge has been characterized by brick Gothic, particularly the style of construction known as travée brugeoise. Maintained until the seventeenth century, this type of construction was also the main inspiration for nineteenth-century restorations.

Brugge (Bruges) is an outstanding example of a medieval settlement that has maintained its historic fabric as it evolved over the centuries, and where original Gothic constructions form part of the town's identity. As one of Europe's commercial and cultural capitals, Brugge developed cultural links to different parts of the world. It is closely associated with the school of Flemish Primitive painting.

Archaeological excavations show evidence of human presence in the area of Brugge from the Iron Age and the Gallo-Roman period. It was the military and administrative centre of the region and commercial links with Scandinavia started at that time. The name of Brugge is first mentioned in the ninth century and is documented in Carolingian coins bearing the name Bruggia. At this time it was part of a defence system against the Normans, and the first fortification existed in 851 at the site of the present-day Bourg. The settlement developed gradually and it became a harbour and commercial centre with European connections.

The Brugge fair was established in 1200 and contacts with Britain were the first to develop, particularly in the wool trade. The city's growing prosperity was reflected in the construction of public buildings such

as the imposing belfry in the Grand Place, and Brugge was quickly established as an economic capital of Europe. Under Philippe le Bon (1419–67), the Duke of Burgundy, who set up his court in Brugge, the city became a centre of court life, of Flemish art, of miniature painting, and printing. Owing to the presence of Italians it soon became a centre of humanism and the Renaissance.

From the late fifteenth century Brugge gradually entered a period of stagnation. The Flemish regions were integrated into the Habsburg Empire, and the discovery of America displaced economic interests from the Atlantic to the Mediterranean. However, from 1600 to 1800, as a result of the construction of canal systems, Brugge re-established its maritime connection, albeit at a modest level. From 1815 to 1830 Brugge was part of the United Kingdom of the Netherlands and since 1830 has been part of Belgium. During the nineteenth century a colony of English aristocrats influenced the cultural life of the city and contributed to a renewed interest in the artistic heritage of Brugge and the restoration of historic buildings.

World Heritage site since

1978 1979 1980 1981 1982 1983 1984 1985 1986 1987 1988 1989 1990 1991 1992 1993 1994 1995 1996 1997 1998 1999 **2000** 2001 2002 2003 2004 2005 2006 2007 200

Neolithic Flint Mines at Spiennes (Mons)
Belgium

Criteria – Human creative genius; Testimony to cultural tradition; Significance in human history

The Neolithic flint mines at Spiennes, covering more than 1 km², are the largest and earliest concentration of ancient mines in Europe. They are also remarkable for the diversity of technological solutions used for extraction and for the fact that they are directly linked to a settlement of the same period. Currently the site appears on the surface as a large area of meadows and fields strewn with millions of scraps of worked flint. Underground the site is an immense network of galleries linked to the surface by vertical shafts dug by Neolithic man. Many of the mines have never been excavated and those which are open to the public are in their original condition, with the exception of some modern shoring and props.

Underground flint mining began around the mid-fifth millennium BC and carried on for centuries. The remains of the mines show the gradual development of technology by prehistoric man to extract a material essential for the production of tools and implements.

World Heritage site since

1978 1979 1980 1981 1982 1983 1984 1985 1986 1987 1988 1989 1990 1991 1992 1993 1994 1995 1996 1997 1998 1999 **2000** 2001 2002 2003 2004 2005 2006 2007 2008

Ancient Villages in Southern Anhui – Xidi and Hongcun
China

Criteria – Interchange of values; Significance in human history; Traditional human settlement

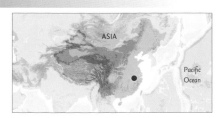

Hongcun (**pictured below**) retains many of its fine buildings and its exceptional water system. The open watercourse runs through the village and forms two ponds, one in the centre (Moon Pond) and the other to the south of the village (South Lake).

The traditional non-urban settlements of China, which have to a very large extent disappeared during the twentieth century, are exceptionally well preserved in the villages of Xidi and Hongcun. The two villages are graphic illustrations of a type of human settlement created during a feudal period and based on a prosperous trading economy. In their buildings and their street patterns, they reflect the socio-economic structure of a long-lived settled period of Chinese history. The streets are all paved with granite and the buildings, which are widely spaced, are timber-framed with brick walls and elegantly carved decoration.

World Heritage site since

1978 1979 1980 1981 1982 1983 1984 1985 1986 1987 1988 1989 1990 1991 1992 1993 1994 1995 1996 1997 1998 1999 **2000** 2001 2002 2003 2004 2005 2006 2007 200

Longmen Grottoes
China

Criteria – Human creative genius; Interchange of values; Testimony to cultural tradition

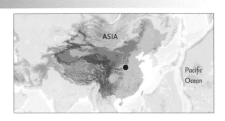

ASIA

Pacific
Ocean

The grottoes and niches of Longmen contain the largest and most impressive collection of Chinese art of the late Northern Wei and Tang dynasties. These works, entirely devoted to the Buddhist religion, represent the high point of Chinese stone carving. This perfection of a long-established art form was to play a highly significant role in the cultural evolution of this region of Asia. Work began on the

Longmen Grottoes in 493, when Emperor Xiaowen of the Northern Wei dynasty moved his capital to Luoyang, and was to continue for the next four centuries. The group of giant statues in Fengxiansi Cave is most fully representative of this phase of Chinese art at Longmen; they are generally acknowledged to be artistic masterpieces of truly global significance.

In total 2,345 niches or grottoes have been recorded on the two sides of the river at Longmen. They house more than 100,000 Buddhist statues, about 2,500 stelae and inscriptions, and over 60 Buddhist pagodas.

Boddhisatvas in the main grotto.
▼

World Heritage site since

1978 1979 1980 1981 1982 1983 1984 1985 1986 1987 1988 1989 1990 1991 1992 1993 1994 1995 1996 1997 1998 1999 **2000** 2001 2002 2003 2004 2005 2006 2007 2008

Notre-Dame Cathedral in Tournai
Belgium

Criteria – Interchange of values; Significance in human history

The Cathedral of Notre-Dame in Tournai was built in the first half of the twelfth century. The cathedral lies at the heart of the old town, not far from the left bank of the river Escaut. In architectural terms, it is the product of three design periods that can still easily be distinguished: the Romanesque nave is of extraordinary dimensions with a wealth of sculpture on its capitals; the choir, rebuilt in the thirteenth century, is in the pure Gothic style; these are linked by a transept in a transitional style featuring an impressive group of five bell towers. The cathedral's front is decorated with

sculptures dating from different periods (fourteenth, sixteenth and seventeenth centuries) depicting Old Testament scenes, episodes from the city's history, and saints.

The cathedral bears witness to an exchange of architectural influence between the Île de France, the Rhineland and Normandy in the short period at the start of the twelfth century that preceded the flowering of Gothic architecture.

Cathedral and Churches of Echmiatsin and the Archaeological Site of Zvartnots
Armenia

Criteria – Interchange of values; Testimony to cultural tradition

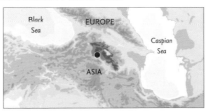

The cathedral and churches of Echmiatsin and the archaeological remains at Zvartnots graphically illustrate the evolution and development of the Armenian central-domed cross-hall type of church, which exerted a profound influence on architectural and artistic development in the region.

The Cathedral of Holy Echmiatsin, built in 301–3, is Armenia's most ancient Christian place of worship. Together with the other religious buildings and archaeological remains in Echmiatsin and Zvartnots, it bears witness to the founding of Christianity in the country.

◄
Ruins of the temple at Zvartnots.

World Heritage site since

1978 1979 1980 1981 1982 1983 1984 1985 1986 1987 1988 1989 1990 1991 1992 1993 1994 1995 1996 1997 1998 1999 **2000** 2001 2002 2003 2004 2005 2006 2007 200

Pantanal Conservation Area
Brazil

Criteria – Natural phenomena or beauty;
Significant ecological and biological processes;
Significant natural habitat for biodiversity

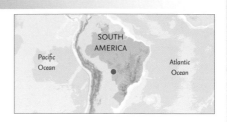

The Pantanal Conservation Complex consists of a cluster of four protected areas located in western central Brazil at the southwest corner of the state of Mato Grosso. The site represents 1.3 per cent of Brazil's Pantanal region, one of the world's largest freshwater wetland ecosystems. The headwaters of the region's two major river systems, the Cuiabá and the Paraguay Rivers, are located here, and the abundance and diversity of its vegetation and animal life are spectacular.

The Pantanal is an immense alluvial plain. Its landscape encompasses a variety of ecological subregions, including river corridors, gallery forests, perennial wetlands and lakes, seasonally inundated grasslands and terrestrial forests.

Surrounded by mountain ridges and plains, the region presents a flat landscape with a small inclination which follows a north-south, east-west direction. The main source of water for the Pantanal is the Cuiabá River. The water spreads out and covers broad expanses, seeking a natural outlet, which lies hundreds of kilometres downstream at the confluence of the river and the Atlantic Ocean. Hydrological studies indicate the presence of a network of underground streams and a degree of subsurface water movement.

The vegetation is located in an area of transition between the dry savanna (cerrado) of central Brazil and the semi-deciduous forest of the south and southeast. The wide range of interacting habitat types produces a remarkable plant diversity.

The fauna of the Pantanal is extremely varied and includes 80 species of mammal, 650 bird, 50 reptile and 400 fish types. Dense populations of species of conservation concern live in the region; these include jaguar, marsh deer, giant anteater and giant otter.

The Pantanal is a sanctuary for birds with many species occurring in large numbers. It is one of the most important breeding grounds for typical wetland birds such as Jabiru stork, as well as several other species of heron, ibis and duck, which are found in enormous flocks.

The Pantanal Conservation Complex consists of a cluster of four protected areas: Pantanal Matogrossense National Park, Dorochê Private Reserve, Acurizal Private Reserve and Penha Private Reserve, with a total area of 1,878 km².

The area is home to twenty-six recorded parrot species including the hyacinth macaw (**pictured right**), the world's largest parrot. A large proportion of the remnant wild population of this species, now endangered, inhabit the region. Habitat destruction and capture for the pet trade are two factors that, in combination, have led to the risk of extinction.

World Heritage site since

1978 1979 1980 1981 1982 1983 1984 1985 1986 1987 1988 1989 1990 1991 1992 1993 1994 1995 1996 1997 1998 1999 2000 2001 2002 2003 2004 2005 2006 2007 20C

Central Suriname Nature Reserve

Suriname

Criteria – Significant ecological and biological processes; Significant natural habitat for biodiversity

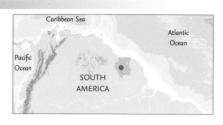

The Central Suriname Nature Reserve comprises 16,000 km² of primary tropical forest of west-central Suriname. It protects the upper watershed of the Coppename River and the headwaters of the Lucie, Oost, Zuid, Saramacca and Gran rivers and covers a range of topography and ecosystems of notable conservation value due to its pristine state. Its montane and lowland forests contain a high diversity of plant life with more than 5,000 vascular plant species collected to date. The Reserve's animals are typical of the region and include the jaguar, giant armadillo, giant river otter, tapir, sloths, eight species of primates and 400 bird species such as harpy eagle, Guiana cock-of-the-rock, and scarlet macaw.

Large parts of the Guyana Shield and Amazon regions are being rapidly transformed by logging, hunting, mining and settlement, but the reserve remains inaccessible, largely unaffected and unthreatened by human activity.

◄
White-faced saki, one of the eight primate species found in the Reserve.

World Heritage site since

1978 1979 1980 1981 1982 1983 1984 1985 1986 1987 1988 1989 1990 1991 1992 1993 1994 1995 1996 1997 1998 1999 2000 2001 2002 2003 2004 2005 2006 2007 200

Aranjuez Cultural Landscape

Spain

Criteria – Interchange of values; Significance in human history

The Aranjuez cultural landscape is an entity of complex relationships: between nature and human activity, between sinuous watercourses and geometric landscape design, between the rural and the urban, between forest landscape and the delicately modulated architecture of its palatial buildings. Three hundred years of royal attention to the development and care of this landscape have seen it express an evolution of concepts from humanism and political centralization, to characteristics such as those found in its eighteenth-century French-style Baroque garden, to the urban lifestyle which developed alongside the sciences of plant acclimatization and stock-breeding during the Age of Enlightenment.

The site incorporates a planned town, large gardens, vegetable gardens and orchards, lagoons, rivers and waterworks, woods and moors. The whole area appears as a green oasis in an otherwise dry, brown and fairly barren sierra-type landscape.

World Heritage site since

1978 1979 1980 1981 1982 1983 1984 1985 1986 1987 1988 1989 1990 1991 1992 1993 1994 1995 1996 1997 1998 1999 2000 **2001** 2002 2003 2004 2005 2006 2007 2008

New Lanark
United Kingdom

Criteria – Interchange of values; Significance in human history; Heritage associated with events of universal significance

Atlantic Ocean

North Sea

EUROPE

New Lanark is a small eighteenth-century village set in a sublime Scottish landscape where the philanthropist and idealist Robert Owen moulded a model industrial community in the early nineteenth century. The first cotton mill at New Lanark went into production in 1786 and Owen began to remodel the village around 1809. The imposing mill buildings, the spacious and well-designed workers' housing, and the dignified educational institute and school all testify to Owen's humanism.

The theme throughout is one of good proportion, good masonry, and simplicity of detail. The success of New Lanark inspired other benevolent industrialists to follow Owen's example and had a profound influence on social developments throughout the nineteenth century and beyond in matters such as progressive education, factory reform, humane working practices, international cooperation, and the concept of the garden city.

Robert Owen's idealistic vision of a society without crime, poverty, and misery had a wide appeal in the years following the Napoleonic Wars. He left New Lanark in 1828, after which new buildings were constructed, and others demolished or destroyed by fire, but the appearance of the village today remains very close to that of its heyday.

World Heritage site since

1978 1979 1980 1981 1982 1983 1984 1985 1986 1987 1988 1989 1990 1991 1992 1993 1994 1995 1996 1997 1998 1999 2000 **2001** 2002 2003 2004 2005 2006 2007 200

Swiss Alps Jungfrau-Aletsch
Switzerland

Criteria – Natural phenomena or beauty; Major stages of Earth's history; Significant ecological and biological processes

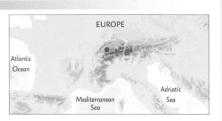

The Jungfrau-Aletsch region is the most glaciated part of the Alps, containing Europe's largest glacier and a range of classic glacial features. It provides an outstanding geological record of the uplift and compression that formed the High Alps. The diversity of flora and wildlife is represented in a range of Alpine and sub-Alpine habitats and plant colonization in the wake of retreating glaciers and provides an outstanding example of plant succession.

The geology of the site derives from the 'Helvetic nappe' (a large body of rock that was thrust over younger rock in Europe during the Miocene epoch). The folding and over-thrusting of rock layers during the formation of the Alps have produced very complex rock formations that have since been exposed by glacial activity. The physiography of the area is characterized by steep north-facing slopes and relatively gentle southern ones.

The area's scenic and aesthetic appeal is one of the most dramatic in the Alps. The impressive northern wall of the site with the panorama of the Eiger, Mönch and Jungfrau mountains provides a classic view of the north face of the High Alps - a view that has played an important role in European art and literature. Nine peaks in the region are

higher than 4,000 m. Classic examples of glacial phenomena in the area include U-shaped valleys, valley glaciers, cirques, horn peaks and moraines. The Aletsch Glacier is the largest and longest glacier in western Eurasia in terms of area (128 km²), length (23 km) and depth (900 m).

Vegetation and fauna vary by slope, aspect and elevation. There is a marked difference in vegetation between the northern and southern slopes. On the north side, forests at lower elevations consist of broadleaved species such as beech, ash, alder, elm and birch. The south side is too dry for beech, which is replaced by Scots pine. On the northern side, the subalpine zone is dominated by Norway spruce with mountain ash, silver birch and stone pine and, on the southern side, by more continental species such as European larch. Above the timberline are extensive areas of rhododendron scrub, alpine grassland and tundra vegetation and, on the dry southern slopes, steppe grassland.

The Aletsch Glacier.

Fauna in the region is typical of the Alps, with a wide variety of species including ibex, lynx, red deer, roe deer, chamois and marmot as well as several reptiles and amphibians. A representative range of alpine birds also occurs in the area, including golden eagle, kestrel, chough ptarmigan, black grouse, snow finch, wallcreeper, lammergeier, pygmy owl and various woodpecker species.

World Heritage site since

1978 1979 1980 1981 1982 1983 1984 1985 1986 1987 1988 1989 1990 1991 1992 1993 1994 1995 1996 1997 1998 1999 2000 **2001** 2002 2003 2004 2005 2006 2007 2008

Alto Douro Wine Region
Portugal

Criteria – Testimony to cultural tradition; Significance in human history; Traditional human settlement

The Alto Douro Region has been producing wine for some 2,000 years. Since the eighteenth century, its main product, port wine, has been world-famous. This long tradition of viticulture has produced a cultural landscape of outstanding beauty that reflects its technological, social and economic evolution. The landscape in the Demarcated Region of the Douro is formed by steep hills and boxed-in valleys that flatten out into plateaux above 400 m. The most dominant feature of the landscape is the terraced vineyards that blanket the countryside. Throughout the centuries, row upon row of terraces have been built according to different techniques. The earliest were narrow, irregular terraces buttressed by walls of stone that were regularly taken down and rebuilt. The long lines of continuous, regularly-shaped terraces date mainly from the end of the nineteenth century.

Vineyard by the river Douro. ▲

The more recent terracing techniques, the *patamares*, have greatly altered the appearance of the landscape. Large plots of slightly sloping earth-banked land were laid out to facilitate mechanization of the vineyard. Trials of other systems are continuing with a view to finding alternatives to the *patamares* and to minimize the impact of the new methods on the landscape.

World Heritage site since

1978 1979 1980 1981 1982 1983 1984 1985 1986 1987 1988 1989 1990 1991 1992 1993 1994 1995 1996 1997 1998 1999 2000 **2001** 2002 2003 2004 2005 2006 2007 2008

Tombs of Buganda Kings at Kasubi
Uganda

Criteria – Human creative genius; Testimony to cultural tradition; Significance in human history; Heritage associated with events of universal significance

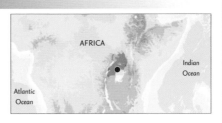

The Tombs of Buganda Kings at Kasubi constitute a site embracing almost 0.3 km² of hillside within Kampala district. Most of the site is agricultural, farmed by traditional methods. At its core on the hilltop is the former palace of the Kabakas of Buganda, built in 1882 and converted into the royal burial ground in 1884. Four royal tombs now lie within the Muzibu Azaala Mpanga, the main building, which is circular and surmounted by a dome. It is a major example of an architectural achievement in organic materials, principally wood, thatch, reed, wattle and daub. The site's main significance lies, however, in its intangible values of belief, spirituality, continuity and identity.

The spatial organization of the Tombs represents the best extant example of a Baganda palace/architectural ensemble. It reflects technical achievements developed over many centuries and is the most active religious place in the kingdom.

◄

The Muzibu Azaala Mpanga.

The Royal Hill of Ambohimanga
Madagascar

Criteria – Testimony to cultural tradition; Significance in human history; Heritage associated with events of universal significance

The Royal Hill of Ambohimanga is the most significant symbol of the cultural identity of the people of Madagascar. Its traditional design, materials and layout are representative of the social and political structure of Malagasy society from at least the sixteenth century. The site consists of a royal city and burial site and an ensemble of sacred places (wood, spring, lake, public meeting place).

Fortifications protected the royal city in an arrangement of banks, ditches, and fourteen stone gateways. The site is associated with strong feelings of national identity, and has maintained its spiritual and sacred character both in ritual practice and in the popular imagination for the past 500 years. It remains a place of worship for pilgrims from Madagascar and elsewhere.

In 1897 the remains of royalty were transferred to Antananarivo by the French authorities in a failed attempt to erase the holiness of the site and the nationalistic legitimacy attached to it. The tombs were demolished and military buildings erected for the garrison on the site.

World Heritage site since

1978 1979 1980 1981 1982 1983 1984 1985 1986 1987 1988 1989 1990 1991 1992 1993 1994 1995 1996 1997 1998 1999 2000 **2001** 2002 2003 2004 2005 2006 2007 200

Central Sikhote-Alin
Russian Federation
Criteria – Significant natural habitat for
biodiversity

The Sikhote-Alin mountain range contains
one the richest and most unusual temperate
forests in the world. The unique
combination of its severe climatic
characteristics, physical isolation, and
traditional resource use by the indigenous
peoples has meant that 80–90 per cent of
the region's vegetation still remains as
dense temperate forest and taiga. Alpine
tundra, coastal shrublands, meadows and
bogs account for the rest of the area. In this

mixed zone between taiga and subtropics,
southern species such as the tiger and
Himalayan bear cohabit with northern
species such as the brown bear and lynx.
The site stretches from the peaks of
Sikhote-Alin to the Sea of Japan and is
important for the survival of many
endangered species such as the Amur tiger.

The Sikhote-Alin
protected areas are
considered to contain
the greatest plant and
animal diversity on
the northwestern
Pacific coastline.
Many of the plants,
such as ginseng and
Siberian ginseng, are
of medicinal value
and are important to
the indigenous people

Fertö/Neusiedlersee
Cultural Landscape
Austria and Hungary
Criteria – Traditional human settlement

The Fertö/Neusiedler Lake and its
surroundings are an outstanding example
of a traditional human settlement and land
use representative of a culture. The lake lies
between the Alps, 70 km distant, and the
lowlands in the territory of two states,
Austria and Hungary. It has been the
meeting place of different cultures for
eight millennia, and this is graphically
demonstrated by its varied landscape.
The present character of the landscape is

the result of ancient land-use forms based
on stockraising and viticulture to an extent
not found in other European lake areas.
The remarkable rural architecture of the
villages surrounding the lake is typified by
the historic centre of the medieval free town
of Rust, which perfectly symbolises a united
society of townspeople and farmers.

Several eighteenth-
and nineteenth-
century palaces add to
the area's considerable
cultural interest.
These include the
Fertöd Esterházy
Palace, the most
important eighteenth-
century palace of
Hungary, built on the
model of Versailles.
Between 1769 and
1790 Josef Haydn's
compositions were
first heard here.

World Heritage site since

1978 1979 1980 1981 1982 1983 1984 1985 1986 1987 1988 1989 1990 1991 1992 1993 1994 1995 1996 1997 1998 1999 2000 **2001** 2002 2003 2004 2005 2006 2007 2008

Saltaire
United Kingdom

Criteria – Interchange of values; Significance in human history

Saltaire in west Yorkshire is a well-preserved industrial village of the second half of the nineteenth century. Its textile mills, public buildings and workers' housing are built in a harmonious style of high architectural quality, giving a vivid impression of Victorian philanthropic paternalism. Saltaire represents an important stage in the development of modern town planning and had a major influence on the 'garden city' movement. When Titus Salt, a wealthy businessman, became mayor of Bradford in

1848, he committed himself to reducing the town's pollution problems. Work on the mill began in 1851 and it was opened in 1853. Salt's new village eventually had over 800 dwellings in wide streets, with a large dining hall and kitchens, baths and wash houses, almshouse for retired workers, hospital and dispensary, educational institute and church, and ample recreational land and allotments.

The houses, built between 1854 and 1868, are fine examples of nineteenth-century hierarchical workers' homes. Each was equipped with its own water and gas supply and an outside lavatory. They vary in size from 'two-up two-down' terraces to much larger managers' houses with gardens.

Saltaire Mill and the river Aire.
▼

World Heritage site since

1978 1979 1980 1981 1982 1983 1984 1985 1986 1987 1988 1989 1990 1991 1992 1993 1994 1995 1996 1997 1998 1999 2000 **2001** 2002 2003 2004 2005 2006 2007 2008

Medina of Essaouira (formerly Mogador)
Morocco

Criteria – Interchange of values; Significance in human history

Essaouira is an outstanding and well-preserved example of a late-eighteenth-century European fortified seaport town translated to a North African context. Since its foundation, it has been a major international trading seaport, linking Morocco and its Saharan hinterland with Europe and the rest of the world. As Morocco increasingly opened up to the rest of the world in the later seventeenth century, the old town needed to expand and a new plan was laid out by a French architect. The town has retained its European appearance to a substantial extent, and is a leading example of building inspired by European architecture – a town unique by virtue of its design.

In the medina of Essaouira a symbiosis was achieved between building techniques from Morocco and elsewhere that gave birth to some unique architectural masterpieces: the Sqalas (fortifications) of the port and of the medina, the Bab Marrakesh bastion, the water gate, mosques, synagogues and churches.

The Sqala du Port, Essaouira.
▼

World Heritage site since

1978 1979 1980 1981 1982 1983 1984 1985 1986 1987 1988 1989 1990 1991 1992 1993 1994 1995 1996 1997 1998 1999 2000 **2001** 2002 2003 2004 2005 2006 2007 2008

Mining Area of the Great Copper Mountain in Falun
Sweden

Criteria – Interchange of values; Testimony to cultural tradition; Traditional human settlement

The Great Copper Mountain (Stora Kopparberget) is the oldest and most important mine working in Sweden and one of the world's most remarkable industrial monuments. Mining here began as early as the ninth century and came to an end in the closing years of the twentieth century. By the seventeenth century, Falun was producing 70 per cent of the world's copper and was the mainstay of Sweden's economy, enabling it to become one of the leading European powers. The seventeenth-century planned town, with its many fine historic buildings, together with the industrial and domestic remains of a number of settlements spread over a wide area of the Dalarna region, provide a vivid picture of what was for centuries one of the world's most important mining areas.

The Great Copper Mountain was a corporate operation in which free miners owned shares in proportion to their interests in copper smelters. It is often referred to as 'the oldest company in the world.'

Zollverein Coal Mine Industrial Complex in Essen
Germany

Criteria – Interchange of values; Testimony to cultural tradition

The Zollverein industrial complex in Land Nordrhein-Westfalen consists of the complete infrastructure of a historical coal-mining site, with some twentieth-century buildings of outstanding architectural merit. It constitutes remarkable material evidence of the evolution and decline of an essential industry over the past 150 years. Mining began in the mid-nineteenth century at a depth of some 120 m and finished at 1,200 m. By the end of mining the underground roadways extended over 120 km; they were accessed by twelve shafts, opened up progressively between 1847 and 1932. The methods of mining evolved as technology developed from hand picks to mechanized coal cutting. Elements of the original pits, the central coking plant, railway lines, associated buildings and housing all survive today.

Zollverein is an exceptional industrial monument: its buildings are outstanding examples of the application of Modern Movement design concepts to wholly industrial architecture. Of particular note are the imposing administrative building (1906), the director's villa (1898), and the mine officials' residence (1878).

World Heritage site since

1978 1979 1980 1981 1982 1983 1984 1985 1986 1987 1988 1989 1990 1991 1992 1993 1994 1995 1996 1997 1998 1999 2000 **2001** 2002 2003 2004 2005 2006 2007 2008

Brazilian Atlantic Islands: Fernando de Noronha and Atol das Rocas Reserves
Brazil

Criteria – Natural phenomena or beauty; Significant ecological and biological processes; Significant natural habitat for biodiversity

Peaks of the Southern Atlantic submarine ridge form the Fernando de Noronha Archipelago and Rocas Atoll off the coast of Brazil. They represent a large proportion of the island surface of the South Atlantic and their rich waters are extremely important for the breeding and feeding of tuna, shark, turtle and marine mammals. The islands are home to the largest concentration of tropical seabirds in the western Atlantic. Baia de Golfinhos has an exceptional population of resident dolphin and at low tide the Rocas Atoll provides a spectacular seascape of lagoons and tidal pools teeming with fish.

There are less than ten oceanic islands in the South Atlantic, and the Fernando de Noronha Archipelago and Rocas Atoll represent more than 50 per cent of such islands' surface area. Consequently, they are important for the biodiversity of the South Atlantic basin.

Tsodilo
Botswana

Criteria – Human creative genius; Testimony to cultural tradition; Heritage associated with events of universal significance

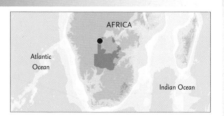

With one of the highest concentrations of rock art in the world, Tsodilo has been called the 'Louvre of the Desert'. For many thousands of years these rocky outcrops in the hostile landscape of the Kalahari Desert have been visited and settled by humans, who have left rich traces of their presence in the form of outstanding rock art. Over 4,500 paintings are preserved in an area of only 10 km². The archaeological record of the area gives a chronological account of human activities and environmental changes over at least 100,000 years.

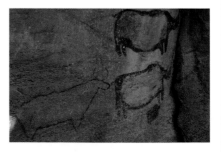

The outcrops had immense symbolic and religious significance for the human communities who survived here, being respected as a place of worship frequented by ancestral spirits. Some of the art is thought to be more than 2,000 years old. Geometric art is regarded as about 1,000 years old while the pictures with cattle date from the animals' introduction to Tsodilo after the sixth century AD.

World Heritage site since

1978 1979 1980 1981 1982 1983 1984 1985 1986 1987 1988 1989 1990 1991 1992 1993 1994 1995 1996 1997 1998 1999 2000 **2001** 2002 2003 2004 2005 2006 2007 2008

Provins, Town of Medieval Fairs
France
Criteria – Interchange of values; Significance in human history

At the beginning of the second millennium AD, Provins was one of several towns in the territory of the Counts of Champagne that became the venues for great annual trading fairs linking northern Europe with the Mediterranean world. The entire town developed in relation to the fairs, either directly serving the fair functions or being indirectly related as an outcome. Of the four towns where medieval fairs were held in the

reign of the Counts of Champagne, Provins is the only one to retain its original medieval fabric. There are two large buildings: the so-called 'Tour de César' or the Big Tower, a stone structure, dating initially from the twelfth-century, and consisting of three large spaces one above the other, covered with a seventeenth-century conical roof; and the Romanesque-Gothic church of Saint-Quiriace.

A characteristic of all the ancient buildings in Provins, whether for mixed or for commercial use, is their system of vaulted cellars, dating from the twelfth to the fourteenth centuries. These are either entirely underground (Upper Town) or partly built-up above ground (Lower Town), and all open out to the street by means of a large door to which access is gained by a wide stone staircase.

World Heritage site since

1978 1979 1980 1981 1982 1983 1984 1985 1986 1987 1988 1989 1990 1991 1992 1993 1994 1995 1996 1997 1998 1999 2000 **2001** 2002 2003 2004 2005 2006 2007 2008

Samarkand – Crossroads of Cultures
Uzbekistan

Criteria – Human creative genius; Interchange of values; Significance in human history

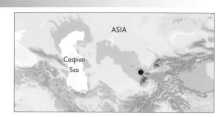

ASIA

Caspian Sea

Located on the crossroads of the great trade routes that traversed central Asia, the historic town of Samarkand illustrates in its art, architecture and urban structure the most important stages of central Asian cultural and political history from the thirteenth century to the present day. Founded in the seventh century BC as ancient Afrasiab, the city emerged as a major centre through the efforts of Timur the Lame (Tamerlane, c. 1336–1405). It was

rebuilt on its present site, southwest of Afrasiab, and became the capital of Timur's powerful state and the repository of the material riches from conquered territories that extended from central Asia to Persia, Afghanistan, and India. The major monuments include the Registan Mosque and madrasas (see photo below), Bibi-Khanum Mosque, the Shakhi-Zinda compound and the Gur-Emir ensemble, as well as Ulugh-Beg's Observatory.

In 1868 the Russians conquered Samarkand, making it a provincial capital (1887) and thus reviving its economy. They constructed schools, churches, and hospitals, and the western part of Samarkand was redeveloped according to current town planning ideas. This period, however, also led to the destruction of the city walls and gates, as well as several monuments.

Registan Mosque.
▼

World Heritage site since

1978 1979 1980 1981 1982 1983 1984 1985 1986 1987 1988 1989 1990 1991 1992 1993 1994 1995 1996 1997 1998 1999 2000 **2001** 2002 2003 2004 2005 2006 2007 2008

Dorset and East Devon Coast
United Kingdom
Criteria – Major stages of Earth's history

Located on the south coast of England, the property comprises eight sections along 155 km of coast. The property has a combination of geological, palaeontological and geomorphological features. These include a variety of fossils, a beach renowned for its pebbles, and textbook examples of common coastal features such as sea stacks and sea caves. The area has been studied for more than 300 years and has contributed to the development of earth sciences in the United Kingdom. The site includes a near-continuous sequence of Triassic, Jurassic and Cretaceous rock exposures, representing much of the Mesozoic era (251–66 million years ago) or approximately 185 million years of the Earth's history. A large number of vertebrate, invertebrate and plant fossils have been discovered. Among the finds are fossil dinosaur footprints, flying reptiles and marine reptiles.

Man o' War Beach in Dorset. ▲

Well-preserved remains of a late-Jurassic fossil forest are exposed on the Isle of Portland and the Purbeck coast. Chesil Beach, stretching from West Bay to Portland, is famous for the volume, type and grading of its pebbles. The Fleet Lagoon is one of the most important saline lagoons in Europe.

World Heritage site since

1978 1979 1980 1981 1982 1983 1984 1985 1986 1987 1988 1989 1990 1991 1992 1993 1994 1995 1996 1997 1998 1999 2000 **2001** 2002 2003 2004 2005 2006 2007 2008

Alejandro de Humboldt National Park
Cuba

Criteria – Significant ecological and biological processes; Significant natural habitat for biodiversity

The park, in southeastern Cuba, includes a complex system of mountains, tablelands, coastal plains, bays and coral reefs. It is the least explored natural area on the island and has locations where the plant life has not yet been assessed.

Complex geology and varied topography have given rise to a diversity of ecosystems and species unmatched in the insular Caribbean and created one of the most biologically diverse tropical island sites on Earth. Many of the underlying rocks are toxic to plants so species have had to adapt to survive in these hostile conditions. This unique process of evolution has resulted in the development of many new species and the park is one of the most important sites in the Western Hemisphere for the conservation of endemic flora. Endemism of vertebrates and invertebrates is also very high.

Yungang Grottoes
China

Criteria – Human creative genius; Interchange of values; Testimony to cultural tradition; Significance in human history

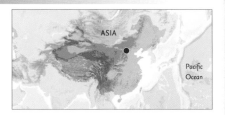

By AD 525 the initial project, sponsored by the court, was mostly completed, but low-ranking officials and monks continued to dig more caves and carve statues. During the Liao dynasty, wooden shelters were built in front of the caves, turning the grottoes into temple buildings, such as the Ten Famous Temples.

The Yungang Grottoes, in Datong city, Shanxi Province, with their 252 caves and 51,000 statues, represent the outstanding achievement of Buddhist cave art in China in the fifth and sixth centuries. The grottoes of the early period (AD 460–5) are composed of five main caves, dug under the direction of the monk Tan Yao and named after him. They have a U-shaped plan and arched roof and each cave has a door and a window. The central images occupy the major part of the caves, while on the outer walls 1,000 Buddhist statues are carved. The Yungang Grottoes also include four groups of twin caves and one group of triple caves; the site extends as much as 1 km east-west.

World Heritage site since

1978 1979 1980 1981 1982 1983 1984 1985 1986 1987 1988 1989 1990 1991 1992 1993 1994 1995 1996 1997 1998 1999 2000 **2001** 2002 2003 2004 2005 2006 2007 2008

Villa d'Este, Tivoli
Italy

Criteria – Human creative genius; Interchange of values; Testimony to cultural tradition; Significance in human history; Heritage associated with events of universal significance

The gardens of the Villa d'Este had a profound influence on the development of garden design throughout Europe. They are among the earliest and finest of the giardini delle meraviglie and symbolize the flowering of Renaissance culture. The ensemble, composed of the palace and gardens, forms an irregular quadrilateral and covers an area of about 45,000 m². The plan of the villa is irregular because the architect was obliged to make use of certain parts of the previous monastic building. On the garden side the architecture of the palace is very simple: a long main body of three storeys, marked by bands, rows of windows, side pavilions and an elegant loggia. The lower level is decorated with the Fountain of Leda. The main rooms of the villa are arranged in rows on two floors and open on to the magnificent garden.

One of the villa's many outstanding features is the Alley of the Hundred Fountains, the waters from which cross the entire garden. There is also the innovative and striking design of the jets of the large cascade which were activated whenever unsuspecting people walked under the arcades. The garden also features its own artificial mountain, with three alcoves holding statues, and a water organ.

The Neptune Fountain (foreground) and the Organ Fountain (background). ▼

World Heritage site since

1978 1979 1980 1981 1982 1983 1984 1985 1986 1987 1988 1989 1990 1991 1992 1993 1994 1995 1996 1997 1998 1999 2000 **2001** 2002 2003 2004 2005 2006 2007 200

Historic Centre of the Town of Goiás
Brazil

Criteria – Interchange of values; Significance in human history

In its layout and architecture the historic town of Goiás is an outstanding example of a European town admirably adapted to the climatic, geographical and cultural constraints of central South America. It represents the evolution of a form of urban structure and architecture characteristic of the colonial settlement of South America, making full use of local materials and

techniques and conserving its exceptional setting. The urban layout is an example of the organic development of a mining town, adapted to the conditions of the site. Although modest, both public and private architecture form a harmonious whole, thanks to the coherent use of local materials and vernacular techniques.

In 1748, Goiás' gold wealth meant it was chosen as the headquarters of a new subdistrict, and its first governor transformed the modest village into a small capital. Its townscape has not been subject to any major changes in modern times, making Goiás a remarkably well-preserved example of a mining town of the eighteenth and nineteenth centuries, including its natural environment, which has remained intact.

World Heritage site since

1978 1979 1980 1981 1982 1983 1984 1985 1986 1987 1988 1989 1990 1991 1992 1993 1994 1995 1996 1997 1998 1999 2000 **2001** 2002 2003 2004 2005 2006 2007 2008

Churches of Peace in Jawor and Swidnica
Poland

Criteria – Testimony to cultural tradition; Significance in human history; Heritage associated with events of universal significance

The Churches of Peace in Jawor and Swidnica, the largest timber-framed religious buildings in Europe, were built in former Silesia in the mid-seventeenth century, at a time of religious strife following the Peace of Westphalia. They are monuments not just to the skill of their designers, but also to the religious tolerance shown by the Catholic Habsburg Emperor towards Protestant communities in Silesia after the Thirty Years'

War. In most of the province Protestants were persecuted but, through the agency of the Lutheran king of Sweden, the Emperor allowed the erection of three Lutheran churches, of which two survive. As a result of conditions imposed by the Emperor, the builders had to employ pioneering constructional techniques of a scale and complexity unknown in wooden architecture.

The churches in Jawor and Swidnica differ in the character of their floor plans. Both have three aisles, and both terminate in a polygonal east end, but whereas in Jawor the eastern end is still a true chancel, in Swidnica it functions as a sacristy.

Derwent Valley Mills
United Kingdom

Criteria – Interchange of values; Significance in human history

The Derwent Valley in central England contains a series of eighteenth- and nineteenth-century cotton mills and an industrial landscape of high historical and technological interest. The modern factory system was effectively born here, when new types of building were erected to house the latest technology for spinning cotton developed by Richard Arkwright. This was large-scale industrial production in a hitherto rural landscape, and the need to provide housing and other facilities for workers and

managers resulted in the creation of the first industrial towns. The workers' housing associated with this and the other mills remains intact and illustrates the socio-economic development of the area. The complete heritage site is a continuous strip, 24 km long, from Matlock Bath to the centre of Derby.

Richard Arkwright developed the village of Cromford to attract the families of the mainly child workforce that he needed. Weavers lived in his houses, the parents weaving calico on their topmost floors and children working in the spinning mills.

World Heritage site since

1978 1979 1980 1981 1982 1983 1984 1985 1986 1987 1988 1989 1990 1991 1992 1993 1994 1995 1996 1997 1998 1999 2000 **2001** 2002 2003 2004 2005 2006 2007 2008

Masada
Israel

Criteria – Testimony to cultural tradition;
Significance in human history; Heritage
associated with events of universal significance

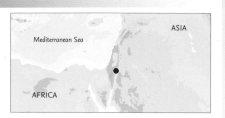

Masada is a rugged natural fortress of
majestic beauty set in the Judaean Desert
overlooking the Dead Sea. It is a symbol of
the ancient kingdom of Israel, its violent
destruction and the last stand of Jewish
patriots against the Roman army at the end
of the First Jewish-Roman War in AD 73.
The camps, fortifications and attack ramp
that encircle the monument constitute the
most complete Roman siege works
surviving to the present day.

The towering hill of Masada overlooks
a natural landscape of savage beauty. To the
west lie the hills and terraces of the Judaean
Desert; to the east is a wildly broken terrain,
running down to the brilliant colours of the
Dead Sea. A giant scarp stretches south to
the horizon and Masada forms part of this.

Masada was built as a palace complex in
the classic style of the early Roman Empire
by King Herod the Great of Judaea, who
ruled from 37–4 BC. The Northern Palace, in
its present form, dates from the main phase
of building in the late first century BC. It was
built on three slightly modified natural rock
terraces. The upper level was mainly
residential while imposing colonnaded
reception halls were situated on the two
lower levels. The lower reception level is the
best preserved of the three.

The nucleus of the Western Palace
comprised a courtyard surrounded by
bedrooms and reception rooms; two
extensive service wings were added in the
main phase of building. Its size, layout and
opulent decoration – mosaic floors and
walls of white plaster painted to imitate
marble panels – suggest that this was the
ceremonial palace, while the Northern
Palace was the private palace for the king
and his family.

The massive defensive wall, built in the
final phase, was 1,290 m long with twenty-
seven towers and about seventy rooms on
its inner side. Three gates pierce the wall:
the Western Gate, the Southern Gate, and
the Snake Path Gate (the eastern gate).
Afarth gate - the Water Gate - provided
access to the Northern Section; this
however was not integrated into the
casement wall. Water was delivered, during
the winter floods in the wadis to the west,
through a network of dams and channels to
the cisterns dug into the rock of Masada.

Most of the buildings on the hilltop were
occupied by around 1,000 people who lived
there in the Zealot period of the first
century AD.

Aerial view of
Masada, with one of
the Roman camps in
the foreground.

There is a network of
eight Roman military
camps around
Masada. Most striking
are the hundreds of
contubernia (messing
units), consisting of
walls of stones
1–1.5 m high on which
the soldiers erected
their leather tents.
The great ramp used
for the final assault
on Masada was built
from soil and stones
braced by an armature
of timber beams.

The remains of a fifth-
century Byzantine
church also stands on
the summit. The floor
was originally covered
with mosaic but much
of this was removed
to the Louvre in the
nineteenth century.

World Heritage site since

1978 1979 1980 1981 1982 1983 1984 1985 1986 1987 1988 1989 1990 1991 1992 1993 1994 1995 1996 1997 1998 1999 2000 **2001** 2002 2003 2004 2005 2006 2007 2008

Old City of Acre
Israel

Criteria – Interchange of values; Testimony to cultural tradition; Traditional human settlement

Acre is an exceptional port-historic town that preserves the substantial remains of the medieval Crusader buildings beneath the existing Ottoman fortified town which dates from the eighteenth and nineteenth centuries. The remains of the Crusader town, dating from 1104–1291, lie almost intact, both above and below today's street level, providing an exceptional picture of the layout and structures of the capital of the medieval Crusader kingdom of Jerusalem.

During the two centuries of Crusader rule, Acre symbolized, better than any other city, the interchange between eastern and western cultures. New neighbourhoods such as Monmizar to the north were built and Acre was given a new double city wall. What remains today is a remarkable mixture of cultural elements from every period of Acre's eventful history between the eleventh and twentieth centuries.

Destroyed in the sixteenth century, Acre was a deserted ghost town until reconstruction began in the mid-eighteenth century, under Daher El Amar and later El Jazar, who rebuilt the port and fortifications. Enjoying renewed economic expansion in the nineteenth century, wealthy merchants settled there, building grand mansions in the eastern neoclassical style.

World Heritage site since

1978 1979 1980 1981 1982 1983 1984 1985 1986 1987 1988 1989 1990 1991 1992 1993 1994 1995 1996 1997 1998 1999 2000 **2001** 2002 2003 2004 2005 2006 2007 2008

Cerrado Protected Areas: Chapada dos Veadeiros and Emas National Parks
Brazil

Criteria – Significant ecological and biological processes; Significant natural habitat for biodiversity

The two sites included in the designated area contain flora and fauna and key habitats that characterize the Cerrado – one of the world's oldest and most diverse tropical ecosystems. For millennia, these sites have served as refuge for several species during periods of climate change and will be vital for maintaining the biodiversity of the Cerrado region during future climate fluctuations.

Comprising the Chapada dos Veadeiros and the Emas National Parks the area is home to many threatened and endemic species, while mammals include giant anteater, giant armadillo, maned wolf, jaguar and pampas deer.

◄
Wild giant anteater.

Tugendhat Villa in Brno
Czech Republic

Criteria – Interchange of values; Significance in human history

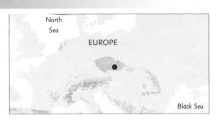

The house lost most of its original furniture and suffered some damage after it was taken over by the German State in 1939. The Tugendhat Villa Fund was established in 1993 and a scientific restoration of the building took place.

The Tugendhat Villa in Brno is a masterpiece of the Modern Movement in architecture in Europe in the 1920s. Its particular value lies in the way the German architect Mies van der Rohe (1886–1969) applied the radical new concepts of the movement to the design of residential buildings and made extensive use of modern industrial capabilities. The architect designed original furniture specifically for this house, such as the steel and leather Tugendhat chair.

The back wall of the living area is made from beautiful onyx, brought from the Atlas Mountains and processed on site. The mechanical equipment designed and built for the house was also exceptional, including electrically operated large steel-frame windows, central heating and an air-conditioning system with a regulated fine-spray humidifying chamber.

World Heritage site since

1978 1979 1980 1981 1982 1983 1984 1985 1986 1987 1988 1989 1990 1991 1992 1993 1994 1995 1996 1997 1998 1999 2000 **2001** 2002 2003 2004 2005 2006 2007 200

Historic Centre of Guimarães
Portugal

Criteria – Interchange of values; Testimony to cultural tradition; Significance in human history

The historic town of Guimarães is associated with the emergence of Portuguese national identity in the twelfth century. It is an exceptionally well-preserved town illustrating the evolution of particular building types from the medieval settlement to the present-day city. The historic centre is formed by a large number of stone constructions (950–1498). The period from Renaissance to neoclassicism is characterized by noble houses and the development of civic facilities, city squares, etc. The residential buildings make use of two construction techniques, a half-timbered one (taipa de rodízio) dating from before the sixteenth century, and another from the nineteenth century (taipa de fasquio) that uses timber alone. Despite some changes during the modern period, the town has maintained its medieval urban layout.

Guimarães is of particular significance by virtue of the fact that the specialized building techniques developed there in the Middle Ages were transmitted to Portuguese colonies in Africa and the New World, becoming their characteristic feature.

Lamu Old Town
Kenya

Criteria – Interchange of values; Significance in human history; Heritage associated with events of universal significance

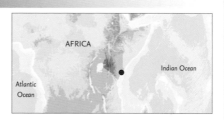

Lamu Old Town is the oldest and best-preserved Swahili settlement in east Africa, retaining its traditional functions. Built in coral stone and mangrove timber, the town is characterized by the simplicity of structural forms enriched by such features as inner courtyards, verandas, and elaborately carved wooden doors. Lamu has hosted major Muslim religious festivals since the nineteenth century, and has become a significant centre for the study of Islamic and Swahili cultures.

The architecture and urban structure of Lamu graphically demonstrate the cultural influences that have come together there over several centuries from Europe, Arabia and India, utilizing traditional Swahili techniques to produce a distinct culture.

World Heritage site since

1978 1979 1980 1981 1982 1983 1984 1985 1986 1987 1988 1989 1990 1991 1992 1993 1994 1995 1996 1997 1998 1999 2000 2001 **2002** 2003 2004 2005 2006 2007 2008

Minaret and Archaeological Remains of Jam
Afghanistan

Criteria – Interchange of values; Testimony to cultural tradition; Significance in human history

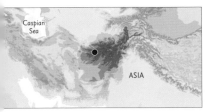

The Minaret of Jam is a graceful, soaring structure, dating back to the twelfth century, believed to have been built to commemorate a major victory of the sultans of the Ghurid dynasty. Rising to 65 m from a 9 m diameter octagonal base, its four tapering cylindrical shafts are constructed of fired brick bonded with lime mortar. The exterior of the minaret is completely covered with geometric decoration in relief laid over the plain structural bricks. The first cylinder is the most decorated: it is divided into eight vertical segments, matching those of the base. Each vertical zone has a narrow band of inscriptions running in an unbroken line around each panel. It is an outstanding example of Islamic architecture and ornamentation in this region and played a significant role in their further dissemination.

The remains of castles and towers of the Ghurid settlement are to be found on the opposite bank of the Hari River, north of the minaret and high on the cliff. There are also the remains of fortifications visible to the east of the minaret, suggesting that the minaret was surrounded not by a settlement but by a military camp.

World Heritage site since

1978 1979 1980 1981 1982 1983 1984 1985 1986 1987 1988 1989 1990 1991 1992 1993 1994 1995 1996 1997 1998 1999 2000 2001 **2002** 2003 2004 2005 2006 2007 200

Upper Middle Rhine Valley
Germany
Criteria – Interchange of values; Significance in human history; Traditional human settlement

The Rhine Valley at Oberwesel.

As one of the most important transport routes in Europe, the Middle Rhine Valley has facilitated the exchange of culture between the Mediterranean region and the north for two millennia. It is an outstanding organic cultural landscape and an excellent example of an evolving traditional way of life and means of communication in a narrow river valley. The terracing of its steep slopes in particular has shaped the landscape. However, this form of land use is under threat from today's socio-economic pressures.

The 65-km-stretch of the Middle Rhine Valley, with its castles, historic towns, and vineyards, graphically illustrates the long history of human involvement with a dramatic and varied natural landscape. It is intimately associated with history and legend, and for centuries has exercised a powerful influence on writers, artists and composers.

The river breaks through the Rhenish Slate Mountains, connecting the broad floodplain of the Oberrheingraben with the lowland basin of the Lower Rhine. At the 5-km-long Bingen Gate the Rhine enters the upper canyon stretch of the river; here, the vineyards of the Rüdesheimer Berg are among the best in the Rheingau. In the 15-km-long Bacharach valley, the small town of Lorch is lined with terraced vineyards, while Bacharach contains many timber-framed houses and retains its medieval appearance. Kaub and its environs contain a number of monuments, including the town wall, the Pfalzgrafenstein castle and terraced vineyards created in the Middle Ages. Oberwesel has preserved some fine early houses, two Gothic churches, the medieval Schönburg castle, and the town wall.

The valley landscape begins to change at Oberwesel with the transition from soft clay-slates to hard sandstone. The result is a series of narrows, the most famous of which is the Lorelei. This stretch of river was once hazardous for shipping and is reputed to be the place where the fabulous treasure of the Niebelungs lies hidden. On the right bank of the river is St Goarshausen, with its castle of Neu-Katzenelnbogen. The design of the fortress of Burg Reichenberg suggests that it may have been inspired by Crusader fortresses in Syria and Palestine. Bad Salzig on the left bank marks the beginning of the section of horseshoe-shaped bends known as the Boppard Loops.

Boppard originated as a Roman way-station and fort. Beyond is Osterspai which has merged into one town with Niederspay. Together they contain more timber-framed houses than anywhere else on the Middle Rhine. On the left bank, Rhens is where the German Emperors were enthroned after their election and coronation.

The fortress of Marksburg, the only surviving medieval fortification on the Middle Rhine, towers above Braubach. The castle of Stolzenfels was restored in 1835 by the Prussians, while at Koblenz is the New Castle, the first and most important classicist building in the Rhineland.

World Heritage site since

1978 1979 1980 1981 1982 1983 1984 1985 1986 1987 1988 1989 1990 1991 1992 1993 1994 1995 1996 1997 1998 1999 2000 2001 **2002** 2003 2004 2005 2006 2007 200

St Catherine Area
Egypt

Criteria – Human creative genius; Testimony to cultural tradition; Significance in human history; Heritage associated with events of universal significance

The Orthodox Monastery of St Catherine stands at the foot of Mount Horeb, of the Old Testament, where Moses received the Tablets of the Law. The entire area is sacred to three world religions: Christianity, Islam, and Judaism. Ascetic monasticism in remote areas prevailed in the early Christian church and resulted in the establishment of monastic communities in such places. St Catherine's Monastery is one of the earliest of these, and the oldest to have survived intact, having been used for its initial function without interruption since the sixth century. Its walls and buildings are very significant in the study of Byzantine architecture and the monastery houses outstanding collections of early Christian manuscripts and icons.

The Christian communities of St Catherine's Monastery have always maintained close relations with Islam. In 623 a document signed by the Prophet himself exempted the monks of St Catherine's from military service and tax and called upon Muslims to give them every help. As a reciprocal gesture the monastic community permitted the conversion of a chapel within the walled enceinte, to a mosque.

Mahabodhi Temple Complex at Bodh Gaya
India

Criteria – Human creative genius; Interchange of values; Testimony to cultural tradition; Significance in human history; Heritage associated with events of universal significance

The Mahabodhi Temple Complex is one of the four holy sites related to the life of the Lord Buddha (566–486 BC) as the place where, in 531 BC, he attained the supreme and perfect insight while seated under the Bodhi Tree. It provides exceptional records for the events associated with his life and for subsequent worship, particularly since Emperor Asoka made a pilgrimage to this spot around 260 BC and built the first temple at the site of the Bodhi Tree. The present temple dates from the fifth or sixth centuries and is one of the earliest Buddhist temples built entirely in brick still standing in India.

The most important of the sacred places is the giant Bodhi Tree (ficus religiosa). This tree is to the west of the main temple and is supposed to be a direct descendant of the original Bodhi Tree under which the Buddha spent his First Week and where he had his enlightenment.

World Heritage site since

1978 1979 1980 1981 1982 1983 1984 1985 1986 1987 1988 1989 1990 1991 1992 1993 1994 1995 1996 1997 1998 1999 2000 2001 2002 **2003** 2004 2005 2006 2007 200

Three Parallel Rivers of Yunnan Protected Areas
China

Criteria – Natural phenomena or beauty; Major stages of Earth's history; Significant ecological and biological processes; Significant natural habitat for biodiversity

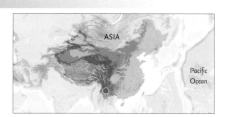

The Protected Areas within the boundaries of the Three Parallel Rivers National Park comprise a 17,000-km² site that features sections of the upper reaches of three of the great rivers of Asia: the Yangtze (Jinsha), the Mekong and the Salween. The rivers run roughly parallel, north to south for over 300 km, through steep gorges which, in places, are 3,000 m deep and bordered by glaciated peaks more than 6,000 m high. The site is an epicentre of Chinese biodiversity. It is also one of the richest temperate regions of the world in terms of biodiversity.

Three Parallel Rivers of Yunnan Protected Areas is situated in the mountainous northwest of Yunnan Province in south-central China. The site consists of fifteen protected areas in eight geographical clusters and extends 310 km from north to south and 180 km from east to west.

The World Heritage site lies over four parallel mountain ranges that reach in excess of 4,000 m above sea level. The ranges are part of the Hengduan Mountains which have been corrugated and uplifted by the pressures of crustal folding.

The land area encompassed by Three Parallel Rivers of Yunnan Protected Areas is one of the world's least-disturbed temperate ecological areas, an epicentre of Chinese endemic species and a natural gene pool of great richness. It supports the richest diversity of higher plants of China, owing to its altitudinal range and its position in a climatic corridor between north and south. It includes the equivalents of seven climatic zones: southern, central and northern subtropical zones, with dry hot valleys; warm, cool and cold temperate zones; and cold zones.

Owing to its function as a refuge during the last Ice Age and its location near the boundaries of three major biogeographic realms (east Asia, southeast Asia and the Tibetan plateau), the park has twenty-two vegetation subtypes and 6,000 plant species. The fauna is a complex mosaic of Palaearctic, oriental and endemic species adapted to almost all the inland climates from southern subtropical to frigid. The area is believed to support over 25 per cent of the world's animal species, including a concentration of rare and endangered animals. There are numerous primitive animals that are relics of the ecological past, alongside animals that have recently adapted to colder conditions.

Snow-capped Baimang Snow Mountain between th Yangtze and Mekong rivers.

The site is dominated by a composite orogenic belt that shows the signs of powerful crustal movements. Notable is the compression of the Eurasian plate edge by the underlying Indian plate. The resulting squeeze created vast thrust nappes, violent shearing and uplift into high mountains, through which pre-existing rivers continue to cut, resulting in the area's characteristic extreme vertical relief.

Alpine landscapes and their evolution are represented in the eastern mountains, where plateaus and valleys are covered with meadows, waterfalls, streams and hundreds of lakes left by glacial erosion.

World Heritage site since

1978 1979 1980 1981 1982 1983 1984 1985 1986 1987 1988 1989 1990 1991 1992 1993 1994 1995 1996 1997 1998 1999 2000 2001 2002 **2003** 2004 2005 2006 2007 20

Wooden Churches of Southern Little Poland

Poland

Criteria – Testimony to cultural tradition;
Significance in human history

North
Sea

Baltic
Sea

EUROPE

The wooden churches of southern Little Poland represent outstanding examples of medieval church-building traditions in Roman Catholic culture. Built using the horizontal log technique, common in eastern and northern Europe since the Middle Ages, these churches offered an alternative to the stone structures erected in urban centres. Churches have been of particular significance in the development of Polish wooden architecture, and an essential element of settlement structures, both as landmarks and as ideological symbols. They were an outward sign of the cultural identity of communities, reflecting the artistic and social aspirations of their patrons and creators. The six sites in southern Little Poland represent different aspects of these developments.

The Church of Archangel Michael of Szalowa, built in 1736–1756, differs from the others because of its architectural form although the same construction techniques were used. The church has a nave and two aisles, and is built in basilica form. The extremely rich polychrome decoration and fittings date from the eighteenth century.

The Church of St Philip and St James the Apostles (Sekowa).

World Heritage site since

1978 1979 1980 1981 1982 1983 1984 1985 1986 1987 1988 1989 1990 1991 1992 1993 1994 1995 1996 1997 1998 1999 2000 2001 2002 **2003** 2004 2005 2006 2007 2008

Franciscan Missions in the Sierra Gorda of Querétaro
Mexico

Criteria – Interchange of values; Testimony to cultural tradition

The five Franciscan missions of Sierra Gorda were built during the last phase of the conversion to Christianity of the interior of Mexico in the mid-eighteenth century and became an important reference for the continuation of the evangelisation of California, Arizona and Texas. Each mission had to erect the church, find the natives, subdue them, and then group them in huts around the church. The missionaries had to learn the native language, supply the population with food, teach them how to behave, and only then evangelise them. The richly decorated church façades are of special interest as they represent an example of the joint creative efforts of the missionaries and the native Indios. The rural settlements that grew around the missions have retained their vernacular character.

The architecture of the missions follows a similar pattern and generally includes an atrium, sacramental doorway, open chapel, processional chapels and a cloister. All five missions share similar elements in relation to their environment, the town and the religious buildings.

Cultural Landscape and Archaeological Remains of the Bamiyan Valley
Afghanistan

Criteria – Human creative genius; Interchange of values; Testimony to cultural tradition; Significance in human history; Heritage associated with events of universal significance

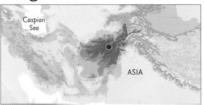

The cultural landscape and archaeological remains of the Bamiyan Valley represent the artistic and religious developments which, from the first to the thirteenth centuries, characterized ancient Bakhtria, integrating various cultural influences into the Gandhara school of Buddhist art. The area contains numerous Buddhist monastic ensembles and sanctuaries, as well as fortified edifices from the Islamic period. The site is also testimony to the tragic destruction by the Taliban of the two standing Buddha statues in March 2001.

The Bamiyan Valley is a high pass (2,500 m) that formed one of the branches of the Silk Road. Its beautiful landscape is associated with legendary figures which contributed to its development as a major religious and cultural centre.

◄
One of the Buddha statues destroyed in 2001.

World Heritage site since

1978 1979 1980 1981 1982 1983 1984 1985 1986 1987 1988 1989 1990 1991 1992 1993 1994 1995 1996 1997 1998 1999 2000 2001 2002 **2003** 2004 2005 2006 2007 200

Monte San Giorgio
Switzerland

Criteria – Major stages of Earth's history

The pyramid-shaped, wooded mountain 1,096 m above sea level, to the south of Lake Lugano in Canton Ticino, is regarded as the best fossil record of marine life from the Triassic Period, 245–230 million years ago. The sequence records life in a tropical lagoon environment, sheltered and partially separated from the open sea by an offshore reef. Diverse marine life flourished within this lagoon, including reptiles, fish, bivalves, ammonites, echinoderms and crustaceans. Because the lagoon was near to land, the fossil remains also include some land-based

fossils including reptiles, insects and plants. The result is a fossil resource of great richness.

Fossils from the mountain have been known to science for over 150 years. The vertebrate material includes particularly spectacular specimens, with articulated skeletons up to 6 m in length. The site's record of marine life during a critical period in vertebrate evolution on Earth provides a global reference point for comparative studies of evolution.

◀

Monte San Giorgio on Lake Lugano.

Jewish Quarter and St Procopius' Basilica in Třebíč
Czech Republic

Criteria – Interchange of values; Testimony to cultural tradition

The ensemble of the Jewish Quarter, the old Jewish cemetery and the Basilica of St Procopius in Třebíč, are reminders of the co-existence of Jewish and Christian cultures from the Middle Ages to the twentieth century. The Jewish Quarter bears outstanding testimony to the different aspects of the life of this community. The area has preserved all essential social functions, including synagogues and

schools, as well as a leather factory. St Procopius' Basilica is situated in a good position on the hill with a view over the whole of Třebíč. Built in the thirteenth century as a monastic church, it is a mixture of Romanesque and early Gothic styles and presents a remarkable example of the influence of western European architectural heritage in this region.

The Jewish Quarter rises up the hillside from the river to which its two main streets are linked by small mediaeval alleys; some of these run through the houses. All the Jewish residents were deported during the Second World War.

orld Heritage site since

8 1979 1980 1981 1982 1983 1984 1985 1986 1987 1988 1989 1990 1991 1992 1993 1994 1995 1996 1997 1998 1999 2000 2001 2002 **2003** 2004 2005 2006 2007 2008

Quebrada de Humahuaca
Argentina

riteria – Interchange of values; Significance in uman history; Traditional human settlement

Quebrada de Humahuaca follows the line f a major cultural route, the Camino Inca, long the spectacular valley of the Rio rande, from its source in the cold High ndean desert plateau to its confluence with ne Rio Leone some 150 km to the south. he valley shows substantial evidence of its se as a major trade route over the past 0,000 years. Scattered along the valley are

extensive remains of successive settlements whose inhabitants created and used these linear routes. It features visible traces of prehistoric hunter-gatherer communities, of the Inca Empire (fifteenth–sixteenth centuries) and of the fight for independence in the sixteenth and twentieth centuries.

Of particular note are the extensive remains of stone-walled agricultural terraced fields at Coctaca, thought to have originated around 1,500 years ago and still in use today. The field system makes a dramatic impact on the landscape that is unrivalled in South America.

Seven Colours Mountain in Quebrada de Humahuaca.
▼

World Heritage site since

1978 1979 1980 1981 1982 1983 1984 1985 1986 1987 1988 1989 1990 1991 1992 1993 1994 1995 1996 1997 1998 1999 2000 2001 2002 **2003** 2004 2005 2006 2007 200

Phong Nha-Ke Bang National Park

Vietnam

Criteria – Major stages of Earth's history

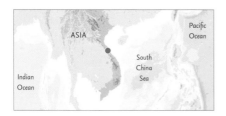

The karst formation of Phong Nha-Ke Bang National Park has evolved since the Palaeozoic era, some 400 million years ago, and is the oldest major karst area in Asia. Subject to massive tectonic changes, the park's karst landscape is extremely complex with many geomorphic features of considerable significance. The vast area, extending to the border with the Lao People's Democratic Republic, contains spectacular formations including 65 km of caves and underground rivers. The Phong Nha Cave is the most famous in the system, with a currently surveyed length of 44.5 km. Its entrance is part of an underground river and tour boats can penetrate inside to a distance of 1,500 m. Other extensive caves include the Vom cave and the Hang Khe Rhy cave.

Stalactites and pillar formations in Phong Nha-Ke Bang National Park. ▲

Some 92 per cent of the park is covered by tropical forest. A total of 568 vertebrate species have been recorded in the site, comprising 113 mammals, 81 reptiles and amphibians, 302 birds and 72 fish.

The site is particularly rich in primates, with ten species and subspecies forming 45 per cent of the total number of species in Vietnam.

World Heritage site since

1978 1979 1980 1981 1982 1983 1984 1985 1986 1987 1988 1989 1990 1991 1992 1993 1994 1995 1996 1997 1998 1999 2000 2001 2002 **2003** 2004 2005 2006 2007 2008

Mausoleum of Khoja Ahmed Yasawi
Kazakhstan

Criteria – Human creative genius; Testimony to cultural tradition; Significance in human history

Black Sea
Caspian Sea
ASIA

The Mausoleum of Khoja Ahmed Yasawi, a distinguished Sufi master of the twelfth century, is situated in the city of Turkestan (Yasi) in southern Kazakhstan. The mausoleum is in the area of the former citadel, in the northeastern part of the ancient town, now an open archaeological site. To the south is a nature protection area; on the other sides the modern city of Turkestan surrounds the site.

The mausoleum was built at the time of Timur (Tamerlane), from 1389 to 1405. In this partly unfinished building, Persian master builders experimented with architectural and structural solutions later used in the construction of Samarkand, the capital of the Timurid Empire. Today it is one of the largest and best-preserved constructions of the Timurid period.

The Mausoleum's Main Hall is covered with a conic-spherical dome which is the largest in central Asia (18.2 m in diameter). The building also features a mosque, which is the only room where fragments of the original wall paintings are preserved.

World Heritage site since

1978 1979 1980 1981 1982 1983 1984 1985 1986 1987 1988 1989 1990 1991 1992 1993 1994 1995 1996 1997 1998 1999 2000 2001 2002 2003 **2004** 2005 2006 2007 2008

Liverpool – Maritime Mercantile City
United Kingdom

Criteria – Interchange of values; Testimony to cultural tradition; Significance in human history

Atlantic Ocean
North Sea
EUROPE

Six areas in the historic centre and docklands of the maritime mercantile city of Liverpool bear witness to the development of one of the world's major trading centres in the eighteenth and nineteenth centuries. Liverpool played an important role in the growth of the British Empire and became the major port for the mass movement of people, e.g. slaves, and emigrants from northern Europe to America. Liverpool was a pioneer in the development of modern dock technology, transport systems and port management. It features a great number of significant commercial, civic and public buildings.

The site stretches along the waterfront from Albert Dock to Pier Head and Stanley Dock and takes in the historic, commercial and cultural districts of the city centre.

World Heritage site since

1978 1979 1980 1981 1982 1983 1984 1985 1986 1987 1988 1989 1990 1991 1992 1993 1994 1995 1996 1997 1998 1999 2000 2001 2002 2003 **2004** 2005 2006 2007 2008

Ilulissat Icefjord
Greenland (Denmark)

Criteria – Natural phenomena or beauty; Major stages of Earth's history

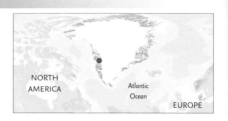

Located on the west coast of Greenland, 250 km north of the Arctic Circle, Ilulissat Icefjord (402 km²) is the sea mouth of Sermeq Kujalleq, one of the few glaciers through which the Greenland icecap reaches the sea. Sermeq Kujalleq is one of the fastest moving (19 m per day) and most active glaciers in the world. It annually calves over 35 km³ of ice, i.e. 10 per cent of the production of all Greenland calf ice and more than any other glacier outside

Antarctica. Studied for over 250 years, it has helped to develop our understanding of climate change and icecap glaciology. The combination of a huge ice-sheet and the dramatic sounds of a fast-moving glacial ice-stream calving into a fjord covered by icebergs makes for a dramatic and awe-inspiring natural phenomenon.

Norsemen inhabited southwest Greenland between AD 985 and 1450. During the sixteenth–eighteenth centuries explorers, followed by whalers, inhabited the area. The World Heritage area includes the archaeologically valuable sites of Sermermuit, abandoned in 1850, and Qajaa on the south side of the fjord, abandoned earlier. The early settler summered in tents but used stone and turf hovels in winter.

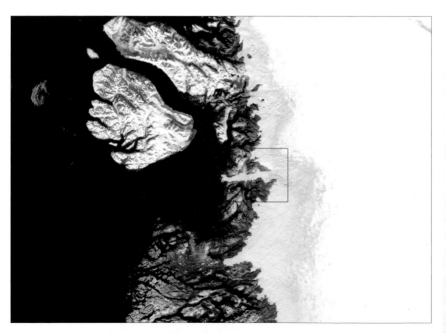

◄

Ilulissat Icefjord as seen from space.

World Heritage site since

978 1979 1980 1981 1982 1983 1984 1985 1986 1987 1988 1989 1990 1991 1992 1993 1994 1995 1996 1997 1998 1999 2000 2001 2002 2003 **2004** 2005 2006 2007 2008

Dresden Elbe Valley
Germany

Criteria – Interchange of values; Testimony to cultural tradition; Significance in human history; Traditional human settlement

The Dresden Elbe Valley has been the crossroads of Europe in culture, science and technology. The cultural landscape extends some 18 km along the river from Übigau Palace and Ostragehege fields in the northwest to the Pillnitz Palace and the Elbe River Island in the southeast. It is crowned by the Pillnitz Palace and the centre of Dresden with its numerous monuments and parks dating from the sixteenth to twentieth centuries. The landscape also features nineteenth- and twentieth-century suburban villas and gardens and valuable natural features. Some old villages have retained their historic structure, and elements remain from the industrial revolution, notably the 147 m Blue Wonder steel bridge (1891–3), the single-rail suspension cable railway (1898–1901), and the funicular (1894–5). The passenger steamships (the oldest from 1879) and shipyard (c. 1900) are still in use.

The fortified city of Dresden developed in the Middle Ages with its main part on the south side of the river Elbe. After a fire in the late seventh century, the city was modernized in Baroque and Rococo styles. The north bank became known as Neustadt (New Town) and the German town on the south bank as Altstadt (Old Town).

World Heritage site since

1978 1979 1980 1981 1982 1983 1984 1985 1986 1987 1988 1989 1990 1991 1992 1993 1994 1995 1996 1997 1998 1999 2000 2001 2002 2003 **2004** 2005 2006 2007 2008

Þingvellir National Park
Iceland

Criteria – Testimony to cultural tradition; Heritage associated with events of universal significance

Þingvellir (Thingvellir) is the National Park where the Althing – an open-air assembly, which represented the whole of Iceland – was established in 930 and continued to meet until 1798. Over two weeks a year, the assembly set laws – seen as a covenant between free men – and settled disputes. The Althing has deep historical and symbolic associations for the people of Iceland. The property includes the Þingvellir National Park and the remains of the Althing itself: fragments of around fifty booths built from turf and stone. Remains from the tenth century are thought to be buried underground. The site also includes

remains of agricultural use from the eighteenth and nineteenth centuries. The park shows evidence of the way the landscape was husbanded over 1,000 years.

The Althing and its hinterland, Þingvellir National Park, represent a unique reflection of medieval Norse/Germanic culture which persisted from its foundation in AD 980 until the eighteenth century.

Bam and its Cultural Landscape
Islamic Republic of Iran

Criteria – Interchange of values; Testimony to cultural tradition; Significance in human history; Traditional human settlement

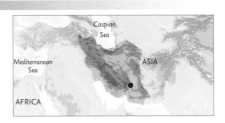

Bam is situated in a desert environment on the southern edge of the Iranian high plateau. The origins of Bam can be traced back to the Achaemenid period, sixth to fourth centuries BC. Its heyday was from the seventh to eleventh centuries, being at the crossroads of important trade routes and known for the production of silk and cotton

garments. The existence of life in the oasis was based on the underground irrigation canals, the qanāts, of which Bam has preserved some of the earliest evidence in Iran. Arg-e Bam is the most representative example of a fortified medieval town built in vernacular technique using mud layers (Chineh).

Bam is an outstanding expression of the interaction of man and nature in a desert environment. The civilisation depended on a strict social system with precise tasks and responsibilities, which have been maintained in use until the present.

World Heritage site since

78 1979 1980 1981 1982 1983 1984 1985 1986 1987 1988 1989 1990 1991 1992 1993 1994 1995 1996 1997 1998 1999 2000 2001 2002 2003 **2004** 2005 2006 2007 2008

Pitons Management Area
Saint Lucia

Criteria – Natural phenomena or beauty; Major stages of Earth's history

Dominating the mountainous landscape of St Lucia are the Pitons, two steep-sided volcanic spires rising side by side from the sea. Gros Piton (770 m) is 3 km in diameter at its base, and Petit Piton (743 m) is 1 km in diameter and linked to the former by the Piton Mitan ridge. The Pitons are part of a volcanic complex, known to geologists as the Soufrière Volcanic Centre, which is the remnant of one, or more, huge collapsed stratovolcano. The Marine Management Area is a coastal strip 11 km long and about 1 km wide. The coral reefs, which cover almost 60 per cent of the marine area, are healthy and diverse. The area is a multiple-use management system where agriculture, artisan fishing, human settlement (1,500 residents) and tourism (four large hotel developments) were present at the time of inscription.

At least 148 plant species have been recorded on Gros Piton and 97 on Petit Piton. Among these are several endemic or rare plants, including eight rare species of tree. Some bird species, including five endemics, are known from Gros Piton, along with indigenous rodents, opossum, bats, reptiles and amphibians.

World Heritage site since

1978 1979 1980 1981 1982 1983 1984 1985 1986 1987 1988 1989 1990 1991 1992 1993 1994 1995 1996 1997 1998 1999 2000 2001 2002 2003 2004 2005 2006 2007 200

Tropical Rainforest Heritage of Sumatra
Indonesia

Criteria – Natural phenomena or beauty;
Significant ecological and biological processes;
Significant natural habitat for biodiversity

ASIA
Pacific Ocean
Indian Ocean
OCEANIA

The 25,000 km² Tropical Rainforest Heritage of Sumatra comprises three widely-separated national parks along the Bukit Barisan mountain range: Gunung Leuser National Park, Kerinci Seblat National Park and Bukit Barisan Selatan National Park. The site holds the greatest potential for long-term conservation of the distinctive biodiversity of Sumatra, including many endangered species. The protected area is home to an estimated 10,000 plant species, including 17 endemic genera; more than 200 mammal species; and some 580 bird species, of which 465 are resident and 21 are endemic. Of the mammal species, twenty-two are Asian, not found elsewhere in the archipelago, and fifteen are confined to the Indonesian region, including the endemic Sumatran orang-utan. The site also provides biogeographic evidence of the evolution of the island.

Kerinci Seblat National Park contains the magnificent, active volcano Gunung Kerinci – at 3,805 m, the highest peak in Sumatra and the highest volcano in Indonesia. Nearby Gunung Tujuh is an outstandingly beautiful crater lake at 1,996 m.

A Sumatran orang-utan.
▼

World Heritage site since

1978 1979 1980 1981 1982 1983 1984 1985 1986 1987 1988 1989 1990 1991 1992 1993 1994 1995 1996 1997 1998 1999 2000 2001 2002 2003 **2004** 2005 2006 2007 2008

Um er-Rasas
(Kastrom Mefa'a)
Jordan

Criteria – Human creative genius; Significance in human history; Heritage associated with events of universal significance

Most of this archaeological site, which started as a Roman military camp and grew to become a town from the fifth century onwards, has not been excavated. It contains remains from the Roman, Byzantine and Early Muslim periods, the end of the third to ninth centuries AD, and a fortified Roman military camp. The site also has sixteen churches, some with well-preserved mosaic floors. Particularly noteworthy is the mosaic floor of the Church of St Stephen. Two square towers are probably the only remains of the practice, well known in this part of the world, of the stylites (ascetic monks who spent time in isolation atop a column or tower). Um er-Rasas is surrounded by, and dotted with, remains of ancient agricultural cultivation in an arid area. It is here that the Prophet Muhammad, travelling as a tradesman, met a monk who convinced him of the virtue of monotheism.

The mosaic floor of the church of St Stephen shows an incredible representation of towns in Palestine, Jordan and Egypt, including their identification. Its artistic and technical qualities justify describing Um er-Rasas as a masterpiece of human creative genius.

Southern area of Um er-Rasas.
▼

World Heritage site since

1978 1979 1980 1981 1982 1983 1984 1985 1986 1987 1988 1989 1990 1991 1992 1993 1994 1995 1996 1997 1998 1999 2000 2001 2002 2003 **2004** 2005 2006 2007 200

Royal Exhibition Building and Carlton Gardens
Australia
Criteria – Interchange of values

The aesthetic significance of the Carlton Gardens lies in its representation of the nineteenth-century Gardenesque style. This includes parterre garden beds, significant avenues including the southern carriage drive and Grande Allée, the path system clusters of trees, two small lakes and three fountains.

The Royal Exhibition Building, in its original setting of the Carlton Gardens, is the only substantially intact example in the world of a Great Hall from a major international exhibition. The building and gardens were designed for the great exhibitions of 1880 and 1888 in Melbourne. The building is constructed of brick and timber, steel and slate. It combines elements from the Byzantine, Romanesque, Lombardic and Italian Renaissance styles. The property is typical of the international exhibition movement, which aimed to showcase technological innovation, and boasts many of the important features that made the expositions so dramatic, including a dome, a great hall, giant entry portals and complementary gardens and viewing areas. Unlike the structures at many international exhibitions, the building was conceived as a permanent construction that would have a future role in the cultural activities of the growing city of Melbourne.

World Heritage site since

1978 1979 1980 1981 1982 1983 1984 1985 1986 1987 1988 1989 1990 1991 1992 1993 1994 1995 1996 1997 1998 1999 2000 2001 2002 2003 **2004** 2005 2006 2007 2008

Orkhon Valley Cultural Landscape
Mongolia
Criteria – Interchange of values; Testimony to cultural tradition; Significance in human history

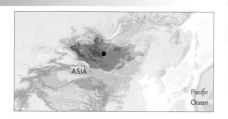

The 1,220-km² Orkhon Valley Cultural Landscape encompasses an extensive area of pastureland on both banks of the Orkhon River and includes numerous archaeological remains dating back to the sixth century. The site also includes Kharkhorum, the thirteenth- and fourteenth-century capital of Chingis (Genghis) Khan's vast Empire. Collectively the remains in the site reflect the symbiotic links between nomadic, pastoral societies and their administrative and religious centres, and the importance of the Orkhon valley in the history of central Asia. The grassland is still grazed by Mongolian nomadic pastoralists.

In Mongolia, nomadic pastoralism is revered and glorified as the heart of Mongolian culture, and in the Orkhon Valley Cultural Landscape the links between such nomadic pastoralism and its associated settlements can be seen clearly.

Muskauer Park / Park Muzakowski
Germany and Poland
Criteria – Human creative genius; Significance in human history

A landscaped park of 5.6 km² astride the Neisse River and the border between Poland and Germany, Muskauer Park was created by Prince Hermann von Pückler-Muskau from 1815 to 1844. Blending seamlessly with the surrounding farmed landscape, the park pioneered new approaches to landscape design and influenced the development of landscape architecture in Europe and America. Designed as a 'painting with plants', it did not seek to evoke classical landscapes, paradise, or some lost perfection, instead using local plants to enhance the inherent qualities of the existing landscape. This integrated landscape extends into the town of Muskau with green passages that formed urban parks framing areas for development. The town thus became a design component in a utopian landscape. The site also features a reconstructed castle, bridges and an arboretum.

The site is the centre of a landscape park which extended around Muskau and into the countryside. After the Second World War the Neisse became the international border, leaving 3.5 km² of the park within Poland and 2.1 km² in Germany.

World Heritage site since

1978 1979 1980 1981 1982 1983 1984 1985 1986 1987 1988 1989 1990 1991 1992 1993 1994 1995 1996 1997 1998 1999 2000 2001 2002 2003 **2004** 2005 2006 2007 2008

Cape Floral Region Protected Areas
South Africa

Criteria – Significant ecological and biological processes; Significant natural habitat for biodiversity

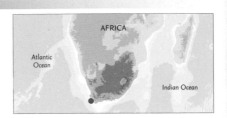

AFRICA

Atlantic Ocean

Indian Ocean

Fynbos, Afrikaans for 'fine bush' is the natural, primarily evergreen, shrubland vegetation of the Cape Floral region.

Characteristics of the Cape Floral Region that are of global scientific interest are the responses of its plants to fire; seed dispersal by ants and termites; the high level (83 per cent) of plant pollination by insects (mainly beetles and flies); and its links to the ancient continent of Gondwanaland, which allowed reconstruction of the flora's ancient connections.

Some of the species actually require fire for germination. Pollination and nutrient-cycling by termites are notable and the region also has a very high number of plants that are pollinated by birds and mammals.

Eight protected areas covering 5,530 km² make up the Cape Floral Region, one of the richest areas for plants in the world. It represents less than 0.5 per cent of the area of Africa but is home to nearly 20 per cent of the continent's flora. The site displays outstanding ecological and biological processes associated with the fynbos (fine bush) vegetation which is unique to the Cape Floral Region. The outstanding diversity, density and endemism of the flora are among the highest anywhere in the world. Unique plant-reproductive strategies, adaptive to fire, patterns of seed dispersal by insects and patterns of endemism and adaptive radiation found in the flora, are of outstanding value to science.

The region is located in the southwest corner of South Africa in Cape Province. Elevations range from 2,077 m in the Groot Winterhoek to sea level in the De Hoop Nature Reserve. A great part of the area is characterized by rugged mountain passes, rivers, rapids, cascades and pools.

The area has been designated as one of the World Centres of Plant Diversity. It has 44 per cent of the subcontinental flora of 20,367 species, including endemic and subendemic families and threatened species. The Cape Peninsula contains almost half of these species, with 25 per cent of the

flora of the whole region. The richness is due to the wide variety of macrohabitats and microhabitat mosaics resulting from the range of elevations, soils and climatic conditions, including the co-existence of winter-rainfall species with summer-rainfall species from further east.

Eight centres of endemism have been distinguished in the Cape Floral Region. The distinctive flora of the area, comprising 80 per cent of its richness, is the fynbos, fine-leaved vegetation adapted to both the Mediterranean type of climate and to periodic fires, and defined by the location or dominant species. Plant variety is based on soil types that range from predominantly coarse, sandy, acidic, nutrient-poor soils, to alkaline marine sands and slightly richer alluvials. There are pockets of evergreen forest in fire-protected gorges and on deeper soils; in the east are valley thickets and succulent thickets, which are less fire-dependent, and in the drier north, low succulent Karoo shrubland.

World Heritage site since

1978 1979 1980 1981 1982 1983 1984 1985 1986 1987 1988 1989 1990 1991 1992 1993 1994 1995 1996 1997 1998 1999 2000 2001 2002 2003 **2004** 2005 2006 2007 20

Luis Barragán House and Studio
Mexico

Criteria – Human creative genius; Interchange of values

Built in 1948, the House and Studio of architect Luis Barragán in the suburbs of Mexico City represents an outstanding example of the architect's creative work in the post-Second World War period. The concrete building, totalling 1,161 m², consists of a ground floor and two upper storeys, as well as a small private garden.

Barragán's work integrated modern and traditional artistic and vernacular currents and elements into a new synthesis, which has been greatly influential, especially in the contemporary design of gardens, plazas and landscapes.

Luis Barragán began work on the house fo a client but in 1948 decided to take the house for himself. Th plans were gradually developed over the construction period and the house remained his studio and residence until h death.

Capital Cities and Tombs of the Ancient Koguryo Kingdom
China

Criteria – Human creative genius; Interchange of values; Testimony to cultural tradition; Significance in human history; Traditional human settlement

The site includes archaeological remains of three cities – Wunu Mountain City, Guonei City and Wandu Mountain City – and forty tombs. Fourteen tombs are imperial and twenty-six are of nobles. All belong to the Koguryo culture, named after the dynasty that ruled over parts of northern China and the northern half of the Korean Peninsula from 277 BC to AD 668. Wunu Mountain City is only partly excavated. Guonei City, within the modern city of Ji'an, played the role of a 'supporting capital' after

the main Koguryo capital moved to Pyongyang (in present day Democratic People's Republic of Korea). Wandu Mountain City, one of the capitals of the Koguryo Kingdom, contains many vestiges including a large palace and thirty-seven tombs. Some of the tombs show great ingenuity in their elaborate ceilings, which were designed to roof wide spaces without columns and carry the heavy load of a stone or earth tumulus mound placed above them.

The site represents exceptional testimony to the vanished Koguryo civilization. The capital cities are early examples of mountain cities and were later imitated by neighbouring cultures. Conversely, the Koguryo cities and tombs also show evidence of strong impact from other cultures.

World Heritage site since

1978 1979 1980 1981 1982 1983 1984 1985 1986 1987 1988 1989 1990 1991 1992 1993 1994 1995 1996 1997 1998 1999 2000 2001 2002 2003 **2004** 2005 2006 2007 2008

Chhatrapati Shivaji Terminus formerly Victoria Terminus
India

Criteria – Interchange of values; Significance in human history

The Chhatrapati Shivaji Terminus, formerly known as Victoria Terminus Station, in Mumbai (Bombay), is an outstanding example of Victorian Gothic Revival architecture in India. British architects worked closely with Indian craftsmen on the building to include local architectural traditions. The resulting structure, with its remarkable stone dome, turrets, pointed arches and eccentric ground plan is close to traditional Indian palace architecture, making it an outstanding example of the meeting of two cultures. The unique new style of the building soon came to symbolize Bombay as the 'Gothic City' and major international mercantile port of India.

The main structure is built from a judicious blend of Indian sandstone and limestone, while high-quality Italian marble was used for the key decorative elements. The main interiors are lavishly decorated: the ground floor of the North Wing, known as the Star Chamber, which is still the booking office, is embellished with Italian marble and polished Indian blue stone.

World Heritage site since

1978 1979 1980 1981 1982 1983 1984 1985 1986 1987 1988 1989 1990 1991 1992 1993 1994 1995 1996 1997 1998 1999 2000 2001 2002 2003 **2004** 2005 2006 2007 2008

Val d'Orcia
Italy

Criteria – Significance in human history;
Heritage associated with events of universal
significance

EUROPE

Mediterranean Sea

Ionian
Sea

The World Heritage
site is significant in
that the large
farmhouses assume
a dominant position
in the landscape and
are enriched by
prominent
architectural elements
such as loggias,
belvederes, porches
and avenues of trees
bordering the
approach roads.

Val d'Orcia is an exceptional reflection
of the way the landscape was rewritten in
Renaissance times to reflect the ideals of
good governance and to create an
aesthetically pleasing picture. The landscape
is 25 km from Siena's centre and was, in
effect, colonized by the city's merchants in
the fourteenth and fifteenth centuries. They
aimed to create an area of efficient

agricultural units that was also pleasing to
the eye. The landscape that resulted was one
of careful and conscious planning and
design and led to the beginning of the
concept of 'landscape' as a man-made
creation. The landscape's distinctive
aesthetics of fortified settlements on conical
hills rising out of flat chalk plains have
inspired many important artists.

World Heritage site since

1978 1979 1980 1981 1982 1983 1984 1985 1986 1987 1988 1989 1990 1991 1992 1993 1994 1995 1996 1997 1998 1999 2000 2001 2002 2003 **2004** 2005 2006 2007 2008

Varberg Radio Station
Sweden

Criteria – Interchange of values; Significance in human history

Varberg Radio Station was in regular service until the 1960s. It has been partly open to the public since 1997 although some equipment is still used by the Swedish Navy.

The Varberg Radio Station at Grimeton in southern Sweden built 1922–4 is an exceptionally well-preserved monument to early wireless transatlantic communication. It consists of the transmitter equipment, including the aerial system of six 127-m-high steel towers. Although no longer in regular use, the equipment has been maintained in operating condition. The 1.1 km² site comprises buildings housing the original Alexanderson transmitter, including the towers with their antennae, short-wave transmitters with their antennae, and a residential area with staff housing. The architect Carl Åkerblad designed the main buildings in the neoclassical style and the structural engineer Henrik Kreüger was responsible for the antenna towers, the tallest built structures in Sweden at that time. The site is an outstanding example of the development of telecommunications and is the only surviving example of a major transmitting station based on pre-electronic technology.

Tomb of Askia
Mali

Criteria – Interchange of values; Testimony to cultural tradition; Significance in human history

The tomb's builder, Askia Mohamed, was founder of the Askia dynasty. It is said that on passing through Egypt on his way to Mecca, Askia Mohamed was impressed by the pyramids and decided to construct a pyramidal tomb for himself.

The dramatic 17-m pyramidal structure of the Tomb of Askia was built by Askia Mohamed, the Emperor of Songhai, in 1495 in his capital Gao. It bears testimony to the power and riches of the empire that flourished in the fifteenth and sixteenth centuries through its control of the trans-Saharan trade, notably in salt and gold. It is also a fine example of the monumental mud-building traditions of the Sahel.

World Heritage site since

1978 1979 1980 1981 1982 1983 1984 1985 1986 1987 1988 1989 1990 1991 1992 1993 1994 1995 1996 1997 1998 1999 2000 2001 2002 2003 2004 2005 2006 2007 2008

Koutammakou, the Land of the Batammariba

Togo

Criteria – Traditional human settlement; Heritage associated with events of universal significance

AFRICA

Atlantic Ocean

Whether hand-modelled or built from mud brick, the variety of architectural forms in West Africa illustrates the many ways in which the simple elements of earth and water are brought together to create works of striking artistic sophistication and interest.

The Koutammakou landscape in northeastern Togo, which extends into neighbouring Benin, is home to the Batammariba whose remarkable mud tower-houses, Takienta, have come to be seen as a symbol of Togo. In this landscape, nature is strongly associated with the rituals and beliefs of society. The 500 km² cultural landscape is remarkable due to the architecture of its tower-houses which are a reflection of social structure; its farmland and forest; and the associations between people and landscape. Many of the buildings are two-storeys high and those with granaries feature an almost spherical form above a cylindrical base. Some of the buildings have flat roofs, others have conical thatched roofs. They are grouped in villages, which also include ceremonial spaces, springs, rocks and sites reserved for initiation ceremonies.

Sacred Sites and Pilgrimage Routes in the Kii Mountain Range

Japan

Criteria – Interchange of values; Testimony to cultural tradition; Significance in human history; Heritage associated with events of universal significance

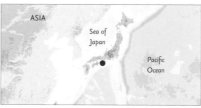

ASIA

Sea of Japan

Pacific Ocean

The shrines contain both buildings and objects, such as temples and statues, as well as such revered natural elements as trees and waterfalls. The journey to the shrines over arduous mountain routes was intended as part of the religious experience.

Set in the dense forests of the Kii Mountains, south of Osaka and overlooking the Pacific Ocean, these three sacred sites – Yoshino and Omine, Kumano Sanzan, Koyasan – linked by pilgrimage routes to the ancient capital cities of Nara and Kyoto, reflect the fusion of Shinto, rooted in the ancient tradition of nature worship in Japan, and Buddhism, which was introduced from China and the Korean Peninsula. The sites (5 km²) and their surrounding forest landscape reflect a persistent and extraordinarily well-documented tradition of sacred mountains over 1,200 years. The area, with its abundance of streams, rivers and waterfalls, is still part of the living culture of Japan and is much visited for ritual purposes and hiking, with up to fifteen million visitors annually. Each of the three sites contains shrines, some of which were founded as early as the ninth century.

World Heritage site since

1978 1979 1980 1981 1982 1983 1984 1985 1986 1987 1988 1989 1990 1991 1992 1993 1994 1995 1996 1997 1998 1999 2000 2001 2002 2003 **2004** 2005 2006 2007 2008

Town Hall and Roland on the Marketplace of Bremen
Germany

Criteria – Testimony to cultural tradition; Significance in human history; Heritage associated with events of universal significance

Bremen is in northwestern Germany, on the river Weser. The medieval town was oblong, limited by the river on the south side and by the moat of the ancient defence system on the north side. The Town Hall and Roland statue are an outstanding ensemble representing civic autonomy and market freedom, as developed in the Holy Roman Empire. The old town hall was built in the Gothic style in the early fifteenth century, after Bremen joined the Hanseatic League. The building was expertly renovated in the so-called Weser Renaissance style in the early seventeenth century. A new town hall was built next to the old one in the early twentieth century as part of an ensemble that survived bombardment during the Second World War.

The stone statue of Roland (a legendary figure in medieval Europe) is about 5.5 m tall, and it was initially erected in 1404, replacing an earlier wooden statue. It is considered the oldest Roland statue still in place in Germany and symbolizes the rights and privileges of the free and imperial city of Bremen.

World Heritage site since

1978 1979 1980 1981 1982 1983 1984 1985 1986 1987 1988 1989 1990 1991 1992 1993 1994 1995 1996 1997 1998 1999 2000 2001 2002 2003 **2004** 2005 2006 2007 20

Madriu-Perafita-Claror Valley
Andorra

Criteria – Traditional human settlement

The cultural landscape of Madriu-Perafita-Claror Valley offers a microcosmic perspective of the way people have harvested the resources of the high Pyrenees over millennia. Its dramatic glacial landscapes of craggy cliffs and glaciers, with high open pastures and steep wooded valleys, covers an area of 42.5 km², 9 per cent of the total area of the principality. It reflects past changes in climate, economic fortune and social systems, as well as the persistence of pastoralism and a strong mountain culture, notably the survival of a communal land-ownership system dating back to the thirteenth century. The site features houses, notably summer settlements, terraced fields, stone tracks and evidence of iron smelting.

The valley has maintained its structures of organization and management since medieval times, surviving as a living witness to the history of Andorra, the culture of the men of the mountains and their coexistence with an extraordinary natural environment.

World Heritage site since

1978 1979 1980 1981 1982 1983 1984 1985 1986 1987 1988 1989 1990 1991 1992 1993 1994 1995 1996 1997 1998 1999 2000 2001 2002 2003 2004 **2005** 2006 2007 200

Historic Centres of Berat and Gjirokastra
Albania

Criteria – Testimony to cultural tradition; Significance in human history

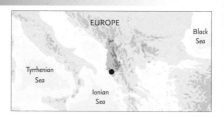

The historic towns of Gjirokastra and Berat, in the Drinos River valley in southern Albania, are rare examples of well-preserved towns of an architectural character typical of the Ottoman period. The thirteenth-century citadel of Gjirokastra provides the focal point of the town with its typical tower houses (Turkish kule). Characteristic of the Balkan region, Gjirokastra contains a series of outstanding examples of kule, a type of building that crystallized in the seventeenth century. But Gjirokastra also features some more elaborate examples from the early nineteenth century. The kule has a tall basement, a first floor for use in the cold season, and a second floor for the warm season. Interiors feature rich decorative details and painted floral patterns, particularly in the zones reserved for the reception of visitors. The town also retains a bazaar, an eighteenth-century mosque and two churches of the same period.

Berat, which was inscribed as a World Heritage Site in 2008 features a kala, or castle, most of which was built in the thirteenth century. The citadel area contains many thirteenth-century Byzantine churches and several mosques of the Ottoman era.

orld Heritage site since

78 1979 1980 1981 1982 1983 1984 1985 1986 1987 1988 1989 1990 1991 1992 1993 1994 1995 1996 1997 1998 1999 2000 2001 2002 2003 2004 **2005** 2006 2007 2008

slands and Protected Areas of the Gulf of California

Mexico

Criteria – Natural phenomena or beauty;
significant ecological and biological processes;
significant natural habitat for biodiversity

he site comprises 244 islands, islets and coastal areas located in the Gulf of California in northwestern Mexico. The Sea of Cortez and its islands have been called a natural laboratory for the investigation of speciation. Moreover, almost all major oceanographic processes are present in the property, giving it extraordinary importance. The site is one of striking natural beauty characterized by rugged islands with high cliffs and sandy beaches. It is home to 695 vascular plant species, more than in any other marine and insular property on the World Heritage List. Equally exceptional is the number of fish species: 891, of which 90 are endemic. The site, moreover, contains 39 per cent of the world's total number of species of marine mammals and a third of the world's marine cetacean species.

Rock formation near Cabo San Lucas at the southern tip of the Baja California peninsula. ▲

The site is unique in that, in a very short distance, there are simultaneously 'bridge islands' (accessible by land) and 'oceanic islands' (accessible by sea and air).

The diversity and abundance of the marine life and the high water transparency make this a diver's paradise.

World Heritage site since

1978 1979 1980 1981 1982 1983 1984 1985 1986 1987 1988 1989 1990 1991 1992 1993 1994 1995 1996 1997 1998 1999 2000 2001 2002 2003 2004 **2005** 2006 2007 20

Old Bridge Area of the Old City of Mostar

Bosnia-Herzegovina

Criteria – Heritage associated with events of universal significance

Mostar has long been known for its old Turkish houses and Old Bridge, Stari Most. In the conflict of the 1990s, however, most of the historic town and the Old Bridge, designed by the renowned architec Sinan (1489–1588), was destroyed. The Old Bridge was recently rebuilt and many of the buildings in the Old Town have been restored or rebuil with the contribution of an international scientific committee established by UNESCO.

The Old Bridge area of Mostar, with its medieval, Ottoman, Mediterranean and Western European architectural features, is an outstanding example of a multicultural urban settlement. The Old Bridge and Old City of Mostar, reconstructed after conflict in 1990, is a symbol of reconciliation, international co-operation and of the coexistence of diverse cultural, ethnic and religious communities.

The historic town of Mostar spans a deep valley of the Neretva River. The area has been settled since prehistoric times and there is evidence of Roman occupation. Little is known of its medieval period, although Christian churches were established in the fifth–sixth centuries. The name of Mostar is first mentioned in 1474; its name came from the bridge-keepers (mostari) of the wooden bridge that crossed from the market town on the left bank of the river. Mostar's key position on the trade route between the Adriatic and mineral-rich central Bosnia led to the settlement's growth across the river. It became the leading town in Herzegovina and, after invasion by the Ottomans in 1468, the centre of Turkish rule in the area.

Mostar was an Ottoman frontier town and fortifications were built up in the sixteenth century; the bridge was also rebuilt in stone.

Religious and public buildings were constructed in a religious complex on the left bank, while private and commercial buildings, organized in distinct quarters, were also built. Several Ottoman inns survive, along with other buildings from this period, such as fountains and schools. Surviving late-Ottoman houses demonstrate the component features of this form of architecture – a hall, a residential upper storey, a paved courtyard and a verandah on one or two storeys.

Some early trading and craft buildings are still extant, notably low shops in wood or stone, stone storehouses and a group of former tanneries around an open courtyard. A number of elements of the early fortifications are visible. The Hercegusa Tower dates from the medieval period, whereas the Ottoman defences are represented by the Halebinovka and Tara Towers, the watchtowers over the ends of the Old Bridge and a stretch of the ramparts.

The city became part of Austro-Hungary in 1878 and all the administrative buildings of that period have neoclassical and Secessionist features. The nineteenth-century houses and commercial buildings are also predominantly neoclassical.

World Heritage site since

1978 1979 1980 1981 1982 1983 1984 1985 1986 1987 1988 1989 1990 1991 1992 1993 1994 1995 1996 1997 1998 1999 2000 2001 2002 2003 2004 **2005** 2006 2007 2008

West Norwegian Fjords – Geirangerfjord and Nærøyfjord

Norway

Criteria – Natural phenomena or beauty;
Major stages of Earth's history

The two fjords, Geirangerfjord and Nærøyfjord, are considered archetypical fjord landscapes and among the most scenically outstanding anywhere in the world. Their exceptional natural beauty is derived from their narrow and steep-sided crystalline rock walls that rise up to 1,400 m from the Norwegian Sea and extend 500 m below sea level. The sheer walls of the fjords have numerous waterfalls, while free-flowing rivers cross their deciduous and coniferous forests from glacial lakes, glaciers and rugged mountains. The landscape features a range of supporting natural phenomena, both terrestrial and marine, such as submarine moraines and marine mammals. Remnants of old and now mostly abandoned transhumant farms add a cultural aspect to the dramatic natural landscape that complements and adds human interest to the area.

Situated in southwestern Norway, 120 km from one another, Geirangerfjord and Nærøyfjord are part of the west Norwegian fjord landscape that stretches from Stavanger in the south to Andalsnes, 500 km to the northeast. The two fjords are among the world's longest and deepest, and are considered distinctive in a country of spectacular fjords. 'Fjord' is a word of Norwegian origin, meaning a glacially over-deepened valley, usually narrow and steep-sided and extending below sea level. Norway's fjords are among the most extensive on Earth and are considered the type locality for study of fjord landscapes.

Both the fjords developed along faults and fracture zones at right angles, giving them a characteristic zigzag form. They are submarine hanging valleys with floors between 300–500 m deep in ice-scoured basins, and between 1 and 2 km wide. They are surrounded by mountains with old transhumance farms in the hanging valleys, and high glacial lakes.

Climate is transitional between oceanic and continental and varies markedly with aspect and altitude. The vegetation is moderately diverse due to the range of gradients from coast to inland, from north to south, from sea level to 1800 m and to the consequent variety of terrain and microclimates.

Wildlife includes four species of deer, arctic fox, otter, and many marine species such as Atlantic salmon, seals, porpoise, dolphins and whales. Over 100 bird species have been recorded.

The two areas that comprise the property complement each other in their characteristics. The more southerly Nærøyfjord is located 100 km inland near the end of Sognefjord. Its surrounding mountains are smooth-topped with high glacial lakes and a plateau glacier. The uplands preserve much of the rounded landforms of the pre-glacial landscape.

Geirangerfjord (pictured on the right) lies to the north, 60 km inland on the upper end of Storfjord. Its mountains are more Alpine in character. Block fields are more prevalent and there is still permafrost and several small glaciers on the highest summits.

World Heritage site since

1978 1979 1980 1981 1982 1983 1984 1985 1986 1987 1988 1989 1990 1991 1992 1993 1994 1995 1996 1997 1998 1999 2000 2001 2002 2003 2004 **2005** 2006 2007 2008

Soltaniyeh
Islamic Republic of Iran

Criteria – Interchange of values; Testimony to
cultural tradition; Significance in human history

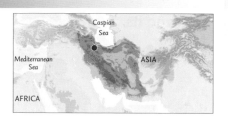

The mausoleum of Oljaytu was constructed
in 1302–12 in the city of Soltaniyeh, the capital
of the Ilkhanid dynasty, which was founded
by the Mongols. Situated in the province of
Zanjān, Soltaniyeh is one of the outstanding
examples of the achievements of Persian
architecture and a key monument in the
development of its Islamic architecture. The
octagonal building is crowned with a 50-m-
tall dome covered in turquoise-blue faïence
and surrounded by eight slender minarets. It
is the earliest existing example of the double-
shelled dome in Iran. The mausoleum's

interior decoration is also outstanding and
scholars such as A.U. Pope have described the
building as 'anticipating the Taj Mahal'.

When Oljaytu came
to power in 1304 he
made the existing
town his capital,
enlarging it and
renaming it
Soltaniyeh (Imperial).
The city was a major
trading centre
between Asia and
Europe but gradually
declined in the
sixteenth–seventeenth
centuries.

◀
Mausoleum of
Oljaytu in Saltaniyeh

Qal'at al-Bahrain – Ancient Harbour and Capital of Dilmun
Bahrain

Criteria – Interchange of values; Testimony to
cultural tradition; Significance in human history

Qal'at al-Bahrain is a typical tell – an artificial
mound created by many successive layers
of human occupation. The strata of the
300–600 m tell testify to continuous human
presence from about 2300 BC to the
sixteenth century AD. About 25 per cent of
the site has been excavated, revealing
structures of different types: residential,
public, commercial, religious and military.
They testify to the importance of the site, a

trading port, over the centuries. On the top
of the 12-m mound there is the impressive
Portuguese fort, which gave the whole site
its name, qal'a (fort). The site was the capital
of the Dilmun, one of the most important
ancient civilizations of the region. It
contains the richest remains inventoried
of this civilization, which was hitherto only
known from written Sumerian references.

The once thriving and
important city of
Qal'at al-Bahrain was
finally abandoned
when its access
channel through the
coral reef silted up,
bringing about
the gradual
transformation from
a 4,500-year-old
settlement to an
archaeological site.

World Heritage site since

1978 1979 1980 1981 1982 1983 1984 1985 1986 1987 1988 1989 1990 1991 1992 1993 1994 1995 1996 1997 1998 1999 2000 2001 2002 2003 2004 **2005** 2006 2007 2008

Struve Geodetic Arc
Belarus, Estonia, Finland, Latvia, Lithuania, Norway, Moldova, Russian Fed., Sweden and Ukraine

Criteria – Interchange of values; Testimony to cultural tradition; Heritage associated with events of universal significance

The first accurate measuring of a long segment of a meridian, helping to establish the exact size and shape of the world, exhibits an important step in the development of earth sciences. The Struve Arc is a chain of survey triangulations stretching from Hammerfest in Norway to the Black Sea over 2,820 km away. These are points of a survey, carried out between 1816 and 1855 by the astronomer Friedrich Georg Wilhelm Struve, which represented the first accurate measuring of a long segment of a meridian. It is an extraordinary example of scientific collaboration among scientists from different countries, and of collaboration between monarchs for a scientific cause.

Thirty-four of the original station points established by Struve and his colleagues, in ten countries, exist today. These are commemorated by different marks: a drilled hole in rock, an iron cross, cairns, or built obelisks.

Vredefort Dome
South Africa

Criteria – Major stages of Earth's history

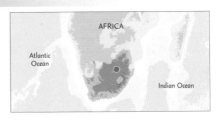

The site contains high quality and accessible geological (outcrop) sites that demonstrate a range of geological evidences of a complex meteorite impact structure.

Vredefort Dome, approximately 120 km southwest of Johannesburg, is a representative part of a larger meteorite impact structure, or astrobleme. Dating back 2,023 million years, it is the oldest astrobleme yet found on Earth. With a radius of 190 km, it is also the largest and the most deeply eroded. Vredefort Dome bears witness to the world's greatest known single energy release event, which had devastating global effects including, according to some scientists, major evolutionary changes. It provides critical evidence of the Earth's geological history and is crucial to the understanding of the evolution of the planet. Despite the importance of impact sites to the planet's history, geological activity on the Earth's surface has led to the disappearance of evidence from most of them, and Vredefort is the only example to provide a full geological profile of an astrobleme below the crater floor.

World Heritage site since

1978 1979 1980 1981 1982 1983 1984 1985 1986 1987 1988 1989 1990 1991 1992 1993 1994 1995 1996 1997 1998 1999 2000 2001 2002 2003 2004 **2005** 2006 2007 200

Shiretoko
Japan

Criteria – Significant ecological and biological processes; Significant natural habitat for biodiversity

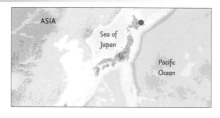

Shiretoko Peninsula has a central spine of volcanoes that produce geothermal features such as geysers and thermally heated pools. It is one of Japan's most unspoiled areas.

Shiretoko Peninsula is located in the northeast of Hokkaido, the northernmost island of Japan. The site includes the land from the central part of the peninsula to its tip, Shiretoko Cape, and the surrounding marine area. It provides an outstanding example of the interaction of marine and terrestrial ecosystems as well as extraordinary ecosystem productivity, largely influenced by the formation of seasonal sea ice at the lowest latitude in the northern hemisphere. It has particular importance for a number of marine and terrestrial species, some of them endangered and endemic, such as Blackiston's fish owl and the Viola kitamiana plant. The site is globally important for threatened seabirds and migratory birds, a number of salmonid species, and for marine mammals including Steller's sea lion and some cetacean species.

Architectural, Residential and Cultural Complex of the Radziwill Family at Nesvizh
Belarus

Criteria – Interchange of values; Significance in human history, Heritage associated with events of universal significance

The ten interconnected buildings of the castle include the palace, the galleries, the residence and the arsenal, all of which are set within the remains of the sixteenth-century fortifications. The castle is connected via a dam to the Church of Corpus Christi.

The Architectural, Residential and Cultural Complex of the Radziwill Family at Nesvizh is located in central Belarus. The Radziwill dynasty, who built and kept the ensemble from the sixteenth century until 1939, gave birth to some of the most important personalities in European history and culture. Due to their efforts, the town of Nesvizh came to exercise great influence in the sciences, arts, crafts and architecture. The complex consists of the residential castle and the mausoleum Church of Corpus Christi with their setting. The castle has ten interconnected buildings, which developed as an architectural whole around a six-sided courtyard. The palaces and church became important prototypes marking the development of architecture throughout central Europe and Russia.

World Heritage site since

1978 1979 1980 1981 1982 1983 1984 1985 1986 1987 1988 1989 1990 1991 1992 1993 1994 1995 1996 1997 1998 1999 2000 2001 2002 2003 2004 **2005** 2006 2007 2008

Urban Historic Centre of Cienfuegos
Cuba

Criteria – Interchange of values; Significance in human history

Cienfuegos was founded in 1819 on the Caribbean coast of southern-central Cuba, at the heart of the country's sugar cane, mango, tobacco and coffee production. Trading powered the city's growth, with wax production, as well as timber and sugar, becoming increasingly important in the nineteenth century. This historic town exhibits an important interchange of cultural and social influences based on the Spanish Enlightenment. It is also the first and finest example of an architectural ensemble representing the new ideas of modernity, hygiene and order in urban planning as developed in Latin America from the nineteenth century.

In its two centuries of existence, Cienfuegos has always been a particularly important trading city. Despite its success, the heritage area has retained its historic fabric without the drastic changes common in many other cities.

The cathedral of Cienfuegos. ▶

World Heritage site since

1978 1979 1980 1981 1982 1983 1984 1985 1986 1987 1988 1989 1990 1991 1992 1993 1994 1995 1996 1997 1998 1999 2000 2001 2002 2003 2004 **2005** 2006 2007 200

Kunya-Urgench
Turkmenistan
Criteria – Interchange of values; Testimony to cultural tradition

The monuments of Kunya-Urgench testify to outstanding achievements in architecture and craftsmanship. Their influence reached as far south and west as Iran and Afghanistan, and later extended to the architecture of the Mogul Empire of sixteenth-century India.

The origins of Kunya-Urgench go back to the sixth or fifth centuries, the early Achaemenid period. Situated at the crossing of trade routes, it was the capital of Khorezm from the twelfth century and the second city after Bukhara in central Asia. The site has three distinct sections: the southern section contains a series of monuments dating mainly from the eleventh–sixteenth centuries, including a mosque, the gates of a caravanserai, fortresses, mausoleums and a 60-m-high minaret; the northern section consists of a large Muslim graveyard with a group of three mausoleums at its centre; the western section is a small area in the western part of the old town containing the monument of Ibn Khajib.

Syracuse and the Rocky Necropolis of Pantalica
Italy
Criteria – Interchange of values; Testimony to cultural tradition; Significance in human history; Heritage associated with events of universal significance

Situated on the Mediterranean coast in southeastern Sicily and enjoying a favourable climate, the sites have been inhabited since protohistoric times. The Syracuse-Pantalica area is remarkable for its cultural diversity.

The site consists of two separate elements, containing outstanding vestiges dating back to Greek and Roman times: The Necropolis of Pantalica contains over 5,000 tombs cut into the rock near open stone quarries, most of them dating from the thirteenth to seventh centuries BC. Vestiges of the Byzantine era also remain in the area, notably the foundations of the Anaktoron Prince's Palace. The other part of the property, Ancient Syracuse, includes the nucleus of the city's foundation as Ortygia by Greeks from Corinth in the eighth century BC. The site of the city, which Cicero described as 'the greatest Greek city and the most beautiful of all', retains vestiges such as the Temple of Athena, fifth century BC, later transformed to serve as a cathedral, a Greek theatre, a Roman amphitheatre, a fort and more. Many remains bear witness to the troubled history of Sicily, from the Byzantines to the Bourbons, interspersed with the Arabo-Muslims, the Normans, Frederick II of the Hohenstaufen dynasty 1197–1250, the Aragons and the Kingdom of the Two Sicilies. Historic Syracuse offers a unique testimony to the development of Mediterranean civilization over three millennia.

World Heritage site since

1978 1979 1980 1981 1982 1983 1984 1985 1986 1987 1988 1989 1990 1991 1992 1993 1994 1995 1996 1997 1998 1999 2000 2001 2002 2003 2004 2005 **2006** 2007 2008

Sichuan Giant Panda Sanctuaries - Wolong, Mt Siguniang and Jiajin Mountains
China

Criteria – Significant natural habitat for biodiversity

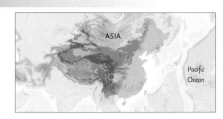

Sichuan Giant Panda Sanctuaries, home to more than 30 per cent of the world's pandas which are classed as highly endangered, covers 9,245 km² with seven nature reserves and nine scenic parks in the Qionglai and Jiajin Mountains. The sanctuaries constitute the largest remaining contiguous habitat of the giant panda, a relic from the palaeo-tropic forests of the Tertiary Era. It is also the species' most important site for captive breeding. The sanctuaries are home to other globally endangered animals such as the red panda, the snow leopard and clouded leopard. They are among the botanically richest sites of any region in the world outside the tropical rainforests, with between 5,000 and 6,000 species of flora in over 1,000 genera.

The giant panda is recognized as a 'National Treasure' of China and is the flagship for global conservation efforts. In the wild it feeds almost exclusively on bamboo, and its preferred habitat is between altitudes of 2,200 m and 3,200 m. As a unique single species and family, the giant panda is very important for studying mammal classification and evolution.

A young panda cub in the Sichuan Giant Panda Sanctuaries. ▼

World Heritage site since

1978 1979 1980 1981 1982 1983 1984 1985 1986 1987 1988 1989 1990 1991 1992 1993 1994 1995 1996 1997 1998 1999 2000 2001 2002 2003 2004 2005 **2006** 2007 200

Centennial Hall in Wroclaw
Poland

Criteria – Human creative genius; Interchange of values; Significance in human history

The Centennial Hall, a landmark in the history of reinforced-concrete architecture, was erected in 1911–3 by the architect Max Berg as a multi-purpose recreational building, situated in the Exhibition Grounds. In form it is a symmetrical quatrefoil with a vast circular central space that can seat some 6,000 persons. The 23-m-high dome is topped with a lantern in steel and glass. The Centennial Hall is a pioneering work of modern engineering and architecture,

which exhibits an important interchange of influences in the early twentieth century, becoming a key reference in the later development of reinforced-concrete structures.

With its diameter of 65 m, the Centennial Hall dome was at the time the largest ever built. Before then, the largest dome was that of the Pantheon in Rome, completed b AD 126. The Centennia Hall dome was twice as big: this stunning achievement was made possible by the new material (ferroconcrete) and Berg's innovative approach to structura design.

Harar Jugol, the Fortified
Historic Town
Ethiopia

Criteria – Interchange of values; Testimony to cultural tradition; Significance in human history; Traditional human settlement

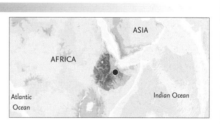

The fortified historic town of Harar is located in the eastern part of the Ethiopia on a plateau with deep gorges surrounded by desert and savanna. The walls surrounding this sacred Muslim city were built between the thirteenth and sixteenth centuries. Harar Jugol, said to be the fourth holiest city of Islam, has eighty-two mosques, three of which date from the tenth century, and 102 shrines, but the townhouses with their exceptional interior design constitute the most spectacular part

of Harar's cultural heritage. The impact of African and Islamic traditions on the development of the town's building types and urban layout make for its particular character and uniqueness.

Harar Jugol is the centre of an Islamic region within otherwise Christian Ethiopia. The World Heritage site comprises the entire historic walled city of Jugol, the name 'Jugol referring to the defensive wall as well as to the fortified town.

◄
One of Harar Jugol's city gates.

World Heritage site since

1978 1979 1980 1981 1982 1983 1984 1985 1986 1987 1988 1989 1990 1991 1992 1993 1994 1995 1996 1997 1998 1999 2000 2001 2002 2003 2004 2005 **2006** 2007 2008

Old town of Regensburg with Stadtamhof

Germany

Criteria – Interchange of values; Testimony to cultural tradition; Significance in human history

A feature of Regensburg are the towers built by patrician families, for which there are no other comparable examples north of the Alps. Similar in form to north Italian towers, they were built more for the purpose of display than for protection. The Goldene Turm (twelfth-century) is nearly 50 m high while the seven-storey Baumburgerturm was built in 1270.

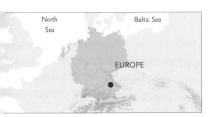

Located on the Danube River in Bavaria, this medieval town contains many buildings of exceptional quality that testify to its history as a trading centre and to its influence on the region from the ninth century. A notable number of historic structures span some two millennia and include ancient Roman, Romanesque and Gothic buildings. Regensburg's eleventh- to thirteenth-century architecture – including the market, city hall and cathedral – still defines the character of the town marked by tall buildings, dark and narrow lanes, and strong fortifications. The buildings include medieval patrician houses and towers, a large number of churches and monastic ensembles as well as the twelfth-century Old Bridge. The town is also remarkable for the vestiges testifying to its rich history as one of the centres of the Holy Roman Empire that turned to Protestantism.

▶ Regensburg Cathedral.

World Heritage site since

1978 1979 1980 1981 1982 1983 1984 1985 1986 1987 1988 1989 1990 1991 1992 1993 1994 1995 1996 1997 1998 1999 2000 2001 2002 2003 2004 2005 **2006** 2007 200

Chongoni Rock-Art Area
Malawi

Criteria – Testimony to cultural tradition;
Heritage associated with events of universal
significance

The dense and
extensive collection
of rock-art shelters
reflects a remarkable
persistence of cultura
traditions over many
centuries, connected
to the role of rock art
in women's initiations,
in rain making and in
funeral rites. These
traditions make the
Chongoni landscape
a powerful force in
Chewa society and
a significant place for
the whole of southern
Africa.

Situated within a cluster of forested granite
hills and covering an area of 126.4 km²,
high up the plateau of central Malawi, the
127 sites of this area feature the richest
concentration of rock art in central Africa.
They reflect the comparatively scarce
tradition of farmer rock art, as well as
paintings by BaTwa hunter-gatherers who
inhabited the area from the late Stone Age.
The Chewa agriculturalists, whose ancestors
lived there from the late Iron Age, practised
rock painting until well into the twentieth
century. The symbols in the rock art, which
are strongly associated with women, still
have cultural relevance amongst the Chewa,
and the sites are actively associated with
ceremonies and rituals.

Malpelo Fauna and Flora
Sanctuary
Colombia

Criteria – Natural phenomena or beauty;
Significant ecological and biological processes

Malpelo Island is
thought to be a
geological hotspot –
a product of the
welling up of the
Earth's mantle. Its
rocky surface supports
a sparse vegetation
of ferns, lichen,
mosses and algae
which are sustained
by guano.

Located some 506 km off the coast of
Colombia, the site includes Malpelo Island
(3.5 km²) and the surrounding marine
environment (8,572 km²). This vast marine
park, the largest no-fishing zone in the
eastern tropical Pacific, provides a critical
habitat for internationally threatened
marine species, and is a major source of
nutrients resulting in large aggregations
of marine biodiversity. It is in particular
a 'reservoir' for sharks, giant grouper and
billfish. Widely recognized as one of the top
diving sites in the world, due to the steep
walls and caves of outstanding natural
beauty, these deep waters support
important populations of large predators
and pelagic species in an undisturbed
environment where they maintain natural
behavioural patterns.

World Heritage site since

1978 1979 1980 1981 1982 1983 1984 1985 1986 1987 1988 1989 1990 1991 1992 1993 1994 1995 1996 1997 1998 1999 2000 2001 2002 2003 2004 2005 **2006** 2007 2008

Cornwall and West Devon Mining Landscape
United Kingdom

Criteria – Interchange of values; Testimony to cultural tradition; Significance in human history

The substantial remains of the copper- and tin-mining industries are a testimony to the contribution of Cornwall and West Devon to the Industrial Revolution in Britain and to the area's fundamental influence on the mining world at large. The area was the heartland from which mining technology rapidly spread and its expertise and technology, in the form of engines, engine houses and mining equipment, was exported around the world.

Much of the local landscape was transformed in the eighteenth and early nineteenth centuries as a result of the rapid growth of mining. The deep underground mines, engine houses, foundries, new towns, great houses and estates, smallholdings, ports and harbours, together with ancillary industries such as smelting, canals and railways, reflect the prolific innovation which led to the region producing two-thirds of the world's copper supply in the early nineteenth century.

The success of the copper, tin and arsenic mines of Cornwall and West Devon was based on deep-shaft mining made possible by technological innovations; these included the safety fuse for blasting and developments in steam-driven pumping.

When mining in the area declined in the 1860s, many miners emigrated, taking Cornish traditions to South Africa, Australia and Central and South America.

Tonwanroath pumping engine house at Wheal Cotes mine, St Agnes in Cornwall. ▼

World Heritage site since

1978 1979 1980 1981 1982 1983 1984 1985 1986 1987 1988 1989 1990 1991 1992 1993 1994 1995 1996 1997 1998 1999 2000 2001 2002 2003 2004 2005 **2006** 2007 200

Aflaj Irrigation Systems of Oman
Oman
Criteria – Traditional human settlement

The word 'aflaj' is the plural of 'falaj', which means 'to divide into shares'. Equitable sharing of a scarce resource to ensure sustainability remains the hallmark of this ancient irrigation system.

The property includes five aflaj irrigation systems and is representative of some 3,000 such systems still in use in Oman. The origins of this system of irrigation may date back to AD 500, but archaeological evidence suggests that irrigation systems existed in this extremely arid area as early as 2500 BC. Using gravity, water is channelled from underground sources or springs to support agriculture and domestic use. The fair and effective management and sharing of water in villages and towns is still underpinned by mutual dependence and communal values and guided by astronomical observations. Numerous watchtowers built to defend the water systems form part of the site, reflecting the historic dependence of communities on the aflaj system. Threatened by the falling level of the underground water table, the aflaj represent an exceptionally well-preserved form of land use.

Aapravasi Ghat
Mauritius
Criteria – Heritage associated with events of universal significance

The immigration depot that received the Indian immigrants is a complex of buildings in Port Louis. Of the original facility founded in the mid-nineteenth century, only around 15 per cent remains today.

In the district of Port Louis, lies the 1,640 m² site where the modern indentured labour diaspora began. In 1834, the British Government selected the island of Mauritius to be the first site for what it called 'the great experiment' in the use of 'free' labour to replace slaves. Between 1834 and 1920, almost half a million indentured labourers arrived from India at Aapravasi Ghat to work in the sugar plantations of Mauritius, or to be transferred to Réunion Island, Australia, southern and eastern Africa or the Caribbean. The buildings of Aapravasi Ghat are among the earliest explicit manifestations of what was to become a global economic system and one of the greatest migrations in history.

World Heritage site since

978 1979 1980 1981 1982 1983 1984 1985 1986 1987 1988 1989 1990 1991 1992 1993 1994 1995 1996 1997 1998 1999 2000 2001 2002 2003 2004 2005 **2006** 2007 2008

Crac des Chevaliers and Qal'at Salah El-Din
Syrian Arab Republic

Criteria – Interchange of values; Significance in human history

Crac des Chevaliers and Qal'at Salah El-Din are among the foremost examples of medieval fortified castles in the world. Both were vital Crusader strongholds. The Crac des Chevaliers (Fortress of the Knights) or Qala'at al-Hosn was built on the site of an existing fortification by the Order of St John of Jerusalem, the Knights Hospitaller, who held it from 1142, turning it into the largest Crusader fortress in the Holy Land. It finally fell to a Mameluke siege in 1271. Still towering over the surrounding landscape, between Homs and Tartous, the magnificent Crac is largely restored to its original state.

The Qal'at Salah El-Din (Fortress of Saladin) is partly in ruins but retains features from its tenth-century Byzantine construction, its reinforcement by the Crusaders in the twelfth century, and modifications by the Ayyubids who captured it under Saladin, Sultan of Egypt and Syria, in 1188.

Crac des Chevaliers guarded the route from Syria to the Holy Land and was one of a chain of Crusader castles commanding the eastern Mediterranean. British soldier and scholar T.E. Lawrence called it 'perhaps the best preserved and most wholly admirable castle in the world'.

Qal'at Salah El-Din was renamed in 1957 in honour of its conqueror, Muslim leader Saladin.

Crac des Chevaliers. ▼

World Heritage site since

1978 1979 1980 1981 1982 1983 1984 1985 1986 1987 1988 1989 1990 1991 1992 1993 1994 1995 1996 1997 1998 1999 2000 2001 2002 2003 2004 2005 2006 **2007** 200

Iwami Ginzan Silver Mine and its Cultural Landscape
Japan

Criteria – Interchange of values; Testimony to cultural tradition; Traditional human settlement

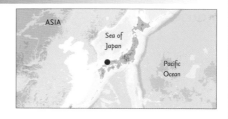

Japan was a major silver producer. By the first half of the seventeenth century, when production at Iwami Ginzan was at its peak, it is estimated that its output accounted for almost a third of all world silver production.

The Iwami Ginzan Silver Mine, in the southwest of Honshu Island, is a cluster of mountains, rising to 600 m and interspersed with deep river valleys featuring the archaeological remains of large-scale mines, smelting and refining sites and mining settlements worked between the sixteenth and twentieth centuries. The mines contributed substantially to the overall economic development of Japan and southeast Asia in the sixteenth and seventeenth centuries, prompting the mass production of silver and gold in Japan. The mining area is now heavily wooded. Included in the site are fortresses, shrines, parts of Kaidô transport routes to the coast, and three port towns, Tomogaura, Okidomari and Yunotsu, from where the ore was shipped to Korea and China.

Richtersveld Cultural and Botanical Landscape
South Africa

Criteria – Significance in human history; Traditional human settlement

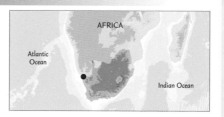

The extensive communal grazed lands are a testimony to land management processes that have ensured the protection of the succulent Karoo vegetation. The seasonal pastoral grazing regimes of the Nama, which sustain the extensive biodiversity of the area, were once much more widespread and are now vulnerable.

The 1,600 km² Richtersveld Cultural and Botanical Landscape of dramatic mountainous desert in northwestern South Africa, constitutes a cultural landscape communally owned and managed. This site sustains the semi-nomadic pastoral livelihood of the Nama people, reflecting seasonal patterns that may have persisted for as much as two millennia in southern Africa. It is the only area where the Nama still construct portable rush-mat houses (haru om), and includes seasonal migrations and grazing grounds, together with stock posts. The pastoralists collect medicinal and other plants and have a strong oral tradition associated with different places and attributes of the landscape.

World Heritage site since

1978 1979 1980 1981 1982 1983 1984 1985 1986 1987 1988 1989 1990 1991 1992 1993 1994 1995 1996 1997 1998 1999 2000 2001 2002 2003 2004 2005 2006 **2007** 2008

Red Fort Complex
India

Criteria – Interchange of values; Testimony to cultural tradition; Heritage associated with events of universal significance

The Red Fort, completed in 1648, represents the zenith of Mughal creativity. The palace plan is based on standard Islamic designs but each pavilion displays architectural elements typical of Mughal building – a fusion of Persian, Timurid, Hindu and Islamic traditions.

Emperor Shah Jahan established his capital at Shahjahanabad and built the Red Fort Complex as his palace fort, enclosing it in ornate red sandstone walls that stretch for 2.5 km. The innovative planning, gardens and architectural style of the fort complex strongly influenced later building and garden design in Rajasthan, Delhi and Agra.

The Red Fort has been a powerful symbol for the Indian nation since its construction. The British Army captured it after the Indian Mutiny of 1857–8 and held it until India gained independence in 1947. It has remained at the centre of national independence celebrations ever since.

The Mughal Empire in India lasted from 1526 until the mid-eighteenth century. It reached the height of its power around 1700, when it encompassed almost the whole subcontinent. The empire's most notable existing legacy is its architecture: Shah Jahan, who built the Red Fort, was also the builder of the Taj Mahal.

World Heritage site since

1978 1979 1980 1981 1982 1983 1984 1985 1986 1987 1988 1989 1990 1991 1992 1993 1994 1995 1996 1997 1998 1999 2000 2001 2002 2003 2004 2005 2006 **2007** 200

Teide National Park
Spain

Criteria – Natural phenomena or beauty;
Major stages of Earth's history

Situated on the island of Tenerife, Teide National Park features the Teide-Pico Viejo stratovolcano – at 3,718 m, the highest peak on Spanish soil. Rising 7,500 m above the ocean floor, it is regarded as the world's third-tallest volcanic structure and stands in a spectacular environment. The visual impact of the site is all the greater due to atmospheric conditions that create constantly changing textures and tones in the landscape and a 'sea of clouds' that forms a visually impressive backdrop to the mountain. Teide is of global importance in providing evidence of the geological processes that underpin the evolution of oceanic islands.

Mount Teide is an exceptional example of a relatively old, slow moving, geologically complex and mature volcanic system. With diverse and accessible features in a relatively limited area, it has long been a major centre for international research with a significant history of influence on the study of geology and geomorphology.

World Heritage site since

1978 1979 1980 1981 1982 1983 1984 1985 1986 1987 1988 1989 1990 1991 1992 1993 1994 1995 1996 1997 1998 1999 2000 2001 2002 2003 2004 2005 2006 **2007** 2008

Samarra Archaeological City
Iraq

Criteria – Interchange of values; Testimony to cultural tradition; Significance in human history

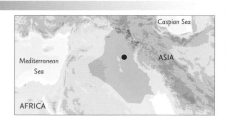

Samarra Archaeological City is the site of a powerful Islamic capital city that ruled for a century over the provinces of the Abbasid Empire, extending from Tunisia to central Asia. Located on both sides of the river Tigris, 130 km north of Baghdad, the length of the site from north to south is 41.5 km; its width varying from 8 km to 4 km. It testifies to the architectural and artistic innovations that developed there and spread to other regions of the Islamic world and beyond. The ninth-century Great Mosque and its spiral minaret are among the numerous remarkable architectural monuments of the site, 80 per cent of which remain to be excavated.

Samarra contains two of the largest mosques and the largest palaces in the Islamic world. Carved stucco, known as the Samarra style, was developed there and spread to other parts of the region. A new type of ceramic, known as lustre ware, was also developed in Samarra, imitating utensils made of precious metals such as gold and silver.

Gobustan Rock Art Cultural Landscape
Azerbaijan

Criteria – Testimony to cultural tradition

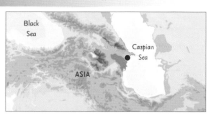

Gobustan Rock Art Cultural Landscape covers three areas of a plateau of rocky boulders rising out of the semi-desert of east-central Azerbaijan, with an outstanding collection of more than 6,000 rock engravings bearing testimony to 40,000 years of rock art. The site also features the remains of inhabited caves, settlements and burials, all reflecting an intensive human use by the inhabitants of the area during the wet period that followed the last Ice Age, from the Upper Paleolithic to the Middle Ages. The site, which covers an area of 5.4 km², is part of the larger Gobustan Reservation.

The rock art at Gobustan is detailed and wide-ranging: it includes plants and animals, and human figures dancing, hunting, fighting and travelling in boats. Nearby are inscriptions left by troops of Alexander the Great and Roman soldiers of the Emperor Trajan.

◄
Animal rock-art engraving, Gobustan.

World Heritage site since

1978 1979 1980 1981 1982 1983 1984 1985 1986 1987 1988 1989 1990 1991 1992 1993 1994 1995 1996 1997 1998 1999 2000 2001 2002 2003 2004 2005 2006 **2007** 200

South China Karst
China
Criteria – Natural phenomena or beauty;
Major stages of Earth's history

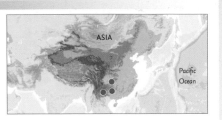

Shilin stone forest in Yunnan province.

Four types of karst landscape in the area are considered outstanding. These are: 'fengcong' karst (cone karst), characterised by linked conical hills and depressions, valleys and gorges; 'fenglin' karst (tower karst), comprising isolated cones or towers on broad plains; stone forests, with a wide diversity of closely spaced pinnacles and towers, and 'tiankeng' karst (giant dolines) – massive circular collapse structures often in close proximity to spectacular gorges and decorated caves, where cave or doline collapse can create natural rock bridges.

The South China Karst region extends over a surface of 500,000 km² lying mainly in Yunnan, Guizhou and Guangxi provinces. The region displays a series of karst landforms in a variety of humid, subhumid, tropical and subtropical climate conditions, and geographical settings. It comprises three clusters: Shilin Karst, Libo Karst and Wulong Karst.

Shilin, in Yunnan province, contains classic examples of stone-forest landscapes, noted for high limestone pinnacles and towers decorated with deep, sharp karren (see photo on the right). They formed over some 270 million years during four major geological time periods from the Permian to present, illustrating the episodic nature of the evolution of these karst features. The stone forests of Shilin are considered superlative natural phenomena and a world reference with a wider range of pinnacle shapes than other karst landscapes, and a higher diversity of shapes and changing colours.

Libo contains carbonate outcrops of different ages that erosive processes shaped over millions of years into impressive cone and tower karsts. It contains a combination of numerous tall karst peaks, deep dolines, sinking streams and long river caves. Libo is also noted for its biodiversity, with over 314 vertebrate species and 1,532 plant species, including several endemics and a number of plants and animals that are globally or nationally endangered.

Wulong represents high inland karst plateaus that have experienced considerable uplift, and includes giant collapse depressions and high natural bridges between stretches of deep, unroofed caves. Its giant dolines and bridges are representative of south China's Tiankeng landscapes. Wulong's landscapes contain evidence for the history of one of the world's great river systems, the Yangtze and its tributaries.

Minority peoples live in two of the karst areas and make up most of the population. There is a strong relationship between karst and the cultural identity and traditions of the people. In Shilin, the Yi people have developed a lifestyle adapted to the karst environment and the stone forests are reflected in every aspect of their culture. The Shui people in Libo have managed their lands for at least a thousand years and provide an exemplary example of sustainable forest management.

World Heritage site since

1978 1979 1980 1981 1982 1983 1984 1985 1986 1987 1988 1989 1990 1991 1992 1993 1994 1995 1996 1997 1998 1999 2000 2001 2002 2003 2004 2005 2006 **2007** 200

Central University City Campus of the Universidad Nacional Autónoma de México (UNAM)
Mexico

Criteria – Human creative genius; Interchange of values; Significance in human history

The ensemble of buildings, sports facilities and open spaces of the Central University City Campus of the Universidad Nacional Autónoma de México (UNAM), was built from 1949 to 1952 by more than sixty architects, engineers and artists who were involved in the project. As a result, the campus constitutes a unique example of twentieth-century modernism integrating urbanism, architecture, engineering, landscape design and fine arts with references to local traditions, especially to Mexico's pre-Hispanic past. The ensemble embodies social and cultural values of universal significance and is one of the most significant icons of modernity in Latin America.

The Biblioteca Centra (Central Library) of UNAM is the university's most famous and iconic building. Tiled murals, the work of Mexican artist Juan O'Gorman, cover the four walls and represent historic and modern Mexico and the university.

Twyfelfontein or /Ui-//aes
Namibia

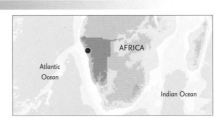

Criteria – Testimony to cultural tradition; Traditional human settlement

Twyfelfontein or /Ui-//aes has one of the largest concentrations of rock petroglyphs (engravings) in Africa. Most of these well-preserved engravings represent rhinoceros, elephant, ostrich and giraffe, as well as drawings of human and animal footprints. The site also includes six painted rock-shelters with motifs of human figures in red ochre. The objects excavated from two sections, date from the Late Stone Age. The site forms a coherent, extensive and high-quality record of ritual practices relating to hunter-gatherer communities in this part of southern Africa over at least 2,000 years.

Over 2,000 figures have been documented, most of which are recognisable depictions of animals together with engravings of their tracks. It is thought the imagery was linked to the beliefs of hunter-gatherers who dominated this area until around AD 1000.

World Heritage site since

1978 1979 1980 1981 1982 1983 1984 1985 1986 1987 1988 1989 1990 1991 1992 1993 1994 1995 1996 1997 1998 1999 2000 2001 2002 2003 2004 2005 2006 **2007** 2008

Old Town of Corfu
Greece
Criteria – Significance in human history

The Old Town of Corfu, on the island of Corfu off the western coasts of Albania and Greece, is located in a strategic position at the entrance to the Adriatic Sea, and has its roots in the eighth century BC. The three forts of the town, designed by renowned Venetian engineers, were used for four centuries to defend the maritime trading interests of the Republic of Venice against the Ottoman Empire. In the course of time, the forts were repaired and partly rebuilt several times, more recently under British rule in the nineteenth century. The mainly neoclassical housing stock of the Old Town is partly from the Venetian period, partly of later construction, notably the nineteenth century. As a fortified Mediterranean port, Corfu's urban and port ensemble is notable for its high level of integrity and authenticity.

The fortifications of Corfu have been actively involved in many conflicts between the west and the east Mediterranean from the fifteenth to the twentieth centuries. During rebuilding they have been altered to allow for developments in weapons of attack and principles of defence, successively by the Venetians and by the British.

World Heritage site since

1978 1979 1980 1981 1982 1983 1984 1985 1986 1987 1988 1989 1990 1991 1992 1993 1994 1995 1996 1997 1998 1999 2000 2001 2002 2003 2004 2005 2006 **2007** 200

Primeval Beech Forests of the Carpathians
Slovakia and Ukraine

Criteria – Significant ecological and biological processes

North Sea

EUROPE

Black Sea

The Primeval Beech Forests of the Carpathians are an outstanding example of undisturbed, complex temperate forests. Together they comprise the largest area of virgin forests of European beech (fagus sylvatica) in existence. They represent all stages of beech forest in age and development, and they contain the largest and tallest beech specimens in the world. In all, the forests constitute an invaluable genetic reservoir of beech and many species associated with, and dependent on, these forest habitats.

The areas contain complete naturally functioning ecosystems. Flora and fauna are rich and include rare and unique plants and animals. Some species, such as black stork, are associated with, and depend on, undisturbed forest habitats.

The forests are also an outstanding example of the recolonization and development of terrestrial ecosystems and communities after the last Ice Age, a process which is still ongoing.

The Primeval Beech Forests of the Carpathians are comprised of ten separate sites strung along a 185-km axis at altitudes ranging from 210 m to 1700 m. Six of the sites are in the Ukraine and four in Slovakia.

World Heritage site since

978 1979 1980 1981 1982 1983 1984 1985 1986 1987 1988 1989 1990 1991 1992 1993 1994 1995 1996 1997 1998 1999 2000 2001 2002 2003 2004 2005 2006 **2007** 2008

Rideau Canal
Canada

Criteria – Human creative genius; Significance in human history

The Rideau Canal, built by the British to defend Canada against the USA, was one of the first canals designed for steam-powered vessels. It is the only canal from the great North American canal-building era that remains operational along its original line with most of its structures intact. The site also features an ensemble of fortifications in Kingston, a reminder of the period when Britain and the USA vied for control of the region.

The Rideau was one of several canals the British built after the war of 1812, when Britain defended Upper Canada against American invasion. It was intended as an alternative route to the St Lawrence River, which was vulnerable to American blockade. In the event, the canal was never used for military purposes but instead has served important commercial and recreational purposes since its construction.

The Rideau Canal was completed in 1832 and covers 202 km of the Rideau and Cataraqui rivers, from Ottawa south to Kingston on Lake Ontario. It is the best-preserved example of a slack-water canal in North America, demonstrating the use of this European technology on a large scale.

Rideau Canal and The Houses of Parliament, Parliament Hill, Ottawa.
▼

World Heritage site since

1978 1979 1980 1981 1982 1983 1984 1985 1986 1987 1988 1989 1990 1991 1992 1993 1994 1995 1996 1997 1998 1999 2000 2001 2002 2003 2004 2005 2006 **2007** 2008

Kaiping Diaolou and Villages
China

Criteria – Interchange of values; Testimony to cultural tradition; Significance in human history

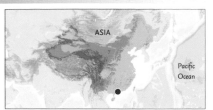

ASIA

Pacific Ocean

Kaiping Diaolou and Villages feature the diaolou, multi-storeyed defensive village houses in Kaiping, which display a complex and elaborate fusion of Chinese and Western structural and decorative forms. The main towers, with their settings and through their flamboyant display of wealth, are a type of building that reflects the significant role of émigré Kaiping people in the development of several countries in south Asia, Australasia and North America, during the late nineteenth and early twentieth centuries. There are four groups

of diaolou and twenty of the most symbolic ones are inscribed on the World Heritage List. These buildings take three forms: communal towers built by several families and used as temporary refuge, residential towers built by individual rich families and used as fortified residences, and watch towers. They are built of stone, pisé (rammed earth or clay), brick or concrete and retain a harmonious relationship with the surrounding landscape.

The building of defensive towers was a local tradition in the Kaiping area since Ming times. The conspicuous wealth of the returning Chinese émigrés contributed to the spread of banditry in the area and the 'diaolou' were an extreme response to the threat. Those in the site represent the final flourishing of a centuries-old tradition.

World Heritage site since

.978 1979 1980 1981 1982 1983 1984 1985 1986 1987 1988 1989 1990 1991 1992 1993 1994 1995 1996 1997 1998 1999 2000 2001 2002 2003 2004 2005 2006 **2007** 2008

Bordeaux, Port of the Moon
France
Criteria – Interchange of values; Significance in human history

The Port of the Moon, port city of Bordeaux in southwest France, is inscribed on the World Heritage List as an inhabited historic city, an outstanding urban and architectural ensemble, created in the age of the Enlightenment, whose values continued up to the first half of the twentieth century, and with more protected buildings than any other French city except Paris. It is also recognized for its historic role as a place of exchange of cultural values over more than 2,000 years, particularly since the twelfth century, due to commercial links with Britain and the Low Countries. Urban plans and architectural ensembles of the early eighteenth century onwards place the city as an outstanding example of innovative classical and neoclassical trends and give it an exceptional urban and architectural unity and coherence. Its urban form represents the success of philosophers who wanted to make towns into melting pots of humanism, universality and culture.

The age of Enlightenment produced Bordeaux's best known architectural and urban features. Louis-Urbain Aubert, Marquis de Tourny, arrived in Bordeaux in 1743, staying there until 1757. He undertook major projects for the renovation and opening up the medieval city, especially to the façades of buildings on the quays along the river Garonne, the vital commercial artery of the community.

The water mirror and Place de la Bourse. ▼

World Heritage site since

1978 1979 1980 1981 1982 1983 1984 1985 1986 1987 1988 1989 1990 1991 1992 1993 1994 1995 1996 1997 1998 1999 2000 2001 2002 2003 2004 2005 2006 **2007** 2008

Gamzigrad-Romuliana, Palace of Galerius
Serbia

Criteria – Testimony to cultural tradition;
Significance in human history

Gamzigrad-Romuliana, the Palace of Galerius in the east of Serbia, is one of the most important Late Roman sites. A fortified palace compound and memorial complex, it was built by the Tetrarch Galerius Maximianus (c. AD 260–311), between the late third and early fourth centuries.

The association of rulers with the divine hierarchy was one of the characteristics of the tetrarchy, and the palace is representative of Late Roman imperial and religious symbolism: the glorification of the emperor as all-powerful ruler and as a god underlies its construction.

The site consists of fortifications, a palace, basilicas, temples, hot baths, a memorial complex and a tetrapylon.

The buildings' intertwining of ceremonial and memorial functions is unique, as are the spatial and visual relationships between the palace and the memorial complex, where the mausoleums of Galerius and his mother Romula are located.

Gamzigrad-Romuliana was built as a place to which Galerius could retire, much like the palace built at Split by his successor, the Emperor Diocletian, who had instituted the tetrarchy in the Roman Empire. The palace was known as Felix Romuliana after the emperor's mother.

World Heritage site since

978 1979 1980 1981 1982 1983 1984 1985 1986 1987 1988 1989 1990 1991 1992 1993 1994 1995 1996 1997 1998 1999 2000 2001 2002 2003 2004 2005 2006 **2007** 2008

Lavaux Vineyard Terraces
Switzerland
Criteria – Significance in human history

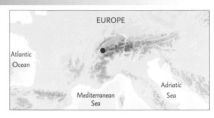

EUROPE

Atlantic
Ocean

Mediterranean
Sea

Adriatic
Sea

The Lavaux Vineyard Terraces are an outstanding example of a centuries-long interaction between people and their environment, developed to optimize local resources and so produce a highly valued wine that has long been important to the economy.

Although there is evidence that vines were grown in the area in Roman times, the present vine terraces can be traced back to the eleventh century when monks from Benedictine and Cistercian monasteries cultivated the area. There are more than 10,000 terraces together with buildings – churches, castles, cellars and houses – that reflect the local history from monastic times to the twentieth century, when the landscape took on its present appearance.

The area is a thriving cultural landscape that demonstrates its millennium-long evolution and development through its well-preserved landscape and buildings, and in the continuation and adaptation of long-standing local cultural traditions.

The Lavaux Vineyard Terraces stretch for about 30 km along the south-facing northern shores of Lake Geneva from the Chateau de Chillon to the eastern outskirts of Lausanne in the Vaud region, covering the lower slopes of the mountainside between the villages and the lake. Conditions in the area are ideal for the production of the Chasselas grape.

World Heritage site since

1978 1979 1980 1981 1982 1983 1984 1985 1986 1987 1988 1989 1990 1991 1992 1993 1994 1995 1996 1997 1998 1999 2000 2001 2002 2003 2004 2005 2006 **2007** 2008

Parthian Fortresses of Nisa
Turkmenistan
Criteria – Interchange of values; Testimony to
cultural tradition

Nisa was the capital
of a powerful empire.
Old Nisa, the royal
citadel, is a 0.14 km²
tell, or mound,
surrounded by a
rampart with towers;
while the 0.25 km² tell
of the city, known as
New Nisa, has walls
up to 9 m high.

The Parthian Fortresses of Nisa consist of
two tells (prehistoric settlement mounds) of
Old and New Nisa, indicating the site of one
of the earliest and most important cities of
the Parthian Empire, a major power from
mid-third century BC to the third century AD.
They conserve the unexcavated remains of
an ancient civilization which skilfully
combined its own traditional cultural
elements with those of the Hellenistic and
Roman west. Archaeological excavations in
two parts of the site have revealed richly
decorated architecture, illustrative of
domestic, state and religious functions.
Situated at the crossroads of important
commercial and strategic axes, this powerful
empire formed a barrier to Roman
expansion while serving as an important
communication and trading centre between
east and west, north and south.

Rainforests of the Atsinanana
Madagascar
Criteria – Significant ecological and biological
processes; Significant natural habitat for
biodiversity

Madagascar is one
of the foremost
countries for
megadiversity of flora
and fauna. The island
has around 12,000
endemic plants and is
globally significant for
its wildlife, especially
primates. The site
protects key areas of
their habitat in
original forests.

The Rainforests of the Atsinanana comprise
six national parks distributed along the
eastern part of the island. These relict
forests are critically important for
maintaining ongoing ecological processes
necessary for the survival of Madagascar's
unique biodiversity, which reflects the
island's geological history. Having
completed its separation from all other land
masses more than sixty million years ago,
Madagascar's plant and animal life evolved
in isolation. The rainforests are inscribed in
the World Heritage List for their importance
to both ecological and biological processes
as well as their biodiversity and the
threatened species they support. Many
species are rare and threatened, especially
primates and lemurs.

World Heritage site since

1978 1979 1980 1981 1982 1983 1984 1985 1986 1987 1988 1989 1990 1991 1992 1993 1994 1995 1996 1997 1998 1999 2000 2001 2002 2003 2004 2005 2006 **2007** 2008

Mehmed Paša Sokolović Bridge in Višegrad
Bosnia-Herzegovina

Criteria – Interchange of values; Significance in human history

The Mehmed Paša Sokolović Bridge of Višegrad crosses the Drina River and was built between 1571 and 1577. It is considered to be characteristic of the height of Ottoman monumental architecture and civil engineering, and a representative masterpiece of the renowned Ottoman architect Sinan.

The 179.5-m-long bridge has eleven masonry arches with spans of 10.7 m to 15 m, and an access ramp at right angles with four arches on the left bank of the river. The arches are enhanced by architectural features typical of the classical Ottoman period.

The bridge was important in allowing control of the inner Balkans by the Ottoman Empire in Istanbul. It forms a highlight of the route linking the plains of the Danube to Sarajevo and the Adriatic coast, particularly to the port of Dubrovnik.

The Višegrad bridge was commissioned by the Grand Vizier Mehmed Paša Sokolović (1505–79), the foremost minister or adviser to the sultan, who was a Bosnian. The building of the bridge was primarily a tribute to his native region.
▼

World Heritage site since

1978 1979 1980 1981 1982 1983 1984 1985 1986 1987 1988 1989 1990 1991 1992 1993 1994 1995 1996 1997 1998 1999 2000 2001 2002 2003 2004 2005 2006 **2007** 200

Jeju Volcanic Island and Lava Tubes
Korea, Republic of
Criteria – Natural phenomena or beauty;
Major stages of Earth's history

Jeju Volcanic Island and Lava Tubes together display a range and quality of accessible volcanic features that offer a distinctive and important contribution to the understanding of volcanic activity and lava cave formation.

The volcanic island of Jeju is an area of outstanding natural beauty and the site covers an area of 188 km². It comprises three locations: Geomunoreum lava-tube system of caves; Seongsan Ilchulbong tuff cone, rising out of the ocean; and the shield or shallow-sided volcano of Mount Hallasan, at 1,950 m, the highest mountain in the Republic of Korea. Hallasan, with its waterfalls, multi-shaped rock formations and lake-filled crater, is the primary volcano, with the Hallasan Natural Reserve at its summit. There are around 360 subsidiary cones around the island, many of which, on cooling, formed the lava-tube caves.

Columnar jointing in the volcanic rock, Jeju. ▼

A lava tube is a conduit that develops under a lava flow and through which lava is expelled during a volcanic eruption. After the rock cools, the empty, elongated channel or lava tube remains.

Geomunoreum, with its multicoloured carbonate roofs and floors and dark lava walls, is regarded as the finest lava-tube system of caves in the world.

World Heritage site since

1978 1979 1980 1981 1982 1983 1984 1985 1986 1987 1988 1989 1990 1991 1992 1993 1994 1995 1996 1997 1998 1999 2000 2001 2002 2003 2004 2005 2006 **2007** 2008

Ecosystem and Relict Cultural Landscape of Lopé-Okanda
Gabon

Criteria – Testimony to cultural tradition; Significance in human history; Significant ecological and biological processes; Significant natural habitat for biodiversity

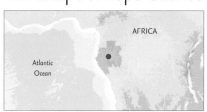

The Ecosystem and Relict Cultural Landscape of Lopé-Okanda demonstrates an unusual interface between dense and well-conserved tropical rainforest and relict savanna environments with a great diversity of species, including endangered large mammals, and habitats. The site illustrates ecological and biological processes in terms of species and habitat adaptation to post-glacial climatic changes. It contains evidence of the successive passages of different peoples who have left extensive and comparatively well-preserved remains of habitation around hilltops, caves and shelters, evidence of iron-working and a remarkable collection of some 1,800 petroglyphs (rock carvings). The property's collection of Neolithic and Iron Age sites, together with the rock art found there, reflects a major migration route of Bantu and other peoples from west Africa along the river Ogooué valley to the north of the dense evergreen Congo forests and to central, east and southern Africa, that has shaped the development of the whole of sub-Saharan Africa.

Over 1,550 plant species have been recorded so far in the park and it is anticipated that the final total could reach over 3,000, making Lopé-Okanda one of the most outstanding areas for floral diversity in the Congo rainforest ecoregion.

World Heritage site since

1978 1979 1980 1981 1982 1983 1984 1985 1986 1987 1988 1989 1990 1991 1992 1993 1994 1995 1996 1997 1998 1999 2000 2001 2002 2003 2004 2005 2006 2007 **2008**

Protective town of San Miguel and the Sanctuary of Jesús Nazareno de Atotonilco
Mexico

Criteria – Interchange of values; Significance in human history

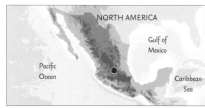

This fortified town, established in the sixteenth century, reached its apogee in the eighteenth century, when many of its outstanding religious and civic buildings were built in the Mexican Baroque style. It was a melting pot of Spanish, Creole and Amerindian cultures. The Jesuit sanctuary, 14 km away, is one of the finest examples of eighteenth-century Baroque art and architecture in Latin America.

During its long history the town has also played a part in the struggle for Mexican independence. It was the birthplace of national hero General Ignacio Allende and was renamed San Miguel de Allende in 1926 in his honour.

World Heritage site since

1978 1979 1980 1981 1982 1983 1984 1985 1986 1987 1988 1989 1990 1991 1992 1993 1994 1995 1996 1997 1998 1999 2000 2001 2002 2003 2004 2005 2006 2007 **2008**

San Marino Historic Centre and Mount Titano
San Marino

Criteria – Testimony to cultural tradition

San Marino is one of the world's oldest republics and the only Italian city-state that survives today. With its unique, uninterrupted continuity as the capital of an independent republic since the thirteenth century, San Marino exemplifies the establishment of a representative democracy based on civic autonomy and self-governance and so represents an important stage in the development of democratic models in Europe and around the world.

The San Marino site covers 0.6 km². The Historic Centre is strategically sited on the top of Mount Titano and its many monuments include fortification towers, walls, gates and bastions; a neoclassical nineteenth-century basilica; fourteenth- and sixteenth-century convents; the nineteenth-century Palazzo Publico; and the Titano Theatre, dating from the eighteenth century.

First tower (Guaita) on Mount Titano.

The state of San Marino is an enclave in the Appenine Mountains and is landlocked and surrounded by Italy. It was the only city-state not to be brought into union with the rest of Italy by Garibaldi during the Risorgimento, the nineteenth-century national unification movement, and maintains its independence today.

Sacred Mijikenda Kaya Forests
Kenya

Criteria – Testimony to cultural tradition; Traditional human settlement; Heritage associated with events of universal significance

Spread out along around 200 km of the Kenyan coast are a number of separate forested sites, mostly on low hills, in which are the remains of fortified villages, kayas, of the Mijikenda people. Tradition tells how kayas were created from the sixteenth century onwards as the Mijikenda migrated south from Somalia. The kayas began to fall out of use in the early twentieth century and are now revered as the repositories of

spiritual beliefs of the Mijikenda people and are seen as the sacred abode of their ancestors. The forests around the kayas have been nurtured by the Mijikenda community to protect the sacred graves and groves and are now almost the only remains of the once extensive coastal lowland forest.

A typical kaya consisted of a circular stockade in a clearing in the forest, approached by well-defined paths. Houses were arranged around the edge of the stockade and within the centre of the village there would be either a grove of trees or a large thatched structure called a moro, places for meetings of the council of elders.

World Heritage site since

1978 1979 1980 1981 1982 1983 1984 1985 1986 1987 1988 1989 1990 1991 1992 1993 1994 1995 1996 1997 1998 1999 2000 2001 2002 2003 2004 2005 2006 2007 **2008**

Rhaetian Railway in the Albula / Bernina Landscapes
Italy and Switzerland

Criteria – Interchange of values; Significance in human history

The railway crosses the Swiss Alps, to the south of the upper valley of the Rhine, by two passes. It follows the valley and the Albula Pass, and then crosses the upper valley of the Engadin (Saint-Moritz), before crossing the Bernina Pass and descending to the river Adda, in the Italian Veltin.

The Rhaetian railway brings together two historic mountain railway lines that cross the Swiss Alps through two passes. Opened in 1904, the Albula line is 67 km long. It features an impressive set of structures including forty-two tunnels and covered galleries and 144 viaducts and bridges and reaches a height of 1,819 m. The 61 km Bernina Pass line features thirteen tunnels and galleries and fifty-two viaducts and bridges and reaches a height of 2,253 m. These railways overcame the isolation of settlements in the central Alps early in the twentieth century, with a major and lasting socio-economic impact on life in the mountains. They display outstanding architectural and civil engineering achievements, built in harmony with the landscapes through which they pass.

Berlin Modernism Housing Estates
Germany

Criteria – Interchange of values; Significance in human history

The idea of these Berlin estates was to create housing for all income levels, of equal standard and varying size, with dedicated bathrooms and kitchens and generous loggias and balconies, which faced the sun. The designers not only aimed at creating a new social and spatial order; they also wanted to create beautiful facilities and make the inhabitants happy.

The site consists of six housing estates in Berlin that testify to innovative housing policies from 1910 to 1933, especially during the Weimar Republic, when the city of Berlin was particularly progressive socially, politically and culturally. The estates are an outstanding example of the building reform movement that contributed to improving housing and living conditions for people through novel approaches to town planning, architecture and garden design. The estates also provide exceptional examples of Modernism, the new architectural style, featuring fresh design solutions, as well as technical and aesthetic innovations. Bruno Taut, Martin Wagner and Walter Gropius were among the leading architects of these projects which exercised considerable influence on the development of housing around the world.

World Heritage site since

1978 1979 1980 1981 1982 1983 1984 1985 1986 1987 1988 1989 1990 1991 1992 1993 1994 1995 1996 1997 1998 1999 2000 2001 2002 2003 2004 2005 2006 2007 **2008**

Socotra Archipelago
Yemen

Criteria – Significant natural habitat for biodiversity

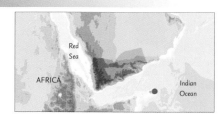

Socotra is globally important for biodiversity conservation because of its exceptionally rich and distinct flora and fauna; 37 per cent of its plant species, 90 per cent of its reptile species and 95 per cent of its land snail species do not occur anywhere else in the world. Socotra, one of the most biodiverse and distinct islands in the world, has been termed the 'Galápagos of the Indian Ocean'.

▲ Socotra osprey.

The site represents all the terrestrial and marine features and processes essential for the long-term conservation of the archipelago's rich and distinct biodiversity. It covers about 75 per cent of the total land area, protecting all the major vegetation types, areas of high floral and faunal values and important bird areas, and the most significant areas of marine biodiversity.

Historic Centre of Camagüey
Cuba

Criteria – Significance in human history; Traditional human settlement

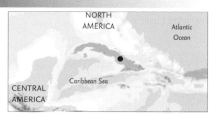

Camagüey is an exceptional example of a traditional urban settlement relatively isolated from main trade routes. It was one of the first villages founded by the Spaniards in Cuba and was the centre of a region dedicated to cattle breeding and the sugar industry. Settled in its current location in 1528, its irregular urban pattern developed with large and small squares, winding streets, alleys and irregular blocks of buildings, highly unusual for Latin American colonial towns located on flat land. The

Spanish colonizers followed medieval European influences in terms of urban layout and traditional construction techniques brought to the Americas by their masons and construction masters. The centre reflects the influence of numerous styles including neoclassical, neocolonial, Art Nouveau and Art Deco.

Religious architecture reached its peak in the eighteenth century. Churches are compact, usually with symmetrical façades with scant decoration. Some adjoin convents, hospitals or cemeteries. As a group, the religious buildings show extreme simplicity, but their historical, artistic and symbolic value has contributed to the naming of Camagüey as the 'City of Churches'.

World Heritage site since

1978 1979 1980 1981 1982 1983 1984 1985 1986 1987 1988 1989 1990 1991 1992 1993 1994 1995 1996 1997 1998 1999 2000 2001 2002 2003 2004 2005 2006 2007 **2008**

Kuk Early Agricultural Site
Papua New Guinea

Criteria – Testimony to cultural tradition;
Significance in human history

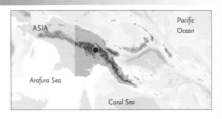

This site consists of 1.2 km² of swamps in the southern highlands of New Guinea, 1,500 m above sea level. Archaeological excavation has established Kuk as the site of the earliest, independent agriculture in Oceania and could indicate that it contributed to the spread of domesticated plants, and of settlement, culture and societies across the region. It contains well-preserved archaeological remains demonstrating the technological leap from plant exploitation to agriculture, initially with taro and yam on wetland margins around 7,000 years ago, and then with organised domestication and cultivation of bananas on drained ground some 4,000 years ago. It is an excellent example of transformation of agricultural practices over time, from cultivation mounds to draining the wetlands through the digging of ditches with wooden tools.

Kuk is one of the few places in the world where archaeological evidence suggests independent agricultural development and changes in agricultural practice over more than 7,000 years. Modern farming activities at Kuk remain relatively low-key and do not intrude upon the archaeological features of the site.

Saryarka – Steppe and Lakes of Northern Kazakhstan
Kazakhstan

Criteria – Significant ecological and biological processes; Significant natural habitat for biodiversity

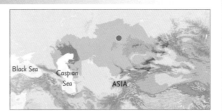

This site features wetlands of outstanding importance for migratory birds, including the globally threatened Siberian white crane, Dalmatian pelican and Pallas's fish eagle, that travel from Africa, Europe and south Asia to their breeding grounds in western and eastern Siberia. The central Asian steppe provides a valuable refuge for over half the species of the region's steppe flora, a number of threatened bird species and the critically endangered Saiga antelope.

The wetlands of the site are globally important. The Korgalzhyn-Tengiz lakes provide feeding grounds for up to 15–16 million birds, including up to 2.5 million geese. They also support 350,000 nesting waterfowl, while the Naurzum lakes support 500,000 nesting waterfowl.

◄

Dalmation pelican.

World Heritage site since

1978 1979 1980 1981 1982 1983 1984 1985 1986 1987 1988 1989 1990 1991 1992 1993 1994 1995 1996 1997 1998 1999 2000 2001 2002 2003 2004 2005 2006 2007 **2008**

Swiss Tectonic Arena Sardona
Switzerland
Criteria – Major stages of Earth's history

The Swiss Tectonic Arena Sardona lies in the Alps in northeastern Switzerland and covers a mountainous area of 329 km²; it includes seven peaks that rise above 3,000 m.

The Swiss Tectonic Arena Sardona is an exceptional and dramatic example of the process of mountain building through continental collision. The structures and processes that characterize the phenomenon are displayed in a mountain setting and the area has been a key site for the study of geological sciences since the eighteenth century.

The Glarus Alps are glaciated mountains rising dramatically above narrow river valleys and are the site of the largest post-glacial landslide in the central Alpine region.

The Glarus Overthrust is a key part of the whole site. This is a break in the earth's crust where compression forces the upper side of two tectonic plates further upwards over the lower one. The exposures of the rocks below and above this feature are visible in three dimensions and they have made substantial contributions to the understanding of mountain-building tectonics.

Klöntaler lake and the Glarus Alps.
▼

World Heritage site since

1978 1979 1980 1981 1982 1983 1984 1985 1986 1987 1988 1989 1990 1991 1992 1993 1994 1995 1996 1997 1998 1999 2000 2001 2002 2003 2004 2005 2006 2007 **2008**

Melaka and George Town, Historic Cities of the Straits of Malacca
Malaysia

Criteria – Interchange of values; Testimony to cultural tradition, Significance in human history

The historic cities of Melaka and George Town, on the Straits of Malacca, are remarkable examples of historic colonial towns, demonstrating a succession of cultural influences that derive from their former function as trading ports linking East and West.

The towns complement one another in illustrating different stages of development and successive changes over a period of time. Melaka demonstrates their early history: originating in the fifteenth-century Malay sultanate and the Portuguese and Dutch periods beginning in the early sixteenth century, its notable monuments are its government buildings, churches, squares and fortifications. Residential and commercial buildings feature strongly in George Town, which represents the British era from the end of the eighteenth century until the twentieth century. Together, the two towns constitute a unique architectural and cultural townscape without parallel anywhere in east and southeast Asia.

Melaka and George Town were forged from the mercantile interaction of Malay, Chinese, and Indian cultures and of three successive European colonial powers over almost 500 years. Each has left its imprint on the local architecture and urban form, technology and monumental art.

Stari Grad Plain
Croatia

Criteria – Interchange of values; Testimony to cultural tradition; Traditional human settlement

The fertile plain near the port, now known as Stari Grad, on the Adriatic island of Hvar, was colonized by Ionian Greeks from Paros in the fourth century BC. It quickly became one of the most important Greek colonies of the Adriatic because of its flourishing agriculture. This agricultural activity, based on grapes and olives, has been maintained since Greek times to the present. The site is also a nature reserve. The landscape features ancient stone walls and trims, or small stone shelters, and bears testimony to the regular geometrical system of land division used by the ancient Greeks, the chora, which has remained virtually intact over twenty-four centuries. This system was completed with tanks of varying sizes for the retention of rainwater.

The agricultural plain of Stari Grad and its environment are an example of very ancient traditional human settlement, which is today under threat from modern economic development, particularly rural depopulation and the abandonment of traditional farming practices.

World Heritage site since

1978 1979 1980 1981 1982 1983 1984 1985 1986 1987 1988 1989 1990 1991 1992 1993 1994 1995 1996 1997 1998 1999 2000 2001 2002 2003 2004 2005 2006 2007 **2008**

Monarch Butterfly Biosphere Reserve
Mexico
Criteria – Natural phenomena or beauty

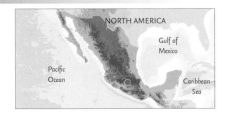

The millions of monarch butterflies that return to the property every year colour the trees orange, bend branches with their massed weight, fill the sky when they take flight, and sound like light rain with the beating of their wings. To witness this unique phenomenon is an exceptional experience of nature.

A monarch butterfly feeding on a flower's nectar.
▼

The Monarch Butterfly Biosphere Reserve protects key overwintering sites for the monarch butterfly whose concentration in the property is a superlative natural phenomenon.

The 563 km² biosphere lies within rugged forested mountains about 100 km northwest of Mexico City. In a dramatic manifestation of insect migration, up to a billion monarch butterflies return annually from northern breeding areas to land in close-packed clusters within fourteen

overwintering colonies in central Mexico's oyamel fir forests. The biosphere reserve protects eight of these colonies and an estimated 70 per cent of the total overwintering population of the monarch butterfly's eastern population.

In the spring the butterflies begin an eight-month round-trip migration that takes them as far as eastern Canada and back, during which four successive generations are born and die.

World Heritage site since

1978 1979 1980 1981 1982 1983 1984 1985 1986 1987 1988 1989 1990 1991 1992 1993 1994 1995 1996 1997 1998 1999 2000 2001 2002 2003 2004 2005 2006 2007 **2008**

Le Morne Cultural Landscape
Mauritius

Criteria – Testimony to cultural tradition; Heritage associated with events of universal significance

Le Morne, a rugged mountain that juts into the Indian Ocean in the southwest of Mauritius, was used as a shelter by runaway slaves, called 'maroons', through the eighteenth and early years of the nineteenth centuries. Mauritius was a major transhipment place for slaves between Africa, India and the Americas and became known as the 'Maroon republic' because of the comparatively large number of escaped slaves who were in hiding on Le Morne. Protected by the mountain's isolated, wooded and almost inaccessible cliffs, it offered a retreat against those who hunted them down. Many only survived as free men for a few weeks; others managed to set up small communities on the mountain, around its base, in caves on its sides or on its summit.

Since the abolition of slavery in 1835 the 'maroons' have achieved legendary status as heroic resistance fighters and Le Morne has become the symbol of their suffering, their bid for freedom and their sacrifice. The mountain, together with its surrounding foothills and lagoons, is a place of great scenic beauty.

Armenian Monastic Ensembles of Iran
Islamic Republic of Iran

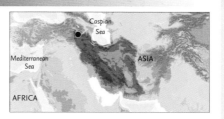

Criteria – Interchange of values; Testimony to cultural tradition; Heritage associated with events of universal significance

The fortified Armenian monasteries of northwest Iran bear testimony, since the origins of Christianity and certainly since the seventh century, to Armenian culture and its contact with Persian and later Iranian civilizations. The monasteries of St Thaddeus and St Stepanos and the Chapel of Dzordzor are outstanding examples of Armenian architectural and decorative traditions. Situated on the south-eastern fringe of Armenian influence, the monasteries constituted a major centre for the dissemination of that culture. The monastery of St Thaddeus, the presumed location of the tomb of St Thaddeus, the apostle of Jesus Christ, has always been a place of high spiritual significance for Christians and other inhabitants in the region. It is still a living place of pilgrimage for the Armenian Church.

The monasteries have survived human and natural destruction over the course of 2,000 years and have been rebuilt several times in keeping with Armenian cultural traditions. Today they are the only important vestiges of Armenian culture in this region.

World Heritage site since

1978 1979 1980 1981 1982 1983 1984 1985 1986 1987 1988 1989 1990 1991 1992 1993 1994 1995 1996 1997 1998 1999 2000 2001 2002 2003 2004 2005 2006 2007 **2008**

Bahá'í Holy Places in Haifa and the Western Galilee
Israel

Criteria – Testimony to cultural tradition; Heritage associated with events of universal significance

The Bahá'í Holy Places in Haifa and the Western Galilee are inscribed for their profound spiritual meaning and the testimony they bear to the strong tradition of pilgrimage in the Bahá'í Faith.

The sites include the two most holy places in the Bahá'í religion associated with its founders, the Shrine of Bahá'u'lláh in Acre and the Shrine of the Báb in Haifa, together with their surrounding gardens, associated buildings and monuments. These two shrines are part of a larger complex of buildings, monuments and sites at seven distinct locations in Haifa and the Western Galilee that draw large numbers of pilgrims from around the world.

The two holy Bahá'í shrines are tangible places of great meaning for one of the world's religions.

The Bahá'í Faith is a monotheistic religion that emphasizes the spiritual unity of all peoples. It was founded by Bahá'u'lláh in Persia in the nineteenth century and it is estimated that there are over five million Bahá'ís around the world.

The Shrine of the Báb and the Bahá'í Gardens, Haifa.
▼

World Heritage site since

1978 1979 1980 1981 1982 1983 1984 1985 1986 1987 1988 1989 1990 1991 1992 1993 1994 1995 1996 1997 1998 1999 2000 2001 2002 2003 2004 2005 2006 2007 **2008**

Fortifications of Vauban
France

Criteria – Human creative genius; Interchange of values; Significance in human history

Vauban played a major role in the history of fortification His work crystallized earlier theories of strategy into a rationa system of fortifications based on a concrete relationship to the territory to be defended. His theories and models were studied and usec across the world and made a major contribution to military architecture.

The Fortifications of Vauban comprise twelve groups of fortified buildings and sites that together represent the peak of classic bastioned fortification typical of Western military architecture. They are the finest examples of the work of Sébastien Le Prestre de Vauban (1633–1707), the renowned military engineer of King Louis XIV.

The twelve properties in the site form a ring around France's borders. They are at Arras, Besançon, Blaye-Cussac-Fort-Médoc, Briançon, Camaret-sur-Mer, Longwy, Mont-Dauphin, Mont-Louis, Neuf-Brisach, Saint-Martin-de-Ré, Sant-Vaast-la-Hougue/Tatihou and Villefranche-de-Conflent. The fortifications include towns built from scratch by Vauban, citadels built on plains, urban bastion walls and bastion towers. There are also mountain forts, sea forts, a mountain battery and two mountain communication structures.

Fort at Briançon.
▼

World Heritage site since

1978 1979 1980 1981 1982 1983 1984 1985 1986 1987 1988 1989 1990 1991 1992 1993 1994 1995 1996 1997 1998 1999 2000 2001 2002 2003 2004 2005 2006 2007 **2008**

Wooden Churches of the Slovak part of the Carpathian Mountain Area
Slovakia

Criteria – Testimony to cultural tradition; Significance in human history

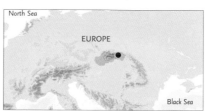

North Sea

EUROPE

Black Sea

The wooden churches are an outstanding testimony to the inter-ethnic and inter-cultural character of this small area in the Carpathian Mountains, where Latin and Byzantine cultures met and overlapped, and of tolerance at a time of religious and political upheaval in the Habsburg Empire.

These wooden churches, two Roman Catholic, three Protestant and three Greek Orthodox, were built between the sixteenth and eighteenth centuries, mostly in quite isolated villages, using wood as the main material and traditional construction techniques. They are good examples of a rich local tradition of religious architecture, marked by the meeting of Latin and Byzantine cultures. The buildings exhibit some variations in their floor plans, interior spaces and external appearance due to their respective religious practices. They bear testimony to the development of major architectural and artistic trends during the period of construction, adapted to a specific geographical and cultural context. Interiors are decorated with paintings on the walls and ceilings and other works of art that enrich their cultural significance.

Chief Roi Mata's Domain
Vanuatu

Criteria – Testimony to cultural tradition; Traditional human settlement; Heritage associated with events of universal significance

OCEANIA

Melanesia

Pacific Ocean

Coral Sea

The landscape reflects continuing Pacific chiefly systems and respect for this authority through tapu (tabu) prohibitions on the use of Roi Mata's residence and burial that have been observed for over 400 years and that have structured the local landscape and social practices. Roi Mata still lives for many in contemporary Vanuatu as a source of power and inspiration.

The domain comprises three early seventeenth-century sites on the islands of Efate, Lelepa and Artok associated with the life and death of the last paramount chief, or Roi Mata, of what is now central Vanuatu. It includes Roi Mata's residence, the site of his death and Roi Mata's mass burial site. It is closely associated with the oral traditions surrounding the chief and the moral values he espoused.

World Heritage site since

1978 1979 1980 1981 1982 1983 1984 1985 1986 1987 1988 1989 1990 1991 1992 1993 1994 1995 1996 1997 1998 1999 2000 2001 2002 2003 2004 2005 2006 2007 **2008**

Joggins Fossil Cliffs
Canada

Criteria – Major stages of Earth's history

Upright fossil trees are preserved at a series of levels in the cliffs together with animal, plant and trace fossils. These provide environmental context and enable a complete reconstruction to be made of the extensive fossil forests that dominated land at this time, and which are now the source of most of the world's coal deposits.

The Joggins Fossil Cliffs have been described as the 'coal age Galápagos' due to their wealth of fossils from the Carboniferous (354 to 290 million years ago). These include the remains and tracks of the first known reptiles, and the rainforest in which they lived, left intact and undisturbed. With its 14.7 km of sea cliffs, low bluffs, rock platforms and beaches, the site contains remains of three ecosystems: estuarine bay, floodplain rainforest and fire-prone forested alluvial plain with freshwater pools. Joggins offers the richest assemblage known of the

fossil life in these three ecosystems with 96 genera and 148 species of fossils and 20 footprint groups. It played a vital role in the development of geological and evolutionary principles.

Mount Sanqingshan
National Park
China

Criteria – Natural phenomena or beauty

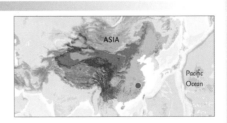

The park is well known for its Taoist cultural relics, stone carvings and temples. Mount Sanqingshan has been a Taoist shrine since a priest, Ge Hong, came to the mountain 400 years ago. The ancient religion of Taoism is based on worship in and of nature, a philosophy much in keeping with conservation ethics now practised here.

Mount Sanqingshan National Park contains a unique array of fantastically-shaped granite pillars and peaks, concentrated in a relatively small area. The looming, intricate rock formations intermingled with delicate forest cover and combined with ever-shifting weather patterns create a landscape of arresting beauty. It is located in the west of the Huyaiyu mountain range in the northeast of Jiangxi Province. The area is

subject to a combination of subtropical, monsoonal and maritime influences and forms an island of temperate forest above the surrounding subtropical landscape. It also features numerous waterfalls, some of them 60 m in height, lakes and springs. The access afforded by suspended walking trails in the park permits visitors to appreciate the park's stunning scenery and enjoy its serene atmosphere.

World Heritage site since

1978 1979 1980 1981 1982 1983 1984 1985 1986 1987 1988 1989 1990 1991 1992 1993 1994 1995 1996 1997 1998 1999 2000 2001 2002 2003 2004 2005 2006 2007 **2008**

Mantua and Sabbioneta
Italy

Criteria – Interchange of values; Testimony to cultural tradition

Mantua and Sabbioneta in the Po valley of northern Italy, are important both for the value of their architecture and also for their prominent role in the dissemination of Renaissance culture. They represent two aspects of Renaissance town planning: Mantua shows the renewal and extension of an already existing city, while 30 km away, sixteenth-century Sabbioneta represents the expression of contemporary theories on the planning of the ideal city.

Typically, Mantua's layout is irregular with regular parts showing different stages of its growth since the Roman period. It includes many medieval buildings, among them an eleventh-century rotunda and a Baroque theatre. Sabbioneta, on the other hand, is a single-period city. Together they offer exceptional testimony to the urban, architectural and artistic realizations of the Renaissance, linked by the vision, planning and work of the local ruling family of Gonzaga.

Mantua was renovated in the fifteenth–sixteenth centuries with hydrological engineering, urban and architectural works. The participation of renowned architects and painters made it a prominent capital of the Renaissance.

Sabbioneta was a new town built according to Renaissance ideals. Its defensive walls, grid-patterned streets, public spaces and monuments make it one of the best examples of European ideal cities.

Mantua from Lago di Mezzo. ▼

World Heritage site since

1978 1979 1980 1981 1982 1983 1984 1985 1986 1987 1988 1989 1990 1991 1992 1993 1994 1995 1996 1997 1998 1999 2000 2001 2002 2003 2004 2005 2006 2007 **2008**

Al-Hijr Archaeological Site (Madâin Sâlih)
Saudi Arabia

Criteria – Interchange of values; Testimony to cultural tradition

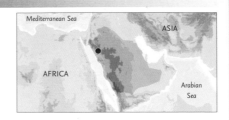

The remarkable archaeological site of Al-Hijr – formerly known as Hegra – is the largest conserved site of the Nabataean civilization south of Petra in modern-day Jordan. It includes a major ensemble of tombs whose architecture and decorations are cut directly into the sandstone of the local landscape.

The site has in total 111 well-preserved monumental tombs, ninety-four of which have decorated façades dating from the first century BC to the first century AD. There are also fifty inscriptions of the pre-Nabataean period and some cave drawings. It is an outstanding illustration of the Nabataeans' specific architectural style and accomplishment.

The site includes a set of wells, most of which were sunk into the rock, demonstrating the Nabataeans' mastery of water techniques for agricultural purposes. Some wells are still in use today.

Al-Hijr is located at a meeting point between various civilizations of late Antiquity, on a trade route between the Arabian Peninsula, the Mediterranean world and Asia. It bears outstanding witness to important cultural exchanges in architecture, decoration, language use and the caravan route. Although abandoned during the pre-Islamic period, the route continued to play its international role for caravans and then for the pilgrimage to Mecca.

World Heritage and UNESCO's World Heritage Mission

Heritage is our legacy from the past, what we live with today, and what we pass on to future generations. Our cultural and natural heritage are both irreplaceable sources of life and inspiration. Places as unique and diverse as the wilds of East Africa's Serengeti, the Pyramids of Egypt, the Great Barrier Reef in Australia and the Baroque cathedrals of Latin America make up our world's heritage.

The United Nations Educational, Scientific and Cultural Organization (UNESCO) seeks to encourage the identification, protection and preservation of cultural and natural heritage around the world considered to be of outstanding universal value to humanity. This is embodied in a unique international treaty, called the Convention Concerning the Protection of the World Cultural and Natural Heritage adopted by UNESCO in 1972 (see http://whc.unesco.org/en/conventiontext).

One of the world's most successful conservation instruments, the World Heritage Convention is exceptional in that it links together in a single document the concepts of nature conservation and the preservation of cultural properties. It is also significant in its universal application – World Heritage sites belong to all the peoples of the world, irrespective of the territory on which they are located. By regarding heritage as both cultural and natural, the Convention recognizes the ways in which people interact with nature, and of the fundamental need to preserve the balance between the two.

UNESCO's World Heritage mission is to:

- encourage countries to sign the World Heritage Convention and to ensure the protection of their natural and cultural heritage;
- encourage States Parties to the Convention to nominate sites within their national territory for inclusion on the World Heritage List;
- encourage States Parties to establish management plans and set up reporting systems on the state of conservation of their World Heritage sites;
- help States Parties safeguard World Heritage properties by providing technical assistance and professional training;
- provide emergency assistance for World Heritage sites in immediate danger;
- support States Parties' public awareness-building activities for World Heritage conservation;
- encourage participation of the local population in the preservation of their cultural and natural heritage;
- encourage international cooperation in the conservation of our world's cultural and natural heritage.

THE CRITERIA FOR SELECTION

The World Heritage Convention stipulates the creation of a World Heritage List. In a detailed process, properties are inscribed by an intergovernmental twenty-one member elected Committee, only after a preselection, nomination, and evaluation process. Two leading international Non-Governmental Organizations, the International Union for the Conservation of Nature (IUCN) and the International Council on Monuments and Sites (ICOMOS), review and advise on the natural and cultural nominations respectively. The International Centre for the Study of the Preservation and Restoration of Cultural Property provides the Committee with expert advice on conservation of cultural sites. To be included, sites must be of outstanding universal value and meet at least one out of ten selection criteria:

Human creative genius

i. to represent a masterpiece of human creative genius;

Interchange of values

ii. to exhibit an important interchange of human values, over a span of time or within a cultural area of the world, on developments in architecture or technology, monumental arts, town-planning or landscape design;

Testimony to cultural tradition

iii. to bear a unique or at least exceptional testimony to a cultural tradition or to a civilization which is living or which has disappeared;

Significance in human history

iv. to be an outstanding example of a type of building, architectural or technological ensemble or landscape which illustrates (a) significant stage(s) in human history;

Traditional human settlement

v. to be an outstanding example of a traditional human settlement, land-use, or sea-use which is representative of a culture (or cultures), or human interaction with the environment especially when it has become vulnerable under the impact of irreversible change;

Heritage associated with events of universal significance

vi. to be directly or tangibly associated with events or living traditions, with ideas, or with beliefs, with artistic and literary works of outstanding universal significance. The Committee considers that this criterion should preferably be used in conjunction with other criteria);

Natural phenomena or beauty

vii. to contain superlative natural phenomena or areas of exceptional natural beauty and aesthetic importance;

Major stages of Earth's history

viii. to be outstanding examples representing major stages of Earth's history, including the record of life, significant on-going geological processes in the development of landforms, or significant geomorphic or physiographic features;

Significant ecological and biological processes

ix. to be outstanding examples representing significant on-going ecological and biological processes in the evolution and development of terrestrial, fresh water, coastal and marine ecosystems and communities of plants and animals;

Significant natural habitat for biodiversity

x. to contain the most important and significant natural habitats for in-situ conservation of biological diversity, including those containing threatened species of outstanding universal value from the point of view of science or conservation.

The protection, management, authenticity and integrity of properties are also important considerations. The process is further detailed in the Operational Guidelines for the Implementation of the World Heritage Convention which, along with the actual text of the Convention, is the main working tool on World Heritage.

UNESCO **Worldwide**

Africa

Algeria	Libyan Arab Jamahiriya
Angola	Madagascar
Benin	Malawi
Botswana	Mali
Burkina Faso	Mauritania
Burundi	Mauritius
Cameroon	Morocco
Cape Verde	Mozambique
Central African Republic	Namibia
Chad	Niger
Comoros	Nigeria
Congo	Rwanda
Côte d'Ivoire	Sao Tome and Principe
Democratic Republic of the Congo	Senegal
Djibouti	Seychelles
Egypt	Sierra Leone
Equatorial Guinea	Somalia
Eritrea	South Africa
Ethiopia	Sudan
Gabon	Swaziland
Gambia	Togo
Ghana	Tunisia
Guinea	Uganda
Guinea-Bissau	United Republic of Tanzania
Kenya	Zambia
Lesotho	Zimbabwe
Liberia	

Asia and the Pacific

Afghanistan	New Zealand
Australia	Niue
Bangladesh	Pakistan
Bhutan	Palau
Brunei Darussalam	Papua New Guinea
Cambodia	Philippines
China	Republic of Korea
Cook Islands	Russian Federation
Democratic People's Republic of Korea	Samoa
Fiji	Singapore
India	Solomon Islands
Indonesia	Sri Lanka
Iran, Islamic Republic of	Tajikistan
Japan	Thailand
Kazakhstan	Timor-Leste
Kiribati	Tonga
Kyrgyzstan	Turkey
Lao People's Democratic Republic	Turkmenistan
Malaysia	Tuvalu
Maldives	Uzbekistan
Marshall Islands	Vanuatu
Micronesia (Federated States of)	Viet Nam
Mongolia	
Myanmar	
Nauru	
Nepal	

The regions presented here follow specific UNESCO definitions which do not forcibly reflect geography. They refer to the execution of regional activities of the Organization.

Europe and North America

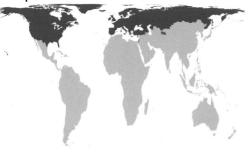

Albania	Luxembourg
Andorra	Malta
Armenia	Monaco
Austria	Montenegro
Azerbaijan	Netherlands
Belarus	Norway
Belgium	Poland
Bosnia and Herzegovina	Portugal
Bulgaria	Republic of Moldova
Canada	Romania
Croatia	Russian Federation
Cyprus	San Marino
Czech Republic	Serbia
Denmark	Slovakia
Estonia	Slovenia
Finland	Spain
France	Sweden
Georgia	Switzerland
Germany	Tajikistan
Greece	The former Yugoslav Republic of
Hungary	Macedonia
Iceland	Turkey
Ireland	Ukraine
Israel	United Kingdom of Great Britain
Italy	and Northern Ireland
Kazakhstan	United States of America
Latvia	
Lithuania	

Arab States

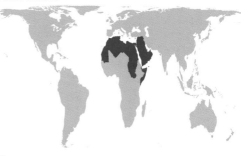

Algeria	Morocco
Bahrain	Oman
Djibouti	Qatar
Egypt	Saudi Arabia
Iraq	Somalia
Jordan	Sudan
Kuwait	Syrian Arab Republic
Lebanon	Tunisia
Libyan Arab Jamahiriya	United Arab Emirates
Malta	Yemen
Mauritania	

Latin America and the Caribbean

Antigua and Barbuda	Guatemala
Argentina	Guyana
Bahamas	Haiti
Barbados	Honduras
Belize	Jamaica
Bolivia	Mexico
Bolivarian Republic of Venezuela	Nicaragua
Brazil	Panama
Chile	Paraguay
Colombia	Peru
Costa Rica	Saint Kitts and Nevis
Cuba	Saint Lucia
Dominica	Saint Vincent and the Grenadines
Dominican Republic	Suriname
Ecuador	Trinidad and Tobago
El Salvador	Uruguay
Grenada	

Country index

Afghanistan
Cultural Landscape and Archaeological
  Remains of the Bamiyan Valley 703
Minaret and Archaeological Remains of
  Jam 683

Albania
Butrint 389
Historic Centres of Berat and Gjirokastra 732

Algeria
Al Qal'a of Beni Hammad 109
Djémila 148
Kasbah of Algiers 379
M'Zab Valley 145
Tassili n'Ajjer 145
Timgad 148
Tipasa 142

Andorra
Madriu-Perafita-Claror Valley 732

Argentina
Cueva de las Manos, Río Pinturas 607
Iguazu National Park 188
Ischigualasto / Talampaya Natural Parks 628
Jesuit Block and Estancias of Córdoba 647
Jesuit Missions of the Guaranis 159
Los Glaciares 111
Península Valdés 590
Quebrada de Humahuaca 705

Armenia
Cathedral and Churches of Echmiatsin and
  the Archaeological Site of Zvartnots 655
Monasteries of Haghpat and Sanahin 500
Monastery of Geghard and the Upper Azat
  Valley 630

Australia
Australian Fossil Mammal Sites
  (Riversleigh / Naracoorte) 429
Fraser Island 393
Gondwana Rainforests of Australia 260
Great Barrier Reef 122
Greater Blue Mountains Area 609
Heard and McDonald Islands 522
Kakadu National Park 127
Lord Howe Island Group 136
Macquarie Island 506
Purnululu National Park 699
Royal Exhibition Building and Carlton
  Gardens 718
Shark Bay 368
Sydney Opera House 763
Tasmanian Wilderness 143
Uluṟu-Kata Tjuṯa National Park 306
Wet Tropics of Queensland 311
Willandra Lakes Region 125

Austria
City of Graz – Historic Centre 585
Fertö / Neusiedlersee Cultural Landscape 664
Hallstatt-Dachstein / Salzkammergut
  Cultural Landscape 531
Historic Centre of the City of Salzburg 498
Historic Centre of Vienna 678
Palace and Gardens of Schönbrunn 502
Semmering Railway 551
Wachau Cultural Landscape 641

Azerbaijan
Gobustan Rock Art Cultural Landscape 767
Walled City of Baku with the Shirvanshah's
  Palace and Maiden Tower 607

Bahrain
Qal'at al-Bahrain – Ancient Harbour and
  Capital of Dilmun 746

Bangladesh
Historic Mosque City of Bagerhat 228
Ruins of the Buddhist Vihara at Paharpur 222
The Sundarbans 512

Belarus
Architectural, Residential and Cultural
  Complex of the Radziwill Family at
  Nesvizh 748
Belovezhskaya Pushcha / Białowieża Forest 61
Mir Castle Complex 635
Struve Geodetic Arc 747

Belgium
Belfries of Belgium and France 575
Flemish Béguinages 562
Historic Centre of Brugge 650
La Grand-Place, Brussels 555
Major Town Houses of the Architect Victor
  Horta (Brussels) 648
Neolithic Flint Mines at Spiennes (Mons) 652
Notre-Dame Cathedral in Tournai 655
Plantin-Moretus House-Workshops-Museum
  Complex 745
The Four Lifts on the Canal du Centre and
  their Environs, La Louvière and Le Roeulx
  (Hainault) 554

Belize
Belize Barrier Reef Reserve System 483

Benin
Royal Palaces of Abomey 232

Bolivia
City of Potosí 289
Fuerte de Samaipata 556
Historic City of Sucre 367
Jesuit Missions of the Chiquitos 348
Noel Kempff Mercado National Park 636
Tiwanaku: Spiritual and Political Centre of the
  Tiwanaku Culture 634

Bosnia-Herzegovina
Mehmed Paša Sokolović Bridge in
  Višegrad 779
Old Bridge Area of the Old City of
  Mostar 734

Botswana
Tsodilo 668

Brazil
Atlantic Forest South-East Reserves 606
Brasilia 280
Brazilian Atlantic Islands: Fernando de
  Noronha and Atol das Rocas Reserves 668
Central Amazon Conservation Complex 624
Cerrado Protected Areas: Chapada dos
  Veadeiros and Emas National Parks 681

Discovery Coast Atlantic Forest
  Reserves 579
Historic Centre of Salvador de Bahia 208
Historic Centre of São Luís 537
Historic Centre of the Town of
  Diamantina 588
Historic Centre of the Town of Goiás 674
Historic Centre of the Town of Olinda 150
Historic Town of Ouro Preto 108
Iguaçu National Park 235
Jesuit Missions of the Guaranis 159
Pantanal Conservation Area 656
Sanctuary of Bom Jesus do Congonhas 223
Serra da Capivara National Park 374

Bulgaria
Ancient City of Nessebar 165
Boyana Church, Sofia 60
Madara Rider 69
Pirin National Park 179
Rila Monastery 178
Rock-Hewn Churches of Ivanovo 69
Srebarna Nature Reserve 168
Thracian Tomb of Kazanlak 49
Thracian Tomb of Sveshtari 234

Cambodia
Angkor 390
Temple of Preah Vihear 789

Cameroon
Dja Faunal Reserve 308

Canada
Canadian Rocky Mountain Parks 194
Dinosaur Provincial Park 48
Gros Morne National Park 302
Head-Smashed-In Buffalo Jump 118
Historic District of Old Québec 206
Joggins Fossil Cliffs 796
Kluane / Wrangell-St Elias / Glacier Bay /
  Tatshenshini-Alsek 52
L'Anse aux Meadows National Historic
  Site 37
Miguasha National Park 599
Nahanni National Park 38
Old Town Lunenburg 463
Rideau Canal 773
SGang Gwaay 133

Waterton Glacier International Peace
  Park 450
Wood Buffalo National Park 170

Central African Republic
Manovo-Gounda St Floris National Park 313

Chile
Churches of Chiloé 620
Historic Quarter of the Seaport City of
  Valparaíso 696
Humberstone and Santa Laura Saltpeter
  Works 736
Rapa Nui National Park 454
Sewell Mining Town 762

China
Ancient Building Complex in the Wudang
  Mountains 429
Ancient City of Ping Yao 540
Ancient Villages in Southern Anhui–Xidi and
  Hongcun 653
Capital Cities and Tombs of the Ancient
  Koguryo Kingdom 722
Classical Gardens of Suzhou 518
Dazu Rock Carvings 584
Fujian Tulou 788
Historic Centre of Macao 737
Historic Ensemble of the Potala Palace,
  Lhasa 425
Huanglong Scenic and Historic Interest
  Area 395
Imperial Palaces of the Ming and Qing
  Dynasties in Beijing and Shenyang 277
Imperial Tombs of the Ming and Qing
  Dynasties 616
Jiuzhaigou Valley Scenic and Historic Interest
  Area 380
Kaiping Diaolou and Villages 774
Longmen Grottoes 654
Lushan National Park 484
Mausoleum of the First Qin Emperor 300
Mogao Caves 298
Mount Emei Scenic Area, including Leshan
  Giant Buddha Scenic Area 481
Mount Huangshan 357
Mount Qingcheng and the Dujiangyan
  Irrigation System 648
Mount Sanqingshan National Park 796
Mount Taishan 297

Mount Wuyi 598
Mountain Resort and its Outlying Temples,
  Chengde 436
Old Town of Lijiang 510
Peking Man Site at Zhoukoudian 298
Sichuan Giant Panda Sanctuaries – Wolong,
  Mt Siguniang and Jiajin Mountains 751
South China Karst 768
Summer Palace and Imperial Garden in
  Beijing 552
Temple and Cemetery of Confucius and the
  Kong Family Mansion in Qufu 433
Temple of Heaven: an Imperial Sacrificial Altar
  in Beijing 559
The Great Wall 294
Three Parallel Rivers of Yunnan Protected
  Areas 700
Wulingyuan Scenic and Historic Interest
  Area 382
Yin Xu 759
Yungang Grottoes 672

Colombia
Historic Centre of Santa Cruz de Mompox 473
Los Katíos National Park 426
Malpelo Fauna and Flora Sanctuary 756
National Archeological Park of
  Tierradentro 473
Port, Fortresses and Group of Monuments,
  Cartagena 184
San Agustín Archeological Park 458

Congo, Democratic Republic of the
Garamba National Park 94
Kahuzi-Biega National Park 97
Okapi Wildlife Reserve 478
Salonga National Park 186
Virunga National Park 45

Costa Rica
Area de Conservación Guanacaste 596
Cocos Island National Park 516
Talamanca Range-La Amistad Reserves /
  La Amistad National Park 162

Côte d'Ivoire
Comoé National Park 153
Mount Nimba Strict Nature Reserve 126
Taï National Park 139

Croatia

Cathedral of St James in Šibenik **627**
Episcopal Complex of the Euphrasian Basilica
  in the Historic Centre of Poreč **516**
Historic City of Trogir **533**
Historical Complex of Split with the Palace of
  Diocletian **82**
Old City of Dubrovnik **86**
Plitvice Lakes National Park **57**
Stari Grad Plain **790**

Cuba

Alejandro de Humboldt National Park **672**
Archaeological Landscape of the First Coffee
  Plantations in the South-East of Cuba **649**
Desembarco del Granma National Park **583**
Historic Centre of Camagüey **785**
Old Havana and its Fortifications **146**
San Pedro de la Roca Castle, Santiago de
  Cuba **548**
Trinidad and the Valley de los Ingenios **323**
Urban Historic Centre of Cienfuegos **749**
Viñales Valley **593**

Cyprus

Choirokoitia **554**
Painted Churches in the Troodos
  Region **210**
Paphos **101**

Czech Republic

Gardens and Castle at Kroměříž **565**
Historic Centre of Český Krumlov **386**
Historic Centre of Prague **384**
Historic Centre of Telč **392**
Holašovice Historical Village
  Reservation **551**
Holy Trinity Column in Olomouc **637**
Jewish Quarter and St Procopius' Basilica in
  Třebíč **704**
Kutná Hora: Historical Town Centre with the
  Church of St Barbara and the Cathedral of
  Our Lady at Sedlec **469**
Lednice-Valtice Cultural Landscape **493**
Litomyšl Castle **588**
Pilgrimage Church of St John of Nepomuk at
  Zelená Hora **443**
Tugendhat Villa in Brno **681**

Denmark

Ilulissat Icefjord **708**
Jelling Mounds, Runic Stones and Church **426**
Kronborg Castle **608**
Roskilde Cathedral **452**

Dominica

Morne Trois Pitons National Park **515**

Dominican Republic

Colonial City of Santo Domingo **343**

Ecuador

City of Quito **32**
Galápagos Islands **26**
Historic Centre of Santa Ana de los Ríos de
  Cuenca **576**
Sangay National Park **171**

Egypt

Abu Mena **49**
Ancient Thebes with its Necropolis **74**
Historic Cairo **78**
Memphis and its Necropolis – the Pyramid
  Fields from Giza to Dahshur **70**
Nubian Monuments from Abu Simbel to
  Philae **56**
St Catherine Area **688**
Wadi Al-Hitan (Whale Valley) **738**

El Salvador

Joya de Cerén Archaeological Site **416**

Estonia

Historic Centre (Old Town) of Tallinn **509**
Struve Geodetic Arc **747**

Ethiopia

Aksum **98**
Fasil Ghebbi, Gondar Region **60**
Harar Jugol, the Fortified Historic Town **752**
Lower Valley of the Awash **102**
Lower Valley of the Omo **106**
Rock-Hewn Churches, Lalibela **32**
Simien National Park **37**
Tiya **106**

Finland

Bronze Age Burial Site of
  Sammallahdenmäki **599**
Fortress of Suomenlinna **378**
Kvarken Archipelago **630**
Old Rauma **368**
Petäjävesi Old Church **436**
Struve Geodetic Arc **747**
Verla Groundwood and Board Mill **486**

Former Yugoslav Republic of Macedonia

Natural and Cultural Heritage of the
  Ohrid region **77**

France

Abbey Church of Saint-Savin sur Gartempe **168**
Amiens Cathedral **117**
Arles, Roman and Romanesque
  Monuments **131**
Belfries of Belgium and France **575**
Bordeaux, Port of the Moon **775**
Bourges Cathedral **381**
Canal du Midi **480**
Cathedral of Notre-Dame, Former Abbey of
  Saint-Rémi and Palace of Tau, Reims **372**
Chartres Cathedral **65**
Cistercian Abbey of Fontenay **121**
Fortifications of Vauban **794**
Gulf of Porto: Calanche of Piana, Gulf of
  Girolata, Scandola Reserve **174**
Historic Centre of Avignon: Papal Palace,
  Episcopal Ensemble and Avignon Bridge **444**
Historic Fortified City of Carcassonne **514**
Historic Site of Lyons **561**
Jurisdiction of Saint-Emilion **600**
Lagoons of New Caledonia: Reef Diversity
  and Associated Ecosystems **788**
Le Havre, the city rebuilt by Auguste Perret **738**
Loire Valley between Sully-sur-Loire and
  Chalonnes **622**
Mont-Saint-Michel and its Bay **58**
Palace and Park of Fontainebleau **113**
Palace and Park of Versailles **62**
Paris, Banks of the Seine **360**
Place Stanislas, Place de la Carrière and Place
  d'Alliance in Nancy **169**
Pont du Gard (Roman Aqueduct) **229**
Prehistoric Sites and Decorated Caves of the
  Vézère Valley **44**
Provins, Town of Medieval Fairs **669**
Pyrénées - Mont Perdu **508**

Roman Theatre and its Surroundings and the 'Triumphal Arch' of Orange **120**
Routes of Santiago de Compostela in France **557**
Royal Saltworks of Arc-et-Senans **153**
Strasbourg – Grande Île **334**
Vézelay, Church and Hill **66**

Gabon
Ecosystem and Relict Cultural Landscape of Lopé-Okanda **781**

Georgia
Bagrati Cathedral and Gelati Monastery **435**
Historical Monuments of Mtskheta **437**
Upper Svaneti **481**

Germany
Aachen Cathedral **33**
Abbey and Altenmünster of Lorsch **379**
Bauhaus and its Sites in Weimar and Dessau **484**
Berlin Modernism Housing Estates **784**
Castles of Augustusburg and Falkenlust at Brühl **193**
Classical Weimar **549**
Collegiate Church, Castle, and Old Town of Quedlinburg **441**
Cologne Cathedral **474**
Dresden Elbe Valley **709**
Frontiers of the Roman Empire **273**
Garden Kingdom of Dessau-Wörlitz **610**
Hanseatic City of Lübeck **281**
Historic Centres of Stralsund and Wismar **690**
Luther Memorials in Eisleben and Wittenberg **503**
Maulbronn Monastery Complex **407**
Messel Pit Fossil Site **456**
Mines of Rammelsberg and Historic Town of Goslar **387**
Monastic Island of Reichenau **640**
Museumsinsel (Museum Island), Berlin **578**
Muskauer Park / Park Muzakowski **719**
Old Town of Regensburg with Stadtamhof **753**
Palaces and Parks of Potsdam and Berlin **346**
Pilgrimage Church of Wies **176**
Roman Monuments, Cathedral of St Peter and Church of Our Lady in Trier **251**
St Mary's Cathedral and St Michael's Church at Hildesheim **225**

Speyer Cathedral **128**
Town of Bamberg **406**
Town Hall and Roland on the Marketplace of Bremen **727**
Upper Middle Rhine Valley **684**
Völklingen Ironworks **441**
Wartburg Castle **589**
Würzburg Residence with the Court Gardens and Residence Square **132**
Zollverein Coal Mine Industrial Complex in Essen **667**

Ghana
Asante Traditional Buildings **94**
Forts and Castles, Volta, Greater Accra, Central and Western Regions **44**

Greece
Acropolis, Athens **282**
Archaeological Site of Aigai (modern name Vergina) **504**
Archaeological Site of Delphi **262**
Archaeological Site of Mystras **341**
Archaeological Site of Olympia **336**
Archaeological Sites of Mycenae and Tiryns **574**
Delos **353**
Historic Centre (Chorá) with the Monastery of Saint John 'the Theologian' and the Cave of the Apocalypse on the Island of Pátmos **603**
Medieval City of Rhodes **324**
Meteora **332**
Monasteries of Daphni, Hosios Loukas and Nea Moni of Chios Greece **342**
Mount Athos **309**
Old Town of Corfu **771**
Paleochristian and Byzantine Monuments of Thessalonika **313**
Pythagoreion and Heraion of Samos **388**
Sanctuary of Asklepios at Epidaurus **315**
Temple of Apollo Epicurius at Bassae **240**

Guatemala
Antigua Guatemala **73**
Archaeological Park and Ruins of Quirigua **130**
Tikal National Park **88**

Guinea
Mount Nimba Strict Nature Reserve **126**

Haiti
National History Park – Citadel, Sans Souci, Ramiers **139**

Holy See
Historic Centre of Rome, the Properties of the Holy See in that City Enjoying Extraterritorial Rights and San Paolo Fuori le Mura **92**
Vatican City **180**

Honduras
Maya Site of Copán **90**
Río Plátano Biosphere Reserve **142**

Hungary
Budapest, including the Banks of the Danube, the Buda Castle Quarter and Andrássy Avenue **274**
Caves of Aggtelek Karst and Slovak Karst **456**
Early Christian Necropolis of Pécs (Sopianae) **619**
Fertö / Neusiedlersee Cultural Landscape **664**
Hortobágy National Park - the Puszta **583**
Millenary Benedictine Abbey of Pannonhalma and its Natural Environment **479**
Old Village of Hollókő and its Surroundings **293**
Tokaj Wine Region Historic Cultural Landscape **686**

Iceland
Surtsey **789**
Þingvellir National Park **710**

India
Agra Fort **161**
Ajanta Caves **156**
Buddhist Monuments at Sanchi **342**
Champaner-Pavagadh Archaeological Park **730**
Chhatrapati Shivaji Terminus (formerly Victoria Terminus) **723**
Churches and Convents of Goa **237**
Elephanta Caves **271**
Ellora Caves **160**
Fatehpur Sikri **247**
Great Living Chola Temples **276**
Group of Monuments at Hampi **252**

India (continued)

Group of Monuments at Mahabalipuram **204**
Group of Monuments at Pattadakal **308**
Humayun's Tomb, Delhi **403**
Kaziranga National Park **224**
Keoladeo National Park **216**
Khajuraho Group of Monuments **249**
Mahabodhi Temple Complex at Bodh
  Gaya **688**
Manas Wildlife Sanctuary **215**
Mountain Railways of India **576**
Nanda Devi and Valley of Flowers
  National Parks **333**
Qutb Minar and its Monuments, Delhi **413**
Red Fort Complex **765**
Rock Shelters of Bhimbetka **694**
Sun Temple, Konârak **183**
Sundarbans National Park **271**
Taj Mahal **154**

Indonesia

Borobudur Temple Compounds **376**
Komodo National Park **370**
Lorentz National Park **604**
Prambanan Temple Compounds **373**
Sangiran Early Man Site **503**
Tropical Rainforest Heritage of Sumatra **712**
Ujung Kulon National Park **374**

Iran, Islamic Republic of

Armenian Monastic Ensembles of Iran **792**
Bam and its Cultural Landscape **710**
Bisotun **762**
Meidan Emam, Esfahan **38**
Pasargadae **714**
Persepolis **68**
Soltaniyeh **746**
Takht-e Soleyman **691**
Tchogha Zanbil **88**

Iraq

Ashur (Qal'at Sherqat) **697**
Hatra **215**
Samarra Archaeological City **767**

Ireland

Archaeological Ensemble of the Bend of the
  Boyne **401**
Skellig Michael **492**

Israel

Bahá'í Holy Places in Haifa and the Western
  Galilee **793**
Biblical Tels – Megiddo, Hazor,
  Beer Sheba **739**
Incense Route – Desert Cities in the Negev **740**
Masada **676**
Old City of Acre **680**
White City of Tel-Aviv – The Modern
  Movement **694**

Italy

Archaeological Area of Agrigento **542**
Archaeological Area and the Patriarchal
  Basilica of Aquileia **565**
Archaeological Areas of Pompei, Herculaneum
  and Torre Annunziata **524**
Assisi, the Basilica of San Francesco and
  Other Franciscan Sites **632**
Botanical Garden (Orto Botanico),
  Padua **532**
Castel del Monte **506**
Cathedral, Torre Civica and Piazza Grande,
  Modena **520**
Church and Dominican Convent of Santa
  Maria delle Grazie with 'The Last Supper'
  by Leonardo da Vinci **95**
Cilento and Vallo di Diano National Park with
  the Archeological sites of Paestum and Velia,
  and the Certosa di Padula **558**
City of Verona **618**
City of Vicenza and the Palladian Villas of the
  Veneto **428**
Costiera Amalfitana **538**
Crespi d'Adda **465**
Early Christian Monuments of Ravenna **487**
Eighteenth-Century Royal Palace at Caserta
  with the Park, the Aqueduct of Vanvitelli, and
  the San Leucio Complex **527**
Etruscan Necropolises of Cerveteri and
  Tarquinia **729**
Ferrara, City of the Renaissance, and its Po
  Delta **453**
Genoa: Le Strade Nuove and the system of the
  Palazzi dei Rolli **759**
Historic Centre of the City of Pienza **495**
Historic Centre of Florence **134**
Historic Centre of Naples **448**
Historic Centre of Rome, the Properties of the
  Holy See in that City Enjoying
  Extraterritorial Rights and San Paolo Fuori
  le Mura **92**

Historic Centre of San Gimignano **358**
Historic Centre of Siena **460**
Historic Centre of Urbino **567**
Isole Eolie (Aeolian Islands) **625**
Late Baroque Towns of the Val di Noto
  (South-Eastern Sicily) **690**
Mantua and Sabbioneta **797**
Piazza del Duomo, Pisa **265**
Portovenere, Cinque Terre, and the Islands
  (Palmaria, Tino and Tinetto) **513**
Residences of the Royal House of Savoy **541**
Rhaetian Railway in the Albula / Bernina
  Landscapes **784**
Rock Drawings in Valcamonica **76**
Sacri Monti of Piedmont and
  Lombardy **695**
Su Nuraxi di Barumini **546**
Syracuse and the Rocky Necropolis of
  Pantalica **750**
The Sassi and the park of the Rupestrian
  Churches of Matera **399**
The Trulli of Alberobello **501**
Val d'Orcia **724**
Venice and its Lagoon **268**
Villa Adriana, Tivoli **595**
Villa d'Este, Tivoli **673**
Villa Romana del Casale **546**

Japan

Buddhist Monuments in the
  Horyu-ji Area **415**
Gusuku Sites and Related Properties of the
  Kingdom of Ryukyu **643**
Himeji-jo **409**
Hiroshima Peace Memorial
  (Genbaku Dome) **477**
Historic Monuments of Ancient Kyoto
  (Kyoto, Uji and Otsu Cities) **438**
Historic Monuments of Ancient Nara **560**
Historic Villages of Shirakawa-go and
  Gokayama **462**
Itsukushima Shinto Shrine **486**
Iwami Ginzan Silver Mine and its Cultural
  Landscape **764**
Sacred Sites and Pilgrimage Routes in the Kii
  Mountain Range **726**
Shirakami-Sanchi **415**
Shiretoko **748**
Shrines and Temples of Nikko **604**
Yakushima **400**

Jordan
Petra 212
Quseir Amra 221
Um er-Rasas (Kastrom Mefa'a) 713

Kazakhstan
Mausoleum of Khoja Ahmed Yasawi 707
Petroglyphs within the Archaeological
    Landscape of Tamgaly 729
Saryarka – Steppe and Lakes of Northern
    Kazakhstan 786

Kenya
Lake Turkana National Parks 517
Lamu Old Town 682
Mount Kenya National Park / Natural
    Forest 508
Sacred Mijikenda Kaya Forests 782

Korea, Democratic People's Republic of
Complex of Koguryo Tombs 715

Korea, Republic of
Changdeokgung Palace Complex 521
Gochang, Hwasun and Ganghwa
    Dolmen Sites 626
Gyeongju Historic Areas 642
Haeinsa Temple Janggyeong Panjeon, the
    Depositories for the Tripitaka Koreana
    Woodblocks 462
Hwaseong Fortress 536
Jeju Volcanic Island and Lava Tubes 780
Jongmyo Shrine 468
Seokguram Grotto and Bulguksa
    Temple 457

Lao People's Democratic Republic (Laos)
Town of Luang Prabang 464
Vat Phou and Associated Ancient Settlements
    within the Champasak Cultural
    Landscape 679

Latvia
Historic Centre of Riga 544
Struve Geodetic Arc 747

Lebanon
Anjar 205
Baalbek 191
Byblos 192
Ouadi Qadisha (the Holy Valley) and the
    Forest of the Cedars of God
    (Horsh Arz el-Rab) 569
Tyre 197

Libyan Arab Jamahiriya (Libya)
Archaeological Site of Cyrene 136
Archaeological Site of Leptis Magna 144
Archaeological Site of Sabratha 149
Old Town of Ghadamès 261
Rock-Art Sites of Tadrart Acacus 234

Lithuania
Curonian Spit 615
Kernavė Archaeological Site (Cultural Reserve
    of Kernavė) 715
Struve Geodetic Arc 747
Vilnius Historic Centre 421

Luxembourg
City of Luxembourg: its Old Quarters and
    Fortifications 424

Madagascar
Rainforests of the Atsinanana 778
Royal Hill of Ambohimanga 663
Tsingy de Bemaraha Strict
    Nature Reserve 352

Malawi
Chongoni Rock-Art Area 756
Lake Malawi National Park 190

Malaysia
Gunung Mulu National Park 638
Kinabalu Park 631
Melaka and George Town, Historic Cities of
    the Straits of Malacca 790

Mali
Cliff of Bandiagara (Land of the Dogons) 340
Old Towns of Djenné 310

Timbuktu 321
Tomb of Askia 725

Malta
City of Valletta 93
Hal Saflieni Hypogeum 98
Megalithic Temples of Malta 96

Mauritania
Ancient Ksour of Ouadane, Chinguetti, Tichitt
    and Oualata 478
Banc d'Arguin National Park 340

Mauritius
Aapravasi Ghat 760
Le Morne Cultural Landscape 792

Mexico
Agave Landscape and Ancient Industrial
    Facilities of Tequila 754
Ancient Maya City of Calakmul,
    Campeche 687
Archaeological Monuments Zone of
    Xochicalco 605
Archaeological Zone of Paquimé, Casas
    Grandes 563
Central University City Campus of the
    Universidad Nacional Autónoma de México
    (UNAM) 770
Earliest 16th-Century Monasteries on the
    Slopes of Popocatepetl 432
Franciscan Missions in the Sierra Gorda of
    Querétaro 703
Historic Centre of Mexico City and
    Xochimilco 284
Historic Centre of Morelia 369
Historic Centre of Oaxaca and Archaeological
    Site of Monte Albán 286
Historic Centre of Puebla 284
Historic Centre of Zacatecas 419
Historic Fortified Town of Campeche 577
Historic Monuments Zone of Querétaro 496
Historic Monuments Zone of Tlacotalpan 548
Historic Town of Guanajuato and Adjacent
    Mines 318
Hospicio Cabañas, Guadalajara 534
Islands and Protected Areas of the Gulf of
    California 733
Luis Barragán House and Studio 722
Monarch Butterfly Biosphere Reserve 791

Mexico (continued)
Pre-Hispanic City of Chichen-Itza **330**
Pre-Hispanic City of El Tajin **394**
Pre-Hispanic City and National Park of
Palenque **299**
Pre-Hispanic City of Teotihuacan **288**
Pre-Hispanic Town of Uxmal **489**
Protective Town of San Miguel and the
Sanctuary of Jesús Nazareno de
Atotonilco **781**
Rock Paintings of the Sierra de San
Francisco **419**
Sian Ka'an **267**
Whale Sanctuary of El Vizcaino **401**

Moldova, Republic of
Struve Geodetic Arc **747**

Mongolia
Orkhon Valley Cultural Landscape **719**
Uvs Nuur Basin **693**

Montenegro
Durmitor National Park **107**
Natural and Culturo-Historical Region of
Kotor **84**

Morocco
Archaeological Site of Volubilis **543**
Historic City of Meknes **494**
Ksar of Ait-Ben-Haddou **305**
Medina of Essaouira (formerly Mogador) **666**
Medina of Fez **119**
Medina of Marrakesh **218**
Medina of Tétouan (formerly known as
Titawin) **522**
Portuguese City of Mazagan
(El Jadida) **717**

Mozambique
Island of Mozambique **365**

Namibia
Twyfelfontein or /Ui-//aes **770**

Nepal
Kathmandu Valley **50**
Lumbini, the Birthplace of the Lord
Buddha **512**
Royal Chitwan National Park **196**
Sagarmatha National Park **85**

Netherlands
Defence Line of Amsterdam **488**
Droogmakerij de Beemster
(Beemster Polder) **580**
Historic Area of Willemstad, Inner City and
Harbour, Netherlands Antilles **528**
Ir. D.F. Woudagemaal (D.F. Wouda Steam
Pumping Station) **566**
Mill Network at Kinderdijk-Elshout **523**
Rietveld Schröderhuis (Rietveld Schröder
House) **636**
Schokland and Surroundings **449**

New Zealand
New Zealand Sub-Antarctic Islands **556**
Te Wahipounamu – South West New
Zealand **354**
Tongariro National Park **349**

Nicaragua
Ruins of León Viejo **619**

Niger
Aïr and Ténéré Natural Reserves **367**
W National Park of Niger **485**

Nigeria
Osun-Osogbo Sacred Grove **741**
Sukur Cultural Landscape **587**

Norway
Bryggen **43**
Røros Mining Town **104**
Rock Art of Alta **232**
Struve Geodetic Arc **747**
Urnes Stave Church **81**
Vegaøyan – the Vega Archipelago **716**
West Norwegian Fjords – Geirangerfjord
and Nærøyfjord **742**

Oman
Aflaj Irrigation Systems of Oman **760**
Archaeological Sites of Bat, Al-Khutm and
Al-Ayn **322**
Bahla Fort **303**
Land of Frankincense **614**

Pakistan
Archaeological Ruins at Moenjodaro **102**
Buddhist Ruins at Takht-i-Bahi and
Neighbouring City Remains at
Sahr-i-Bahlol **110**
Fort and Shalamar Gardens in Lahore **124**
Historic Monuments of Thatta **118**
Rohtas Fort **532**
Taxila **109**

Panama
Archaeological Site of Panamá Viejo and
Historic District of Panamá **529**
Coiba National Park and its Special Zone of
Marine Protection **745**
Darien National Park **130**
Fortifications on the Caribbean Side of
Panama: Portobelo-San Lorenzo **99**
Talamanca Range-La Amistad Reserves /
La Amistad National Park **162**

Papua New Guinea
Kuk Early Agricultural Site **786**

Paraguay
Jesuit Missions of La Santísima
Trinidad de Paraná and Jesús de
Tavarangue **407**

Peru
Chan Chan Archaeological Zone **238**
Chavín (Archaeological site) **210**
City of Cuzco **158**
Historic Centre of Lima **326**
Historic Sanctuary of Machu Picchu **172**
Historical Centre of the City of Arequipa **617**
Huascarán National Park **214**
Lines and Geoglyphs of Nasca and Pampas de
Jumana **434**
Manú National Park **296**
Río Abiseo National Park **356**

Philippines

Baroque Churches of the Philippines 402
Historic Town of Vigan 582
Puerto-Princesa Subterranean River National
   Park 601
Rice Terraces of the Philippine Cordilleras 466
Tubbataha Reef Marine Park 412

Poland

Auschwitz Birkenau, German Nazi
   Concentration and Extermination Camp
   (1940-1945) 40
Belovezhskaya Pushcha / Białowieża Forest 61
Castle of the Teutonic Order in Malbork 547
Centennial Hall in Wrocław 752
Churches of Peace in Jawor and Swidnica 675
Cracow's Historic Centre 30
Historic Centre of Warsaw 89
Kalwaria Zebrzydowska: the Mannerist
   Architectural and Park Landscape Complex
   and Pilgrimage Park 594
Medieval Town of Toruń 526
Muskauer Park / Park Muzakowski 719
Old City of Zamość 382
Wieliczka Salt Mine 36
Wooden Churches of Southern Little
   Poland 702

Portugal

Alto Douro Wine Region 662
Central Zone of the Town of Angra do
   Heroismo in the Azores 162
Convent of Christ in Tomar 175
Cultural Landscape of Sintra 447
Historic Centre of Évora 258
Historic Centre of Guimarães 682
Historic Centre of Oporto 482
Landscape of the Pico Island Vineyard
   Culture 731
Laurisilva of Madeira 602
Monastery of Alcobaça 337
Monastery of Batalha 177
Monastery of the Hieronymites and Tower of
   Belém in Lisbon 163
Prehistoric Rock-Art Sites in the
   Côa Valley 563

Romania

Churches of Moldavia 412
Dacian Fortresses of the Orastie Mountains 597

Danube Delta 362
Historic Centre of Sighişoara 592
Monastery of Horezu 408
Villages with Fortified Churches in
   Transylvania 404
Wooden Churches of Maramureş 581

Russian Federation

Architectural Ensemble of the Trinity Sergius
   Lavra in Sergiev Posad 420
Central Sikhote-Alin 664
Church of the Ascension, Kolomenskoye 427
Citadel, Ancient City and Fortress Buildings
   of Derbent 698
Cultural and Historic Ensemble of the
   Solovetsky Islands 392
Curonian Spit 615
Ensemble of the Ferrapontov Monastery 643
Ensemble of the Novodevichy Convent 728
Golden Mountains of Altai 557
Historic and Architectural Complex of the
   Kazan Kremlin 646
Historic Centre of Saint Petersburg and
   Related Groups of Monuments 344
Historic Monuments of Novgorod and
   Surroundings 383
Historical Centre of the City of Yaroslavl 744
Kizhi Pogost 356
Kremlin and Red Square, Moscow 350
Lake Baikal 476
Natural System of Wrangel Island Reserve 716
Struve Geodetic Arc 747
Uvs Nuur Basin 693
Virgin Komi Forests 449
Volcanoes of Kamchatka 490
Western Caucasus 601
White Monuments of Vladimir and
   Suzdal 396

Saint Kitts and Nevis

Brimstone Hill Fortress National Park 596

Saint Lucia

Pitons Management Area 711

San Marino

San Marino Historic Centre and
   Mount Titano 782

Saudi Arabia

Al-Hijr Archaeological Site (Madâin Sâlih) 798

Senegal

Djoudj National Bird Sanctuary 121
Island of Gorée 29
Island of Saint-Louis 626
Niokolo-Koba National Park 116
Stone Circles of Senegambia 755

Serbia

Gamzigrad-Romuliana, Palace of Galerius 776
Medieval Monuments in Kosovo 717
Stari Ras and Sopoćani 76
Studenica Monastery 250

Seychelles

Aldabra Atoll 133
Vallée de Mai Nature Reserve 171

Slovakia

Bardejov Town Conservation Reserve 645
Caves of Aggtelek Karst and Slovak Karst 456
Historic Town of Banská Štiavnica and the
   Technical Monuments in its Vicinity 414
Primeval Beech Forests of the Carpathians 772
Spišský Hrad and its Associated Cultural
   Monuments 417
Vlkolínec 418
Wooden Churches of the Slovak part of the
   Carpathian Mountain Area 795

Slovenia

Škocjan Caves 261

Solomon Islands

East Rennell 558

South Africa

Cape Floral Region Protected Areas 720
Fossil Hominid Sites of Sterkfontein,
   Swartkrans, Kromdraai, and Environs 593
iSimangaliso Wetland Park 575
Mapungubwe Cultural Landscape 693
Richtersveld Cultural and Botanical
   Landscape 764

South Africa (continued)
Robben Island 569
uKhahlamba / Drakensberg Park 611
Vredefort Dome 747

Spain
Alhambra, Generalife and Albayzín,
    Granada 202
Aranjuez Cultural Landscape 658
Archaeological Ensemble of Mérida 410
Archaeological Ensemble of Tárraco 621
Archaeological Site of Atapuerca 647
Burgos Cathedral 200
Catalan Romanesque Churches of the
    Vall de Boí 644
Cathedral, Alcázar and Archivo de Indias
    in Seville 266
Cave of Altamira and Paleolithic Cave
    Art of Northern Spain 220
Doñana National Park 430
Garajonay National Park 250
Historic Centre of Córdoba 198
Historic City of Toledo 242
Historic Walled Town of Cuenca 505
Ibiza Biodiversity and Culture 572
La Lonja de la Seda de Valencia 485
Las Médulas 535
Monastery and Site of the Escurial,
    Madrid 193
Monuments of Oviedo and the Kingdom
    of the Asturias 217
Mudéjar Architecture of Aragon 256
Old City of Salamanca 328
Old Town of Ávila with its Extra-Muros
    Churches 233
Old Town of Cáceres 253
Old Town of Segovia and its Aqueduct 209
Palau de la Música Catalana and Hospital de
    Sant Pau, Barcelona 528
Palmeral of Elche 639
Poblet Monastery 366
Pyrénées - Mont Perdu 508
Renaissance Monumental Ensembles of
    Úbeda and Baeza 697
Rock Art of the Mediterranean Basin on the
    Iberian Peninsula 562
Roman Walls of Lugo 649
Route of Santiago de Compostela 397
Royal Monastery of Santa María de
    Guadalupe 416
San Cristóbal de La Laguna 587
San Millán Yuso and Suso Monasteries 535

Santiago de Compostela (Old Town) 211
Teide National Park 766
University and Historic Precinct of Alcalá
    de Henares 568
Vizcaya Bridge 758
Works of Antoni Gaudí 182

Sri Lanka
Ancient City of Polonnaruwa 140
Ancient City of Sigiriya 151
Golden Temple of Dambulla 363
Old Town of Galle and its Fortifications 320
Sacred City of Anuradhapura 138
Sacred City of Kandy 314
Sinharaja Forest Reserve 316

Sudan
Gebel Barkal and the Sites of the
    Napatan Region 695

Suriname
Central Suriname Nature Reserve 658
Historic Inner City of Paramaribo 687

Sweden
Agricultural Landscape of Southern
    Öland 640
Birka and Hovgården 404
Church Village of Gammelstad, Luleå 504
Engelsberg Ironworks 408
Hanseatic Town of Visby 459
High Coast 630
Laponian Area 497
Mining Area of the Great Copper Mountain
    in Falun 667
Naval Port of Karlskrona 553
Rock Carvings in Tanum 432
Royal Domain of Drottningholm 371
Skogskyrkogården 435
Struve Geodetic Arc 747
Varberg Radio Station 725

Switzerland
Benedictine Convent of St John at Müstair 159
Convent of St Gall 176
Lavaux Vineyard Terraces 777
Monte San Giorgio 704
Old City of Berne 157

Rhaetian Railway in the Albula / Bernina
    Landscapes 784
Swiss Alps Jungfrau-Aletsch 660
Swiss Tectonic Arena Sardona 787
Three Castles, Defensive Wall and Ramparts
    of the Market-Town of Bellinzone 610

Syrian Arab Republic (Syria)
Ancient City of Aleppo 236
Ancient City of Bosra 105
Ancient City of Damascus 39
Crac des Chevaliers and Qal'at Salah
    El-Din 761
Site of Palmyra 100

Tanzania
Kilimanjaro National Park 278
Kondoa Rock-Art Sites 755
Ngorongoro Conservation Area 42
Ruins of Kilwa Kisiwani and Ruins of
    Songo Mnara 116
Selous Game Reserve 137
Serengeti National Park 112
Stone Town of Zanzibar 612

Thailand
Ban Chiang Archaeological Site 388
Dong Phayayen-Khao Yai Forest
    Complex 741
Historic City of Ayutthaya 359
Historic Town of Sukhothai and Associated
    Historic Towns 375
Thungyai-Huai Kha Khaeng Wildlife
    Sanctuaries 364

The Gambia
James Island and Related Sites 698
Stone Circles of Senegambia 755

Togo
Koutammakou, the Land of the
    Batammariba 726

Tunisia
Amphitheatre of El Jem 80
Dougga / Thugga 530
Ichkeul National Park 97

Kairouan 335
Medina of Sousse 317
Medina of Tunis 72
Punic Town of Kerkuane and
    its Necropolis 220
Site of Carthage 46

Turkey
Archaeological Site of Troy 550
City of Safranbolu 442
Göreme National Park and the Rock Sites of
    Cappadocia 230
Great Mosque and Hospital of Divriği 225
Hattusha: the Hittite Capital 237
Hierapolis-Pamukkale 329
Historic Areas of Istanbul 226
Nemrut Dağ 292
Xanthos-Letoon 333

Turkmenistan
Kunya-Urgench 750
Parthian Fortresses of Nisa 778
State Historical and Cultural Park
    Ancient Merv 573

Uganda
Bwindi Impenetrable National Park 422
Rwenzori Mountains National Park 430
Tombs of Buganda Kings at Kasubi 663

Ukraine
Kiev: Saint-Sophia Cathedral and Related
    Monastic Buildings, Kiev-Pechersk
    Laura 347
L'viv – the Ensemble of the Historic
    Centre 564
Primeval Beech Forests of the
    Carpathians 772
Struve Geodetic Arc 747

United Kingdom
Blaenavon Industrial Landscape 620
Blenheim Palace 270
Canterbury Cathedral, St Augustine's Abbey,
    and St Martin's Church 327
Castles and Town Walls of King Edward in
    Gwynedd 257
City of Bath 290

Cornwall and West Devon Mining
    Landscape 757
Derwent Valley Mills 675
Dorset and East Devon Coast 671
Durham Castle and Cathedral 246
Frontiers of the Roman Empire 273
Giant's Causeway and Causeway Coast 244
Gough and Inaccessible Islands 468
Heart of Neolithic Orkney 570
Henderson Island 321
Historic Town of St George and Related
    Fortifications, Bermuda 621
Ironbridge Gorge 239
Liverpool – Maritime Mercantile City 707
Maritime Greenwich 507
New Lanark 659
Old and New Towns of Edinburgh 470
Royal Botanic Gardens, Kew 692
St Kilda 241
Saltaire 665
Stonehenge, Avebury and Associated Sites 254
Studley Royal Park including the Ruins of
    Fountains Abbey 248
Tower of London 312
Westminster Palace, Westminster Abbey and
    Saint Margaret's Church 272

Uruguay
Historic Quarter of the City of Colonia del
    Sacramento 472

USA
Cahokia Mounds State Historic Site 151
Carlsbad Caverns National Park 446
Chaco Culture 285
Everglades National Park 53
Grand Canyon National Park 54
Great Smoky Mountains National Park 164
Hawaii Volcanoes National Park 264
Independence Hall 64
Kluane / Wrangell-St Elias / Glacier Bay /
    Tatshenshini-Alsek 52
La Fortaleza and San Juan National Historic
    Site in Puerto Rico 166
Mammoth Cave National Park 126
Mesa Verde National Park 28
Monticello and the University of Virginia in
    Charlottesville 304
Olympic National Park 128
Pueblo de Taos 380
Redwood National Park 103

Statue of Liberty 187
Waterton Glacier International Peace Park 450
Yellowstone National Park 34
Yosemite National Park 184

Uzbekistan
Historic Centre of Bukhara 405
Historic Centre of Shakhrisyabz 629
Itchan Kala 348
Samarkand – Crossroads of Cultures 670

Vanuatu
Chief Roi Mata's Domain 795

Venezuela
Canaima National Park 431
Ciudad Universitaria de Caracas 644
Coro and its Port 398

Vietnam
Complex of Hué Monuments 411
Ha Long Bay 440
Hoi An Ancient Town 586
My Son Sanctuary 579
Phong Nha-Ke Bang National Park 706

Yemen
Historic Town of Zabid 398
Old City of Sana'a 245
Old Walled City of Shibam 152
Socotra Archipelago 785

Zambia
Mosi-oa-Tunya / Victoria Falls 338

Zimbabwe
Great Zimbabwe National Monument 240
Khami Ruins National Monument 257
Mana Pools National Park, Sapi and Chewore
    Safari Areas 186
Matobo Hills 696
Mosi-oa-Tunya / Victoria Falls 338

Index

A

Aachen Cathedral Germany 33
Aapravasi Ghat Mauritius 760
Abbey and Altenmünster of Lorsch
Germany 379
Abbey Church of Saint-Savin sur Gartempe
France 168
Abomey, Royal Palaces of Benin 232
Abu Mena Egypt 49
Abu Simbel, Nubian Monuments Egypt 56
Acre, Old City of Israel 680
Acropolis, Athens Greece 282
Aeolian Islands Italy 625
Aflaj Irrigation Systems of Oman Oman 760
Agave Landscape and Ancient Industrial
Facilities of Tequila Mexico 754
Aggtelek Karst and Slovak Karst, Caves of
Hungary and Slovakia 456
Agra Fort India 161
Agricultural Landscape of Southern Öland
Sweden 640
Agrigento, Archaeological Area of Italy 542
Aigai, Archaeological Site of (modern Vergina)
Greece 504
Aïr et Ténéré Natural Reserves Niger 367
Aït-Ben-Haddou, Ksar of Morocco 305
Ajanta Caves India 156
Aksum Ethiopia 98
Al Qal'a of Beni Hammad Algeria 109
Al-Ayn Oman 322
Al-Hijr Archaeological Site (Madâin Sâlih)
Saudi Arabia 798
Al-Khutm Oman 322
Albayzín, Granada Spain 202
Alberobello, The Trulli of Italy 501
Albula / Bernina Landscape, Rhaetian Railway
in the Italy and Switzerland 784
Alcalá de Henares, University and Historic
Precinct of Spain 568
Alcázar, Seville Spain 266
Alcobaça Monastery of Portugal 337
Aldabra Atoll Seychelles 133
Alejandro de Humboldt National Park
Cuba 672
Aleppo, Ancient City of Syrian Arab
Republic (Syria) 236
Algiers, Kasbah of Algeria 379
Alhambra, Generalife and Albayzín, Granada
Spain 202

Alta, Rock Art of Norway 232
Altai, Golden Mountains of Russian
Federation 557
Altamira, Cave of Spain 220
Alto Douro Wine Region Portugal 662
Ambohimanga, Royal Hill of
Madagascar 663
Amiens Cathedral France 117
Amphitheatre of El Jem Tunisia 80
Amsterdam, Defence Line of
Netherlands 488
Ancient Building Complex in the Wudang
Mountains China 429
Ancient City of Aleppo Syrian Arab
Republic (Syria) 236
Ancient City of Bosra Syrian Arab
Republic (Syria) 105
Ancient City of Damascus Syrian Arab
Republic (Syria) 39
Ancient City of Nessebar Bulgaria 165
Ancient City of Ping Yao China 540
Ancient City of Polonnaruwa Sri Lanka 140
Ancient City of Sigiriya Sri Lanka 151
Ancient Ksour of Ouadane, Chinguetti, Tichitt
and Oualata Mauritania 478
Ancient Maya City of Calakmul, Campeche
Mexico 687
Ancient Merv Turkmenistan 573
Ancient Thebes with its Necropolis Egypt 74
Ancient Villages in Southern Anhui–Xidi and
Hongcun China 653
Andrássy Avenue, Budapest Hungary 274
Angkor Cambodia 390
Angra do Heroismo in the Azores, Central
Zone of the Town of Portugal 162
Anjar Lebanon 205
Antigua Guatemala Guatemala 73
Anuradhapura, Sacred City of Sri Lanka 138
Apocalypse, Cave of the Greece 603
Aquileia, Archaeological Area and the
Patriarchal Basilica of Italy 565
Aragon, Mudéjar Architecture of Spain 256
Aranjuez Cultural Landscape Spain 658
Arc-et-Senans, Royal Saltworks of France 153
Archaeological Area of Agrigento Italy 542
Archaeological Area and the Patriarchal
Basilica of Aquileia Italy 565
Archaeological Areas of Pompei, Herculaneum
and Torre Annunziata Italy 524
Archaeological Ensemble of the Bend of the
Boyne Ireland 401
Archaeological Ensemble of Mérida Spain 410
Archaeological Ensemble of Tárraco Spain 621

Archaeological Landscape of the First Coffee
Plantations in the South-East Cuba 649
Archaeological Monuments Zone of
Xochicalco Mexico 605
Archaeological Park and Ruins of Quirigua
Guatemala 130
Archaeological Ruins at Moenjodaro
Pakistan 102
Archaeological Site of Aigai (modern name
Vergina) Greece 504
Archaeological Site of Atapuerca Spain 647
Archaeological Site of Cyrene Libyan Arab
Jamahiriya (Libya) 136
Archaeological Site of Delphi Greece 262
Archaeological Site of Leptis Magna Libyan
Arab Jamahiriya (Libya) 144
Archaeological Site of Mystras Greece 341
Archaeological Site of Panamá Viejo and
Historic District of Panamá Panama 529
Archaeological Site of Olympia Greece 336
Archaeological Site of Sabratha Libyan Arab
Jamahiriya (Libya) 149
Archaeological Site of Troy Turkey 550
Archaeological Site of Volubilis Morocco 543
Archaeological Sites of Bat, Al-Khutm and
Al-Ayn Oman 322
Archaeological Sites of Mycenae and Tiryns
Greece 574
Archaeological Zone of Paquimé, Casas
Grandes Mexico 563
Architectural Ensemble of the Trinity Sergius
Lavra in Sergiev Posad Russian
Federation 420
Architectural, Residential and Cultural
Complex of the Radziwill Family at
Nesvizh Belarus 748
Archivo de Indias, Seville Spain 266
Area de Conservación Guanacaste
Costa Rica 596
Arequipa, Historical Centre of the City of
Peru 617
Arles, Roman and Romanesque Monuments
France 131
Armenian Monastic Ensembles of Iran
Islamic Republic of Iran 792
Asante Traditional Buildings Ghana 94
Ashur (Qal'at Sherqat) Iraq 697
Askia, Tomb of Mali 725
Asklepios at Epidaurus, Sanctuary of
Greece 315
Assisi, the Basilica of San Francesco and
Other Franciscan Sites Italy 632
Asturias, Kingdom of the Spain 217

Atapuerca, Archaeological Site of Spain **647**
Athens, Acropolis Greece **282**
Athos, Mount Greece **309**
Atlantic Forest South-East Reserves
   Brazil **606**
Atol das Rocas Reserve Brazil **668**
Atsinanana, Rainforests of the
   Madagascar **778**
Augustusburg and Falkenlust at Brühl,
   Castles of Germany **193**
Auschwitz Birkenau, German Nazi
   Concentration and Extermination Camp
   (1940-1945) Poland **40**
Australian Fossil Mammal Sites
   (Riversleigh / Naracoorte) Australia **429**
Avebury United Kingdom **254**
Avignon, Historic Centre of France **444**
Ávila with its Extra-Muros Churches,
   Old Town of Spain **233**
Awash, Lower Valley of the Ethiopia **102**
Ayutthaya, Historic City of Thailand **359**

B
Baalbek Lebanon **191**
Baeza, Renaissance Monumental Ensembles
   of Spain **697**
Bagerhat, Historic Mosque City of
   Bangladesh **228**
Bagrati Cathedral and Gelati Monastery
   Georgia **435**
Bahá'í Holy Places in Haifa and the Western
   Galilee Israel **793**
Bahla Fort Oman **303**
Baikal, Lake Russian Federation **476**
Baku, Walled City of Azerbaijan **607**
Bam and its Cultural Landscape Islamic
   Republic of Iran **710**
Bamberg, Town of Germany **406**
Bamiyan Valley, Cultural Landscape and
   Archaeological Remains of the
   Afghanistan **703**
Ban Chiang Archaeological Site Thailand **388**
Banc d'Arguin National Park
   Mauritania **340**
Banská Štiavnica, Historic Town of
   Slovakia **414**
Barcelona, Palau de la Música Catalana and
   Hospital de Sant Pau Spain **528**
Bardejov Town Conservation Reserve
   Slovakia **645**
Baroque Churches of the Philippines
   Philippines **402**
Barumini, Su Nuraxi di Italy **546**

Bassae, Temple of Apollo Epicurius at
   Greece **240**
Bat Oman **322**
Batalha, Monastery of Portugal **177**
Batammariba, The Land of the Togo **726**
Bath, City of United Kingdom **290**
Bauhaus and its Sites in Weimar and Dessau
   Germany **484**
Beemster Polder Netherlands **580**
Beer Sheba, Tel Israel **739**
Béguinages, Flemish Belgium **562**
Belfries of Belgium and France Belgium and
   France **575**
Belize Barrier Reef Reserve System Belize **483**
Bellinzone Switzerland **610**
Belovezhskaya Pushcha / Białowieża Forest
   Belarus and Poland **61**
Benedictine Convent of St John at Müstair
   Switzerland **159**
Berat, Historic Centre of Albania **732**
Berlin Modernism Housing Estates
   Germany **784**
Berlin, Palaces and Parks of Germany **346**
Bermuda, Historic Town of St George and
   Related Fortifications United Kingdom **621**
Berne, Old City of Switzerland **157**
Bernina Landscape, Rhaetian Railway in the
   Italy and Switzerland **784**
Bhimbetka, Rock Shelters of India **694**
Białowieża Forest Belarus and Poland **61**
Biblical Tels – Megiddo, Hazor, Beer Sheba
   Israel **739**
Birka and Hovgården Sweden **404**
Bisotun Islamic Republic of Iran **762**
Blaenavon Industrial Landscape United
   Kingdom **620**
Blenheim Palace United Kingdom **270**
Bodh Gaya, Mahabodhi Temple Complex at
   India **688**
Bom Jesus do Congonhas, Sanctuary of
   Brazil **223**
Bordeaux, Port of the Moon France **775**
Borobudur Temple Compounds
   Indonesia **376**
Bosra, Ancient City of Syrian Arab
   Republic (Syria) **105**
Botanical Garden (Orto Botanico), Padua
   Italy **532**
Bourges Cathedral France **381**
Boyana Church, Sofia Bulgaria **60**
Boyne, Archaeological Ensemble
   of the Bend of the Ireland **401**
Brasilia Brazil **280**

Brazilian Atlantic Islands: Fernando de
   Noronha and Atol das Rocas Reserves
   Brazil **668**
Bremen, Town Hall and Roland on the
   Marketplace of Germany **727**
Brimstone Hill Fortress National Park
   Saint Kitts and Nevis **596**
Brno, Tugendhat Villa in Czech Republic **681**
Bronze Age Burial Site of Sammallahdenmäki
   Finland **599**
Brugge, Historic Centre of Belgium **650**
Bryggen Norway **43**
Buda Castle Quarter, Budapest
   Hungary **274**
Budapest, including the Banks of the Danube,
   the Buda Castle Quarter and Andrássy
   Avenue Hungary **274**
Buddhist Monuments at Sanchi India **342**
Buddhist Monuments in the Horyu-ji Area
   Japan **415**
Buddhist Ruins at Takht-i-Bahi and
   Neighbouring City Remains at
   Sahr-i-Bahlol Pakistan **110**
Buganda Kings, Tombs of Uganda **663**
Bukhara, Historic Centre of Uzbekistan **405**
Bulguksa Temple Republic of Korea **457**
Burgos Cathedral Spain **200**
Butrint Albania **389**
Bwindi Impenetrable National Park
   Uganda **422**
Byblos Lebanon **192**

C
Cáceres, Old Town of Spain **253**
Cahokia Mounds State Historic Site USA **151**
Cairo, Historic Egypt **78**
Calakmul, Ancient Maya City of Mexico **687**
Camagüey, Historic Centre of Cuba **785**
Campeche, Historic Fortified Town of
   Mexico **577**
Canadian Rocky Mountain Parks Canada **194**
Canaima National Park Venezuela **431**
Canal du Centre, The Four Lifts on the
   Belgium **554**
Canal du Midi France **480**
Canterbury Cathedral, St Augustine's Abbey,
   and St Martin's Church
   United Kingdom **327**
Cape Floral Region Protected Areas
   South Africa **720**
Capital Cities and Tombs of the Ancient
   Koguryo Kingdom China **722**
Cappadocia, Rock Sites of Turkey **230**

Caracas, Ciudad Universitaria de
Venezuela **644**
Carcassonne, Historic Fortified City of
France **514**
Carlsbad Caverns National Park
USA **446**
Carlton Gardens, Melbourne Australia **718**
Carpathian Mountain Area, Wooden Churches
of the Slovak part of the Slovakia **795**
Carpathians, Primeval Beech Forests of the
Slovakia and Ukraine **772**
Cartagena Colombia **184**
Carthage, Site of Tunisia **46**
Casas Grandes, Archaeological Zone of
Paquimé Mexico **563**
Caserta, Eighteenth-Century Royal Palace at
Italy **527**
Castel del Monte Italy **506**
Castle of the Teutonic Order in Malbork
Poland **547**
Castles and Town Walls of King Edward in
Gwynedd United Kingdom **257**
Castles of Augustusburg and Falkenlust at
Brühl Germany **193**
Catalan Romanesque Churches of the Vall de
Boí Spain **644**
Cathedral and Churches of Echmiatsin and
the Archaeological Site of Zvartnots
Armenia **655**
Cathedral of Notre-Dame, Former Abbey of
Saint-Rémi and Palace of Tau, Reims
France **372**
Cathedral of St James in Šibenik
Croatia **627**
Cathedral, Alcázar and Archivo de Indias
in Seville Spain **266**
Cathedral, Torre Civica and Piazza Grande,
Modena Italy **520**
Caucasus, Western Russian Federation **601**
Causeway Coast United Kingdom **244**
Cave of Altamira and Paleolithic Cave Art of
Northern Spain Spain **220**
Cave of the Apocalypse on the Island of
Pátmos, Historic Centre (Chorá) with the
Monastery of Saint John 'the Theologian'
and the Greece **603**
Caves of Aggtelek Karst and Slovak Karst
Hungary and Slovakia **456**
Cedars of God, Forest of the Lebanon **569**
Centennial Hall in Wroclaw Poland **752**
Central Amazon Conservation Complex
Brazil **624**
Central Sikhote-Alin Russian Federation **664**

Central Suriname Nature Reserve
Suriname **658**
Central University City Campus of the
Universidad Nacional Autónoma de México
(UNAM) Mexico **770**
Central Zone of the Town of Angra do
Heroismo in the Azores Portugal **162**
Cerrado Protected Areas: Chapada dos
Veadeiros and Emas National Parks
Brazil **681**
Certosa di Padula Italy **558**
Cerveteri and Tarquinia, Etruscan
Necropolises of Italy **729**
Český Krumlov, Historic Centre of
Czech Republic **386**
Chaco Culture USA **285**
Champaner-Pavagadh Archaeological Park
India **730**
Champasak Cultural Landscape Lao People's
Democratic Republic (Laos) **679**
Chan Chan Archaeological Zone Peru **238**
Changdeokgung Palace Complex
Republic of Korea **521**
Chapada dos Veadeiros National Park
Brazil **681**
Chartres Cathedral France **65**
Chavín (Archaeological site) Peru **210**
Chengde, Mountain Resort and its Outlying
Temples China **436**
Chewore Safari Area Zimbabwe **186**
Chhatrapati Shivaji Terminus (formerly
Victoria Terminus), Mumbai India **723**
Chichen-Itza, Pre-Hispanic City of
Mexico **330**
Chief Roi Mata's Domain Vanuatu **795**
Chiloé, Churches of Chile **620**
Chinguetti, Ancient Ksour of
Mauritania **478**
Chiquitos, Jesuit Missions of the
Bolivia **348**
Choirokoitia Cyprus **554**
Chola Temples, Great Living India **276**
Chongoni Rock-Art Area Malawi **756**
Chorá Greece **603**
Church and Dominican Convent of Santa
Maria delle Grazie with 'The Last Supper'
by Leonardo da Vinci Italy **95**
Church of the Ascension, Kolomenskoye
Russian Federation **427**
Church Village of Gammelstad, Luleå
Sweden **504**
Churches and Convents of Goa India **237**
Churches of Chiloé Chile **620**

Churches of Moldavia Romania **412**
Churches of Peace in Jawor and Swidnica
Poland **675**
Cienfuegos, Urban Historic Centre of
Cuba **749**
Cilento and Vallo di Diano National Park with
the Archeological sites of Paestum and Velia,
and the Certosa di Padula Italy **558**
Cinque Terre Italy **513**
Cistercian Abbey of Fontenay France **121**
Citadel, Ancient City and Fortress Buildings of
Derbent Russian Federation **698**
City of Bath United Kingdom **290**
City of Cuzco Peru **158**
City of Graz – Historic Centre Austria **585**
City of Luxembourg: its Old Quarters and
Fortifications Luxembourg **424**
City of Potosí Bolivia **289**
City of Quito Ecuador **32**
City of Safranbolu Turkey **442**
City of Valletta Malta **93**
City of Verona Italy **618**
City of Vicenza and the Palladian Villas of the
Veneto Italy **428**
Ciudad Universitaria de Caracas
Venezuela **644**
Classical Gardens of Suzhou China **518**
Classical Weimar Germany **549**
Cliff of Bandiagara (Land of the Dogons)
Mali **340**
Côa Valley, Prehistoric Rock-Art Sites in the
Portugal **563**
Cocos Island National Park Costa Rica **516**
Coffee Plantations in the South-East of Cuba,
Archaeological Landscape of the First
Cuba **649**
Coiba National Park and its Special Zone of
Marine Protection Panama **745**
Collegiate Church, Castle, and Old Town of
Quedlinburg Germany **441**
Cologne Cathedral Germany **474**
Colonia del Sacramento, Historic Quarter of
the City of Uruguay **472**
Colonial City of Santo Domingo
Dominican Republic **343**
Comoé National Park Côte d'Ivoire **153**
Complex of Hué Monuments Vietnam **411**
Complex of Koguryo Tombs Democratic
People's Republic of Korea **715**
Confucius, Temple and Cemetery of China **433**
Convent of Christ in Tomar Portugal **175**
Convent of St Gall Switzerland **176**
Copan, Maya Site of Honduras **90**

Córdoba, Historic Centre of Spain **198**
Córdoba, Jesuit Block and Estancias of
    Argentina **647**
Corfu, Old Town of Greece **771**
Cornwall and West Devon Mining Landscape
    United Kingdom **757**
Coro and its Port Venezuela **398**
Costiera Amalfitana Italy **538**
Crac des Chevaliers and Qal'at Salah El-Din
    Syrian Arab Republic (Syria) **761**
Cracow Historic Centre Poland **30**
Crespi d'Adda Italy **465**
Cuenca, Historic Walled Town of Spain **505**
Cueva de las Manos, Río Pinturas
    Argentina **607**
Cultural Landscape and Archaeological
    Remains of the Bamiyan Valley
    Afghanistan **703**
Cultural and Historic Ensemble of the
    Solovetsky Islands Russian Federation **392**
Cultural Landscape of Sintra
    Portugal **447**
Curonian Spit Lithuania and Russian
    Federation **615**
Cuzco, City of Peru **158**
Cyrene, Archaeological Site of Libyan Arab
    Jamahiriya (Libya) **136**

D

Dacian Fortresses of the Orastie Mountains
    Romania **597**
Dahshur Egypt **70**
Damascus, Ancient City of Syrian Arab
    Republic (Syria) **39**
Dambulla, Golden Temple of
    Sri Lanka **363**
Danube, Banks of the, Budapest
    Hungary **274**
Danube Delta Romania **362**
Daphni, Monastery of Greece **342**
Darien National Park Panama **130**
Dazu Rock Carvings China **584**
Defence Line of Amsterdam
    Netherlands **488**
Delos Greece **353**
Delphi, Archaeological Site of Greece **262**
Derbent, Ancient City of
    Russian Federation **698**
Derwent Valley Mills United Kingdom **675**
Desembarco del Granma National Park
    Cuba **583**
Dessau, Bauhaus and its Sites in
    Germany **484**

Dessau-Wörlitz, Garden Kingdom of
    Germany **610**
Diamantina, Historic Centre of the Town of
    Brazil **588**
Dilmun, Ancient Harbour and Capital of
    Bahrain **746**
Dinosaur Provincial Park Canada **48**
Diocletian, Palace of Croatia **82**
Discovery Coast Atlantic Forest Reserves
    Brazil **579**
Divriği, Great Mosque and Hospital of
    Turkey **225**
Dja Faunal Reserve Cameroon **308**
Djémila Algeria **148**
Djenné, Old Towns of Mali **310**
Djoudj National Bird Sanctuary
    Senegal **121**
Dogons, Land of the Mali **340**
Doñana National Park Spain **430**
Dong Phayayen-Khao Yai Forest Complex
    Thailand **741**
Dorset and East Devon Coast United
    Kingdom **671**
Dougga / Thugga Tunisia **530**
Drakensberg Park South Africa **611**
Dresden Elbe Valley Germany **709**
Droogmakerij de Beemster (Beemster Polder)
    Netherlands **580**
Drottningholm, Royal Domain of
    Sweden **371**
Dubrovnik, Old City of Croatia **86**
Dujiangyan Irrigation System
    China **648**
Durham Castle and Cathedral United
    Kingdom **246**
Durmitor National Park
    Montenegro **107**

E

Earliest 16th-Century Monasteries on the
    Slopes of Popocatepetl Mexico **432**
Early Christian Monuments of Ravenna
    Italy **487**
Early Christian Necropolis of Pécs (Sopianae)
    Hungary **619**
East Devon Coast United Kingdom **671**
East Rennell Solomon Islands **558**
Echmiatsin, Cathedral and Churches of
    Armenia **655**
Ecosystem and Relict Cultural Landscape of
    Lopé-Okanda Gabon **781**
Edinburgh, Old and New Towns of
    United Kingdom **470**

Eighteenth-Century Royal Palace at Caserta
    with the Park, the Aqueduct of Vanvitelli, and
    the San Leucio Complex Italy **527**
Eisleben and Wittenberg, Luther Memorials in
    Germany **503**
El Jadida Morocco **717**
El Jem, Amphitheatre of Tunisia **80**
El Tajin, Pre-Hispanic City of Mexico **394**
El Vizcaino, Whale Sanctuary of
    Mexico **401**
Elche, Palmeral of Spain **639**
Elephanta Caves India **271**
Ellora Caves India **160**
Emas National Park Brazil **681**
Emei, Mount China **481**
Engelsberg Ironworks Sweden **408**
Ensemble of the Ferrapontov Monastery
    Russian Federation **643**
Ensemble of the Novodevichy Convent Russian
    Federation **728**
Epidaurus, Sanctuary of Asklepios at
    Greece **315**
Episcopal Complex of the Euphrasian Basilica
    in the Historic Centre of Poreč
    Croatia **516**
Escurial, Monastery and site of the
    Spain **193**
Essaouira (formerly Mogador), Medina of
    Morocco **666**
Essen, Zollverein Coal Mine Industrial
    Complex in Germany **667**
Etruscan Necropolises of Cerveteri and
    Tarquinia Italy **729**
Euphrasian Basilica, Episcopal Complex
    of the Croatia **516**
Everest, Mount Nepal **85**
Everglades National Park USA **53**
Évora, Historic Centre of Portugal **258**

F

Falkenlust Castle at Brühl Germany **193**
Falun, Mining Area of the Great Copper
    Mountain in Sweden **667**
Fasil Ghebbi, Gondar Region Ethiopia **60**
Fatehpur Sikri India **247**
Fernando de Noronha and Atol das Rocas
    Reserves Brazil **668**
Ferrapontov Monastery, Ensemble of the
    Russian Federation **643**
Ferrara, City of the Renaissance, and its Po
    Delta Italy **453**
Fertö / Neusiedlersee Cultural Landscape
    Austria and Hungary **664**

Fez, Medina of Morocco **119**
Flemish Béguinages Belgium **562**
Flint Mines, Neolithic Belgium **652**
Florence, Historic Centre of Italy **134**
Fontainebleau, Palace and Park of
   France **113**
Fontenay, Cistercian Abbey of France **121**
Fort and Shalamar Gardens in Lahore
   Pakistan **124**
Forts and Castles, Volta, Greater Accra,
   Central and Western Regions Ghana **44**
Fortifications of Vauban France **794**
Fortifications on the Caribbean Side of
   Panama: Portobelo-San Lorenzo
   Panama **99**
Fortress of Suomenlinna Finland **378**
Fossil Cliffs, Joggins Canada **796**
Fossil Hominid Sites of Sterkfontein,
   Swartkrans, Kromdraai, and Environs South
   Africa **593**
Fossil Mammal Sites (Riversleigh /
   Naracoorte) Australia **429**
Fountains Abbey, Ruins of
   United Kingdom **248**
Four Lifts on the Canal du Centre and their
   Environs, La Louvière and Le Roeulx
   (Hainault), The Belgium **554**
Franciscan Missions in the Sierra Gorda of
   Querétaro Mexico **703**
Frankincense, Land of Oman **614**
Fraser Island Australia **393**
Frontiers of the Roman Empire
   United Kingdom and Germany **273**
Fuerte de Samaipata Bolivia **556**
Fujian Tulou China **788**

G

Galápagos Islands Ecuador **26**
Galerius, Palace of Serbia **776**
Galle and its Fortifications, Old Town of
   Sri Lanka **320**
Gammelstad, Church Village of, Luleå
   Sweden **504**
Gamzigrad-Romuliana, Palace of Galerius
   Serbia **776**
Ganghwa Dolmen Republic of Korea **626**
Garajonay National Park Spain **250**
Garamba National Park Dem. Rep.
   of the Congo **94**
Garden Kingdom of Dessau-Wörlitz
   Germany **610**
Gardens and Castle at Kroměříž
   Czech Republic **565**

Gaudí, Antoni Spain **182**
Gebel Barkal and the Sites of the Napatan
   Region Sudan **695**
Geghard, Monastery of Armenia **630**
Geirangerfjord Norway **742**
Gelati Monastery Georgia **435**
Genbaku Dome Japan **477**
Generalife, Granada Spain **202**
Genoa: Le Strade Nuove and the system of the
   Palazzi dei Rolli Italy **759**
Geoglyphs and Lines of Nasca and Pampas de
   Jumana Peru **434**
George Town, Historic City of the Straits of
   Malacca Malaysia **790**
Ghadamès, Old Town of Libyan Arab
   Jamahiriya (Libya) **261**
Giant's Causeway and Causeway Coast
   United Kingdom **244**
Girolata, Gulf of France **174**
Giza Egypt **70**
Gjirokastra, Historic Centre of
   Albania **732**
Glacier Bay Canada and USA **52**
Goa, Churches and Convents of
   India **237**
Gobustan Rock Art Cultural Landscape
   Azerbaijan **767**
Gochang, Hwasun and Ganghwa Dolmen
   Sites Republic of Korea **626**
Goiás, Historic Centre of the Town of
   Brazil **674**
Gokayama, Historic Villages of
   Japan **462**
Golden Mountains of Altai Russian
   Federation **557**
Golden Temple of Dambulla
   Sri Lanka **363**
Gondwana Rainforests of Australia
   Australia **260**
Gorée, Island of Senegal **29**
Göreme National Park and the Rock Sites of
   Cappadocia Turkey **230**
Goslar, Historic Town of Germany **387**
Gough and Inaccessible Islands United
   Kingdom **468**
Granada, Spain **202**
Grand Canyon National Park USA **54**
Graz, Historic Centre of the City of
   Austria **585**
Great Barrier Reef Australia **122**
Great Copper Mountain in Falun, Mining
   Area of the Sweden **667**
Great Living Chola Temples India **276**

Great Mosque and Hospital of Divriği
   Turkey **225**
Great Smoky Mountains National Park
   USA **164**
Great Wall, The China **294**
Great Zimbabwe National Monument
   Zimbabwe **240**
Greater Blue Mountains Area
   Australia **609**
Greenwich, Maritime United Kingdom **507**
Grimeton, Varberg Radio Station
   Sweden **725**
Gros Morne National Park Canada **302**
Group of Monuments at Hampi India **252**
Group of Monuments at Mahabalipuram
   India **204**
Group of Monuments at Pattadakal India **308**
Guadalajara, Hospicio Cabañas Mexico **534**
Guanacaste, Area de Conservación
   Costa Rica **596**
Guanajuato, Historic Town of Mexico **318**
Guaranis, Jesuit Missions of the Argentina
   and Brazil **159**
Guimarães, Historic Centre of Portugal **682**
Gulf of California, Islands and Protected Areas
   of the Mexico **733**
Gulf of Porto: Calanche di Piana, Gulf of
   Girolata, Scandola Reserve France **174**
Gunung Mulu National Park Malaysia **638**
Gusuku Sites and Related Properties of the
   Kingdom of Ryukyu Japan **643**
Gwynedd, Castles and Town Walls of King
   Edward in United Kingdom **257**
Gyeongju Historic Areas
   Republic of Korea **642**

H

Ha Long Bay Vietnam **440**
Haeinsa Temple Janggyeong Panjeon, the
   Depositories for the Tripitaka Koreana
   Woodblocks Republic of Korea **462**
Haghpat and Sanahin, Monasteries of
   Armenia **500**
Haifa, Bahá'í Holy Places in Israel **793**
Hal Saflieni Hypogeum Malta **98**
Hallstatt-Dachstein / Salzkammergut
   Cultural Landscape Austria **531**
Hampi, Group of Monuments at India **252**
Hanseatic City of Lübeck Germany **281**
Hanseatic Town of Visby Sweden **459**
Harar Jugol, the Fortified Historic Town
   Ethiopia **752**
Hatra Iraq **215**

Hattusha: the Hittite Capital Turkey **237**
Havana, Old Cuba **146**
Hawaii Volcanoes National Park USA **264**
Hazor, Tel Israel **739**
Head-Smashed-In Buffalo Jump Canada **118**
Heard and McDonald Islands Australia **522**
Heart of Neolithic Orkney
    United Kingdom **570**
Henderson Island United Kingdom **321**
Heraion of Samos Greece **388**
Herculaneum, Archaeological Area of
    Italy **524**
Hierapolis-Pamukkale Turkey **329**
Hieronymites, Monastery of the Portugal **163**
High Coast Sweden **630**
Hildesheim, St Mary's Cathedral and
    St Michael's Church Germany **225**
Himeji-jo Japan **409**
Hiroshima Peace Memorial (Genbaku Dome)
    Japan **477**
Historic and Architectural Complex of the
    Kazan Kremlin Russian Federation **646**
Historic Area of Willemstad, Inner City and
    Harbour, Netherlands Antilles
    Netherlands **528**
Historic Areas of Istanbul Turkey **226**
Historic Cairo Egypt **78**
Historic Centre (Chorá) with the Monastery of
    Saint John 'the Theologian' and the Cave of
    the Apocalypse on the Island of Pátmos
    Greece **603**
Historic Centre (Old Town) of Tallinn
    Estonia **509**
Historic Centre of the City of Pienza Italy **495**
Historic Centre of the City of Salzburg
    Austria **498**
Historic Centre of the Town of Diamantina
    Brazil **588**
Historic Centre of the Town of Goiás
    Brazil **674**
Historic Centre of the Town of Olinda
    Brazil **150**
Historic Centre of Avignon: Papal Palace,
    Episcopal Ensemble and Avignon
    Bridge France **444**
Historic Centre of Brugge Belgium **650**
Historic Centre of Bukhara Uzbekistan **405**
Historic Centre of Camagüey Cuba **785**
Historic Centre of Český Krumlov
    Czech Republic **386**
Historic Centre of Córdoba Spain **198**
Historic Centre of Évora Portugal **258**
Historic Centre of Florence Italy **134**

Historic Centre of Guimarães Portugal **682**
Historic Centre of Lima Peru **326**
Historic Centre of Macao China **737**
Historic Centre of Mexico City and Xochimilco
    Mexico **284**
Historic Centre of Morelia Mexico **369**
Historic Centre of Naples Italy **448**
Historic Centre of Oaxaca and Archaeological
    Site of Monte Albán Mexico **286**
Historic Centre of Oporto Portugal **482**
Historic Centre of Prague
    Czech Republic **384**
Historic Centre of Puebla Mexico **284**
Historic Centre of Riga Latvia **544**
Historic Centre of Rome, the Properties of the
    Holy See in that City Enjoying
    Extraterritorial Rights and San Paolo Fuori
    le Mura Italy **92**
Historic Centre of Saint Petersburg and
    Related Groups of Monuments
    Russian Federation **344**
Historic Centre of Salvador de Bahia
    Brazil **208**
Historic Centre of San Gimignano Italy **358**
Historic Centre of Santa Ana de los Ríos de
    Cuenca Ecuador **576**
Historic Centre of Santa Cruz de
    Mompox Colombia **473**
Historic Centre of São Luís Brazil **537**
Historic Centre of Shakhrisyabz
    Uzbekistan **629**
Historic Centre of Siena Italy **460**
Historic Centre of Sighişoara Romania **592**
Historic Centre of Telč Czech Republic **392**
Historic Centre of Urbino Italy **567**
Historic Centre of Vienna Austria **678**
Historic Centre of Warsaw Poland **89**
Historic Centre of Zacatecas Mexico **419**
Historic Centres of Berat and Gjirokastra
    Albania **732**
Historic Centres of Stralsund and Wismar
    Germany **690**
Historic City of Ayutthaya Thailand **359**
Historic City of Meknes Morocco **494**
Historic City of Sucre Bolivia **367**
Historic City of Toledo Spain **242**
Historic City of Trogir Croatia **533**
Historic District of Old Québec Canada **206**
Historic District of Panamá Panamá **529**
Historic Ensemble of the Potala Palace, Lhasa
    China **425**
Historic Fortified City of Carcassonne
    France **514**

Historic Fortified Town of Campeche
    Mexico **577**
Historic Inner City of Paramaribo
    Suriname **687**
Historic Monuments of Ancient Kyoto
    (Kyoto, Uji and Otsu Cities) Japan **438**
Historic Monuments of Ancient Nara
    Japan **560**
Historic Monuments of Novgorod and
    Surroundings Russian Federation **383**
Historic Monuments of Thatta Pakistan **118**
Historic Monuments Zone of Querétaro
    Mexico **496**
Historic Monuments Zone of Tlacotalpan
    Mexico **548**
Historic Mosque City of Bagerhat
    Bangladesh **228**
Historic Quarter of the City of Colonia del
    Sacramento Uruguay **472**
Historic Quarter of the Seaport City of
    Valparaíso Chile **696**
Historic Sanctuary of Machu Picchu
    Peru **172**
Historic Site of Lyons France **561**
Historic Town of Banská Štiavnica and the
    Technical Monuments in its Vicinity
    Slovakia **414**
Historic Town of Guanajuato and Adjacent
    Mines Mexico **318**
Historic Town of Ouro Preto
    Brazil **108**
Historic Town of St George and Related
    Fortifications, Bermuda United
    Kingdom **621**
Historic Town of Sukhothai and Associated
    Historic Towns Thailand **375**
Historic Town of Vigan Philippines **582**
Historic Town of Zabid Yemen **398**
Historic Villages of Shirakawa-go and
    Gokayama Japan **462**
Historic Walled Town of Cuenca
    Spain **505**
Historical Centre of the City of Arequipa
    Peru **617**
Historical Centre of the City of Yaroslavl
    Russian Federation **744**
Historical Complex of Split with the Palace of
    Diocletian Croatia **82**
Historical Monuments of Mtskheta
    Georgia **437**
Hoi An Ancient Town Vietnam **586**
Holašovice Historical Village Reservation
    Czech Republic **551**

Mausoleum of the First Qin Emperor
China **300**

Mausoleum of Khoja Ahmed Yasawi,
Turkestan Kazakhstan **707**

Maya Site of Cópan Honduras **90**

Mazagan (El Jadida), Portuguese City of
Morocco **717**

McDonald Island Australia **522**

Medieval City of Rhodes Greece **324**

Medieval Monuments in Kosovo
Serbia **717**

Medieval Town of Toruń Poland **526**

Medina of Essaouira (formerly Mogador)
Morocco **666**

Medina of Fez Morocco **119**

Medina of Marrakesh Morocco **218**

Medina of Sousse Tunisia **317**

Medina of Tétouan (formerly known as
Titawin) Morocco **522**

Medina of Tunis Tunisia **72**

Megalithic Temples of Malta Malta **96**

Megiddo, Tel Israel **739**

Mehmed Paša Sokolović Bridge in Višegrad
Bosnia-Herzegovina **779**

Meidan Emam, Esfahan Islamic Republic
of Iran **38**

Meknes, Historic City of Morocco **494**

Melaka and George Town, Historic Cities of
the Straits of Malacca Malaysia **790**

Melbourne, Royal Exhibition Building and
Carlton Gardens Australia **718**

Memphis and its Necropolis – the Pyramid
Fields from Giza to Dahshur Egypt **70**

Mérida, Archaeological Ensemble of
Spain **410**

Meru, Ancient Turkmenistan **573**

Mesa Verde National Park USA **28**

Messel Pit Fossil Site Germany **456**

Meteora Greece **332**

Miguasha National Park Canada **599**

Mijikenda Kaya Forests, Sacred
Kenya **782**

Mill Network at Kinderdijk-Elshout
Netherlands **523**

Millenary Benedictine Abbey of Pannonhalma
and its Natural Environment Hungary **479**

Minaret and Archaeological Remains of Jam
Afghanistan **683**

Mines of Rammelsberg and Historic Town of
Goslar Germany **387**

Ming and Qing Dynasties in Beijing and
Shenyang, Imperial Palaces of the
China **277**

Ming and Qing Dynasties, Imperial Tombs
of the China **616**

Mining Area of the Great Copper Mountain
in Falun Sweden **667**

Mir Castle Complex Belarus **635**

Modena Italy **520**

Moenjodaro, Archaeological Ruins at
Pakistan **102**

Mogador (Medina of Essaouira)
Morocco **666**

Mogao Caves China **298**

Moldavia, Churches of Romania **412**

Monarch Butterfly Biosphere Reserve
Mexico **791**

Monasteries of Daphni, Hosios Loukas and
Nea Moni of Chios Greece **342**

Monasteries of Haghpat and Sanahin
Armenia **500**

Monastery and Site of the Escurial, Madrid
Spain **193**

Monastery of Alcobaça Portugal **337**

Monastery of Batalha Portugal **177**

Monastery of Geghard and the Upper Azat
Valley Armenia **630**

Monastery of Horezu Romania **408**

Monastery of the Hieronymites and Tower of
Belém in Lisbon Portugal **163**

Monastic Island of Reichenau Germany **640**

Mons, Neolithic Flint Mines at
Belgium **652**

Mont-Saint-Michel and its Bay France **58**

Monte Albán, Archaeological Site of
Mexico **286**

Monte San Giorgio Switzerland **704**

Monticello and the University of Virginia in
Charlottesville USA **304**

Monuments of Oviedo and the Kingdom
of the Asturias Spain **217**

Morelia, Historic Centre of Mexico **369**

Morne Trois Pitons National Park
Dominica **515**

Mosi-oa-Tunya / Victoria Falls Zambia and
Zimbabwe **338**

Mostar, Old Bridge Area of the Old City of
Bosnia-Herzegovina **734**

Mount Athos Greece **309**

Mount Emei Scenic Area, including Leshan
Giant Buddha Scenic Area China **481**

Mount Huangshan China **357**

Mount Kenya National Park / Natural Forest
Kenya **508**

Mount Nimba Strict Nature Reserve Côte
d'Ivoire and Guinea **126**

Mount Qingcheng and the Dujiangyan
Irrigation System China **648**

Mount Sanqingshan National Park
China **796**

Mount Taishan China **297**

Mount Wuyi China **598**

Mountain Railways of India **576**

Mountain Resort and its Outlying Temples,
Chengde China **436**

Mozambique, Island of Mozambique **365**

Mtskheta, Historical Monuments of
Georgia **437**

Mudéjar Architecture of Aragon
Spain **256**

Mumbai, Chhatrapati Shivaji Terminus
(formerly Victoria Terminus) India **723**

Museumsinsel (Museum Island), Berlin
Germany **578**

Muskauer Park / Park Muzakowski Germany
and Poland **719**

Müstair, Benedictine Convent of St John at
Switzerland **159**

My Son Sanctuary Vietnam **579**

Mycenae, Archaeological Site of
Greece **574**

Mystras, Archaeological Site of
Greece **341**

M'Zab Valley Algeria **145**

N

Nærøyfjord Norway **742**

Nahanni National Park Canada **38**

Nancy, Place Stanislas, Place de la Carrière
and Place d'Alliance in France **169**

Nanda Devi and Valley of Flowers National
Parks India **333**

Napatan Region, Gebel Barkal and the Sites
of the Sudan **695**

Naples, Historic Centre Italy **448**

Nara, Historic Monuments of Ancient
Japan **560**

Naracoorte, Fossil Mammal Site
Australia **429**

Nasca, Lines and Geoglyphs of Peru **434**

National Archeological Park of Tierradentro
Colombia **473**

National History Park – Citadel, Sans Souci,
Ramiers Haiti **139**

Natural and Cultural Heritage of the Ohrid
region Former Yugoslav Republic
of Macedonia **77**

Natural and Culturo-Historical Region of
Kotor Montenegro **84**

Natural System of Wrangel Island Reserve Russian Federation **716**

Naval Port of Karlskrona Sweden **553**

Nea Moni of Chios, Monastery of Greece **342**

Negev, Desert Cities in the Israel **740**

Nemrut Dağ Turkey **292**

Neolithic Flint Mines at Spiennes (Mons) Belgium **652**

Nessebar, Ancient City of Bulgaria **165**

Nesvizh, Architectural, Residential and Cultural Complex of the Radziwill Family at Belarus **748**

New Caledonia, Lagoons of France **788**

New Lanark United Kingdom **659**

New Zealand Sub-Antarctic Islands New Zealand **556**

Ngorongoro Conservation Area Tanzania **42**

Niger, W National Park of Niger **485**

Nikko, Shrines and Temples of Japan **604**

Nimba Strict Nature Reserve, Mount Côte d'Ivoire and Guinea **126**

Niokolo-Koba National Park Senegal **116**

Nisa, Parthian Fortresses of Turkmenistan **778**

Noel Kempff Mercado National Park Bolivia **636**

Notre-Dame, Cathedral of, Reims France **372**

Notre-Dame Cathedral in Tournai Belgium **655**

Novgorod and Surroundings, Historic Monuments of Russian Federation **383**

Novodevichy Convent, Ensemble of the Russian Federation **728**

Nubian Monuments from Abu Simbel to Philae Egypt **56**

O

Oaxaca, Historic Centre of Mexico **286**

Ohrid region, Natural and Cultural Heritage of the Former Yugoslav Republic of Macedonia **77**

Okapi Wildlife Reserve Dem. Rep. of the Congo **478**

Öland, Agricultural Landscape of Southern Sweden **640**

Old Bridge Area of the Old City of Mostar Bosnia-Herzegovina **734**

Old and New Towns of Edinburgh United Kingdom **470**

Old City of Acre Israel **680**

Old City of Berne Switzerland **157**

Old City of Dubrovnik Croatia **86**

Old City of Jerusalem and its Walls **114**

Old City of Salamanca Spain **328**

Old City of Sana'a Yemen **245**

Old City of Zamość Poland **382**

Old Havana and its Fortifications Cuba **146**

Old Rauma Finland **368**

Old Town of Ávila with its Extra-Muros Churches Spain **233**

Old Town of Cáceres Spain **253**

Old Town of Corfu Greece **771**

Old Town of Galle and its Fortifications Sri Lanka **320**

Old Town of Ghadamès Libyan Arab Jamahiriya (Libya)**261**

Old Town of Lijiang China **510**

Old Town of Regensburg with Stadtamhof Germany **753**

Old Town of Segovia and its Aqueduct Spain **209**

Old Town Lunenburg Canada **463**

Old Towns of Djenné Mali **310**

Old Village of Hollókő and its Surroundings Hungary **293**

Old Walled City of Shibam Yemen **152**

Olinda, Historic Centre of the Town of Brazil **150**

Olomouc, Holy Trinity Column in Czech Republic **637**

Olympia, Archaeological Site of Greece **336**

Olympic National Park USA **128**

Omo, Lower Valley of the Ethiopia **106**

Oporto, Historic Centre of Portugal **482**

Orange, Roman Theatre and its Surroundings and the 'Triumphal Arch' of France **120**

Orastie Mountains, Dacian Fortresses of the Romania **597**

Orkhon Valley Cultural Landscape Mongolia **719**

Orkney, Heart of Neolithic United Kingdom **570**

Osun-Osogbo Sacred Grove Nigeria **741**

Otsu Japan **438**

Ouadane, Ancient Ksour of Mauritania **478**

Ouadi Qadisha (the Holy Valley) and the Forest of the Cedars of God (Horsh Arz el-Rab) Lebanon **569**

Oualata, Ancient Ksour of Mauritania **478**

Our Lady at Sedlec, Cathedral of Czech Republic **469**

Ouro Preto, Historic Town of Brazil **108**

Oviedo and the Kingdom of the Asturias, Monuments of Spain **217**

P

Padua, Botanical Garden (Orto Botanico) Italy **532**

Padula, Certosa di Italy **558**

Paestum Italy **558**

Paharpur, Ruins of the Buddhist Vihara at Bangladesh **222**

Painted Churches in the Troodos Region Cyprus **210**

Palace and Gardens of Schönbrunn Austria **502**

Palace and Park of Fontainebleau France **113**

Palace and Park of Versailles France **62**

Palaces and Parks of Potsdam and Berlin Germany **346**

Palau de la Música Catalana and Hospital de Sant Pau, Barcelona Spain **528**

Palazzi dei Rolli, Genoa Italy **759**

Palenque, Pre-Hispanic City and National Park of Mexico **299**

Paleochristian and Byzantine Monuments of Thessalonika Greece **313**

Paleolithic Cave Art of Northern Spain Spain **220**

Palladian Villas of The Veneto Italy **428**

Palmaria Italy **513**

Palmeral of Elche Spain **639**

Palmyra, Site of Syrian Arab Republic (Syria) **100**

Pampas de Jumana Peru **434**

Panamá Viejo, Archaeological Site of Panama **529**

Pannonhalma and its Natural Environment, Millenary Benedictine Abbey of Hungary **479**

Pantalica, Rocky Necropolis of Italy **750**

Pantanal Conservation Area Brazil **656**

Papal Palace, Avignon France **444**

Paphos Cyprus **101**

Paquimé, Archaeological Zone of Mexico **563**

Paramaribo, Historic Inner City of Suriname **687**

Paris, Banks of the Seine France **360**

Parthian Fortresses of Nisa Turkmenistan **778**

Pasargadae Islamic Republic of Iran **714**

Pátmos, Island of Greece **603**

Pattadakal, Group of Monuments at India **308**

Pécs (Sopianae), Early Christian Necropolis of Hungary **619**

Peking Man Site at Zhoukoudian China **298**

Península Valdés Argentina **590**

Perdu, Mont France and Spain **508**

Persepolis Islamic Republic of Iran **68**

Petäjävesi Old Church Finland **436**

Petra Jordan **212**

Petroglyphs within the Archaeological Landscape of Tamgaly Kazakhstan **729**

Philae, Nubian Monuments Egypt **56**

Philippine Cordilleras, Rice Terraces of the Philippines **466**

Phong Nha-Ke Bang National Park Vietnam **706**

Piana, Calanche of France **174**

Piazza del Duomo, Pisa Italy **265**

Piazza Grande, Modena Italy **520**

Pico Island Vineyard Culture, Landscape of the Portugal **731**

Piedmont and Lombardy, Sacri Monti of Italy **695**

Pienza, Historic Centre of the City of Italy **495**

Pilgrimage Church of St John of Nepomuk at Zelená Hora Czech Republic **443**

Pilgrimage Church of Wies Germany **176**

Pilgrimage Park Poland **594**

Ping Yao, Ancient City of China **540**

Pirin National Park Bulgaria **179**

Pisa, Piazza del Duomo Italy **265**

Pitons Management Area Saint Lucia **711**

Place Stanislas, Place de la Carrière and Place d'Alliance in Nancy France **169**

Plantin-Moretus House-Workshops-Museum Complex, Antwerp Belgium **745**

Plitvice Lakes National Park Croatia **57**

Po Delta Italy **453**

Poblet Monastery Spain **366**

Polonnaruwa, Ancient City of Sri Lanka **140**

Pompei, Archaeological Area of Italy **524**

Pont du Gard (Roman Aqueduct) France **229**

Popocatepetl, Earliest 16th-Century Monasteries on the Slopes of Mexico **432**

Poreč, Episcopal Complex of the Euphrasian Basilica in the Historic Centre of Croatia **516**

Port, Fortresses and Group of Monuments, Cartagena Colombia **184**

Porto, Gulf of: Calanche of Piana, Gulf of Girolata, Scandola Reserve France **174**

Portobelo-San Lorenzo Panama **99**

Portovenere, Cinque Terre, and the Islands (Palmaria, Tino and Tinetto) Italy **513**

Portuguese City of Mazagan (El Jadida) Morocco **717**

Potala Palace, Historic Ensemble of the China **425**

Potosí, City of Bolivia **289**

Potsdam and Berlin, Palaces and Parks of Germany **346**

Prague, Historic Centre of Czech Republic **384**

Prambanan Temple Compounds Indonesia **373**

Preah Vihear, Temple of Cambodia **789**

Pre-Hispanic City of Chichen-Itza Mexico **330**

Pre-Hispanic City of El Tajin Mexico **394**

Pre-Hispanic City and National Park of Palenque Mexico **299**

Pre-Hispanic City of Teotihuacan Mexico **288**

Pre-Hispanic Town of Uxmal Mexico **489**

Prehistoric Rock-Art Sites in the Côa Valley Portugal **563**

Prehistoric Sites and Decorated Caves of the Vézère Valley France **44**

Primeval Beech Forests of the Carpathians Slovakia and Ukraine **772**

Protective Town of San Miguel and the Sanctuary of Jesús Nazareno de Atotonilco Mexico **781**

Provins, Town of Medieval Fairs France **669**

Puebla, Historic Centre of Mexico **284**

Pueblo de Taos USA **380**

Puerto Rico, La Fortaleza and San Juan National Historic Site in USA **166**

Puerto-Princesa Subterranean River National Park Philippines **601**

Punic Town of Kerkuane and its Necropolis Tunisia **220**

Pyramid Fields from Giza to Dahshur Egypt **70**

Pyrénées - Mont Perdu France and Spain **508**

Pythagoreion and Heraion of Samos Greece **388**

Q

Qal'at al-Bahrain – Ancient Harbour and Capital of Dilmun Bahrain **746**

Qal'at Salah El-Din Syrian Arab Republic (Syria) **761**

Qal'at Sherqat (Ashur) Iraq **697**

Qin Emperor, Mausoleum of the First China **300**

Qing Dynasty, Imperial Tombs of the China **616**

Qingcheng, Mount, and the Dujiangyan Irrigation System China **648**

Québec, Old, Historic District of Canada **206**

Quebrada de Humahuaca Argentina **705**

Quedlinburg, Old Town of Germany **441**

Queensland, Wet Tropics of Australia **311**

Querétaro, Franciscan Missions in the Sierra Gorda of Mexico **703**

Querétaro, Historic Monuments Zone of Mexico **496**

Qufu, Temple and Cemetery of Confucius and the Kong Family Mansion in China **433**

Quirigua, Archaeological Park and Ruins of Guatemala **130**

Quito, City of Ecuador **32**

Quseir Amra Jordan **221**

Qutb Minar and its Monuments, Delhi India **413**

R

Radziwill Family at Nesvizh, Architectural, Residential and Cultural Complex of the Belarus **748**

Rainforests of the Atsinanana Madagascar **778**

Ramiers Haiti **139**

Rammelsberg, Mines of Germany **387**

Rapa Nui National Park Chile **454**

Ravenna, Early Christian Monuments of Italy **487**

Red Fort Complex India **765**

Red Square, Moscow Russian Federation **350**

Redwood National Park USA **103**

Regensburg with Stadtamhof, Old Town of Germany **753**

Reichenau, Monastic Island of Germany **640**

Reims, Cathedral of Notre-Dame France **372**

Renaissance Monumental Ensembles of Úbeda and Baeza Spain **697**

Residences of the Royal House of Savoy
Italy **541**
Rhaetian Railway in the Albula / Bernina
Landscapes Italy and Switzerland **784**
Rhine Valley, Upper Middle Germany **684**
Rhodes, Medieval City of Greece **324**
Rice Terraces of the Philippine Cordilleras
Philippines **466**
Richtersveld Cultural and Botanical
Landscape South Africa **764**
Rideau Canal Canada **773**
Rietveld Schröderhuis (Rietveld Schröder
House), Utrecht Netherlands **636**
Riga, Historic Centre of Latvia **544**
Rila Monastery Bulgaria **178**
Río Abiseo National Park Peru **356**
Río Pinturas, Cueva de las Manos,
Argentina **607**
Río Plátano Biosphere Reserve
Honduras **142**
Riversleigh, Fossil Mammal Site
Australia **429**
Røros Mining Town Norway **104**
Robben Island South Africa **569**
Rock Art of Alta Norway **232**
Rock Art Cultural Landscape, Gobustan
Azerbaijan **767**
Rock Art of the Mediterranean Basin on the
Iberian Peninsula Spain **562**
Rock Carvings in Tanum Sweden **432**
Rock Drawings in Valcamonica Italy **76**
Rock Paintings of the Sierra de San Francisco
Mexico **419**
Rock Shelters of Bhimbetka India **694**
Rock-Art Area, Chongoni Malawi **756**
Rock-Art Sites, Kondoa Tanzania **755**
Rock-Art Sites in the Côa Valley, Prehistoric
Portugal **563**
Rock-Art Sites of Tadrart Acacus Libyan Arab
Jamahiriya (Libya) **234**
Rock-Hewn Churches, Lalibela Ethiopia **32**
Rohtas Fort Pakistan **532**
Roman Empire, Frontiers of the
United Kingdom and Germany **273**
Roman Monuments, Cathedral of St Peter and
Church of Our Lady in Trier Germany **251**
Roman Theatre and its Surroundings and the
'Triumphal Arch' of Orange France **120**
Roman Walls of Lugo Spain **649**
Rome, Historic Centre of Italy **92**
Roskilde Cathedral Denmark **452**
Route of Santiago de Compostela Spain **397**
Routes of Santiago de Compostela France **557**

Royal Botanic Gardens, Kew United
Kingdom **692**
Royal Chitwan National Park Nepal **196**
Royal Domain of Drottningholm
Sweden **371**
Royal Exhibition Building and Carlton
Gardens, Melbourne Australia **718**
Royal Hill of Ambohimanga
Madagascar **663**
Royal Monastery of Santa María de
Guadalupe Spain **416**
Royal Palaces of Abomey Benin **232**
Royal Saltworks of Arc-et-Senans
France **153**
Ruins of the Buddhist Vihara at Paharpur
Bangladesh **222**
Ruins of Kilwa Kisiwani and Ruins of Songo
Mnara Tanzania **116**
Ruins of León Viejo Nicaragua **619**
Rwenzori Mountains National Park
Uganda **430**
Ryukyu, Gusuku Sites and Related Properties
of the Kingdom of Japan **643**

S
Sabbioneta Italy **797**
Sabratha, Archaeological Site of Libyan Arab
Jamahiriya (Libya) **149**
Sacred City of Anuradhapura Sri Lanka **138**
Sacred City of Kandy Sri Lanka **314**
Sacred Mijikenda Kaya Forests Kenya **782**
Sacred Sites and Pilgrimage Routes in the Kii
Mountain Range Japan **726**
Sacri Monti of Piedmont and Lombardy
Italy **695**
Safranbolu, City of Turkey **442**
Sagarmatha National Park Nepal **85**
Sahr-i-Bahlol Pakistan **110**
St Augustine's Abbey United Kingdom **327**
St Barbara, Church of Czech Republic **469**
St Catherine Area Egypt **688**
St Gall, Convent of Switzerland **176**
St George, Historic Town of,
Bermuda United Kingdom **621**
St James, Cathedral of Croatia **627**
St John of Nepomuk at Zelená Hora,
Pilgrimage Church of Czech Republic **443**
Saint John 'the Theologian', Monastery of,
Greece **603**
St Kilda United Kingdom **241**
Saint Margaret's Church United Kingdom **272**
St Mary's Cathedral and St Michael's Church
at Hildesheim Germany **225**

St Martin's Church United Kingdom **327**
St Peter, Cathedral of Germany **251**
Saint Petersburg and Related Groups of
Monuments, Historic Centre of
Russian Federation **344**
St Procopius' Basilica in Třebíč Czech
Republic **704**
Saint-Emilion, Jurisdiction of France **600**
Saint-Louis, Island of Senegal **626**
Saint-Rémi, Former Abbey of France **372**
Saint-Savin sur Gartempe, Abbey Church of
France **168**
Saint-Sophia Cathedral and Related Monastic
Buildings, Kiev-Pechersk Lavra
Ukraine **347**
Salamanca, Old City of Spain **328**
Salonga National Park Dem. Rep.
of the Congo **186**
Saltaire United Kingdom **665**
Salvador de Bahia, Historic Centre of
Brazil **208**
Salzburg, Historic Centre of the City of
Austria **498**
Salzkammergut Cultural Landscape
Austria **531**
Samaipata, Fuerte de Bolivia **556**
Samarkand – Crossroads of Cultures
Uzbekistan **670**
Samarra Archaeological City Iraq **767**
Sammallahdenmäki, Bronze Age Burial Site of
Finland **599**
Samos, Pythagoreion and Heraion of
Greece **388**
San Agustín Archeological Park
Colombia **458**
San Cristóbal de La Laguna Spain **587**
San Francesco, Basilica of Italy **632**
San Gimignano, Historic Centre of
Italy **358**
San Juan National Historic Site, Puerto Rico
USA **166**
San Leucio Complex Italy **527**
San Lorenzo Panama **99**
San Marino Historic Centre and Mount
Titano San Marino **782**
San Miguel and the Sanctuary of Jesús
Nazareno de Atotonilco, Protective Town of
Mexico **781**
San Millán Yuso and Suso Monasteries
Spain **535**
San Paolo Fuori le Mura Italy **92**
San Pedro de la Roca Castle, Santiago de
Cuba Cuba **548**

Sana'a, Old City of Yemen **245**
Sanahin, Monastery of Armenia **500**
Sanchi, Buddhist Monuments at India **342**
Sanctuary of Asklepios at Epidaurus
  Greece **315**
Sanctuary of Bom Jesus do Congonhas
  Brazil **223**
Sangay National Park Ecuador **171**
Sangiran Early Man Site Indonesia **503**
Sanqingshan, Mount, National Park
  China **796**
Sans Souci Haiti **139**
Sant Pau, Hospital de, Barcelona
  Spain **528**
Santa Ana de los Ríos de Cuenca, Historic
  Centre of Ecuador **576**
Santa Cruz de Mompox, Historic Centre of
  Colombia **473**
Santa María de Guadalupe, Royal Monastery
  of Spain **416**
Santa Maria delle Grazie with 'The Last
  Supper' by Leonardo da Vinci, Church and
  Dominican Convent of Italy **95**
Santiago de Compostela, Route of
  Spain **397**
Santiago de Compostela in France, Routes of
  France **557**
Santiago de Compostela (Old Town)
  Spain **211**
Santiago de Cuba, San Pedro de la Roca
  Castle Cuba **548**
Santo Domingo, Colonial City of Dominican
  Republic **343**
Saõ Miguel das Missões, Ruins of Brazil **159**
Sapi Safari Area Zimbabwe **186**
São Luís, Historic Centre of Brazil **537**
Saryarka – Steppe and Lakes of Northern
  Kazakhstan Kazakhstan **786**
Sassi and the park of the Rupestrian Churches
  of Matera, The Italy **399**
Savoy, Residences of the Royal House of
  Italy **541**
Scandola Reserve France **174**
Schokland and Surroundings
  Netherlands **449**
Schönbrunn, Palace and Gardens of
  Austria **502**
Segovia, Old Town of Spain **209**
Seine, Banks of the France **360**
Selous Game Reserve Tanzania **137**
Semmering Railway Austria **551**
Senegambia, Stone Circles of The Gambia and
  Senegal **755**

Seokguram Grotto and Bulguksa Temple
  Republic of Korea **457**
Serengeti National Park Tanzania **112**
Serra da Capivara National Park Brazil **374**
Seuthopolis, Thracian Tomb of Kazanlak
  Bulgaria **49**
Seville, Cathedral, Alcázar and Archivo de
  Indias in Spain **266**
Sewell Mining Town Chile **762**
SGang Gwaay Canada **133**
Shakhrisyabz, Historic Centre of
  Uzbekistan **629**
Shark Bay, Western Australia
  Australia **368**
Shenyang, Imperial Palaces of the
  Ming and Qing Dynasties China **277**
Shibam, Old Walled City of Yemen **152**
Shirakami-Sanchi Japan **415**
Shirakawa-go and Gokayama,
  Historic Villages of Japan **462**
Shiretoko Japan **748**
Shirvanshah's Palace, Baku Azerbaijan **607**
Shrines and Temples of Nikko Japan **604**
Sian Ka'an Mexico **267**
Šibenik, The Cathedral of St James in
  Croatia **627**
Sichuan Giant Panda Sanctuaries – Wolong,
  Mt Siguniang and Jiajin Mountains
  China **751**
Siena, Historic Centre of Italy **460**
Sierra de San Francisco, Rock Paintings of the
  Mexico **419**
Sierra Gorda of Querétaro, Franciscan
  Missions in the Mexico **703**
Sierra Maestra, Archaeological Landscape of
  the First Coffee Plantations Cuba **649**
Sighişoara, Historic Centre of Romania **592**
Sigiriya, Ancient City of Sri Lanka **151**
Siguniang, Mt, Giant Panda Sanctuary
  China **751**
Sikhote-Alin, Central Russian Federation **664**
Simien National Park Ethiopia **37**
Sinharaja Forest Reserve Sri Lanka **316**
Sintra, Cultural Landscape of Portugal **447**
Site of Carthage Tunisia **46**
Site of Palmyra Syrian Arab Republic
  (Syria) **100**
Skellig Michael Ireland **492**
Škocjan Caves Slovenia **261**
Skogskyrkogården Sweden **435**
Slovak Karst, Caves of Aggtelek Karst and
  Hungary and Slovakia **456**
Socotra Archipelago Yemen **785**

Sofia, Boyana Church Bulgaria **60**
Solovetsky Islands, Cultural and Historic
  Ensemble of the Russian Federation **392**
Soltaniyeh Islamic Republic of Iran **746**
Songo Mnara, Ruins of Tanzania **116**
Sopianae (Pécs), Early Christian Necropolis of
  Hungary **619**
Sopoćani Serbia **76**
Sousse, Medina of Tunisia **317**
South China Karst China **768**
Southern Little Poland, Wooden Churches of
  Poland **702**
Speyer Cathedral Germany **128**
Spiennes (Mons), Neolithic Flint Mines at
  Belgium **652**
Spišský Hrad and its Associated Cultural
  Monuments Slovakia **417**
Split, Historical Complex of Croatia **82**
Srebarna Nature Reserve Bulgaria **168**
Stari Grad Plain Croatia **790**
Stari Ras and Sopoćani Serbia **76**
State Historical and Cultural Park Ancient
  Merv Turkmenistan **573**
Statue of Liberty USA **187**
Sterkfontein, Fossil Hominid Site of South
  Africa **593**
Stone Circles of Senegambia The Gambia and
  Senegal **755**
Stone Town of Zanzibar Tanzania **612**
Stonehenge, Avebury and Associated Sites
  United Kingdom **254**
Stralsund, Historic Centre of
  Germany **690**
Strasbourg – Grande Île France **334**
Struve Geodetic Arc Belarus, Estonia, Finland,
  Latvia, Lithuania, Norway, Republic of
  Moldova, Russian Federation, Sweden
  and Ukraine **747**
Studenica Monastery Serbia **250**
Studley Royal Park including the Ruins of
  Fountains Abbey United Kingdom **248**
Su Nuraxi di Barumini Italy **546**
Sucre, Historic City of Bolivia **367**
Sukhothai, Historic Town of Thailand **375**
Sukur Cultural Landscape Nigeria **587**
Sumatra, Tropical Rainforest Heritage of
  Indonesia **712**
Summer Palace and Imperial Garden in
  Beijing China **552**
Sun Temple, Konârak India **183**
Sundarbans, The Bangladesh **512**
Sundarbans National Park India **271**
Suomenlinna, Fortress of Finland **378**

Surtsey Iceland **789**

Suso Monastery Spain **535**

Southern Anhui, Ancient Villages in Xidi and
Hongcun China **653**

Suzdal, White Monuments of Russian
Federation **396**

Suzhou, Classical Gardens of China **518**

Sveshtari, Thracian Tomb of Bulgaria **234**

Swartkrans, Fossil Hominid Site South
Africa **593**

Swidnica, Church of Peace Poland **675**

Swiss Alps Jungfrau-Aletsch
Switzerland **660**

Swiss Tectonic Arena Sardona
Switzerland **787**

Sydney Opera House Australia **763**

Syracuse and the Rocky Necropolis of
Pantalica Italy **750**

T

Tadrart Acacus, Rock-Art Sites of Libyan Arab
Jamahiriya (Libya) **234**

Taï National Park Côte d'Ivoire **139**

Taishan, Mount China **297**

Taj Mahal India **154**

Takht-e Soleyman Islamic Republic
of Iran **691**

Takht-i-Bahi, Buddhist Ruins at
Pakistan **110**

Talamanca Range-La Amistad Reserves / La
Amistad National Park Costa Rica and
Panama **162**

Talampaya Natural Park Argentina **628**

Tallinn, Historic Centre (Old Town) of
Estonia **509**

Tamgaly, Petroglyphs within the Archaeological
Landscape of Kuzukhstan **729**

Tanum, Rock Carvings in Sweden **432**

Taos, Pueblo de USA **380**

Tarquinia, Etruscan Necropolises of Italy **729**

Tárraco, Archaeological Ensemble of
Spain **621**

Tasmanian Wilderness Australia **143**

Tassili n'Ajjer Algeria **145**

Tatshenshini-Alsek Canada and USA **52**

Tau, Palace of France **372**

Taxila Pakistan **109**

Tchogha Zanbil Islamic Republic of Iran **88**

Te Wahipounamu – South West New Zealand
New Zealand **354**

Teide National Park Spain **766**

Tel-Aviv, White City of – The Modern
Movement Israel **694**

Telč, Historic Centre of Czech Republic **392**

Tels, Bibilical – Megiddo, Hazor, Beer Sheba
Israel **739**

Temple of Apollo Epicurius at Bassae
Greece **240**

Temple and Cemetery of Confucius and the
Kong Family Mansion in Qufu China **433**

Temple of Heaven: an Imperial Sacrificial Altar
in Beijing China **559**

Temple of Preah Vihear Cambodia **789**

Ténéré Natural Reserve Niger **367**

Teotihuacan, Pre-Hispanic City of Mexico **288**

Tequila, Agave Landscape and Ancient
Industrial Facilities of Mexico **754**

Tétouan (formerly known as Titawin),
Medina of Morocco **522**

Thatta, Historic Monuments of
Pakistan **118**

The Four Lifts on the Canal du Centre and
their Environs, La Louvière and Le Roeulx
(Hainault) Belgium **554**

The Great Wall China **294**

The Sassi and the park of the Rupestrian
Churches of Matera Italy **399**

The Sundarbans Bangladesh **512**

The Trulli of Alberobello Italy **501**

Thebes, Ancient Egypt **74**

Thessalonika, Paleochristian and Byzantine
Monuments of Greece **313**

Þingvellir National Park Iceland **710**

Thracian Tomb of Kazanlak Bulgaria **49**

Thracian Tomb of Sveshtari Bulgaria **234**

Three Castles, Defensive Wall and Ramparts
of the Market-Town of Bellinzone
Switzerland **610**

Three Parallel Rivers of Yunnan Protected
Areas China **700**

Thugga Tunisia **530**

Thungyai-Huai Kha Khaeng Wildlife
Sanctuaries Thailand **364**

Tichitt, Ancient Ksour of
Mauritania **478**

Tierradentro, National Archeological Park of
Colombia **473**

Tikal National Park Guatemala **88**

Timbuktu Mali **321**

Timgad Algeria **148**

Tinetto Italy **513**

Tino Italy **513**

Tipasa Algeria **142**

Tiryns, Archaeological Site of Greece **574**

Titano, Mount San Marino **782**

Titawin, Medina of Morocco **522**

Tiwanaku: Spiritual and Political Centre of the
Tiwanaku Culture Bolivia **634**

Tiya Ethiopia **106**

Tlacotalpan, Historic Monuments Zone of
Mexico **548**

Tokaj Wine Region Historic Cultural
Landscape Hungary **686**

Toledo, Historic City of Spain **242**

Tomar, Convent of Christ in Portugal **175**

Tomb of Askia Mali **725**

Tombs of Buganda Kings at Kasubi
Uganda **663**

Tongariro National Park New Zealand **349**

Torre Annunziata, Archaeological Area of
Italy **524**

Torre Civica, Modena Italy **520**

Toruń, Medieval Town of Poland **526**

Tournai, Notre-Dame Cathedral in
Belgium **655**

Tower of Belém Portugal **163**

Tower of London United Kingdom **312**

Town of Bamberg Germany **406**

Town of Luang Prabang Lao People's
Democratic Republic (Laos) **464**

Town Hall and Roland on the Marketplace of
Bremen Germany **727**

Transylvania, Villages with Fortified Churches
in Romania **404**

Trier, Roman Monuments, Cathedral of
St Peter and Church of Our Lady in
Germany **251**

Trinidad and the Valley de los Ingenios
Cuba **323**

Trinity Sergius Laura in Sergiev Posad,
Architectural Ensemble of the
Russian Federation **420**

Tripitaka Koreana Woodblocks, The
Depositories for Republic of Korea **462**

Trogir, Historic City of Croatia **533**

Troodos Region, Painted Churches in the
Cyprus **210**

Tropical Rainforest Heritage of Sumatra
Indonesia **712**

Troy, Archaeological Site of Turkey **550**

Trulli of Alberobello, The Italy **501**

Třebíč, Jewish Quarter and St Procopius'
Basilica in Czech Republic **704**

Tsingy de Bemaraha Strict Nature Reserve
Madagascar **352**

Tsodilo Botswana **668**

Tubbataha Reef Marine Park
Philippines **412**

Tugendhat Villa in Brno Czech Republic **681**

Tunis, Medina of Tunisia **72**
Twyfelfontein or /Ui-//aes Namibia **770**
Tyre Lebanon **197**

U

Úbeda and Baeza, Renaissance Monumental
　Ensembles of Spain **697**
/Ui-//aes, Twyfelfontein or Namibia **770**
Uji Japan **438**
Ujung Kulon National Park Indonesia **374**
uKhahlamba / Drakensberg Park South
　Africa **611**
Uluru-Kata Tjuta National Park
　Australia **306**
Um er-Rasas (Kastrom Mefa'a) Jordan **713**
Universidad Nacional Autónoma de México
　(UNAM), Central University City Campus
　of the Mexico **770**
Universitaria de Caracas, Ciudad
　Venezuela **644**
University and Historic Precinct of Alcalá de
　Henares Spain **568**
University of Virginia in Charlottesville
　USA **304**
Upper Azat Valley Armenia **630**
Upper Middle Rhine Valley
　Germany **684**
Upper Svaneti Georgia **481**
Urban Historic Centre of Cienfuegos
　Cuba **749**
Urbino, Historic Centre of Italy **567**
Urnes Stave Church Norway **81**
Uvs Nuur Basin Mongolia and Russian
　Federation **693**
Uxmal, Pre-Hispanic Town of
　Mexico **489**

V

Val di Noto (South-Eastern Sicily), Late
　Baroque Towns of the Italy **690**
Val d'Orcia Italy **724**
Valcamonica, Rock Drawings in Italy **76**
Vall de Boí, Catalan Romanesque Churches
　of the Spain **644**
Vallée de Mai Nature Reserve
　Seychelles **171**
Valletta, City of Malta **93**
Valley of Flowers National Parks India **333**
Valparaíso, Historic Quarter of the Seaport
　City of Chile **696**
Vanvitelli, Aqueduct of Italy **527**
Varberg Radio Station, Grimeton
　Sweden **725**

Vat Phou and Associated Ancient Settlements
　within the Champasak Cultural Landscape
　Lao People's Democratic Republic
　(Laos) **679**
Vatican City Holy See **180**
Vauban, Fortifications of France **794**
Vegaøyan – the Vega Archipelago
　Norway **716**
Velia Italy **558**
Veneto, Palladian Villas of the Italy **428**
Venice and its Lagoon Italy **268**
Vergina, Archaeological Site of Greece **504**
Verla Groundwood and Board Mill
　Finland **486**
Verona, City of Italy **618**
Versailles, Palace and Park of France **62**
Vézelay, Church and Hill France **66**
Vézère Valley, Prehistoric Sites and Decorated
　Caves of the France **44**
Vicenza, City of Italy **428**
Victoria Falls Zambia and Zimbabwe **338**
Victoria Terminus, India **723**
Vienna, Historic Centre of Austria **678**
Vigan, Historic Town of Philippines **582**
Villa Adriana, Tivoli Italy **595**
Villa d'Este, Tivoli Italy **673**
Villa Romana del Casale, Piazza Armerina
　Italy **546**
Villages with Fortified Churches in
　Transylvania Romania **404**
Vilnius Historic Centre Lithuania **421**
Viñales Valley Cuba **593**
Virgin Komi Forests Russian Federation **449**
Virunga National Park Dem. Rep.
　of the Congo **45**
Visby, Hanseatic Town of Sweden **459**
Višegrad, Mehmed Paša Sokolović Bridge in
　Bosnia-Herzegovina **779**
Vizcaya Bridge Spain **758**
Vladimir and Suzdal, White Monuments of
　Russian Federation **396**
Vlkolínec Slovakia **418**
Volcanoes of Kamchatka Russian
　Federation **490**
Völklingen Ironworks Germany **441**
Volubilis, Archaeological Site of
　Morocco **543**
Vredefort Dome South Africa **747**

W

W National Park of Niger Niger **485**
Wachau Cultural Landscape
　Austria **641**

Wadi Al-Hitan (Whale Valley) Egypt **738**
Walled City of Baku with the Shirvanshah's
　Palace and Maiden Tower Azerbaijan **607**
Warsaw, Historic Centre of Poland **89**
Wartburg Castle Germany **589**
Waterton Glacier International Peace Park
　Canada and USA **450**
Weimar and Dessau, Bauhaus and its Sites in
　Germany **484**
West Devon Mining Landscape
　United Kingdom **757**
Western Caucasus Russian Federation **601**
Western Galilee, Bahá'í Holy Places in the
　Israel **793**
Westminster Palace, Westminster Abbey and
　Saint Margaret's Church United
　Kingdom **272**
Wet Tropics of Queensland Australia **311**
Whale Sanctuary of El Vizcaino
　Mexico **401**
Whale Valley (Wadi Al-Hitan)
　Egypt **738**
White City of Tel-Aviv – The Modern
　Movement Israel **694**
White Monuments of Vladimir and Suzdal
　Russian Federation **396**
Wieliczka Salt Mine, Poland **36**
Wies, Pilgrimage Church of
　Germany **176**
Willandra Lakes Region Australia **125**
Willemstad, Historic Area of
　Netherlands Antilles **528**
Wismar, Historic Centre of Germany **690**
Wittenberg, Luther Memorial in
　Germany **503**
Wolong Giant Panda Sanctuary China **751**
Wood Buffalo National Park Canada **170**
Wooden Churches of the Slovak part of the
　Carpathian Mountain Area
　Slovakia **795**
Wooden Churches of Maramureş
　Romania **581**
Wooden Churches of Southern Little Poland
　Poland **702**
Works of Antoni Gaudí Spain **182**
Wrangel Island Reserve, Natural System of
　Russian Federation **716**
Wrangell-St Elias Canada and USA **52**
Wroclaw, Centennial Hall in Poland **752**
Wudang Mountains, Ancient Building
　Complex in the China **429**
Wulingyuan Scenic and Historic Interest Area
　China **382**

Würzburg Residence with the Court Gardens
  and Residence Square Germany 132
Wuyi, Mount China 598

X

Xanthos-Letoon Turkey 333
Xidi, Southern Anhui China 653
Xochicalco, Archaeological Monuments Zone
  of Mexico 605
Xochimilco Mexico 284

Y

Yakushima Japan 400
Yaroslavl, Historical Centre of the City of
  Russian Federation 744

Yellowstone National Park USA 34
Yin Xu China 759
Yosemite National Park USA 184
Yungang Grottoes China 672
Yunnan Protected Areas, Three Parallel
  Rivers of China 700

Z

Zabid, Historic Town of Yemen 398
Zacatecas, Historic Centre of Mexico 419
Zamość, Old City of Poland 382
Zanzibar, Stone Town of Tanzania 612
Zelená Hora, Pilgrimage Church of St John of
  Nepomuk at Czech Republic 443
Zhoukoudian, Peking Man Site at China 298

Zollverein Coal Mine Industrial Complex in
  Essen Germany 667
Zvartnots, Archaeological Site of Armenia 655

Acknowledgements

Concept, design, maps, editorial and project management by the staff at Collins Geo, Glasgow.

Text edited by Collins Geo. Based on official information made available by the United Nations Educational, Scientific and Cultural Organization (UNESCO) and its World Heritage Centre.

With thanks to:

The UNESCO World Heritage Centre and its staff.

The freelance editors, copywriters, cartographers and pre-press individuals and organizations.

Image credits

Images supplied by www.shutterstock.com unless noted with an asterisk. All photographers credited unless unknown.

▲ picture located on the upper half of the page
▼ picture located on the lower half of the page

27▲ ©javarman; 27▼ ©javarman; 28 ©Duncan Gilbert; 29 ©faberfoto; 31 ©Wiktor Bubniak; 32▲*©UNESCO/Jim Williams; 32▼ ©Jennifer Stone; 33▲ ©skyfish; 33▼ ©skyfish; 35 ©Katrina Leigh; 36*©Anna Rosinska-Renaud; 37*©Tito Dupret; 39 ©Olga Kolos; 41 ©Wiktor Bubniak; 42▲ ©Kitch Bain; 42▼ ©Kitch Bain; 43 ©PixAchi; 45*©Africa Conservation Fund/www.gorilla.cd; 47▲ ©WitR; 47▼ ©Danijela Pavlovic Markovic; 48 ©Ryan Morgan; 49*©Tito Dupret; 51 ©Om Prakash Yadav; 52 ©Mariusz S. Jurgielewicz; 53 ©John A. Anderson; 55 ©Anton Foltin; 56▲ ©Mirek Hejnicki; 56▼ ©Vladimir Wrangel; 57 ©Maugli; 59 ©Cristina Ciochina; 60*©Tito Dupret; 61 ©Aleksander Bolbot; 63 ©Christophe Robard; 64 ©Tony Strong; 65 ©Gautier Willaume; 67 ©Sofilou; 68 ©steba; 71 ©Piotr Sikora; 71 ©Vladimir Wrangel; 73 ©Daniel Loncarevic; 75 ©Connors Bros.; 77 ©Ljupco Smokovski; 79 ©W H Chow; 80 ©PauleMarjanovic; 81 ©Andrea Seemann; 83 ©Krkr; 84 ©Svetlana Tikhonova; 85 ©Jason Maehl; 87 ©Beneda Miroslav; 88*©Tito Dupret; 89 ©Copestello; 90 ©Sam Chadwick; 91 ©Grigory Kubatyan; 92 ©lexan; 93 ©PixAchi; 94*©Nuria Ortega; 95 ©Belle Momenti Photography; 96 ©Cecilia Lim H M; 97*©UNESCO/Marc Patry; 98*©T6 Ecosystem/UNESCO; 99 ©Matt Ragen; 100 ©OPIS; 101 ©Ioannis Ioannou; 102▲*©Tito Dupret; 102▼*©Tito Dupret; 103 ©Andrew Ferguson; 104 ©Inger Anne Hulbækdal; 105 ©OPIS; 106▲*©Tito Dupret; 106▼*©Tito Dupret; 107 ©Milan Ljubisavljevic; 108 ©David Davis; 109▼*©Tito Dupret; 110*©Tito Dupret; 111 ©David Thyberg; 112 ©Stephane Angue; 113 ©Darja Vorontsova; 115▲ ©Bomshtein; 115▼ ©Stavchansky Yakov; 116*©Tito Dupret; 117 ©Claudio Giovanni Colombo; 118*©Waqas Muhammad/Wikipedia; 119 ©Thomas Cristofoletti; 120 ©Philip Lange; 121 ©Norman Bateman; 123▲ ©tororo reaction; 123▼ ©tororo reaction; 124*©(CC by sa 2.5)Ali Imran; 125 ©Sam DCruz; 126*©Prof. Dr Mark-Oliver Rödel; 127 ©Sam DCruz; 129 ©Natalia Bratslavsky; 131 ©Claudio Giovanni Colombo; 132 ©Khirman Vladimir; 133 ©Holger Ehlers; 135 ©Sailorr; 136*©UNESCO/Giovanni Boccardi; 137 ©faberfoto; 138 ©Valery Shanin; 141 ©Emma Holmwood; 143 ©Ashley Whitworth; 144 ©WitR; 147 ©Leonid Katsyka; 149 ©Clara; 150 ©ostill; 151*©Tito Dupret; 152 ©Vladimir Melnik; 153*©Tito Dupret;

155 ©Luciano Mortula; 156 ©JeremyRichards; 157 ©Marek Slusarczyk; 158 ©Chris Howey; 160 ©JeremyRichards; 161 ©Jarno Gonzalez Zarraonandia; 163 ©Matt Trommer; 164 ©Carolina K. Smith M.D.; 165 ©Andreas Gradin; 167 ©Bryan Busovicki; 169 ©Lazar Mihai-Bogdan; 170 ©Sue Smith; 173 ©Jarno Gonzalez Zarraonandia; 174 ©Bensliman; 175 ©Matt Trommer; 177 ©ultimathule; 178 ©sunnyfrog; 179 ©Ljupco Smokovski; 181 ©Slawomir Kruz; 182 ©WH Chow; 183 ©JeremyRichards; 185 ©Susan McKenzie; 187 ©Marcio Jose Bastos Silva; 189 ©urosr; 190 ©Dennis Albert Richardson; 191 ©javarman; 192 ©Palis Michael; 195▲*©Mark Steward; 195▼ ©Nelu Goia; 196 ©Jason Maehl; 197 ©javarman; 199 ©Brandus Dan Lucian; 200 ©Vladimir Korostysheuskiy; 201 ©quantz; 203 ©Rafael Ramirez Lee; 204 ©omkar.a.v; 205 ©Olga Kolos; 207 ©Kenneth V. Pilon; 208 ©Jose Miguel Hernandez Leon; 209 ©JIS; 210*©Alonzo Addison; 211 ©Francisco Turnes; 213 ©Joseph Calev; 214 ©Jakub Cejpek; 216 ©Larsek; 217 ©Rafael Angel Garcia Dobarganes; 219 ©Dainis Derics; 221 ©Factoria singular fotografia; 222*©Tito Dupret; 223 ©Rui Vale de Sousa; 224 ©JeremyRichards; 227 ©ImageDesign; 228*©Tito Dupret; 229 ©Elena Elisseeva; 231 ©Jaroslaw Grudzinski; 233 ©Oscar F. Chuyn; 235 ©Jason Maehl; 236 ©Holger Mette; 237*©Tito Dupret; 238 ©Chris Howey; 239 ©Jean Frooms; 241 ©Joe Gough; 242 ©aguilarphoto; 243 ©Fotowan; 244 ©Joe Gough; 245 ©Vladimir Melnik; 246 ©Gail Johnson; 247 ©Nikolay Titov; 248 ©pdtnc; 249 ©Asit Jain; 251 ©Jeremy R. Smith Sr.; 252 ©Mikhail Nekrasov; 253 ©LianeM; 255 ©John Evans; 256 ©Lola; 257 ©Gail Johnson; 259 ©inacio pires; 260 ©Ralph Loesche; 263 ©dr. Le Thanh Hung 264 ©Bryan Busovicki; 265 ©edobric; 266 ©sokolovsky; 267 ©Joseph Calev; 269▲*©Mark Steward; 269▼*©Mark Steward; 270 ©Rachael Russell; 271▲*©Tito Dupret; 271▼*©Tito Dupret; 272 ©Vinicius Tupinamba; 273 ©SmarterMedium; 275 ©dundanim; 276 ©Sankar; 277 ©sunxuejun; 279 ©enote; 280 ©ostill; 281 ©rubiphoto; 282 ©Matt Houser; 285 ©Zack Frank; 287 ©Michael Levy; 287 ©Ian D Walker; 288*©Tito Dupret; 289 ©Chris Howey; 291 ©Joop Snijder jr.; 292 ©Nathan Chor; 293 ©Mary Lane; 295 ©Mikhail Nekrasov; 296 ©Joseph Calev; 297 ©emily2k; 298▲*©Tito Dupret; 298▼*©Tito Dupret; 299 ©Ales Liska; 301 ©Jack Cronkhite; 302 ©Duncan de Young; 303 ©Martin Preston; 304 ©PRANAV VORA; 305 ©Bidouze Stéphane; 308*©Tito Dupret; 309 ©Koylias Ioannis; 310 ©Jam.si; 311 ©tororo reaction; 312 ©Mary Lane; 313▲*©UNESCO/APF-José Kalpers - Africa Parks Foundation; 313▼ ©Panos Karapanagiotis; 314 ©Ewen Cameron; 315 ©Ariy; 316 ©Mithila Somasiri; 317 ©LouLouPhotos; 319 ©Bill Perry; 320 ©Eugeniapp; 322*©Tito Dupret; 323 ©Alexey Goosev; 325 ©Jozsef Szasz-Fabian; 326 ©Thomas Barrat; 327 ©hauhu; 328 ©VanHart; 329 ©Clara; 331 ©Joseph Calev; 332 ©Andrew Buckin; 333 ©Bill McKelvie 334 ©Sjoerd van der Wal; 335 ©Aneta Skoczewska; 336 ©Richard Bowden; 337 ©Pedro Pinto; 339 ©Korobanova; 341 ©Ioannis Nousis; 342*©Tito Dupret; 343 ©rj lerich; 345 ©Fast Snail; 346 ©Dainis Derics; 347 ©Dmytro Korolov; 348*©Tito Dupret; 349 ©Sam DCruz; 351 ©Dimon; 352 ©POZZO DI BORGO Thomas; 353 ©rj lerich; 355 ©Sander van Sinttruye; 356 ©Artem Samokhvalov; 357 ©Tan Kim Pin; 358 ©MASSIMO MERLINI; 359 ©akua; 361▲ ©CHRISTOPHE ROLLAND; 361▼*©Mark Steward; 362 ©Mircea BEZERGHEANU; 363 ©Magdalena Bujak; 364*©Tito Dupret; 365 ©Grigory Kubatyan; 366 ©Sofilou; 367*©Dieter Biskamp; 368 ©Alexander Studentschnig; 369 ©Bill Perry; 370 ©Specta; 371 ©Mikael Damkier; 372 ©ultimathule; 373 ©Cristina CIOCHINA; 374*©Tito Dupret; 375 ©John Hemmings; 378 ©Juha Sompinmäki; 380*©Tito Dupret; 381 ©POZZO DI BORGO Thomas; 382*©Tito Dupret; 383 ©Alexander Maksimov; 385 ©Ferenc Cegledi; 386 ©robert paul van beets; 387 ©Philip Lange; 388*©Tito Dupret; 389 ©Netfalls; 391 ©Luciano Mortula; 392 ©Olga Kolos; 394 ©Holger Mette; 394 ©Grigory Kubatyan; 395*©Tito Dupret; 396 ©Dmitriy Bryndin; 397 ©MCales; 398▲*©Tito Dupret; 398▼*©Carmen Daly Schelbert; 399 ©Olga Zaporozhskaya; 400 ©tororo reaction; 402 ©Alan Kraft; 403 ©paul prescott; 404 ©PixAchi; 405 ©enote; 406 ©Khirman Vladimir;

409 ©Martin Mette; 410 ©Jarno Gonzalez Zarraonandia; 411 ©Valery Shanin; 413 ©JeremyRichards; 414 ©Jozef Sedmak; 415 ©Sam DCruz; 417 ©Wiktor Bubniak; 418 ©jpatava; 420 ©krechet; 421 ©tfrisch99; 423 ©Steffen Foerster Photography; 424 ©Fedor Selivanov; 425 ©Pichugin Dmitry; 427 ©Svetlana Tikhonova; 428 ©Thomas M Perkins; 429*©Tito Dupret; 431 ©rm; 433 ©Bill Perry; 434 ©Jarno Gonzalez Zarraonandia; 436*©Tito Dupret; 437 ©jorisvo; 439 ©Huang Yuetao; 440 ©Kris Vandereycken; 442 ©polartern; 443 ©vospalej; 445 ©carlos sanchez pereyra; 446 ©Michael J Thompson; 447 ©Ungor; 448 ©Danilo Ascione; 449 ©Motordigitaal; 451 ©vera bogaerts; 452 ©Alan Kraft; 453 ©Gianluca Figliola Fantini; 455 ©Andrzej Gibasiewicz; 456 ©Falk Kienas; 457 ©Keith Brooks; 458 ©Mark Van Overmeire; 459 ©Daniel Gustavsson; 461 ©edobric; 462*©Tito Dupret; 463 ©Helen & Vlad Filatov; 464 ©Willem Tims; 465 ©Claudio Giovanni Colombo; 467 ©Jonald Morales; 468*©Tito Dupret; 469 ©Jiri Krajicek; 471▲©godrick; 471▼©Ron Urquhart; 472 ©Joel Blit; 475 ©Ian D Walker; 476 ©Kochergin; 477 ©koi88; 478*©Reto Kuster; 479 ©Mary Lane; 480 ©Xavier MARCHANT; 481*©Tito Dupret; 482 ©Jorge Felix Costa; 483 ©Joe Barbarite; 484*©Tito Dupret; 487 ©Valeria73; 489 ©Alex Garaev; 491 ©Pichugin Dmitry; 492 ©Sean Prior; 493 ©BESTWEB; 494 ©kirych; 495 ©L F File; 496 ©Bryan Busovicki; 497 ©Andreas Gradin; 499▲©Tobias Guttmann; 499▼©Marek Slusarczyk; 500 ©Zorik Galstyan; 501 ©Valeria73; 502 ©Aron Brand; 503*©Tito Dupret; 505 ©aguilarphoto; 507 ©Benson HE; 509 ©Veronika Trofer; 511 ©szefei; 512▲©Tito Dupret; 512▼*©Tito Dupret; 513 ©Dan Breckwoldt; 514 ©Lagui; 515 ©Maksym Kalyta; 517*©Tito Dupret; 519 ©Mikhail Nekrasov; 520 ©MASSIMO MERLINI; 521 ©Shapiro Svetlana; 522 ©Angels at Work; 523 ©Floris Slooff; 525 ©Perov Stanislaw; 526 ©Tomasz Szymanski; 527 ©Dino; 528 ©Angels at Work; 529 ©rj lerich; 530 ©WitR; 531 ©Razvan Stroie; 532*©Tito Dupret; 533 ©Slawomir Kruz; 534 ©Bill Perry; 536 ©Chris102; 537 ©ostill; 539 ©ollirg; 540*©Tito Dupret; 541 ©Mauro Bighin; 542 ©Tobias Machhaus; 543 ©Dainis Derics; 545 ©Vladimirs Kuskins; 547 ©Marcin-linfernum; 549 ©Uwe Bumann; 550 ©MaxFX; 551*©Austrian National Tourist Office/Diejun; 552 ©Buddhadl; 553 ©Joerg Hausmann; 555 ©Ivo Brezina; 557▲©Pichugin Dmitry; 557▼©MARTAFR; 559 ©sunxuejun; 560 ©Laitr Keiows; 561 ©lexan; 562▲©Tito Dupret; 564 ©Andy Z.; 566*©Tito Dupret; 567 ©luri; 568 ©photooiasson; 569▲*©Tito Dupret; 569▼©Jethro Lennox; 571 ©David Woods; 572 ©Factoria singular fotografia; 573*©Tito Dupret; 574 ©Martin D. Vonka; 576*©David Barrie/Darjeeling Himalayan Railway Society; 577 ©Alfredo Schaufelberger; 578 ©Philip Lange; 579 ©Dole; 580*©Tito Dupret; 581 ©Tudor Stanica; 582*©Tito Dupret; 584*©Tito Dupret; 585 ©OPIS; 586 ©Guillermo Garcia; 589 ©Joerg Humpe; 591 ©Pablo H Caridad; 592 ©Sorin Popa; 593 ©robert paul van beets; 594 ©Bartlomiej K. Kwieciszewski; 595 ©Valeria73; 597 ©Insuratelu Gabriela Gianina; 598 ©Sam DCruz; 599*©Tito Dupret; 600 ©riekephotos; 601*©Tito Dupret; 602 ©Ales Liska; 603 ©baldovina; 604 ©WH CHOW; 605 ©Tootles; 606 ©kwest; 607 ©Marc C. Johnson; 609*©(CC by 2.0) sridgway; 610 ©Fedor Selivanov; 611 ©EcoPrint; 613 ©Albo; 614*©Tito Dupret; 615 ©Birute Vijeikiene; 616 ©zhouhui8525; 617 ©Jarno Gonzalez Zarraonandia; 618 ©Vladimir Daragan; 621 ©Ismael Montero Verdu; 623 ©St. Nick; 624 ©Joao Virissimo; 625 ©slava_vn; 626*©Tito Dupret; 627 ©Hano Uzeirbegovic; 628 ©rm; 629*©Tito Dupret; 630 ©Alexey Averiyanov; 631 ©Chong Wei Jin; 633 ©Ivonne Wierink; 634 ©javarman; 635 ©Tim Arbaev; 636*©Tito Dupret; 637 ©Karel Gallas; 638 ©Chow Shue Ma; 639 ©Miguel Angel Pallardo del Rio; 640 ©Vladimir Korostyshevskiy; 641 ©Larky; 642*©Tito Dupret; 646 ©Pavel K; 648*©Tito Dupret; 648*©Tito Dupret; 649*©Mark Steward; 652*©Tito Dupret; 653 ©Craig Hanson; 654 ©Buddhadl; 655 ©Zorik Galstyan; 655*©Tito Dupret; 657 ©ecoventurestravel; 658 ©Eric Gevaert; 659 ©rubiphoto; 661 ©Julia R.; 662 ©Sílvia Antunes; 663*©Tito Dupret; 665 ©Gyrohype; 666 ©Seleznev Oleg; 668*©(CC by 2.0) Sara & Joachim; 669 ©Graham Bloomfield; 670 ©javarman; 671 ©Daniel Gilbey;

672*©Tito Dupret; 673 ©Joseph Calev; 674 ©Mark Breck; 677*©Israel Ministry of Tourism www.goisrael.com; 678 ©Alexander Cyliax; 679 ©Juha Sompinmäki; 680 ©Rostislav Glinsky; 681 ©Nestor Noci; 682*©Tito Dupret; 683*©UNESCO/Claudio Margottini; 685 ©LianeM; 686 ©Falk Kienas; 689 ©PavleMarjanovic; 690 ©mirabile; 691*©Tito Dupret; 692 ©jeff gynane; 693*©NASA/GSFC/METI/ERSDAC/JAROS and the U.S./Japan ASTER Science Team and Jesse Allen; 694*©Tito Dupret; 697*©UNESCO/Giovanni Boccardi; 699 ©urosr; 701 ©mastiffliu; 702 ©Wiktor Bubniak; 703*©Tito Dupret; 704 ©AND Inc.; 705 ©rm; 706*©Tito Dupret; 707 ©Andrew Barker; 708*©MODIS/NASA; 709 ©vanille; 710 ©Hugo de Wolf; 711 ©James Beach; 712 ©Michael Steden; 713*©Tito Dupret; 714 ©steba; 715 ©fotique; 718 ©Joern; 721*©Fleur Gayet; 723 ©Holger Mette; 724 ©Stuart Blyth; 725*©(CC by 2.0) Crazy Joe Devola; 727 ©Joerg Humpe; 728 ©Eremin Sergey; 730 ©Plotnikoff; 731 ©Horácio José Lopes dos Santos; 733 ©alysta; 736*©UNESCO/Alessandro Balsamo; 737 ©Ng Wei Keong; 739*©Doron Nissim; 740*©Tsvika Tsuk; 741*©beltsazar; 743 ©Natalia Belotelova; 744 ©Zimins@NET; 745*©Tito; 746*©Tito Dupret; 749 ©Rafael Martin-Gaitero; 751 ©newphotoservice; 752▲©Kate Kotova; 752▼*©Tito Dupret; 753 ©manfredxy; 754 ©Jesus Cervantes; 755▲*©(CC by sa 2.0) shaunamullally; 755▼*©Tito Dupret; 757 ©Richard Griffin; 758 ©Ruta Saulyte-Laurinaviciene; 760*©Tito Dupret; 761 ©Holger Mette; 763 ©Neale Cousland; 765 ©Holger Mette; 766 ©Vlad Zharoff; 767 ©Tonis Valing; 769 ©Andy Lim; 770 ©ECOPRINT; 771 ©Petros Tsonis; 772 ©Brykaylo Yuriy; 773 ©Vlad Ghiea; 774*©Tito Dupret; 775 ©Coquilleau; 776 ©El Choclo; 777 ©lavigne herve; 779 ©PavleMarjanovic; 780 ©Connors Bros.; 783 ©Aleksandrs Jermakovichs; 785 ©Vladimir Melnik; 786 ©Jens Stolt; 787 ©Peter Wey; 791 ©Lori Skelton; 793 ©Tatiana Belova; 794 ©Katarzyna Mazurowska; 795*©(CC by 2.0) PhillipC; 796*©Tito Dupret; 797 ©RookCreations; 798 ©Salem Alforaih; 799 ©Salem Alforaih

CC by 2.0

These works are licensed under the Creative Commons 2.0 Attribution License. To view a copy of this license, visit http://creativecommons.org/licenses/by/2.0/

CC by sa 2.5

These works are licensed under the Creative Commons Attribution ShareAlike 2.5 License. To view a copy of this license, visit http://creativecommons.org/licenses/by-sa/2.5/

Tito Dupret/WHTour.org

(www.worldheritage-tour.org) is a non-profit organization documenting World Heritage sites in panography – 360 degree imaging –thanks to the support of the J. M. Kaplan Fund from NewYork, USA.

How to use this book

The page on which the information on a World Heritage site can be found is accessed in a number of ways – by consulting the continent maps on which all the sites are located, or by reference to either the alphabetical or country by country index. All entries are presented in a similar manner and are arranged chronologically by the year in which they were first inscribed on the World Heritage List.

The diagram below indicates the individual components of each entry and explains the colour coding used to distinguish whether a site is classified as natural, cultural or mixed.

Site title
gives the official UNESCO World Heritage title for each entry.

Red band
represents entries classified as cultural sites.

Locator map
shows the location of the site in a wider region.

Blue band
represents entries classified as mixed sites.

Timeline
on every page highlights the year in which the sites were first inscribed.

Site location
indicates the country where the site can be found.

Green band
represents entries classified as natural sites.

Criteria summary
To be included on the World Heritage List, sites must be of outstanding universal value and meet at least one out of ten selection criteria. Full criteria explanation can be found on pages **800–1**.

Main text
gives concise descriptions and information about each site.

Extra information
about each site supplements the details in the main text.